VOLUME FOUR HUNDRED AND FIFTY-THREE

METHODS IN ENZYMOLOGY

Autophagy in Disease and Clinical Applications, Part C

METHODS IN ENZYMOLOGY

Editors-in-Chief

JOHN N. ABELSON AND MELVIN I. SIMON

Division of Biology
California Institute of Technology
Pasadena, California, USA

Founding Editors

SIDNEY P. COLOWICK AND NATHAN O. KAPLAN

VOLUME FOUR HUNDRED AND FIFTY-THREE

METHODS IN ENZYMOLOGY

Autophagy in Disease and Clinical Applications, Part C

EDITED BY

DANIEL J. KLIONSKY

Life Sciences Institute
University of Michigan
Ann Arbor, Michigan, USA

ELSEVIER

AMSTERDAM • BOSTON • HEIDELBERG • LONDON
NEW YORK • OXFORD • PARIS • SAN DIEGO
SAN FRANCISCO • SINGAPORE • SYDNEY • TOKYO
Academic Press is an imprint of Elsevier

Academic Press is an imprint of Elsevier
525 B Street, Suite 1900, San Diego, California 92101-4495, USA
30 Corporate Drive, Suite 400, Burlington,MA01803, USA
32 Jamestown Road, LondonNW17BY, UK

For information on all Elsevier Academic Press publications
visit our Web site at elsevierdirect.com

ISBN-13: 978-0-12-374936-9

PRINTED IN THE UNITED STATES OF AMERICA
09 10 11 9 8 7 6 5 4 3 2 1

Contents

Contributors

Mona Bains
Department of Pharmacology, University of Colorado at Denver, Anchutz Medical Campus, Aurora, Colorado, USA

Brian T. Chait
Laboratory of Mass Spectrometry and Gaseous Ion Chemistry, Rockefeller University, New York, New York, USA

Zhi-Hua Chen
Division of Pulmonary and Critical Care Medicine Brigham and Women's Hospital, Harvard Medical School, Boston, Massachusetts, USA

Salvatore J. Cherra III
Department of Pathology, Division of Neuropathology, University of Pittsburgh School of Medicine and Center for Neuroscience (CNUP), Pittsburgh, Pennsylvania, USA

Fulvio Chiacchiera
Laboratory of Signal-Dependent Transcription, Department of Translational Pharmacology, Consorzio Mario Negri Sud, Santa Maria Imbaro (Chieti), Italy

Augustine M. K. Choi
Division of Pulmonary and Critical Care Medicine Brigham and Women's Hospital, Harvard Medical School, Boston, Massachusetts, USA

Charleen T. Chu
Department of Pathology Division of Neuropathology, University of Pittsburgh School of Medicine and Center for Neuroscience (CNUP), Pittsburgh, Pennsylvania, USA

Robert S. B. Clark
Departments of Critical Care Medicine and Pediatrics, Safar Center for Resuscitation Research University of Pittsburgh School of Medicine, Pittsburgh, Pennsylvania, USA

John L. Cleveland
Department of Cancer Biology, The Scripps Research Institute, Scripps-Florida, Jupiter, Florida, USA

Charles Conrad
Department of Neuro-Oncology, University of Texas M. D. Anderson Cancer Center, Houston, Texas, USA

Ruben K. Dagda
Department of Pathology, Division of Neuropathology, University of Pittsburgh School of Medicine Center for Neuroscience (CNUP), Pittsburgh, Pennsylvania, USA

Wen-Xing Ding
Department of Pathology, University of Pittsburgh School of Medicine, Pittsburgh, Pennsylvania, USA

Frank C. Dorsey
Department of Cancer Biology, The Scripps Research Institute, Scripps-Florida, Jupiter, Florida, USA

Juan Fueyo
Department of Neuro-Oncology, University of Texas M. D. Anderson Cancer Center, Houston, Texas, USA

Candelaria Gomez-Manzano
Department of Neuro-Oncology, University of Texas M. D. Anderson Cancer Center, Houston, Texas, USA

Roberta A. Gottlieb
San Diego State Research Foundation BioScience Center, San Diego State University, San Diego, California, USA

Nirmala Hariharan
Department of Cell Biology and Molecular Medicine, UMDNJ, New Jersey Medical School, Newark, New Jersey, USA

Kim A. Heidenreich
VA Research Service, Eastern Colorado Health Care System, Denver, Colorado, USA

Robert W. Hickey
Departments of Critical Care Medicine and Pediatrics, Safar Center for Resuscitation Research University of Pittsburgh School of Medicine, Pittsburgh, Pennsylvania, USA

Joseph A. Hill
Departments of Internal Medicine (Cardiology) and Molecular Biology, University of Texas Southwestern Medical Center, Dallas, Texas, USA

Victoria Hill
The Arthritis and Rheumatism Branch, NIAMS, National Institutes of Health, Bethesda, Maryland, USA

Gay R. Holstein
Departments of Neurology and Neuroscience, Mount Sinai School of Medicine, New York, New York, USA

Misun Hwang
Department of Radiation Oncology, Vanderbilt Ingram Cancer Center, Vanderbilt University School of Medicine, Nashville, Tennessee, USA

Jerry J. Jaboin
Department of Radiation Oncology, Vanderbilt Ingram Cancer Center, Vanderbilt University School of Medicine, Nashville, Tennessee, USA

Hong Jiang
Department of Neuro-Oncology, University of Texas M. D. Anderson Cancer Center, Houston, Texas, USA

Vassiliki Karantza-Wadsworth
Division of Medical Oncology, Department of Internal Medicine, University of Medicine and Dentistry of New Jersey, Robert Wood Johnson Medical School, Piscataway, New Jersey, USA, and Cancer Institute of New Jersey, New Brunswick, New Jersey, USA

David Kessel
Department of Pharmacology, Wayne State University School of Medicine, Detroit, Michigan, USA

Hong Pyo Kim
Division of Pulmonary, Allergy and Critical Care Medicine, Department of Medicine, University of Pittsburgh, Pittsburgh, Pennsylvania, USA, and Division of Pulmonary and Critical Care Medicine Brigham and Women's Hospital, Harvard Medical School, Boston, Massachusetts, USA

Michiel Knaapen
ZNA Middelheim, Laboratory for Cardiovascular Pathology, Antwerp, Belgium

Masato Koike
Department of Cell Biology and Neuroscience, Juntendo University Graduate School of Medicine, Tokyo, Japan

Masaaki Komatsu
Laboratory of Frontier Science, Tokyo Metropolitan Institute of Medical Science, Tokyo, Japan, and PRESTO, Japan Science and Technology Corporation, Kawaguchi, Japan

Shiori Kyoi
Department of Cell Biology and Molecular Medicine, UMDNJ, New Jersey Medical School, Newark, New Jersey, USA

Ju-Hyun Lee
Center for Dementia Research, Nathan S. Kline Institute, Orangeburg, New York, USA, and Department of Psychiatry, New York University School of Medicine, New York, USA

Michael J. Lenardo
Laboratory of Immunology, National Institute of Allergy and Infectious Diseases, National Institutes of Health, Bethesda, Maryland, USA

Bo Lu
Department of Radiation Oncology, Vanderbilt Ingram Cancer Center, Vanderbilt University School of Medicine, Nashville, Tennessee, USA

May Christine V. Malicdan
Department of Neuromuscular Research, National Institute of Neurosciences, National Center of Neurology and Psychiatry, Tokyo, Japan

Robin Mathew
University of Medicine and Dentistry of New Jersey, Robert Wood Johnson Medical School, Piscataway, New Jersey, USA, and Center for Advanced Biotechnology and Medicine, Rutgers University, Piscataway, New Jersey, USA

Christina K. McPhee
Department of Cancer Biology, University of Massachusetts Medical School, Worcester, Massachusetts, USA, and Department of Cell Biology and Molecular Genetics, University of Maryland, Maryland, USA

Wilhelm Mistiaen
Department of Healthcare Sciences, University College of Antwerp, Antwerp, Belgium

Ichizo Nishino
Department of Neuromuscular Research, National Institute of Neurosciences, National Center of Neurology and Psychiatry, Tokyo, Japan

Ralph A. Nixon
Center for Dementia Research, Nathan S. Kline Institute, Orangeburg, New York, USA, and Department School of Medicine Cell Biology, New York University, New York, USA

Satoru Noguchi
Department of Neuromuscular Research, National Institute of Neurosciences, National Center of Neurology and Psychiatry, Tokyo, Japan

Nancy L. Oleinick
Department of Radiation Oncology, Case Western Reserve University, Cleveland, Ohio, USA

Cynthia N. Perry
San Diego State Research Foundation BioScience Center, San Diego State University, San Diego, California, USA

Paul Plotz
The Arthritis and Rheumatism Branch, NIAMS, National Institutes of Health, Bethesda, Maryland, USA

Edward D. Plowey
Department of Pathology, Division of Neuropathology, University of Pittsburgh School of Medicine and Center for Neuroscience (CNUP) Pittsburgh, Pennsylvania, USA

Stephanie M. Prater
Department of Cancer Biology, The Scripps Research Institute, Scripps-Florida, Jupiter, Florida, USA

Nina Raben
The Arthritis and Rheumatism Branch, NIAMS, National Institutes of Health, Bethesda, Maryland, USA

Brinda Ravikumar
Department of Medical Genetics, University of Cambridge, Cambridge Institute for Medical Research, Addenbrooke's Hospital, Cambridge, UK

Beverly A. Rothermel
Department of Internal Medicine (Cardiology), University of Texas Southwestern Medical Center, Dallas, Texas, USA

David C. Rubinsztein
Department of Medical Genetics, University of Cambridge, Cambridge Institute for Medical Research, Addenbrooke's Hospital, Cambridge, UK

Stefan W. Ryter
Division of Pulmonary, Allergy and Critical Care Medicine, Department of Medicine, University of Pittsburgh, Pittsburgh, Pennsylvania, USA, and Division of Pulmonary and Critical Care Medicine Brigham and Women's Hospital, Harvard Medical School, Boston, Massachusetts, USA

Junichi Sadoshima
Department of Cell Biology and Molecular Medicine, UMDNJ, New Jersey Medical School, Newark, New Jersey, USA

Sovan Sarkar
Department of Medical Genetics, University of Cambridge, Cambridge Institute for Medical Research, Addenbrooke's Hospital, Cambridge, UK

Mitsuho Sasaki
Department of Cell Biology and Neuroscience, Juntendo University Graduate School of Medicine, Tokyo, Japan

Thomas Schröter
Translational Research Institute, The Scripps Research Institute, Scripps-Florida, Jupiter, Florida, USA

Lauren Shea
The Arthritis and Rheumatism Branch, NIAMS, National Institutes of Health, Bethesda, Maryland, USA

Masahiro Shibata
Division of Gross Anatomy and Morphogenesis, Department of Regenerative and Transplant Medicine, Niigata University Graduate School of Medical and Dental Sciences, Niigata, Japan

Cristiano Simone
Laboratory of Signal-Dependent Transcription, Department of Translational Pharmacology, Consorzio Mario Negri Sud, Santa Maria Imbaro (Chieti), Italy

Meredith A. Steeves
Department of Cancer Biology, The Scripps Research Institute, Scripps-Florida, Jupiter, Florida, USA

Hiromitsu Takagi
Department of Cell Biology and Molecular Medicine, UMDNJ, New Jersey Medical School, Newark, New Jersey, USA

Yasuo Uchiyama
Department of Cell Biology and Neuroscience, Juntendo University Graduate School of Medicine, Tokyo, Japan

Satoshi Waguri
Department of Anatomy and Histology, Fukushima Medical University School of Medicine, Fukushima, Japan

Qing Jun Wang
Departments of Neurology and Neuroscience, Mount Sinai School of Medicine, New York, New York, USA, and Laboratory of Mass Spectrometry and Gaseous Ion Chemistry, Rockefeller University, New York, New York, USA

Eileen White
The Cancer Institute of New Jersey, New Jersey, USA, and Molecular Biology and Biochemistry, Rutgers University, New Jersey,USA

Erin J. White
Department of Neuro-Oncology, University of Texas M. D. Anderson Cancer Center, Houston, Texas, USA

Dun-Sheng Yang
Center for Dementia Research, Nathan S. Kline Institute, Orangeburg, New York, USA, and Department of Psychiatry, New York University School of Medicine, New York, USA

Xiao-Ming Yin
Department of Pathology, University of Pittsburgh School of Medicine, Pittsburgh, Pennsylvania, USA

Li Yu
Laboratory of Immunology, National Institute of Allergy and Infectious Diseases, National Institutes of Health, Bethesda, Maryland, USA

Zhenyu Yue
Departments of Neurology and Neuroscience, Mount Sinai School of Medicine, New York, New York, USA

Hongxin Zhu
Department of Internal Medicine (Cardiology), University of Texas Southwestern Medical Center, Dallas, Texas, USA

Preface

The initial studies on autophagy began in the late 1950s and focused on the response of organisms to glucagon. The methods for analyzing autophagy at that time were limited primarily to electron microscopy and subcellular fractionation. In subsequent decades the analyses were concerned with the regulation of autophagy, and additional biochemical methods including sequestration assays became part of the arsenal for following autophagy activity. Most recently, molecular genetic studies have focused on the function of the autophagy-related proteins, and many of the assays have been designed around these newly identified components.

Now, in part, we are seeing a return to whole organism analyses, bringing to bear some of the increased knowledge we have gained from the recent wide range of molecular studies. These analyses include an examination of autophagy in new model systems, such as planaria, hydra and ticks as described in volume A of this series. Similarly, extensive studies have been carried out in standard organisms, most notably mice, and some of the corresponding techniques are included in volume B. These animal studies have certainly paved the way for the examination of autophagy in humans. In addition, however, the renewed interest in whole organisms has resulted from the many health implications of autophagy. Recent studies have indicated a role for autophagy in tumor suppression and immunity, in preventing certain types of neurodegeneration, in liver and lung disease, in myopathies and heart disease, and in some lysosomal storage disorders.

I think it is fair to say that one goal, even of the most basic research on autophagy, is to find information that may ultimately be of therapeutic use. For example, rapamycin or its derivatives are being used in clinical trials for treating certain types of cancer, and there is a high level of interest in the use of autophagy (either through stimulation or inhibition) for potentiating anti-cancer therapies. Similarly, a number of labs have carried out screens for drugs that can modulate autophagy for the potential mitigation of neurodegenerative diseases or myopathies. Accordingly, this final volume of the series on autophagy focuses on disease connections with autophagy and on methods for monitoring autophagy in clinical settings. Once again, certain techniques are presented in more than one chapter, with the emphasis on details that are essential for successful sample preparation and assay using the particular tissue sample being described. There is no question that these methods should facilitate the application of similar techniques to study new systems and additional diseases that become linked with autophagy in the future.

DANIEL J. KLIONSKY

Methods in Enzymology

Volume I. Preparation and Assay of Enzymes
Edited by Sidney P. Colowick and Nathan O. Kaplan

Volume II. Preparation and Assay of Enzymes
Edited by Sidney P. Colowick and Nathan O. Kaplan

Volume III. Preparation and Assay of Substrates
Edited by Sidney P. Colowick and Nathan O. Kaplan

Volume IV. Special Techniques for the Enzymologist
Edited by Sidney P. Colowick and Nathan O. Kaplan

Volume V. Preparation and Assay of Enzymes
Edited by Sidney P. Colowick and Nathan O. Kaplan

Volume VI. Preparation and Assay of Enzymes *(Continued)*
Preparation and Assay of Substrates
Special Techniques
Edited by Sidney P. Colowick and Nathan O. Kaplan

Volume VII. Cumulative Subject Index
Edited by Sidney P. Colowick and Nathan O. Kaplan

Volume VIII. Complex Carbohydrates
Edited by Elizabeth F. Neufeld and Victor Ginsburg

Volume IX. Carbohydrate Metabolism
Edited by Willis A. Wood

Volume X. Oxidation and Phosphorylation
Edited by Ronald W. Estabrook and Maynard E. Pullman

Volume XI. Enzyme Structure
Edited by C. H. W. Hirs

Volume XII. Nucleic Acids (Parts A and B)
Edited by Lawrence Grossman and Kivie Moldave

Volume XIII. Citric Acid Cycle
Edited by J. M. Lowenstein

Volume XIV. Lipids
Edited by J. M. Lowenstein

Volume XV. Steroids and Terpenoids
Edited by Raymond B. Clayton

Volume XXXV. Lipids (Part B)
Edited by John M. Lowenstein

Volume XXXVI. Hormone Action (Part A: Steroid Hormones)
Edited by Bert W. O'Malley and Joel G. Hardman

Volume XXXVII. Hormone Action (Part B: Peptide Hormones)
Edited by Bert W. O'Malley and Joel G. Hardman

Volume XXXVIII. Hormone Action (Part C: Cyclic Nucleotides)
Edited by Joel G. Hardman and Bert W. O'Malley

Volume XXXIX. Hormone Action (Part D: Isolated Cells, Tissues, and Organ Systems)
Edited by Joel G. Hardman and Bert W. O'Malley

Volume XL. Hormone Action (Part E: Nuclear Structure and Function)
Edited by Bert W. O'Malley and Joel G. Hardman

Volume XLI. Carbohydrate Metabolism (Part B)
Edited by W. A. Wood

Volume XLII. Carbohydrate Metabolism (Part C)
Edited by W. A. Wood

Volume XLIII. Antibiotics
Edited by John H. Hash

Volume XLIV. Immobilized Enzymes
Edited by Klaus Mosbach

Volume XLV. Proteolytic Enzymes (Part B)
Edited by Laszlo Lorand

Volume XLVI. Affinity Labeling
Edited by William B. Jakoby and Meir Wilchek

Volume XLVII. Enzyme Structure (Part E)
Edited by C. H. W. Hirs and Serge N. Timasheff

Volume XLVIII. Enzyme Structure (Part F)
Edited by C. H. W. Hirs and Serge N. Timasheff

Volume XLIX. Enzyme Structure (Part G)
Edited by C. H. W. Hirs and Serge N. Timasheff

Volume L. Complex Carbohydrates (Part C)
Edited by Victor Ginsburg

Volume LI. Purine and Pyrimidine Nucleotide Metabolism
Edited by Patricia A. Hoffee and Mary Ellen Jones

Volume LII. Biomembranes (Part C: Biological Oxidations)
Edited by Sidney Fleischer and Lester Packer

VOLUME 71. Lipids (Part C)
Edited by JOHN M. LOWENSTEIN

VOLUME 72. Lipids (Part D)
Edited by JOHN M. LOWENSTEIN

VOLUME 73. Immunochemical Techniques (Part B)
Edited by JOHN J. LANGONE AND HELEN VAN VUNAKIS

VOLUME 74. Immunochemical Techniques (Part C)
Edited by JOHN J. LANGONE AND HELEN VAN VUNAKIS

VOLUME 75. Cumulative Subject Index Volumes XXXI, XXXII, XXXIV–LX
Edited by EDWARD A. DENNIS AND MARTHA G. DENNIS

VOLUME 76. Hemoglobins
Edited by ERALDO ANTONINI, LUIGI ROSSI-BERNARDI, AND EMILIA CHIANCONE

VOLUME 77. Detoxication and Drug Metabolism
Edited by WILLIAM B. JAKOBY

VOLUME 78. Interferons (Part A)
Edited by SIDNEY PESTKA

VOLUME 79. Interferons (Part B)
Edited by SIDNEY PESTKA

VOLUME 80. Proteolytic Enzymes (Part C)
Edited by LASZLO LORAND

VOLUME 81. Biomembranes (Part H: Visual Pigments and Purple Membranes, I)
Edited by LESTER PACKER

VOLUME 82. Structural and Contractile Proteins (Part A: Extracellular Matrix)
Edited by LEON W. CUNNINGHAM AND DIXIE W. FREDERIKSEN

VOLUME 83. Complex Carbohydrates (Part D)
Edited by VICTOR GINSBURG

VOLUME 84. Immunochemical Techniques (Part D: Selected Immunoassays)
Edited by JOHN J. LANGONE AND HELEN VAN VUNAKIS

VOLUME 85. Structural and Contractile Proteins (Part B: The Contractile Apparatus and the Cytoskeleton)
Edited by DIXIE W. FREDERIKSEN AND LEON W. CUNNINGHAM

VOLUME 86. Prostaglandins and Arachidonate Metabolites
Edited by WILLIAM E. M. LANDS AND WILLIAM L. SMITH

VOLUME 87. Enzyme Kinetics and Mechanism (Part C: Intermediates, Stereo-chemistry, and Rate Studies)
Edited by DANIEL L. PURICH

VOLUME 88. Biomembranes (Part I: Visual Pigments and Purple Membranes, II)
Edited by LESTER PACKER

Volume 89. Carbohydrate Metabolism (Part D)
Edited by Willis A. Wood

Volume 90. Carbohydrate Metabolism (Part E)
Edited by Willis A. Wood

Volume 91. Enzyme Structure (Part I)
Edited by C. H. W. Hirs and Serge N. Timasheff

Volume 92. Immunochemical Techniques (Part E: Monoclonal Antibodies and General Immunoassay Methods)
Edited by John J. Langone and Helen Van Vunakis

Volume 93. Immunochemical Techniques (Part F: Conventional Antibodies, Fc Receptors, and Cytotoxicity)
Edited by John J. Langone and Helen Van Vunakis

Volume 94. Polyamines
Edited by Herbert Tabor and Celia White Tabor

Volume 95. Cumulative Subject Index Volumes 61–74, 76–80
Edited by Edward A. Dennis and Martha G. Dennis

Volume 96. Biomembranes [Part J: Membrane Biogenesis: Assembly and Targeting (General Methods; Eukaryotes)]
Edited by Sidney Fleischer and Becca Fleischer

Volume 97. Biomembranes [Part K: Membrane Biogenesis: Assembly and Targeting (Prokaryotes, Mitochondria, and Chloroplasts)]
Edited by Sidney Fleischer and Becca Fleischer

Volume 98. Biomembranes (Part L: Membrane Biogenesis: Processing and Recycling)
Edited by Sidney Fleischer and Becca Fleischer

Volume 99. Hormone Action (Part F: Protein Kinases)
Edited by Jackie D. Corbin and Joel G. Hardman

Volume 100. Recombinant DNA (Part B)
Edited by Ray Wu, Lawrence Grossman, and Kivie Moldave

Volume 101. Recombinant DNA (Part C)
Edited by Ray Wu, Lawrence Grossman, and Kivie Moldave

Volume 102. Hormone Action (Part G: Calmodulin and Calcium-Binding Proteins)
Edited by Anthony R. Means and Bert W. O'Malley

Volume 103. Hormone Action (Part H: Neuroendocrine Peptides)
Edited by P. Michael Conn

Volume 104. Enzyme Purification and Related Techniques (Part C)
Edited by William B. Jakoby

VOLUME 105. Oxygen Radicals in Biological Systems
Edited by LESTER PACKER

VOLUME 106. Posttranslational Modifications (Part A)
Edited by FINN WOLD AND KIVIE MOLDAVE

VOLUME 107. Posttranslational Modifications (Part B)
Edited by FINN WOLD AND KIVIE MOLDAVE

VOLUME 108. Immunochemical Techniques (Part G: Separation and Characterization of Lymphoid Cells)
Edited by GIOVANNI DI SABATO, JOHN J. LANGONE, AND HELEN VAN VUNAKIS

VOLUME 109. Hormone Action (Part I: Peptide Hormones)
Edited by LUTZ BIRNBAUMER AND BERT W. O'MALLEY

VOLUME 110. Steroids and Isoprenoids (Part A)
Edited by JOHN H. LAW AND HANS C. RILLING

VOLUME 111. Steroids and Isoprenoids (Part B)
Edited by JOHN H. LAW AND HANS C. RILLING

VOLUME 112. Drug and Enzyme Targeting (Part A)
Edited by KENNETH J. WIDDER AND RALPH GREEN

VOLUME 113. Glutamate, Glutamine, Glutathione, and Related Compounds
Edited by ALTON MEISTER

VOLUME 114. Diffraction Methods for Biological Macromolecules (Part A)
Edited by HAROLD W. WYCKOFF, C. H. W. HIRS, AND SERGE N. TIMASHEFF

VOLUME 115. Diffraction Methods for Biological Macromolecules (Part B)
Edited by HAROLD W. WYCKOFF, C. H. W. HIRS, AND SERGE N. TIMASHEFF

VOLUME 116. Immunochemical Techniques
(Part H: Effectors and Mediators of Lymphoid Cell Functions)
Edited by GIOVANNI DI SABATO, JOHN J. LANGONE, AND HELEN VAN VUNAKIS

VOLUME 117. Enzyme Structure (Part J)
Edited by C. H. W. HIRS AND SERGE N. TIMASHEFF

VOLUME 118. Plant Molecular Biology
Edited by ARTHUR WEISSBACH AND HERBERT WEISSBACH

VOLUME 119. Interferons (Part C)
Edited by SIDNEY PESTKA

VOLUME 120. Cumulative Subject Index Volumes 81–94, 96–101

VOLUME 121. Immunochemical Techniques (Part I: Hybridoma Technology and Monoclonal Antibodies)
Edited by JOHN J. LANGONE AND HELEN VAN VUNAKIS

VOLUME 122. Vitamins and Coenzymes (Part G)
Edited by FRANK CHYTIL AND DONALD B. MCCORMICK

Volume 123. Vitamins and Coenzymes (Part H)
Edited by Frank Chytil and Donald B. McCormick

Volume 124. Hormone Action (Part J: Neuroendocrine Peptides)
Edited by P. Michael Conn

Volume 125. Biomembranes (Part M: Transport in Bacteria, Mitochondria, and Chloroplasts: General Approaches and Transport Systems)
Edited by Sidney Fleischer and Becca Fleischer

Volume 126. Biomembranes (Part N: Transport in Bacteria, Mitochondria, and Chloroplasts: Protonmotive Force)
Edited by Sidney Fleischer and Becca Fleischer

Volume 127. Biomembranes (Part O: Protons and Water: Structure and Translocation)
Edited by Lester Packer

Volume 128. Plasma Lipoproteins (Part A: Preparation, Structure, and Molecular Biology)
Edited by Jere P. Segrest and John J. Albers

Volume 129. Plasma Lipoproteins (Part B: Characterization, Cell Biology, and Metabolism)
Edited by John J. Albers and Jere P. Segrest

Volume 130. Enzyme Structure (Part K)
Edited by C. H. W. Hirs and Serge N. Timasheff

Volume 131. Enzyme Structure (Part L)
Edited by C. H. W. Hirs and Serge N. Timasheff

Volume 132. Immunochemical Techniques (Part J: Phagocytosis and Cell-Mediated Cytotoxicity)
Edited by Giovanni Di Sabato and Johannes Everse

Volume 133. Bioluminescence and Chemiluminescence (Part B)
Edited by Marlene DeLuca and William D. McElroy

Volume 134. Structural and Contractile Proteins (Part C: The Contractile Apparatus and the Cytoskeleton)
Edited by Richard B. Vallee

Volume 135. Immobilized Enzymes and Cells (Part B)
Edited by Klaus Mosbach

Volume 136. Immobilized Enzymes and Cells (Part C)
Edited by Klaus Mosbach

Volume 137. Immobilized Enzymes and Cells (Part D)
Edited by Klaus Mosbach

Volume 138. Complex Carbohydrates (Part E)
Edited by Victor Ginsburg

VOLUME 157. Biomembranes (Part Q: ATP-Driven Pumps and Related Transport: Calcium, Proton, and Potassium Pumps)
Edited by SIDNEY FLEISCHER AND BECCA FLEISCHER

VOLUME 158. Metalloproteins (Part A)
Edited by JAMES F. RIORDAN AND BERT L. VALLEE

VOLUME 159. Initiation and Termination of Cyclic Nucleotide Action
Edited by JACKIE D. CORBIN AND ROGER A. JOHNSON

VOLUME 160. Biomass (Part A: Cellulose and Hemicellulose)
Edited by WILLIS A. WOOD AND SCOTT T. KELLOGG

VOLUME 161. Biomass (Part B: Lignin, Pectin, and Chitin)
Edited by WILLIS A. WOOD AND SCOTT T. KELLOGG

VOLUME 162. Immunochemical Techniques (Part L: Chemotaxis and Inflammation)
Edited by GIOVANNI DI SABATO

VOLUME 163. Immunochemical Techniques (Part M: Chemotaxis and Inflammation)
Edited by GIOVANNI DI SABATO

VOLUME 164. Ribosomes
Edited by HARRY F. NOLLER, JR., AND KIVIE MOLDAVE

VOLUME 165. Microbial Toxins: Tools for Enzymology
Edited by SIDNEY HARSHMAN

VOLUME 166. Branched-Chain Amino Acids
Edited by ROBERT HARRIS AND JOHN R. SOKATCH

VOLUME 167. Cyanobacteria
Edited by LESTER PACKER AND ALEXANDER N. GLAZER

VOLUME 168. Hormone Action (Part K: Neuroendocrine Peptides)
Edited by P. MICHAEL CONN

VOLUME 169. Platelets: Receptors, Adhesion, Secretion (Part A)
Edited by JACEK HAWIGER

VOLUME 170. Nucleosomes
Edited by PAUL M. WASSARMAN AND ROGER D. KORNBERG

VOLUME 171. Biomembranes (Part R: Transport Theory: Cells and Model Membranes)
Edited by SIDNEY FLEISCHER AND BECCA FLEISCHER

VOLUME 172. Biomembranes (Part S: Transport: Membrane Isolation and Characterization)
Edited by SIDNEY FLEISCHER AND BECCA FLEISCHER

VOLUME 190. Retinoids (Part B: Cell Differentiation and Clinical Applications)
Edited by LESTER PACKER

VOLUME 191. Biomembranes (Part V: Cellular and Subcellular Transport: Epithelial Cells)
Edited by SIDNEY FLEISCHER AND BECCA FLEISCHER

VOLUME 192. Biomembranes (Part W: Cellular and Subcellular Transport: Epithelial Cells)
Edited by SIDNEY FLEISCHER AND BECCA FLEISCHER

VOLUME 193. Mass Spectrometry
Edited by JAMES A. MCCLOSKEY

VOLUME 194. Guide to Yeast Genetics and Molecular Biology
Edited by CHRISTINE GUTHRIE AND GERALD R. FINK

VOLUME 195. Adenylyl Cyclase, G Proteins, and Guanylyl Cyclase
Edited by ROGER A. JOHNSON AND JACKIE D. CORBIN

VOLUME 196. Molecular Motors and the Cytoskeleton
Edited by RICHARD B. VALLEE

VOLUME 197. Phospholipases
Edited by EDWARD A. DENNIS

VOLUME 198. Peptide Growth Factors (Part C)
Edited by DAVID BARNES, J. P. MATHER, AND GORDON H. SATO

VOLUME 199. Cumulative Subject Index Volumes 168–174, 176–194

VOLUME 200. Protein Phosphorylation (Part A: Protein Kinases: Assays, Purification, Antibodies, Functional Analysis, Cloning, and Expression)
Edited by TONY HUNTER AND BARTHOLOMEW M. SEFTON

VOLUME 201. Protein Phosphorylation (Part B: Analysis of Protein Phosphorylation, Protein Kinase Inhibitors, and Protein Phosphatases)
Edited by TONY HUNTER AND BARTHOLOMEW M. SEFTON

VOLUME 202. Molecular Design and Modeling: Concepts and Applications (Part A: Proteins, Peptides, and Enzymes)
Edited by JOHN J. LANGONE

VOLUME 203. Molecular Design and Modeling: Concepts and Applications (Part B: Antibodies and Antigens, Nucleic Acids, Polysaccharides, and Drugs)
Edited by JOHN J. LANGONE

VOLUME 204. Bacterial Genetic Systems
Edited by JEFFREY H. MILLER

VOLUME 205. Metallobiochemistry (Part B: Metallothionein and Related Molecules)
Edited by JAMES F. RIORDAN AND BERT L. VALLEE

Volume 223. Proteolytic Enzymes in Coagulation, Fibrinolysis, and Complement Activation (Part B: Complement Activation, Fibrinolysis, and Nonmammalian Blood Coagulation Factors)
Edited by Laszlo Lorand and Kenneth G. Mann

Volume 224. Molecular Evolution: Producing the Biochemical Data
Edited by Elizabeth Anne Zimmer, Thomas J. White, Rebecca L. Cann, and Allan C. Wilson

Volume 225. Guide to Techniques in Mouse Development
Edited by Paul M. Wassarman and Melvin L. DePamphilis

Volume 226. Metallobiochemistry (Part C: Spectroscopic and Physical Methods for Probing Metal Ion Environments in Metalloenzymes and Metalloproteins)
Edited by James F. Riordan and Bert L. Vallee

Volume 227. Metallobiochemistry (Part D: Physical and Spectroscopic Methods for Probing Metal Ion Environments in Metalloproteins)
Edited by James F. Riordan and Bert L. Vallee

Volume 228. Aqueous Two-Phase Systems
Edited by Harry Walter and Göte Johansson

Volume 229. Cumulative Subject Index Volumes 195–198, 200–227

Volume 230. Guide to Techniques in Glycobiology
Edited by William J. Lennarz and Gerald W. Hart

Volume 231. Hemoglobins (Part B: Biochemical and Analytical Methods)
Edited by Johannes Everse, Kim D. Vandegriff, and Robert M. Winslow

Volume 232. Hemoglobins (Part C: Biophysical Methods)
Edited by Johannes Everse, Kim D. Vandegriff, and Robert M. Winslow

Volume 233. Oxygen Radicals in Biological Systems (Part C)
Edited by Lester Packer

Volume 234. Oxygen Radicals in Biological Systems (Part D)
Edited by Lester Packer

Volume 235. Bacterial Pathogenesis (Part A: Identification and Regulation of Virulence Factors)
Edited by Virginia L. Clark and Patrik M. Bavoil

Volume 236. Bacterial Pathogenesis (Part B: Integration of Pathogenic Bacteria with Host Cells)
Edited by Virginia L. Clark and Patrik M. Bavoil

Volume 237. Heterotrimeric G Proteins
Edited by Ravi Iyengar

Volume 238. Heterotrimeric G-Protein Effectors
Edited by Ravi Iyengar

VOLUME 239. Nuclear Magnetic Resonance (Part C)
Edited by THOMAS L. JAMES AND NORMAN J. OPPENHEIMER

VOLUME 240. Numerical Computer Methods (Part B)
Edited by MICHAEL L. JOHNSON AND LUDWIG BRAND

VOLUME 241. Retroviral Proteases
Edited by LAWRENCE C. KUO AND JULES A. SHAFER

VOLUME 242. Neoglycoconjugates (Part A)
Edited by Y. C. LEE AND REIKO T. LEE

VOLUME 243. Inorganic Microbial Sulfur Metabolism
Edited by HARRY D. PECK, JR., AND JEAN LEGALL

VOLUME 244. Proteolytic Enzymes: Serine and Cysteine Peptidases
Edited by ALAN J. BARRETT

VOLUME 245. Extracellular Matrix Components
Edited by E. RUOSLAHTI AND E. ENGVALL

VOLUME 246. Biochemical Spectroscopy
Edited by KENNETH SAUER

VOLUME 247. Neoglycoconjugates (Part B: Biomedical Applications)
Edited by Y. C. LEE AND REIKO T. LEE

VOLUME 248. Proteolytic Enzymes: Aspartic and Metallo Peptidases
Edited by ALAN J. BARRETT

VOLUME 249. Enzyme Kinetics and Mechanism (Part D: Developments in Enzyme Dynamics)
Edited by DANIEL L. PURICH

VOLUME 250. Lipid Modifications of Proteins
Edited by PATRICK J. CASEY AND JANICE E. BUSS

VOLUME 251. Biothiols (Part A: Monothiols and Dithiols, Protein Thiols, and Thiyl Radicals)
Edited by LESTER PACKER

VOLUME 252. Biothiols (Part B: Glutathione and Thioredoxin; Thiols in Signal Transduction and Gene Regulation)
Edited by LESTER PACKER

VOLUME 253. Adhesion of Microbial Pathogens
Edited by RON J. DOYLE AND ITZHAK OFEK

VOLUME 254. Oncogene Techniques
Edited by PETER K. VOGT AND INDER M. VERMA

VOLUME 255. Small GTPases and Their Regulators (Part A: Ras Family)
Edited by W. E. BALCH, CHANNING J. DER, AND ALAN HALL

VOLUME 256. Small GTPases and Their Regulators (Part B: Rho Family)
Edited by W. E. BALCH, CHANNING J. DER, AND ALAN HALL

Volume 257. Small GTPases and Their Regulators (Part C: Proteins Involved in Transport)
Edited by W. E. Balch, Channing J. Der, and Alan Hall

Volume 258. Redox-Active Amino Acids in Biology
Edited by Judith P. Klinman

Volume 259. Energetics of Biological Macromolecules
Edited by Michael L. Johnson and Gary K. Ackers

Volume 260. Mitochondrial Biogenesis and Genetics (Part A)
Edited by Giuseppe M. Attardi and Anne Chomyn

Volume 261. Nuclear Magnetic Resonance and Nucleic Acids
Edited by Thomas L. James

Volume 262. DNA Replication
Edited by Judith L. Campbell

Volume 263. Plasma Lipoproteins (Part C: Quantitation)
Edited by William A. Bradley, Sandra H. Gianturco, and Jere P. Segrest

Volume 264. Mitochondrial Biogenesis and Genetics (Part B)
Edited by Giuseppe M. Attardi and Anne Chomyn

Volume 265. Cumulative Subject Index Volumes 228, 230–262

Volume 266. Computer Methods for Macromolecular Sequence Analysis
Edited by Russell F. Doolittle

Volume 267. Combinatorial Chemistry
Edited by John N. Abelson

Volume 268. Nitric Oxide (Part A: Sources and Detection of NO; NO Synthase)
Edited by Lester Packer

Volume 269. Nitric Oxide (Part B: Physiological and Pathological Processes)
Edited by Lester Packer

Volume 270. High Resolution Separation and Analysis of Biological Macromolecules (Part A: Fundamentals)
Edited by Barry L. Karger and William S. Hancock

Volume 271. High Resolution Separation and Analysis of Biological Macromolecules (Part B: Applications)
Edited by Barry L. Karger and William S. Hancock

Volume 272. Cytochrome P450 (Part B)
Edited by Eric F. Johnson and Michael R. Waterman

Volume 273. RNA Polymerase and Associated Factors (Part A)
Edited by Sankar Adhya

Volume 274. RNA Polymerase and Associated Factors (Part B)
Edited by Sankar Adhya

Volume 275. Viral Polymerases and Related Proteins
Edited by Lawrence C. Kuo, David B. Olsen, and Steven S. Carroll

Volume 276. Macromolecular Crystallography (Part A)
Edited by Charles W. Carter, Jr., and Robert M. Sweet

Volume 277. Macromolecular Crystallography (Part B)
Edited by Charles W. Carter, Jr., and Robert M. Sweet

Volume 278. Fluorescence Spectroscopy
Edited by Ludwig Brand and Michael L. Johnson

Volume 279. Vitamins and Coenzymes (Part I)
Edited by Donald B. McCormick, John W. Suttie, and Conrad Wagner

Volume 280. Vitamins and Coenzymes (Part J)
Edited by Donald B. McCormick, John W. Suttie, and Conrad Wagner

Volume 281. Vitamins and Coenzymes (Part K)
Edited by Donald B. McCormick, John W. Suttie, and Conrad Wagner

Volume 282. Vitamins and Coenzymes (Part L)
Edited by Donald B. McCormick, John W. Suttie, and Conrad Wagner

Volume 283. Cell Cycle Control
Edited by William G. Dunphy

Volume 284. Lipases (Part A: Biotechnology)
Edited by Byron Rubin and Edward A. Dennis

Volume 285. Cumulative Subject Index Volumes 263, 264, 266–284, 286–289

Volume 286. Lipases (Part B: Enzyme Characterization and Utilization)
Edited by Byron Rubin and Edward A. Dennis

Volume 287. Chemokines
Edited by Richard Horuk

Volume 288. Chemokine Receptors
Edited by Richard Horuk

Volume 289. Solid Phase Peptide Synthesis
Edited by Gregg B. Fields

Volume 290. Molecular Chaperones
Edited by George H. Lorimer and Thomas Baldwin

Volume 291. Caged Compounds
Edited by Gerard Marriott

Volume 292. ABC Transporters: Biochemical, Cellular, and Molecular Aspects
Edited by Suresh V. Ambudkar and Michael M. Gottesman

Volume 293. Ion Channels (Part B)
Edited by P. Michael Conn

Volume 329. Regulators and Effectors of Small GTPases (Part E: GTPases Involved in Vesicular Traffic)
Edited by W. E. Balch, Channing J. Der, and Alan Hall

Volume 330. Hyperthermophilic Enzymes (Part A)
Edited by Michael W. W. Adams and Robert M. Kelly

Volume 331. Hyperthermophilic Enzymes (Part B)
Edited by Michael W. W. Adams and Robert M. Kelly

Volume 332. Regulators and Effectors of Small GTPases (Part F: Ras Family I)
Edited by W. E. Balch, Channing J. Der, and Alan Hall

Volume 333. Regulators and Effectors of Small GTPases (Part G: Ras Family II)
Edited by W. E. Balch, Channing J. Der, and Alan Hall

Volume 334. Hyperthermophilic Enzymes (Part C)
Edited by Michael W. W. Adams and Robert M. Kelly

Volume 335. Flavonoids and Other Polyphenols
Edited by Lester Packer

Volume 336. Microbial Growth in Biofilms (Part A: Developmental and Molecular Biological Aspects)
Edited by Ron J. Doyle

Volume 337. Microbial Growth in Biofilms (Part B: Special Environments and Physicochemical Aspects)
Edited by Ron J. Doyle

Volume 338. Nuclear Magnetic Resonance of Biological Macromolecules (Part A)
Edited by Thomas L. James, Volker Dötsch, and Uli Schmitz

Volume 339. Nuclear Magnetic Resonance of Biological Macromolecules (Part B)
Edited by Thomas L. James, Volker Dötsch, and Uli Schmitz

Volume 340. Drug–Nucleic Acid Interactions
Edited by Jonathan B. Chaires and Michael J. Waring

Volume 341. Ribonucleases (Part A)
Edited by Allen W. Nicholson

Volume 342. Ribonucleases (Part B)
Edited by Allen W. Nicholson

Volume 343. G Protein Pathways (Part A: Receptors)
Edited by Ravi Iyengar and John D. Hildebrandt

Volume 344. G Protein Pathways (Part B: G Proteins and Their Regulators)
Edited by Ravi Iyengar and John D. Hildebrandt

Volume 345. G Protein Pathways (Part C: Effector Mechanisms)
Edited by Ravi Iyengar and John D. Hildebrandt

Volume 346. Gene Therapy Methods
Edited by M. Ian Phillips

Volume 347. Protein Sensors and Reactive Oxygen Species (Part A: Selenoproteins and Thioredoxin)
Edited by Helmut Sies and Lester Packer

Volume 348. Protein Sensors and Reactive Oxygen Species (Part B: Thiol Enzymes and Proteins)
Edited by Helmut Sies and Lester Packer

Volume 349. Superoxide Dismutase
Edited by Lester Packer

Volume 350. Guide to Yeast Genetics and Molecular and Cell Biology (Part B)
Edited by Christine Guthrie and Gerald R. Fink

Volume 351. Guide to Yeast Genetics and Molecular and Cell Biology (Part C)
Edited by Christine Guthrie and Gerald R. Fink

Volume 352. Redox Cell Biology and Genetics (Part A)
Edited by Chandan K. Sen and Lester Packer

Volume 353. Redox Cell Biology and Genetics (Part B)
Edited by Chandan K. Sen and Lester Packer

Volume 354. Enzyme Kinetics and Mechanisms (Part F: Detection and Characterization of Enzyme Reaction Intermediates)
Edited by Daniel L. Purich

Volume 355. Cumulative Subject Index Volumes 321–354

Volume 356. Laser Capture Microscopy and Microdissection
Edited by P. Michael Conn

Volume 357. Cytochrome P450, Part C
Edited by Eric F. Johnson and Michael R. Waterman

Volume 358. Bacterial Pathogenesis (Part C: Identification, Regulation, and Function of Virulence Factors)
Edited by Virginia L. Clark and Patrik M. Bavoil

Volume 359. Nitric Oxide (Part D)
Edited by Enrique Cadenas and Lester Packer

Volume 360. Biophotonics (Part A)
Edited by Gerard Marriott and Ian Parker

Volume 361. Biophotonics (Part B)
Edited by Gerard Marriott and Ian Parker

Volume 362. Recognition of Carbohydrates in Biological Systems (Part A)
Edited by Yuan C. Lee and Reiko T. Lee

VOLUME 363. Recognition of Carbohydrates in Biological Systems (Part B)
Edited by YUAN C. LEE AND REIKO T. LEE

VOLUME 364. Nuclear Receptors
Edited by DAVID W. RUSSELL AND DAVID J. MANGELSDORF

VOLUME 365. Differentiation of Embryonic Stem Cells
Edited by PAUL M. WASSAUMAN AND GORDON M. KELLER

VOLUME 366. Protein Phosphatases
Edited by SUSANNE KLUMPP AND JOSEF KRIEGLSTEIN

VOLUME 367. Liposomes (Part A)
Edited by NEJAT DÜZGÜNEŞ

VOLUME 368. Macromolecular Crystallography (Part C)
Edited by CHARLES W. CARTER, JR., AND ROBERT M. SWEET

VOLUME 369. Combinational Chemistry (Part B)
Edited by GUILLERMO A. MORALES AND BARRY A. BUNIN

VOLUME 370. RNA Polymerases and Associated Factors (Part C)
Edited by SANKAR L. ADHYA AND SUSAN GARGES

VOLUME 371. RNA Polymerases and Associated Factors (Part D)
Edited by SANKAR L. ADHYA AND SUSAN GARGES

VOLUME 372. Liposomes (Part B)
Edited by NEJAT DÜZGÜNEŞ

VOLUME 373. Liposomes (Part C)
Edited by NEJAT DÜZGÜNEŞ

VOLUME 374. Macromolecular Crystallography (Part D)
Edited by CHARLES W. CARTER, JR., AND ROBERT W. SWEET

VOLUME 375. Chromatin and Chromatin Remodeling Enzymes (Part A)
Edited by C. DAVID ALLIS AND CARL WU

VOLUME 376. Chromatin and Chromatin Remodeling Enzymes (Part B)
Edited by C. DAVID ALLIS AND CARL WU

VOLUME 377. Chromatin and Chromatin Remodeling Enzymes (Part C)
Edited by C. DAVID ALLIS AND CARL WU

VOLUME 378. Quinones and Quinone Enzymes (Part A)
Edited by HELMUT SIES AND LESTER PACKER

VOLUME 379. Energetics of Biological Macromolecules (Part D)
Edited by JO M. HOLT, MICHAEL L. JOHNSON, AND GARY K. ACKERS

VOLUME 380. Energetics of Biological Macromolecules (Part E)
Edited by JO M. HOLT, MICHAEL L. JOHNSON, AND GARY K. ACKERS

VOLUME 381. Oxygen Sensing
Edited by CHANDAN K. SEN AND GREGG L. SEMENZA

VOLUME 382. Quinones and Quinone Enzymes (Part B)
Edited by HELMUT SIES AND LESTER PACKER

VOLUME 383. Numerical Computer Methods (Part D)
Edited by LUDWIG BRAND AND MICHAEL L. JOHNSON

VOLUME 384. Numerical Computer Methods (Part E)
Edited by LUDWIG BRAND AND MICHAEL L. JOHNSON

VOLUME 385. Imaging in Biological Research (Part A)
Edited by P. MICHAEL CONN

VOLUME 386. Imaging in Biological Research (Part B)
Edited by P. MICHAEL CONN

VOLUME 387. Liposomes (Part D)
Edited by NEJAT DÜZGÜNEŞ

VOLUME 388. Protein Engineering
Edited by DAN E. ROBERTSON AND JOSEPH P. NOEL

VOLUME 389. Regulators of G-Protein Signaling (Part A)
Edited by DAVID P. SIDEROVSKI

VOLUME 390. Regulators of G-Protein Signaling (Part B)
Edited by DAVID P. SIDEROVSKI

VOLUME 391. Liposomes (Part E)
Edited by NEJAT DÜZGÜNEŞ

VOLUME 392. RNA Interference
Edited by ENGELKE ROSSI

VOLUME 393. Circadian Rhythms
Edited by MICHAEL W. YOUNG

VOLUME 394. Nuclear Magnetic Resonance of Biological Macromolecules (Part C)
Edited by THOMAS L. JAMES

VOLUME 395. Producing the Biochemical Data (Part B)
Edited by ELIZABETH A. ZIMMER AND ERIC H. ROALSON

VOLUME 396. Nitric Oxide (Part E)
Edited by LESTER PACKER AND ENRIQUE CADENAS

VOLUME 397. Environmental Microbiology
Edited by JARED R. LEADBETTER

VOLUME 398. Ubiquitin and Protein Degradation (Part A)
Edited by RAYMOND J. DESHAIES

VOLUME 399. Ubiquitin and Protein Degradation (Part B)
Edited by RAYMOND J. DESHAIES

VOLUME 400. Phase II Conjugation Enzymes and Transport Systems
Edited by HELMUT SIES AND LESTER PACKER

Volume 436. Globins and Other Nitric Oxide-Reactive Protiens (Part A)
Edited by Robert K. Poole

Volume 437. Globins and Other Nitric Oxide-Reactive Protiens (Part B)
Edited by Robert K. Poole

Volume 438. Small GTPases in Disease (Part A)
Edited by William E. Balch, Channing J. Der, and Alan Hall

Volume 439. Small GTPases in Disease (Part B)
Edited by William E. Balch, Channing J. Der, and Alan Hall

Volume 440. Nitric Oxide, Part F Oxidative and Nitrosative Stress in Redox Regulation of Cell Signaling
Edited by Enrique Cadenas and Lester Packer

Volume 441. Nitric Oxide, Part G Oxidative and Nitrosative Stress in Redox Regulation of Cell Signaling
Edited by Enrique Cadenas and Lester Packer

Volume 442. Programmed Cell Death, General Principles for Studying Cell Death (Part A)
Edited by Roya Khosravi-Far, Zahra Zakeri, Richard A. Lockshin, and Mauro Piacentini

Volume 443. Angiogenesis: *In Vitro* Systems
Edited by David A. Cheresh

Volume 444. Angiogenesis: *In Vivo* Systems (Part A)
Edited by David A. Cheresh

Volume 445. Angiogenesis: *In Vivo* Systems (Part B)
Edited by David A. Cheresh

Volume 446. Programmed Cell Death, The Biology and Therapeutic Implications of Cell Death (Part B)
Edited by Roya Khosravi-Far, Zahra Zakeri, Richard A. Lockshin, and Mauro Piacentini

Volume 447. RNA Turnover in Prokaryotes, Archae and Organelles
Edited by Lynne E. Maquat and Cecilia M. Arraiano

Volume 448. RNA Turnover in Eukaryotes: Nucleases, Pathways and Anaylsis of mRNA Decay
Edited by Lynne E. Maquat and Megerditch Kiledjian

Volume 449. RNA Turnover in Eukaryotes: Analysis of Specialized and Quality Control RNA Decay Pathways
Edited by Lynne E. Maquat and Megerditch Kiledjian

Volume 450. Fluorescence Spectroscopy
Edited by Ludwing Brand and Michael Johnson

CHAPTER ONE

Initiation of Autophagy by Photodynamic Therapy

David Kessel* *and* Nancy L. Oleinick†

Contents

Abstract

Photodynamic therapy (PDT) involves the irradiation of photosensitized cells with light. Depending on localization of the photosensitizing agent, the process can induce photodamage to the endoplasmic reticulum (ER), mitochondria, plasma membrane, and/or lysosomes. When ER or mitochondria are targeted, antiapoptotic proteins of the Bcl-2 family are especially sensitive to photodamage. Both apoptosis and autophagy can occur after PDT, autophagy being associated with enhanced survival at low levels of photodamage to some cells. Autophagy can

* Department of Pharmacology, Wayne State University School of Medicine, Detroit, Michigan, USA
† Department of Radiation Oncology, Case Western Reserve University, Cleveland, Ohio, USA

Methods in Enzymology, Volume 453
ISSN 0076-6879, DOI: 10.1016/S0076-6879(08)04001-9

become a cell-death pathway if apoptosis is inhibited or when cells attempt to recycle damaged constituents beyond their capacity for recovery. While techniques associated with characterization of autophagy are generally applicable, PDT introduces additional factors related to unknown sites of photodamage that may alter autophagic pathways. This chapter discusses issues that may arise in assessing autophagy after cellular photodamage.

1. Introduction

Photodynamic therapy (PDT) is based on the ability of certain photosensitizing agents to localize in malignant cells and tissues. Subsequent irradiation with light corresponding to an absorbance optimum of the photosensitizer leads to an energy-transfer process that results in the conversion of molecular oxygen to a reactive oxygen species (ROS) termed singlet oxygen. The highly reactive singlet oxygen oxidizes cellular molecules, usually nearby lipids and proteins. If sufficient drug and light are provided, this can result in a severe oxidative stress in the cells and a very selective means for tumor eradication (Dougherty *et al.*, 1998). Additional ROS may also be formed, including superoxide anion radical, hydrogen peroxide, and hydroxyl radical, any of which can have adverse effects on cellular functions.

Autophagy has been identified as a mode of cell death after PDT and may predominate where apoptosis is unavailable (Buytaert *et al.*, 2006, 2008; Kessel *et al.*, 2006; Xue *et al.*, 2007). Antiapoptotic proteins of the Bcl-2 family, located in the mitochondria and/or endoplasmic reticulum (ER), are among the most sensitive targets for many commonly used photosensitizing drugs (Kim *et al.*, 1999; Xue *et al.*, 2001a). Loss of such proteins can be a trigger for both apoptosis and autophagy (Pattingre *et al.*, 2005). When lysosomes are targeted, PDT leads to the release of lysosomal enzymes, thereby catalyzing the conversion of Bid to the proapoptotic cleavage product termed t-Bid (Reiners *et al.*, 2002). Because of the involvement of lysosomes in the autophagic process, this can have additional implications (i.e., in the fusion of autophagosomes with lysosomes).

Autophagy can be a factor in PDT at two points in the sequence of events. The first involves the ability of autophagy to repair photodamaged cellular components, including organelles. This can result in a loss of the shoulder on the dose-response curve, generally taken to indicate damage repair (Fig. 1.1). Silencing of the autophagy gene designated Atg7 results in the photosensitization of mouse leukemia L1210 cells to photodynamic effects (Kessel and Arroyo, 2007). A similar finding was recently reported with respect to effects of the silencing of other factors involved in autophagy on the lethality of ionizing radiation (Apel *et al.*, 2008). However, in human

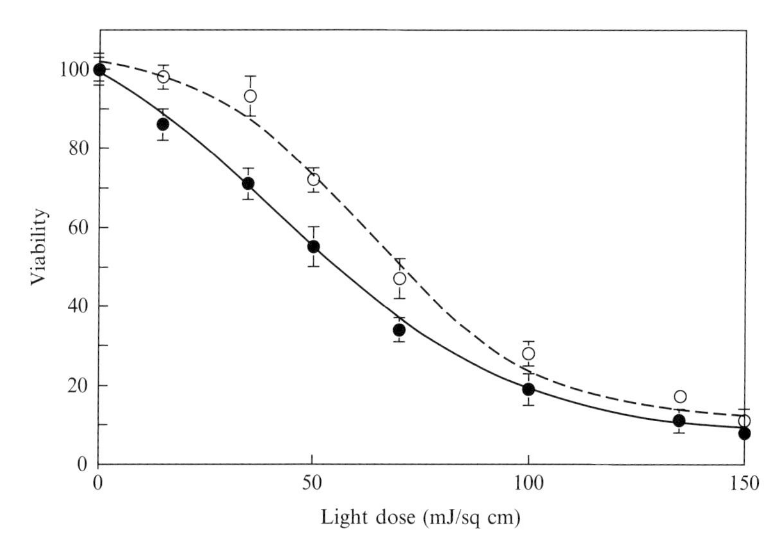

Figure 1.1 Dose-response curve obtained with wild-type L1210 cells (dashed line) and an Atg7 knockdown subline (solid line). Cells were photosensitized with the porphycene CPO and the light dose was varied as indicated. Viability was assessed by clonogenic assays.

breast cancer MCF-7 cells, silencing of Atg7 may make the cells more resistant to PDT (L.Y. Xue, S.M. Chiu, S. Joseph, and N.L. Oleinick, unpublished).

An additional concern in the context of PDT is the propensity of most photosensitizing agents to migrate to relatively hydrophobic intracellular sites (i.e., membranes). Irradiation may therefore result in the inactivation of proteins required for the autophagic process.

2. Photosensitizing Agents

Several photosensitizing agents have been approved for clinical use, and many others are in clinical and preclinical trials. The agents are not always commercially available. Common porphyrins and phthalocyanines can be obtained from the major suppliers (e.g., Sigma-Aldrich). A large selection of porphyrins, phthalocyanines, and related compounds is available from Frontier Science (Logan, Utah; an inventory can be seen at the website www.info@frontiersci.com). Frontier Science can often provide a custom preparation of almost any agent whose structure and synthetic route have been published. Photosensitizers prepared for clinical trials or those received directly from a synthetic chemist are generally supplied at a documented high level of purity; those from commercial sources may require additional purification. The majority of the agents in current use are porphyrins, porphyrin-derived compounds (e.g., benzoporphyrins or pheophorbides), or related types of macrocycle structures (e.g., phthalocyanines). All contain hydrophobic ring systems, even the nonporphyrin photosensitizers (e.g., hypericin). Structures of the latter compound, the porphyrin skeleton and the porphycene CPO are shown in Fig. 1.2.

The concentration and time of incubation will vary as a function of hydrophobicity. Agents at the extreme ends of the solubility spectrum often require extended loading incubation times (Table 1.1). For delivery to cells, the hydrophobic photosensitizers are generally dissolved in a biocompatible organic solvent, such as ethanol, dimethylformamide or DMSO, or in water containing low (0.1%) concentration of a detergent such as Tween 80, and a small aliquot of the stock solution is added to the serum-containing medium bathing the cells. It is necessary to check that the vehicle is not toxic, but if the level of organic solvent remains below 0.1% of the medium volume, we have found no effect from it. Entry of the photosensitizers into the cells involves partitioning of the compounds to intracellular hydrophobic sites; there is little evidence for other mechanisms of cellular uptake.

While there are reports in the literature concerning sites of localization of different photosensitizing agents (e.g., Kessel *et al.*, 1997; Trivedi *et al.*, 2000; Lam *et al.*, 2001; Usuda *et al.*, 2003; Buytaert *et al.*, 2006), it is usually

Table 1.1 Typical incubation conditions for photosensitizing agents

Sensitizer	Relative hydrophobicity	Concentration (μM)	Time (h)
CPO	moderate	2	0.5
NPe6	very low	30	16
mTHPC	very high	2–4	16
Pc 4	high	0.2	1–18

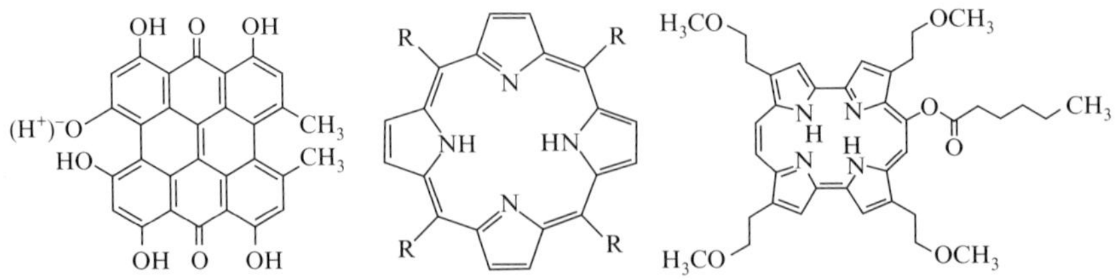

Figure 1.2 Structure of some typical photosensitizing agents. From left: hypericin, the basic porphyrin ring structure and the porphycene termed *CPO*.

necessary for each cell line to be evaluated with regard to these properties. Because the photosensitizers are not directed to specific receptors, and because the structural requirements for localization patterns have not been established for any class of photosensitizers, what applies for one or a few cell lines may not necessarily be pertinent to others.

3. Additional Factors Unique to PDT

Unlike conventional drug-treatment protocols, photodynamic effects occur only when photosensitized cells are irradiated. It may be necessary to incubate cells for several hours so as to obtain a sufficient intracellular concentration of the photosensitizing agent (Table 1.1), but formation of ROS will occur only when irradiation begins. To mimic physiological conditions, irradiation is often carried out at 37 °C, but by manipulating the temperature of irradiation, it is possible to alter the immediate consequences. For example, after Bcl-2 photodamage, insertion of Bax into mitochondria occurs only when the temperature is greater that 15 °C (Pryde *et al.*, 2000). It is therefore possible to carry out irradiation at a temperature $<$15 °C so as to prevent the immediate initiation of apoptosis. If a chilled microscope stage is used for observations, it is then feasible to observe mitochondrial or ER photodamage before the initiation of apoptosis (Kessel and Castelli, 2001). This type of strategy can be important in identifying early events in PDT, as apoptosis causes drastic changes in cell morphology. This may also be true with regard to autophagy.

4. A Typical PDT Protocol

While the precise conditions may vary, a procedure for photosensitization and irradiation of cells in suspension culture using CPO involves the following steps (a modified protocol for cells that grow attached to a substratum is provided in the section labeled "Dose-response data"):

1. Cells are collected in the exponential phase of growth and suspended in fresh growth medium at a density of 3.5×10^5 per ml, with 20 m*M* HEPES, pH 7.0, replacing $NaHCO_3$ to permit maintenance of a near-neutral pH.
2. Cells are loaded with CPO (final concentration = 2 μM) for 30 min at 37 °C, then collected by centrifugation ($100 \times g$, 30 s) and resuspended in fresh HEPES-buffered growth medium containing 10% serum, at 10 °C.

3. A photosensitized cell suspension is placed in a 1-ml glass tube in a thermoelectrically cooled Petri dish (10 °C) and irradiated with light corresponding to the long-wavelength absorbance bands of CPO as defined by a broad-band 600 ± 30 nm interference filter. The precise time will depend on prior experiments that establish a dose-response curve on the basis of clonogenic viability studies. With the murine leukemia L1210 cell line, a 2-min irradiation is required.
4. Cells can then be used directly for an estimate of photodamage by fluorescence microscopy using appropriate probes (Kessel and Reiners, 2007) or incubated for 10–60 min at 37 °C with the cells subsequently used for the determination of autophagic and/or apoptotic effects (e.g., DEVDase activation, LC3 processing, or vacuole formation detected by phase-contrast and electron microscopy).
5. With adhering cells, a similar protocol is followed except that cells are grown on 1.5-cm circular glass disks and incubated in small Petri dishes throughout.
6. We routinely establish viability by plating appropriate dilutions of cells on soft agar and permitting colonies to develop over 6–10 days. When this is done, sterile techniques are employed throughout.
7. Pellets containing 2.5×10^6 cells are lysed in 100 μl of a mixture of 10 m*M* HEPES, pH 7.5, 130 m*M* NaCl, 1% Triton X-100, 10 m*M* NaF, 10 m*M* Na pyrophosphate and 1 m*M* PMSF. After 10 min at 4 °C, the mixture is clarified by brief centrifugation (10,000×*g*, 3 min). The protein content is determined on a 10 μl aliquot and the remainder used for assays (e.g., caspase activity or LC3 processing).

5. Identification and Characterization of Autophagy after PDT

Identification of autophagy can be ambiguous. No kits are available for qualitative or quantitative analysis. Some techniques that have proven useful are outlined subsequently.

5.1. Phase-contrast microscopy

At intervals following irradiation, cells are inspected for vacuole formation using phase-contrast microscopy. While this cannot be considered an unambiguous procedure, the appearance of such vacuoles (compared to a control culture that is irradiated without the photosensitizer) is often associated with an autophagic response (Buytaert *et al.*, 2006; Kessel and Arroyo, 2007; Kessel and Reiners, 2007; Kessel *et al.*, 2006).

5.2. Fluorescence microscopy

Adequate image acquisition can be carried out with any conventional fluorescence microscope. We have successfully used the Nikon Eclipse E600 system with Plan fluor objectives specifically designed for fluorescence detection. This is an upright system but can be used to examine adhering cells if coverslips are placed in the dishes being used for cell culture.

It is helpful to have phase rings incorporated into the objectives so that both phase-contrast and fluorescence images can be obtained. We have also used a Plan Apo 60X water immersion objective. This has the advantage of using water rather than oil so that there is less drag on the cover slip when acquiring a Z-series of images. This procedure is helpful in assessing numbers of lysosomes, or acidic vesicles. Use of a Nikon Apo 60X objective enhances resolution of images. This objective does not contain a phase ring, improving sensitivity to fluorescence, but eliminating the possibility for comparing phase with fluorescence images.

5.3. Techniques in fluorescence microscopy

A qualitative estimate of initiation of autophagy after PDT can be obtained by examining the subsequent appearance of vesicles labeled with monodansylcadaverine (MDC) or any of the LysoTracker or LysoSensor probes provided by Molecular Probes/Invitrogen (Eugene, OR). In a typical study, cells are labeled with 10 μM MDC for 10 min at 37 °C, and patterns of punctate green fluorescence determined using 360–380 nm excitation and 520–560 nm emission (also see the chapter by Vázquez and Colombo in this volume). There is some controversy in the literature as to whether this is (Iwai-Kanai *et al.*, 2008) or is not (Bampton *et al.*, 2005) a suitable test for autophagy.

Enhanced numbers of lysosomes or other acidic vesicles can be visualized using one of the LysoTrackers or LysoSensors provided by Molecular Probes. Incubation for 5 min at 37 °C is usually sufficient for labeling of L1210 cells in suspension culture using 100 nM of LysoTracker Red or 100 nM of the LysoSensors. Of the probes in current use, we have tested LysoTracker Red (λ_{ex} = 570 nm, λ_{em} = 590 nm), LysoSensor Yellow (λ_{ex} = 465 nm, λ_{em} = 535 nm), and LysoSensor Green (λ_{ex} = 440 nm, λ_{em} = 505 nm). Although the latter probes have a greater Stokes shift, any of these probes is satisfactory for assessing relative numbers of lysosomes and other acidic vesicles. It may be feasible to obtain more quantitative data using flow cytometry.

To improve the quantitation of labeling patterns, a fluorescence microscope with a Z-drive is useful. A confocal microscope will eliminate out-of-plane emissions, allowing both sharper images and the ability to visualize a series of optical planes through the cells using the Z-drive. This permits acquisition of a series of stacked images, with subsequent image processing capable of resolving numbers of fluorescent loci. The Metamorph software contains a program for selecting the in-focus pixels from a stack of images. Alternatively, removal of blur from out-of-focus pixels can be better accomplished with AutoQuant software (AutoQuant Imaging, Troy, NY). This can offer an improved assessment of MDC or LysoTracker fluorescence.

To date, we have only limited experience using GFP-labeled LC3, a procedure that has been successful in other contexts (Klionsky *et al.*, 2008). Preliminary studies suggest that a substantial increase in appearance of punctate fluorescence is associated with other indices of autophagy after photodamage.

5.4. Electron microscopy

Electron microscopy provides an estimate of the nature of vacuoles associated with PDT. For this purpose, cells are fixed with 2.5% glutaraldehyde in phosphate-buffered saline, and stored for processing (also see the chapter by Ylä-Anttila *et al.*, in volume 452). It is important to include treatment with uranium acetate and lead citrate for optimal visualization of the double-membrane structure associated with autophagy. A guide to interpretation of data has been reported (Eskelinen, 2008). The major problems to consider are that (1) supralethal light doses can obliterate most cellular structures (something that can also be visualized by phase-contrast microscopy), and (2) it is important to examine a population of control cells to assess the level of autophagy in the normal cell population. In MCF-7 cells, a large increase in the content of autophagosomes was observed following PDT (Xue *et al.*, 2007). Typical results are as shown in Fig. 1.3.

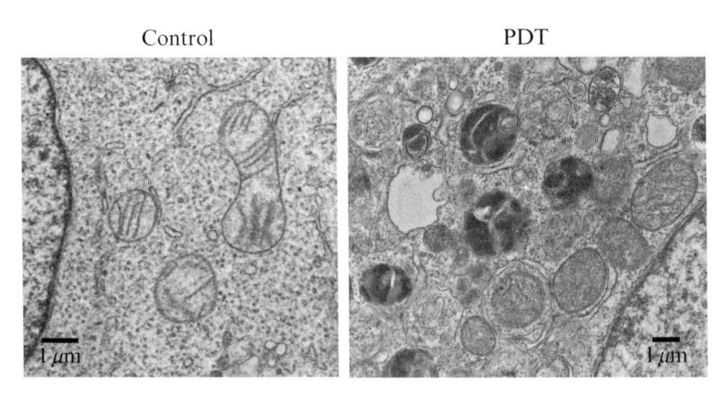

Figure 1.3 Electron micrographs of untreated (left) and PDT-treated (right) MCF-7 cells. Note the presence of numerous autophagosomes after PDT, some containing what appear to be degraded mitochondria.

5.5. Western blots: LC3 processing

Western blots as an index of LC3 processing have proved useful for providing a semiquantitative indication of autophagy, if appropriate precautions are taken (Mizushima and Yoshimori, 2007) (also see the chapter by Kimura *et al.*, in this volume). A useful, but sometimes neglected procedure for assessing the autophagic flux involves carrying out experiments in the presence versus absence of lysosomal protease inhibitors (e.g., E64d and acetyl pepstatin). With the murine L1210 leukemia cells, where autophagy can be detected within 30–60 min after photodamage, these inhibitors can be added during the sensitizer-loading incubation. If longer intervals are required, it will be necessary to determine that the inhibitors are still active during the process and/or have not induced toxic responses by themselves. Perhaps additional supplements will be required. This procedure is found to provide an indication of autophagy after ER photodamage to L1210 cells (Kessel and Arroyo, 2007). However, in MCF-7 cells, the inhibitors cause extensive accumulation of LC3-II independent of PDT and do not alter the level of LC3-II that accumulates in the cells in response to PDT (Xue *et al.*, 2007). Thus, in this latter case, it is possible that both the inhibitors and PDT with the photosensitizer Pc 4 block autophagy at the same late stage of proteolysis following fusion of autophagosomes with lysosomes.

6. Effects of Autophagy on PDT Responses

6.1. Dose-response data

In the context of PDT, the simplest method for altering the dose is to vary the light intensity or duration. Under these conditions, a dose-response curve can be generated and effects of autophagy most readily observed by use of appropriate knockouts. Another simple method for altering the dose is to vary the amount of the photosensitizer in the cell culture medium, allowing sufficient time for cell uptake of the photosensitizer to reach a maximum level. In such a protocol, all of the cultures would then receive the same light dose. In either case, an appropriate measure of overall cell killing is required. A variety of short-term assays of cell viability, such as uptake of vital dyes (e.g., trypan blue) or determination of the ability of cell mitochondria to reduce a tetrazolium dye to a colored product (e.g., MTT assay), are often used because of their relative simplicity and ease. We prefer to avoid these assays whenever possible, as none give a total accounting of cell deaths, both immediate and delayed. Colony-forming assays provide the most reliable assessment of the percentage of cells in a population that are capable of dividing indefinitely and forming a colony. Moreover, short-term assay results can be misleading (Xue *et al.*, 2001b).

There are two types of clonogenic assay used for adhering cells, each of which has advantages and disadvantages. We call these preplate and post-plate protocols. In the former, small numbers of cells (e.g., 100, 300, 1,000) are plated in tissue-culture dishes (60-mm diameter) and allowed to attach to the substratum. This usually requires approximately 4 h. The cells in some of the dishes are exposed to the photosensitizer and light, while other dishes receive photosensitizer only, light only, or no treatment. Then the dishes are returned to the incubator for 1–2 weeks to allow colony formation. In the postplate protocol, cells are treated while in exponential growth or after they have reached confluence. Cultures are exposed to PDT or are designated controls, and then the cells from each plate are recovered by trypsinization, counted, and replated in multiple plates in sufficient number to yield 50–100 colonies. This number is chosen to allow accurate counting of individual colonies while providing statistical significance. The advantage of the preplate protocol is its comparative simplicity. However, a serious disadvantage is that the extent of uptake of the photosensitizer into the small number of cells plated may be quite different from that into each cell in a subconfluent culture. As a result, the toxicity measured with the preplate protocol may not reflect the toxicity under the same conditions in which most molecular studies (e.g., LC3 processing) are carried out. In contrast, in the postplate protocol, identical cultures may be exposed to PDT, then processed for colony formation or molecular measurements. However, it has been observed occasionally that monolayer cultures subjected to some PDT protocols can become so firmly attached to the substrate by PDT that it is not possible to remove them with trypsin (Ball *et al.*, 2001; Uzdensky *et al.*, 2004). If this occurs, the postplate protocol cannot be carried out with that combination of cells, photosensitizer, and dose. In most cases, this is not a problem, and because any number of cells up to 10^6 may be plated to determine colony formation, it is possible to determine if the treatment has left as little as 1 live cell out of a million treated. Because this is the method we advocate, the procedure will be described in more detail.

6.2. A protocol for determining the effect of PDT on the colony-forming ability of adhering cultures

1. Setting up the cultures. Sterile technique is used for steps 1–5. A series of T-25 flasks is inoculated with sufficient cells to yield replicate subconfluent cultures on the planned day of the experiment (day 0). As an example, for MCF-7 cells, we routinely inoculate $\approx 7 \times 10^5$ cells into each of 10–20 T-25 flasks 2 days before the day of PDT (i.e., on day –2). These cultures are allowed to grow in a tissue culture incubator at 37 °C

in a humidified atmosphere of 5% CO_2 in air. When used for the PDT experiment on day 0, there are approximately 2×10^6 cells in each flask.
2. Adding the photosensitizer. On day −1, an aliquot of a photosensitizer stock solution is added to the medium above the cell monolayer. The timing of photosensitizer addition and dose are determined in preliminary experiments. Issues concerning solvents for hydrophobic photosensitizers have been discussed previously.
3. Photoirradiation. Flasks are exposed one at a time to the desired wavelengths of light appropriate for the photosensitizer at the light fluence(s) and fluence rate(s) to be studied. We typically use an LED array with a peak output at 675 nm, which is chosen because it is close to the absorption maximum of phthalocyanines, but broad-band light sources with suitable filters can also be used, as discussed previously.
4. Recovery of cells from the monolayer. Following irradiation, the cells from each culture are released from the monolayer by trypsinization (1 mL of 0.25% trypsin-EDTA purchased from Thermo Scientific, South Logan, UT). The trypsin is blocked by addition of 4 mL of serum-containing medium, and the cell suspension is pipetted several times to break cell clumps. The presence of single cells is checked under a microscope.
5. Processing of cells for colony formation. An aliquot of the cells is counted, either with an electronic cell counter or a hemacytometer, and the initial cell density is calculated. Serial dilutions of the cell stock are made into medium such that the final dilution will result in the desired number of cells being delivered to each of at least 3 60-mm tissue-culture dishes. Calculation of that number requires knowing the plating efficiency (PE; the number of colonies formed per 100 cells plated) of the untreated cells from preliminary tests. For many human cancer cell lines, the PE is 20%–50%, whereas for many rodent cell lines, the PE can be 80% or greater. For example, for a cell line that has a PE of 25%, if 400 cells are plated, 100 colonies should be obtained. For a culture that was exposed to a dose of PDT that is expected to kill 90% of the cells, 4000 cells should be plated in each dish to result in 100 colonies. If the expected survival is not known, it may be necessary to plate, for the example given, 3 dishes of 4000 cells, 3 with half and 3 with double that number. All dishes are placed in an incubator and left undisturbed (generally 1–2 weeks) to allow observable individual colonies to form.
6. Determination of the survival response. The medium is removed, and the monolayer is covered with approximately 2 mL of 0.1% crystal violet in 20% ethanol, and after approximately 5 min to allow staining of the cells, the dishes are washed by gentle immersion into a tub of cool water. The plates are turned upside down to drain and dry. The colonies are then counted by eye under a dissecting microscope or magnifying glass.

For each dish, the number of colonies containing at least 50 cells is divided by the number of cells plated to determine the PE. The PE values of triplicate dishes are averaged and normalized to the PE of the controls.

6.3. Role of autophagy in the sensitivity of cells to PDT

Use of siRNA to produce a significant decrease in Atg7 reveals that autophagy can be a survival process at the lower end of the dose-response curve in mouse leukemia L1210 cells (Kessel and Arroyo, 2007; Kessel and Reiners, 2007). In contrast, MCF-7 human breast cancer cells deficient in Atg7 (Abedin *et al.*, 2006) are more resistant to the lowest doses of PDT, as revealed by an increased shoulder on the clonogenic survival curve (L.Y. Xue, S.M. Chiu, S. Joseph, and N.L. Oleinick, unpublished). Some investigators have advised use of different knockouts (e.g., comparing Atg7 with Atg5) to determine whether a process unique to one of the gene products might skew the results. The cell type– and/or agent-dependence of cell responses to loss of Atg7 is also observed with Atg5 knockdown. Thus, murine embryonic fibroblasts (MEFs) deficient in Atg5 are more sensitive to treatment with staurosporine or tunicamycin but more resistant to treatment with menadione or UVC-radiation than the Atg5-replete MEFs (Wang *et al.*, 2008), and SK-N-SH neuroblastoma cells deficient in Atg5 are sensitized to cell death, as compared with wild-type cells, when treated with ER stressors but not when treated with staurosporine or when pretreated with rapamycin to inhibit mTOR activity (Ogata *et al.*, 2006). Clearly, the mechanisms and roles of autophagy following PDT could also differ markedly in different cell types, depending on a variety of cell characteristics, including their propensity to undergo apoptosis, and in response to agents that produce different types or locations of damage. For PDT, this means that different responses may also follow treatment with photosensitizers that localize to different organelles.

6.4. Sequence of events

Autophagy is expected to occur at a minimum of two points in the cellular response to photodamage. When the ER and/or mitochondria are targets for photodamage, the first response occurs shortly after irradiation of photosensitized L1210 cells (Kessel and Arroyo, 2007). Autophagic vacuoles are observed within 15 min after irradiation, presumably associated with the recycling of photodamaged organelles. A second phase of autophagy occurs after the apoptotic response is ending and may reflect the fate of cells that attempt an excessive amount of recycling and cannot survive. After 24 h, there are a few survivors along with highly vacuolated cells that are freely permeably to propidium iodide. Phase-contrast images showing the appearance of autophagic vacuoles and apoptotic morphology after an LD_{90} PDT dose are shown in Fig. 1.4. As with the many other measures, the same sequence of events is extended over a longer time period in carcinoma cells than in the leukemic cells (Xue *et al.*, 2007). In human breast cancer MCF-7 cells, substantial increases in LC3-II can be observed by 2–24 h after PDT (Fig. 1.5). This may indicate that PDT is promoting the early stage of apoptosis but interfering with the later stages of autophagosome processing or LC3-II degradation.

6.5. Apoptosis versus autophagy

The murine L1210 cell line is useful for the study of apoptosis *vs.* autophagy as cell death mechanisms. Silencing the Atg7 gene appears to essentially abolish autophagy, whereas loss of Bax is sufficient to substantially decrease apoptosis (Kessel and Arroyo, 2007). Figure 1.6 shows the relative response of this cell line to 90% photokilling with the porphycene CPO. One hour after irradiation, a few apoptotic cells are detected with HO33342, and

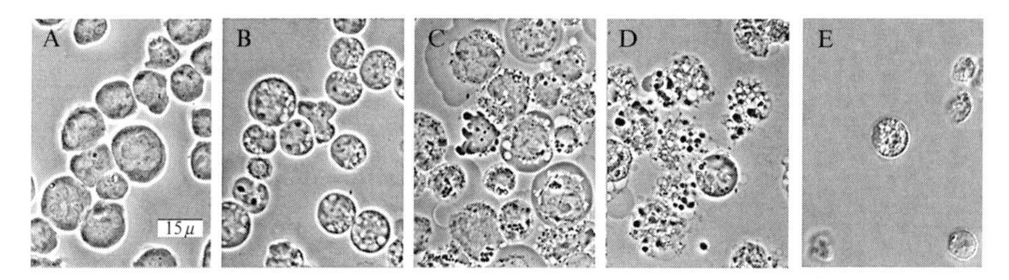

Figure 1.4 Progression of autophagy and apoptosis after photodynamic therapy. Murine leukemia L1210 cells were incubated for 30 min with a 2 μM concentration of the porphycene CPO, resuspended in fresh medium, irradiated (135 mJ/cm^2) with light (600–640 nm) at 10 °C, then warmed to 37 °C. Phase-contrast images were obtained at intervals. (*A*) Before irradiation, (*B*) 15 min after irradiation, (*C*) after 1 h, (*D*) after 4 h, and (*E*) after 24 h.

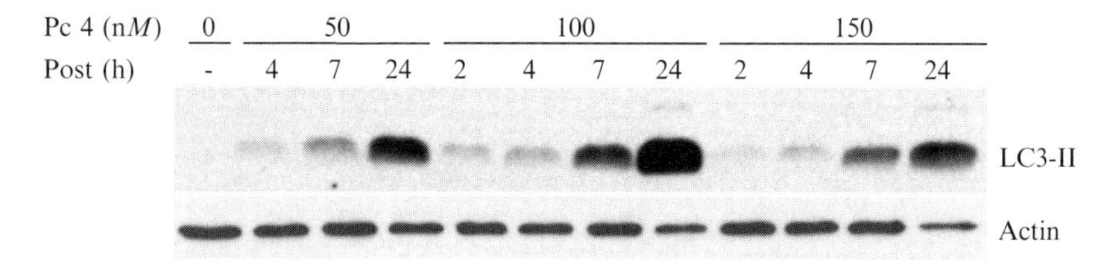

Figure 1.5 Time-course and PDT dose-response for the accumulation of LC3-II in MCF-7c3 cells. Cells were exposed to the indicated concentrations of the photosensitizer Pc 4 overnight, followed by photoirradiation with 200 mJ/cm^2 of red light from a light-emitting photodiode array. Cells were collected at the indicated times post-irradiation and analyzed on Western blots.

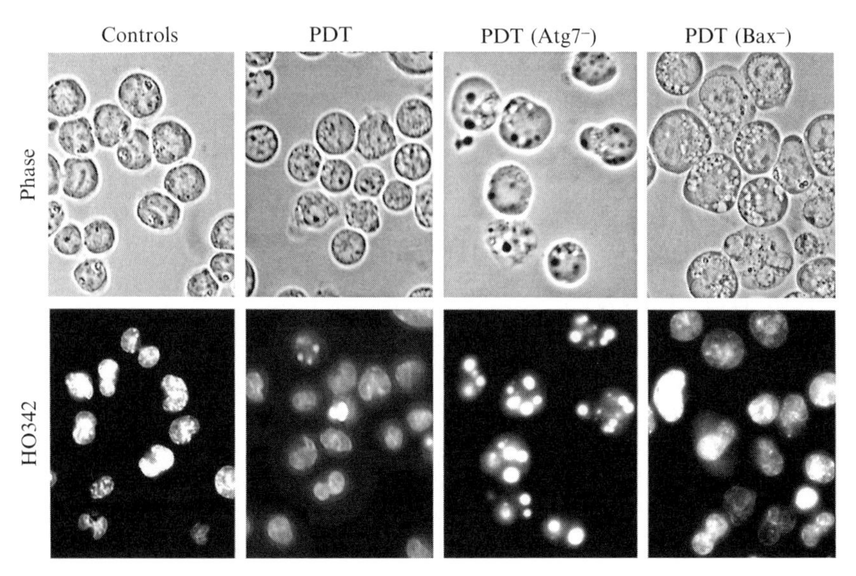

Figure 1.6 Murine leukemia cells were photosensitized and irradiated as shown in Fig. 1.2. Phase-contrast (top) and fluorescence images after treatment with Höchst dye HO33342 [HO342] labeling (bottom) were obtained 1 hr after a subsequent incubation at 37 °C. (A) L1210 controls, (B) L1210 + PDT, (C) L1210/Atg7 + PDT, (D) L1210/Bax + PDT.

some vacuoles are apparent. When Atg7 is silenced, there is a much more substantial apoptotic response, while loss of Bax results in a substantially greater number of vacuoles. In studies from the Kessel laboratory cited previously, vacuolization in this cell line is accompanied by LC3 processing, together with the appearance of double-membranes observed by electron microscopy.

Processing of LC3 appears to occur in all of the cell types that have been examined following PDT (Buytaert *et al.*, 2006; Kessel *et al.*, 2006; Xue *et al.*, 2007). Buytaert *et al.* (2006) found greater levels of PDT-induced LC3-II in Bax/Bak-double knockout MEFs that are deficient in apoptosis than in either the same cells transfected with mitochondrion-directed Bax (where the ability to undergo apoptosis is restored) or in apoptosis-competent wild-type MEFs. Although data for only one PDT dose and one post-PDT time are reported, the results suggest that the autophagic response in this system is greater when apoptosis is defective. In contrast, in a comparison of procaspase-3-deficient versus procaspase-3-overexpressing MCF-7 cells (Xue *et al.*, 2007), there is no marked difference in the rate or extent of accumulation of LC3-II, suggesting that the initiation of autophagy is independent of the ability of the cells to undergo apoptosis. However, apoptosis dominates as a mechanism of cell death in those cells having a fully constituted apoptotic pathway, whereas a non-apoptotic pathway, possibly autophagy, is responsible for cell death when apoptosis is compromised (Xue *et al.*, 2007). Furthermore, as there is the potential for cross talk between the two pathways (e.g., via the interaction of Beclin 1 with Bcl-2), further research is needed to explore more fully the relationship between apoptosis and autophagy following PDT.

ACKNOWLEDGMENTS

The authors' research is supported by NIH grants R01 CA83917, R01 CA106491, P30 CA43703, and CA 23378 from the National Cancer Institute, DHHS, and by the State of Ohio Biomedical Research and Technology Transfer Trust TECH 05-063.

REFERENCES

Abedin, M. J., Wang, D., McDonnell, M. A., Lehmann, U., and Kelekar, A. (2006). Autophagy delays apoptotic death in breast cancer cells following DNA damage. *Cell Death Differ.* **14,** 500–510.

Apel, A., Herr, I., Schwarz, H., Rodemann, H. P., and Mayer, A. (2008). Blocked autophagy sensitizes resistant carcinoma cells to radiation therapy. *Cancer Res.* **68,** 1485–1494.

Ball, D. J., Mayhew, S., Vernon, D. I., Griffin, M., and Brown, S. B. (2001). Decreased efficiency of trypsinization of cells following photodynamic therapy: Evaluation of a role for tissue transglutaminase. *Photochem. Photobiol.* **73,** 47–53.

Bampton, E. T., Goemans, C. G., Niranjan, D., Mizushima, N., and Tolkovsky, A. M. (2005). The dynamics of autophagy visualized in live cells: From autophagosome formation to fusion with endo/lysosomes. *Autophagy* **1,** 23–36.

Buytaert, E., Callewaert, G., Hendrickx, N., Scorrano, L., Hartmann, D., Missiaen, L., Vandenheede, J. R., Heirman, I., Grooten, J., and Agostinis, P. (2006). Role of endoplasmic reticulum depletion and multidomain proapoptotic BAX and BAK proteins in shaping cell death after hypericin- mediated photodynamic therapy. *FASEB J.* **20,** 756–758.

Buytaert, E., Matroule, J. Y., Durinck, S., Close, P., Kocanova, S., Vandenheede, J. R., de Witte, P. A., Piette, J., and Agostinis, P. (2008). Molecular effectors and modulators of hypericin-mediated cell death in bladder cancer cells. *Oncogene* **27,** 1916–1929.

Dougherty, T. J., Gomer, C. J., Henderson, B. W., Jori, G., Kessel, D., Korbelik, M., Moan, J., and Peng, Q. (1998). Photodynamic therapy. *J. Natl. Cancer Inst.* **90,** 889–905.

Eskelinen, E.-L. (2008). To be or not to be? Examples of incorrect identification of autophagic compartments in conventional transmission electron microscopy of mammalian cells. *Autophagy* **4,** 257–260.

Iwai-Kanai, E., Yuan, H., Huang, C., Sayen, M. R., Perry-Garza, C. N., Kim, L., and Gottlieb, R. A. (2008). A method to measure cardiac autophagic flux *in vivo*. *Autophagy* **4,** 322–329.

Kessel, D., and Arroyo, A. S. (2007). Apoptotic and autophagic responses to Bcl-2 inhibition and photodamage. *Photochem. Photobiol. Sci.* **6,** 1290–1295.

Kessel, D., and Castelli, M. (2001). Evidence that bcl-2 is the target of three photosensitizers that induce a rapid apoptotic response. *Photochem. Photobiol.* **74,** 318–322.

Kessel, D., Luo, Y., Deng, Y., and Chang, C. K. (1997). The role of subcellular localization in initiation of apoptosis by photodynamic therapy. *Photochem. Photobiol.* **65,** 422–426.
Kessel, D., and Reiners, J. J. Jr. (2007). Apoptosis and autophagy after mitochondrial or endoplasmic reticulum photodamage. *Photochem. Photobiol.* **83,** 1024–1028.
Kessel, D., Vicente, M. G., and Reiners, J. J. Jr. (2006). Initiation of apoptosis and autophagy by photodynamic therapy. *Lasers Surg. Med.* **38,** 482–488.
Kim, H. R., Luo, Y., Li, G., and Kessel, D. (1999). Enhanced apoptotic response to photodynamic therapy after Bcl-2 transfection. *Cancer Res.* **59,** 3429–3432.
Klionsky, D., Abeliovich, H., Agostinis, P., Agrawal, D. K., Aliev, G., Askew, D. S., Baba, M., Baehrecke, E. H., Bahr, B. A., Ballabio, A., Bamber, B. A., Bassham, D. C., *et al.* (2008). Guidelines for the use and interpretation of assays for monitoring autophagy in higher eukaryotes. *Autophagy* **4,** 151–175.
Lam, M., Oleinick, N. L., and Nieminen, A. L. (2001). Photodynamic therapy-induced apoptosis in epidermoid carcinoma cells: Reactive oxygen species and mitochondrial inner membrane permeabilization. *J. Biol. Chem.* **276,** 47379–47386.
Mizushima, N., and Yoshimori, T. (2007). How to interpret LC3 immunoblotting. *Autophagy* **3,** 542–545.
Ogata, M., Hino, S., Saito, A., Morikawa, K., Kondo, S., Kanemoto, S., Murakami, T., Taniguchi, M., Tanii, I., Yoshinaga, K., Shiosaka, S., Hammarback, J. A., Urano, F., and Imaizumi, K. (2006). Autophagy is activated for cell survival after endoplasmic reticulum stress. *Mol. Cell. Biol.* **26,** 9220–9231.
Pattingre, S., Tassa, A., Qu, X., Garuti, R., Liang, X. H., Mizushima, N., Packer, M., Schneider, M. D., and Levine, B. (2005). Bcl-2 antiapoptotic proteins inhibit Beclin 1-dependent autophagy. *Cell* **122,** 927–939.
Pryde, J. G., Walker, A., Rossi, A. G., Hannah, S., and Haslett, C. (2000). Temperature-dependent arrest of neutrophil apoptosis. Failure of Bax insertion into mitochondria at 15 degrees C prevents the release of cytochrome c. *J. Biol. Chem.* **275,** 33574–33584.
Reiners, J. J. Jr., Caruso, J. A., Mathieu, P., Chelladurai, B., Yin, X. M., and Kessel, D. (2002). Release of cytochrome c and activation of pro-caspase-9 following lysosomal photodamage involves Bid cleavage. *Cell Death Differ.* **9,** 934–944.
Trivedi, N. S., Wang, H.-W., Nieminen, A. L., Oleinick, N. L., and Izatt, J. A. (2000). Quantitative analysis of Pc 4 localization in mouse lymphoma cells via double-label confocal fluorescence microscopy. *Photochem. Photobiol.* **71,** 634–639.
Usuda, J., Chiu, S. M., Murphy, E. S., Lam, M., Nieminen, A. L., and Oleinick, N. L. (2003). Domain-dependent photodamage to Bcl-2: A membrane-anchorage region is needed to form the target of phthalocyanine photosensitization. *J. Biol. Chem.* **278,** 2021–2029.
Uzdensky, A, Juzeniene, A, Ma, L.W, and Moan, J (2004). Photodynamic inhibition of enzymatic detachment of human cancer cells from a substratum. *Biochim. Biophys. Acta* **1670,** 1–11.
Xue, L. Y., Chiu, S. M., Azizuddin, K., Joseph, S., and Oleinick, N. L. (2007). The death of human cancer cells following photodynamic therapy: Apoptosis competence is necessary for Bcl-2 protection but not for induction of autophagy. *Photochem. Photobiol.* **83,** 1016–1023.
Xue, L. Y., Chiu, S. M., and Oleinick, N. L. (2001a). Photochemical destruction of the Bcl-2 oncoprotein during photodynamic therapy with the phthalocyanine photosensitizer Pc 4. *Oncogene* **20,** 3420–3427.
Xue, L. Y., Chiu, S. M., and Oleinick, N. L. (2001b). Photodynamic therapy-induced death of MCF-7 human breast cancer cells: a role for caspase-3 in the late stages of apoptosis but not for the critical lethal event. *Exp. Cell Res.* **263,** 145–155.
Wang, Y., Singh, R., Massey, A. C., Kane, S. S., Kaushik, S., Grant, T., Xiang, Y., Cuervo, A. M., and Czaja, M. J. (2008). Loss of macroautophagy promotes or prevents fibroblast apoptosis depending on the death stimulus. *J. Biol. Chem.* **283,** 4766–4777.

CHAPTER TWO

Autophagic Cell Death

Michael J. Lenardo,* Christina K. McPhee,† *and* Li Yu*

Contents

Abstract

In this chapter we discuss methods to study autophagic cell death. A large body of evidence demonstrates that autophagy is a cell survival mechanism in response to starvation. The role of autophagy in cell death, however, has long been controversial. Recently, molecular approaches have provided direct evidence that autophagy contributes to cell death in certain contexts. We begin this chapter by outlining methods to quantify cell death, for example by assaying for cell viability. Next, we discuss methods to measure processes involved in cell death, such as caspase activation and autophagy. Finally, we discuss methods to genetically or chemically perturb autophagy to test whether autophagy is required for cell death. Together, these approaches provide a guide to investigate the relationship between autophagy and cell death.

* Laboratory of Immunology, National Institute of Allergy and Infectious Diseases, National Institutes of Health, Bethesda, Maryland, USA
† Department of Cancer Biology, University of Massachusetts Medical School, Worcester, Massachusetts, USA, and Department of Cell Biology and Molecular Genetics, University of Maryland, Maryland, USA

Methods in Enzymology, Volume 453
ISSN 0076-6879, DOI: 10.1016/S0076-6879(08)04002-0

1. Introduction

Some time after the discovery of autophagy in the 1950s, researchers reported the accumulation of autophagosomes in certain types of dying cells (Hendy and Grasso, 1972). Observations regarding the morphology of dying cells led to the categorization of programmed cell death into 3 types: type I, or apoptotic cell death; type II, or autophagic cell death; and type III, or nonlysosomal cell death (Clarke, 1990; Schweichel and Merker, 1973). Until recently, the hypothesis that cells undergoing type II cell death use their own machinery to "eat themselves to death" remained highly controversial (Levine and Yuan, 2005). Obtaining direct evidence for the involvement of autophagy in cell death proved difficult. On the one hand, a large body of evidence demonstrates that autophagy is a cell survival mechanism in response to starvation (Lum *et al.*, 2005; Onodera and Ohsumi, 2005). On the other hand, the absence of molecular evidence demonstrating a requirement for autophagy in causing type II cell death made it unclear whether autophagy that occurs in conjunction with dying cells contributes to cell death or is simply a cell's failed attempt to rescue itself from a lethal insult. A turning point came in the 1990s with a now-classic study by Ohsumi's group, which ushered in the molecular era for autophagy investigations (Mizushima *et al.*, 1998; Harding *et al.*, 1995,1996; Thumm *et al.*, 1994; Tsukada *et al.*, 1993). Thirty-one autophagy genes (*ATG*) and many mutants of these genes have been identified in the past decade (summarized in Klionsky, 2007). At the same time, powerful new molecular tools such as RNAi silencing have been utilized in autophagy studies. This has made it feasible to study the molecular role of autophagy in cell death. In 2004, two groups provided evidence that autophagy can contribute to a programmed cell death mechanism (Shimizu *et al.*, 2004; Yu *et al.*, 2004), and in 2007 Berry and Baehrecke (2007) demonstrated *in vivo* that ATG gene function is required for type II cell death occurring in development.

In this chapter, we attempt to clearly define autophagic cell death and provide experimental approaches for further investigation. During the investigation of cell death, complex cellular effects can emerge. In many cases, dead cells bear the hallmarks of different types of cell death pathways, as multiple cell death pathways can be activated in parallel. Prototypical autophagic cell death has features of necrosis, such as swelling, nucleus and organelle degeneration, and rapid loss of membrane integrity (Yu *et al.*, 2006). There are also examples in which it is believed that apoptosis may be the consequence of autophagy (Espert *et al.*, 2006), which can make it challenging to ascribe cell death to one pathway. Moreover, the categorization of a particular variety of cell death as falling into a single type may actually be misleading because this may mask the complexities of a lethal process that has occurred.

Autophagic cell death, by definition, is a type of cell death that requires autophagy, so 3 criteria must be met to certify autophagic cell death: (1) autophagy must be induced; (2) the induction of autophagy must be parallel to or precede cell death; and (3) inhibition of autophagy must block cell death. We consider any type of cell death fulfilling these criteria to be autophagic cell death (ACD), even if other mechanisms may also contribute to a cell's demise.

The methods related to the study of cell death can be divided into three groups. First, there are methods to quantify cell death, such as trypan blue or propidium iodide (PI) staining. These methods are based on the notion that dying cells have altered features, such as the loss of membrane integrity, which can be used to distinguish them from living cells. However, these methods do not offer information about the underlying biochemical events leading to cell death. Second, there are assays that provide evidence of triggering of a specific biochemical pathway. For example, measurements of caspase activation can potentially classify a case of cell death as due to the biochemical pathway leading to apoptosis. Third, there are methods that will reveal the precise genetic pathway leading to cell death. RNAi silencing, for example, or the use of inhibitors or genetic mutants, can be used to identify specific genes involved. So, to classify a particular case of cell death, researchers can extract information using each of these approaches that will lead to a diagnosis. ACD involves both autophagy and cell death and therefore both biochemical processes must be detected together with evidence for causality of the former for the latter. In the following section, we discuss these methods in detail.

2. Methods to Quantify Cell Death

2.1. Trypan blue staining to detect loss of membrane integrity

1. Collect 1×10^6 cells and wash once with phosphate buffered saline (PBS). Suspend cells in 1 ml of PBS. Transfer 100 μl of 0.4% trypan blue solution (w/v) (Invitrogen, 15250061) to a tube.
2. Add 100 μl of the cell suspension (dilution factor = 2) and mix thoroughly.
3. With the coverslip in place, transfer 10 μl of the trypan blue/cell suspension mixture to each chamber of a hemocytometer. Allow each chamber to fill by capillary action.
4. Count all of the cells in the 1-mm center square and in the 4 1-mm corner squares. Nonviable cells will stain blue due to loss of membrane integrity; viable cells will exclude dye and remain unstained (white). Keep a separate count of viable and nonviable cells. The total number of cells is determined using the following calculations: cells/ml = the average count per reference square × dilution factor × 10^4 (refer to the instructions with the hemocytometer to be certain of the boundaries

of the reference square). Count both sides of the hemocytometer and average the cell counts for better accuracy. Variation can be high, so if very accurate counts are required, repeat the procedure and average the cell counts from all determinations. If cell viability is low, count more squares or carry out repeated samplings of the cell suspension until a minimum of 50 cells/corpses (i.e., 50 events) have been quantified.
5. Cell viability (%) = total number of viable cells (unstained)/total number of cells (stained and unstained) × 100.

NOTES:

1. If cells are exposed to Trypan blue for extended periods of time, viable cells may begin to take up dye.
2. Count cells on the top and left touching the middle line of the perimeter of each square. Do not count cells touching the middle line at the bottom or right sides.
3. Make sure most cells are isolated. If more than 10% of the cells appear to be clustered, repeat the entire procedure.
4. The cell number should be between approximately 20–50 cells per square. If cell numbers are out of range, adjust to an appropriate dilution factor.

2.2. Propidium iodide staining for flow cytometry

Similar to trypan blue, propidium iodide (PI) detects the loss of membrane integrity PI penetrates into dead cells and stains DNA, causing strong emission at 620 nm, which can be easily detected by flow cytometry in the FL-3 channel.

1. Dissolve propidium iodide in buffer comprising 1X PBS, pH 7.5, 0.2% bovine serum albumin, Fraction V (BSA; Invitrogen, 11018041), 0.1% sodium azide at a concentration of 1 μg/ml. Use highly purified BSA for the best results.
2. Collect cells by low speed centrifugation (300×*g*, 5 min) and wash once with PBS. After the washing step, resuspend each tube of cells in 2 ml of the PI staining solution and mix well. Let the samples incubate for 5–30 minutes. Keep the samples in this solution at 4 °C, protected from light, until the samples are analyzed on the flow cytometer.
3. Analyze with FACSCalibur or similar instrumentation: the cells with low forward scatter (FSC) and high FL3 (channel for PI) are dead cells.

NOTES:

1. When collecting adherent cells, it is important to collect and include the detached cells in the culture medium, as most of the detached cells are the dead cells.
2. Keep the solution tightly closed at 4 °C, protected from light. Discard after 1 month.

2.3. Lactate dehydrogenase (LDH)-cytotoxicity assay

Lactate dehydrogenase (LDH) is a soluble cytosolic enzyme that is released into the culture medium following the loss of membrane integrity that results from cell death. Various companies have developed cytotoxicity assay kits that are based on the quantification of LDH leakage from dead cells. Most assays have a low control, which is the culture medium by itself, and a high control, which contains cells permeabilized by lysis buffer to enable the maximal release of LDH. Most LDH assay kits measure LDH activity using a two-step reaction. In the first step, LDH oxidizes lactate to generate NADH; in the second step, the newly formed NADH and H^+ catalyze the reduction of a substrate to form a highly colored product. This product can be quantified by changes in optical density (OD). For most kits, reactions and readings are carried out in a 96-well plate for convenience.

1. Collect cells and wash once with corresponding fresh culture medium; then seed 100 μl of cells (with 2–10 $\times$ 10^4 cells per 100 μl) in a 96-well plate as follows:
   a. Background control: 100 μl medium per well, in triplicate, without cells.
   b. Low control: 100 μl of cells in triplicate wells.
   c. High control: 100 μl of cells in triplicate: add 10 μl cell lysis solution to each well and mix.
   d. Test sample: 100 μl of cells in triplicate: add test substances to each well and mix.
2. Incubate the cells in a tissue culture incubator (5% CO_2, 90% humidity, 37 °C) for the appropriate treatment time determined for the test substance. Gently shake the plate at the end of the incubation to ensure that the LDH is evenly distributed throughout the culture medium.
3. Centrifuge the cells at 250×*g* for 10 min in a centrifuge fit with a plate carrier.
4. Accurately transfer 10 μl/well of the clear medium into the corresponding wells of an optically clear 96-well plate.
5. Add 100 μl of LDH reaction mix to each well. Mix and incubate for 30 min at room temperature.
6. Measure the absorbance of all controls and samples with a plate reader equipped with a 450-nm (440-nm to 490-nm) filter. The reference wavelength should be 650 nm.
7. Cytotoxicity (%) = (test sample − low control)/(high control − low control) × 100.

NOTES:

1. The background control will measure background signal from the reagents and background LDH in the culture medium serum. The

background value should be subtracted from all other values prior to calculating the percent cytotoxicity.

2. The quantity of cells to be used per well depends on the cell type. To optimize the assay, a quick test can be performed by using 2, 4, and 8 $\times$ 10^4 cells per well to determine the cell number that you should use. The high control should yield an OD_{450nm} of approximately 2.0 after 30 min of treatment with 10% cell lysis solution, whereas the low control should yield an OD_{450nm} < 0.8. The reaction time should be set to approximately 30 min.
3. A positive control (1 μl LDH) can be used to test whether all reagents are working properly to quantify active LDH enzyme.
4. If the test substances are dissolved in a solvent other than PBS, a solvent control that lacks the testing substrates should be performed by adding the same amount of solvent, in triplicate, without the testing substrates.

2.4. Annexin V staining

Note that an early positive signal with Annexin V staining when membrane integrity is still intact is indicative of apoptosis. This is because phosphatidylserine, which is normally restricted to the inner leaflet of the membrane, is externalized during apoptosis. However, at later times, when the membrane has been compromised, loss of lipid asymmetry allows Annexin to bind indiscriminately and it can no longer be considered a selective marker of apoptosis.

1. Collect cells (approximately 5×10^5 to 1×10^6 cells per tube) by slow centrifugation (300×*g*, 5 min) at room temperature.
2. Wash the cells once in 500 μl of cold 1X PBS buffer, gently resuspend the cells, and then pellet by centrifugation as in step 1.
3. For each sample of cells, prepare 120 μl of Annexin V incubation reagent by combining the following:
   1.2 μl of 10X binding buffer (100 m*M* HEPES, pH 7.4, 1.5 *M* NaCl, 50 m*M* KCl, 10 m*M* $MgCl_2$, and 18 m*M* $CaCl_2$)
   1.2 μl propidium iodide powder (for example, Sigma-Aldrich, P4170)
   1.2 μl Annexin V-FITC (BD Biosciences, 556419)
   9.5 μl distilled H_2O
   120.2 μl Total
   Keep this cocktail in the dark and on ice.
4. Prepare 440 μl of 1X binding buffer per sample to use to dilute the cells after incubation by making a 1:10 dilution of the 10X binding buffer in distilled H_2O.
5. Gently resuspend the washed cells in 100 μl of the Annexin V incubation reagent prepared in step 3.
6. Incubate in the dark for 15 min at room temperature.

7. Add 400 μl of 1X binding buffer to each 100-μl reaction sample.
8. Examine by flow cytometry within 1 h for maximal signal. Measure FL3 for PI and FL1 for Annexin V-fluoroscein.

NOTES:

1. The binding of Annexin V to phosphatidylserine is absolutely dependent on calcium. Therefore, PBS or conventional flow cytometry buffers will not promote binding and should not be used as staining buffers unless 1.8 m*M* calcium has been added. Chelating agents should be avoided.
2. When using adherent cells, both the supernatant and the attached cells should be collected and examined together.

2.5. Caspase activity assay

Caspases are cysteine proteases that hydrolyze target proteins with specific amino acid target sequences. Caspase activation is a major feature of apoptosis. In most cases, the activation of caspases is considered a biochemical hallmark of apoptosis; however, caspase activation also occurs in autophagic cell death. Caspase fluorescent assay kits detect caspase activation by assaying for the cleavage of a fluorescent substrate.

1. Collect $2 - 5 \times 10^6$ cells by centrifugation at 250×*g* for 5 min. Wash once with cold PBS, add 200 μl of PEC lysis buffer (20 m*M* PIPES, pH 7.2, 100 m*M* NaCl, 1 m*M* EDTA, 10 m*M* DTT, 0.1% CHAPS, 10% sucrose), and mix thoroughly.
2. Incubate cells on ice for 30 min (invert samples several times).
3. Centrifuge samples for 5 min at 13,800×*g* and decant the clear supernatant fractions into new tubes to determine caspase activity; discard the pellet fractions. Save an aliquot of each supernatant sample for protein quantification.
4. Prepare N-acetyl-Asp-Glu-Val-Asp-p-Nitroanilide (DEVD-pNA, MBL International, BV-1008-13) stock solution or N-acetyl-Tyr-Val-Ala-Asp-p-Nitroanilide (YVAD-pNA, MBL International, BV-1104-13) stock solution. The protease reactions are carried out with 100–500 μg of protein extract and 160–200 μ*M* substrate in PEC lysis buffer at a final volume of 50–100 μl.
5. Place the reaction mix into a flat-bottomed 96-well microtiter plate and read the absorbance at 405-nm using the kinetics setup of a SpectraMax microtiter plate reader or comparable instrument. Acquire samples every 2–20 s and determine initial rates of substrate hydrolysis; ensure that the measured rate of substrate hydrolysis is linear.

NOTES:

1. Brief vortex, quick freeze/thaw cycles, or short sonication with a microtip can improve cell lysis but should be avoided if the solution becomes

too viscous due to the breakdown of nuclei and the release of DNA molecules into the supernatant.
2. If the samples cannot be tested immediately, they can be stored at −70 °C until measurement.
3. An alternative fluorometric assay can be performed using DEVD- or YVAD-AMC (Calbiochem, 218782-1SET). This gives a 10- to 100-fold increase in sensitivity over the pNA substrates but requires a more sophisticated instrument for quantification.

2.6. TUNEL staining

Various forms of cell death, especially apoptosis, will result in DNA fragmentation. TUNEL is a method for detecting DNA fragmentation. In a typical TUNEL assay, terminal deoxynucleotidyl transferase (TDT) is used to label nicks in the DNA with bromodeoxyuridine triphosphate (BrdUTP); BrdUTP can then be detected with anti-BrdUTP antibody.

The following protocol is from BD Biosciences (556405) with additional comments (see also the chapter by Jaboin *et al.*, in this volume):

1. Suspend the cells in 1% (w/v) paraformaldehyde in PBS, pH 7.4, at a concentration of 1–2 × 10^6 cells/ml.
2. Place the cell suspension on ice for 30–60 min.
3. Centrifuge cells for 5 min at 300×*g* and discard the supernatant fraction.
4. Wash the cells in 5 ml of PBS, and pellet by centrifugation. Repeat the wash and centrifugation step and discard the supernatant fraction.
5. Resuspend the cell pellet in the residual PBS in the tube by gently vortexing the tube.
6. Adjust the cell concentration to 1–2 × 10^6 cells/ml in PBS and add ice-cold ethanol to 70% (v/v) (95% ethanol can be added at a ratio of 3:1 to obtain 70% final concentration). Let the cells stand for a minimum of 30 min on ice or at −20 °C.
7. The cells can be stored in 70% (v/v) ethanol at −20 °C until use for several days.
8. Resuspend the cells by swirling the tubes. Remove 1-ml aliquots of the cell suspensions (approximately 1 × 10^6 cells/ml) and place in 12 × 75-mm flow cytometry centrifuge tubes. Centrifuge the control cell suspensions for 5 min at 300×*g* and remove the 70% ethanol by aspiration, being careful to not disturb the cell pellet.
9. Resuspend each tube of cells with 1.0 ml of wash buffer (6579AZ) for each tube. Centrifuge as before and remove the supernatant fraction by aspiration. Alternatively, touch the lip of the inverted tube onto absorbent paper immediately after decanting to remove as much of the supernatant as possible.
10. Repeat the wash buffer treatment (Step 9).

11. Resuspend each tube of the cell pellets in 50 μl of the DNA labeling solution (prepared according to the manufacturer's protocol).
12. Incubate the cells in the DNA Labeling Solution for 60 min at 37 °C in a temperature-controlled bath. Shake the cells every 15 min to resuspend.
13. At the end of the incubation time, add 1.0 ml of rinse buffer (6583AZ) to each tube and centrifuge each tube at 300×*g* for 5 min. Remove the supernatant fraction by aspiration.
14. Repeat the cell rinsing (as in step 13) with 1.0 ml of the rinse buffer, centrifuge, and remove the supernatant fraction by aspiration.
15. Resuspend the cell pellet in 0.1 ml of the antibody staining solution (using FITC-labeled Anti-BrdU; prepared according to the manufacturer's protocol).
16. Incubate the cells with the FITC-labeled anti-BrdU antibody solution in the dark for 30 min at RT.
17. Add 0.5 ml of the PI/RNase staining buffer (6585AZ) to the tube containing the 0.1 ml antibody staining solution.
18. Incubate the cells in the dark for 30 min at room temperature.
19. Analyze the cells in PI/RNase Staining Buffer by flow cytometry. Analyze the cells within 3 h of staining to obtain optimal results. The cells may begin to deteriorate if left overnight before the start of the analysis.

NOTES:

1. The gradual addition of ethanol while gently vortexing reduces clumping during this fixation step.
2. A higher centrifugation speed (400×*g*) may be required for fixed cells rather than for unfixed cells because ethanol fixation causes cell shrinkage.
3. The best staining is obtained with FITC-conjugated antibody or streptavidin. Other fluorochromes such as phycoerythrin (PE) give much weaker staining despite their inherent brightness. This is probably because higher molecular weight fluorochromes do not penetrate efficiently into fixed cells.
4. Many other companies including MBL, Millipore Corporation, and Promega offer TUNEL assay kits.

2.7. Reactive oxygen species (ROS) detection

Reactive oxygen species (ROS) are byproducts of the normal metabolism of oxygen. ROS are highly reactive due to the presence of unpaired valence shell electrons. High levels of ROS can cause damage to DNA, lipids, and proteins through oxidation, thus ROS are highly toxic. Accumulation of ROS has been detected during various type of cell death including necrosis and autophagic cell death.

ROS can be detected by staining with 2′,7′-dichlorodihydrofluorescein diacetate (H2DCFDA). Oxidation of nonfluorescent H2DCFDA converts it to the highly fluorescent 2′,7′-dichlorofluorescein (DCF), which can be detected by flow cytometry in the FL-1 channel.

1. Collect cells by centrifugation with slow speed (300×*g*, 5 min), and then wash the cells twice with 1X PBS.
2. Stain the cells with 25 μM 2′,7′-dichlorodihydrofluorescein diacetate (H2DCFDA, Invitrogen, D399) in PBS for 30 min.
3. Analyze cells to determine ROS by FACScan. Only analyze live cells (the forward scatter high, side scatter low population) and measure FL1 for ROS.

2.8. Detection of loss of mitochondrial membrane potential by fluorescent probe JC-1 staining

The collapse of the electrochemical gradient across the mitochondrial membrane is an early event of apoptosis and it can be detected by using the JC-1 dye. The JC-1 dye accumulates in healthy mitochondria and produces red fluorescence. However, when the mitochondrial potential collapses during cell death, the JC-1 dye cannot accumulate in mitochondria, and therefore it remains in the cytoplasm in a monomeric form, which emits green fluorescence, that can be easily detected by confocal microscopy or flow cytometry.

1. Harvest cells (at least 2×10^5) by low speed centrifugation (300×*g*, 5 min). Bring the total volume up to 1 mL with corresponding fresh culture medium.
2. Stain the cell suspension with 2.5 mg/mL JC-1 [MBL International, JM-1130-5]. Shake the cell suspension until the dye is well dissolved, giving a uniform red-violet color.
3. Incubate the samples in the dark at room temperature for 15–20 min.
4. Wash twice, centrifuging at 500×*g* for 5 min with a double volume of PBS.
5. Analyze immediately with the flow cytometer; JC-1 can be measured using both the FL1 and the FL2 channels.

3. Methods to Measure Autophagy

A wide array of methods can be used to assess autophagy. Morphological hallmarks of autophagy can be observed by transmission electron microscopy (EM) and confocal microscopy. EM remains the gold standard for assessing autophagy; the high resolution of EM images allows for the detection of distinguishing physical attributes of autophagosomes, such as the double-membrane structures (see the chapter by Ylä-Anttila in volume 452). In addition, engulfed cytosol and organelles can be observed within the double-membrane structure. Another advantage of EM is that it permits the assessment of the signs of other types of cell death, such as nuclear condensation, which is characteristic of apoptosis. EM is also the gold standard for diagnosing apoptosis and necrosis as forms of cell death. However, EM can be cumbersome, expensive, and time consuming. Confocal microscopy, on the other hand, offers a relatively easier method to specifically assess autophagy. Antibody staining for the autophagosomal marker protein LC3 in fixed cells, as well as live imaging for GFP-tagged LC3, have both been widely used (see the chapter by Kimura in this volume). However, researchers must use caution when overexpressing LC3, as this can create artifacts (Klionsky *et al.*, 2008). Alternatively, autophagy can be assessed by monitoring LC3 processing by Western blot.

Assessing the involvement of other cell death mechanisms (e.g., caspase activation) also provides a facile means to establish the timeline of biochemical events.

Other chapters in this book provide detailed protocols concerning these methods, so we will not discuss them in detail here.

4. Methods to Establish Autophagy as the Cause of Cell Death

Observing the accumulation of autophagosomes in dying cells does not prove that the death is caused by autophagy; on the contrary, it may represent a survival attempt on the part of the cell. To distinguish whether autophagy is the cause or consequence of cell death, it is necessary to inhibit autophagy and then assess whether this blocks cell death. Chemical reagents targeting different steps in autophagy are used as autophagy inhibitors (Table 2.1). 3-methyladenine (3-MA), a type III PI 3′-kinase inhibitor, and wortmannin, another PI 3′-kinase inhibitor, inhibit the formation of autophagosomes (Yu *et al.*, 2004). Other inhibitors, such as (1) bafilomycin A_1 (Espert *et al.*, 2006), a specific inhibitor of vacuolar-type H (+)-ATPases; (2) hydroxychloroquine (Boya *et al.*, 2005), a lysosomal lumen alkalizer; and (3) pepstatin A (Yu *et al.*, 2006), an inhibitor of acid proteases, have also been used to inhibit complete autophagy. Whereas these drugs cannot inhibit autophagosome formation, they do inhibit autolysosome maturation by preventing degradation of the engulfed cytosolic contents. The inherent problem with any chemical inhibitor is nonspecific effects. In addition, most autophagy inhibitors do not directly target autophagy genes. For example, as noted previously, 3-MA targets PI 3-kinase, which regulates other pathways in addition to autophagy. Great caution should be taken when interpreting data from chemical inhibitor studies. Whenever possible, molecular approaches should be applied to corroborate the results obtained with chemical inhibitors.

Recently, RNAi silencing for autophagy-essential genes has played a major role in molecularly establishing autophagy as necessary for autophagic cell death (Table 2.2). This approach provides the molecular specificity that chemical reagents lack. However, the precise role of most of the ATG gene products is still not fully understood. In addition, Beclin 1, a gene required for autophagy, regulates functions unrelated to autophagy. As a result, caution must be taken when interpreting data obtained from such experiments. Whenever possible, the expression of multiple ATG genes should be knocked down or otherwise molecularly manipulated to confirm involvement in the death process.

Table 2.1 Autophagy inhibitors

Name	Mechanism	Solvent	Final concentration
3-methyladenine (3-MA)	PI 3-kinase inhibitor	Water	10 m*M*
wortmaninn	PI 3-kinase inhibitor	DMSO	0.1 μM
Bafilomycin A1	Vacuolar-type H(+)-ATPase inhibitor	DMSO	0.1 μM
Hydroxychloroquine	Lysosomal lumen alkalizer	Water	30 μg/ml
Pepstatin A	Acid protease inhibitor	DMSO	1 μM

Table 2.2 Target gene siRNAs

Target gene	Accession number (Human)
Atg5	BC002699
BECN1	NM-003766
Atg7	NM-006395
Atg8	NM_032514
lamp-1	NM-005561
Atg10	NM-8031482
Atg12	NM-004707

5. Autophagy Genes for RNAi Silencing

Various companies provide RNAi design and synthesis services, as well as transfection protocols. Both lipid-based transfection and electroporation have been used for transfecting RNAi into adherent and suspension cell-types. It is worth noting that for some Atg proteins, a very low level may still be sufficient to allow autophagy to proceed, so it is essential to achieve the highest level RNAi transfection efficiency. A new generation of electroporation methods such as the Amaxa nucleofector (http://www.amaxa.com) offer excellent knockdown efficiency. A nonspecific RNAi control should always be included in experimental designs, and verification of knockdown efficiency by quantifying RNA or protein levels is essential.

6. Transfect RNAi by Electroporation (Amaxa Nucleofection)

1. Split 1 dish of confluent cells onto 3 large dishes. Allow the cells to grow overnight to achieve 70% confluence.
2. Use trypsin to release adherent cells. Centrifuge at $300\times g$ for 5 min at 4 °C and resuspend 1×10^6 cells in 100 μl of the appropriate Amaxa transfection buffer (choose the appropriate buffer according to the Amaxa cell line database at http://www.amaxa.com/no_cache/cell-database/).
3. Add RNAi (200 pmol) to this cell mixture and mix gently.
4. Transfer the cell/RNAi mixture into a cuvette.
5. Set up the electroporation program according to the Amaxa database.
6. Remove cells from the cuvette with a pipette and place into 5 ml of corresponding culture medium to grow out in 1 well of 6-well plate.
7. 48 h after transfection, split cells 1 to 3 into 3 wells of 6-well plates and grow overnight.
8. The knockdown efficiency should be verified by Western blot to measure the decrease in the protein level using cells from 1 well; proceed with the experiments using the cells from another 2 wells.

7. Conclusion

Thus far, most methods to quantify cell death and measure cell-death-related parameters were developed for measuring apoptosis and necrosis. However, rapid progress in autophagy research has revealed new aspects of autophagic cell death. This has led to cogent methods to measure parameters specific to autophagic cell death. For example, the degradation of cell survival factors such as catalase has the potential to be a parameter specific for autophagic cell death in certain circumstances. Methods described in this chapter have been used successfully in investigation of autophagic cell death. However, given the complicated nature of autophagic cell death, there is not yet a standard set of methods that can be used universally. Rather, investigators will have to deploy combinations of different methods tailored to the particular biological question of interest.

REFERENCES

Berry, D. L., and Baehrecke, E. H. (2007). Growth arrest and autophagy are required for salivary gland cell degradation in *Drosophila*. *Cell* **131,** 1137–1148.

Boya, P., Gonzalez-Polo, R. A., Casares, N., Perfettini, J. L., Dessen, P., Larochette, N., Metivier, D., Meley, D., Souquere, S., Yoshimori, T., *et al.* (2005). Inhibition of macroautophagy triggers apoptosis. *Mol. Cell Biol.* **25,** 1025–1040.

Clarke, P. G. (1990). Developmental cell death: Morphological diversity and multiple mechanisms. *Anat. Embryol. (Berl.)* **181,** 195–213.

Espert, L., Denizot, M., Grimaldi, M., Robert-Hebmann, V., Gay, B., Varbanov, M., Codogno, P., and Biard-Piechaczyk, M. (2006). Autophagy is involved in T cell death after binding of HIV-1 envelope proteins to CXCR4. *J. Clin. Invest.* **116,** 2161–2172.

Harding, T. M., Hefner-Gravink, A., Thumm, M., and Klionsky, D. J. (1996). Genetic and phenotypic overlap between autophagy and the cytoplasm to vacuole protein. *J. Biol. Chem.* **271,** 17621–17624.

Harding, T. M., Morano, K. A., Scott, S. V., and Klionsky, D. J. (1995). Isolation and characterization of yeast mutants in the cytoplasm to vacuole protein targeting pathway. *J. Cell Bio.* **131,** 591–602.

Hendy, R., and Grasso, P. (1972). Autophagy in acute liver damage produced in the rat by dimethylnitrosamine. *Chem. Biol. Interact.* **5,** 401–413.

Klionsky, D. J. (2007). Autophagy: From phenomenology to molecular understanding in less than a decade. *Nat. Rev. Mol. Cell Biol.* **8,** 931–937.

Klionsky, D. J., Abeliovich, H., Agostinis, P., Agrawal, D. K., Aliev, G., Askew, D. S., Baba, M., Baehrecke, E. H., Bahr, B. A., Ballabio, A., *et al.* (2008). Guidelines for the use and interpretation of assays for monitoring autophagy in higher eukaryotes. *Autophagy* **4,** 151–175.

Levine, B., and Yuan, J. (2005). Autophagy in cell death: An innocent convict? *J. Clin. Invest* **115,** 2679–2688.

Lum, J. J., Bauer, D. E., Kong, M., Harris, M. H., Li, C., Lindsten, T., and Thompson, C. B. (2005). Growth factor regulation of autophagy and cell survival in the absence of apoptosis. *Cell* **120,** 237–248.

Mizushima, N., Noda, T., Yoshimori, T., Tanaka, Y., Ishii, T., George, M. D., Klionsky, D. J., Ohsumi, M., and Ohsumi, Y. (1998). A protein conjugation system essential for autophagy. *Nature* **395,** 395–398.

Onodera, J., and Ohsumi, Y. (2005). Autophagy is required for maintenance of amino acid levels and protein synthesis under nitrogen starvation. *J. Biol. Chem.* **280,** 31582–31586.

Schweichel, J. U., and Merker, H. J. (1973). The morphology of various types of cell death in prenatal tissues. *Teratology* **7,** 253–266.

Shimizu, S., Kanaseki, T., Mizushima, N., Mizuta, T., Arakawa-Kobayashi, S., Thompson, C. B., and Tsujimoto, Y. (2004). Role of Bcl-2 family proteins in a non-apoptotic programmed cell death dependent on autophagy genes. *Nature Cell Biology* **6,** 1221–1228.

Thumm, M., Egner, R., Koch, B., Schlumpberger, M., Straub, M.,Veenhuis, M., and Wolf, D. H. (1994). Isolation of autophagocytosis mutants of *Saccharomyces cerevisiae. FEBS Lett.* **349,** 275–280.

Tsukada, M., and Ohsumi, Y. (1993). Isolation and characterization of autophagy-defective mutants of *Saccharomyces cerevisiae. FEBS Lett.* **1-2,** 169–174.

Yu, L., Alva, A., Su, H., Dutt, P., Freundt, E., Welsh, S., Baehrecke, E. H., and Lenardo, M. J. (2004). Regulation of an *ATG7-beclin 1* program of autophagic cell death by caspase-8. *Science* **304,** 1500–1502.

Yu, L., Wan, F., Dutta, S., Welsh, S., Liu, Z., Freundt, E., Baehrecke, E. H., and Lenardo, M. (2006). Autophagic programmed cell death by selective catalase degradation. *Proc. Natl. Acad. Sci. USA* **103,** 4952–4957.

CHAPTER THREE

Autophagic Neuron Death

Yasuo Uchiyama,* Masato Koike,* Masahiro Shibata,† *and* Mitsuho Sasaki*

Contents

Abstract

Neurons of the central nervous system (CNS) tissue are terminally differentiated cells and have large volumes, unlike cells of peripheral tissues. Such neurons possess abundant lysosomes in which damaged and unneeded intracellular constituents are degraded. A cellular process to bring the unneeded constituents to lysosomes is referred to as macroautophagy (autophagy), which is essential for the maintenance of cellular metabolism under physiological conditions. In fact, mice deficient in Atg7 or Atg5 specifically in CNS tissue have ubiquitin aggregates in neurons and massive loss of cerebral and cerebellar cortical neurons, resulting in neurodegeneration and short life span. In addition, acceleration of autophagy induced by the loss of lysosomal proteinases such as cathepsin D or cathepsins B and L, or by hypoxic/ischemic (H/I) brain injury, causes neurodegeneration. Moreover, lysosomes with undigested materials due to loss of proteinases are enwrapped by double membranes to produce autophagosomes, resulting in the further accumulation of autolysosomes. H/I brain injury at birth that is an important cause of cerebral palsy, mental retardation, and epilepsy causes energy failure, oxidative stress, and unbalanced ion fluxes, leading to a high induction of autophagy in brain

* Department of Cell Biology and Neuroscience, Juntendo University Graduate School of Medicine, Tokyo, Japan
† Division of Gross Anatomy and Morphogenesis, Department of Regenerative and Transplant Medicine, Niigata University Graduate School of Medical and Dental Sciences, Niigata, Japan

Methods in Enzymology, Volume 453
ISSN 0076-6879, DOI: 10.1016/S0076-6879(08)04003-2

neurons. Since mice that are unable to execute autophagy (due to brain-specific deletion of Atg7 or Atg5) die as a result of massive loss of cerebral and cerebellar neurons with accumulation of ubiquitin aggregates, induction of neuronal autophagy after H/I injury is generally considered neuroprotective, as it maintains cellular homeostasis. However, our data showing that H/I injury-induced pyramidal neuron death in the neonatal hippocampus is largely prevented by Atg7 deficiency indicate the presence of autophagic neuron death. In this section, we introduce various methods for the detection of autophagic neuron death in addition to other death modes of CNS neurons.

1. Introduction

Autophagy is a highly regulated process involving the bulk degradation of cytoplasmic macromolecules and organelles in mammalian cells via the lysosomal system. The activity of autophagy is up-regulated under fasting conditions to supply nutrients such as amino acids; however, basal autophagy is also important for the maintenance of cellular metabolic turnover and homeostasis (Mizushima, 2007). In particular, autophagic degradation of proteins with aberrant structures and/or inclusions is essential in achievement of a fine homeostatic balance in the nonproliferating neural cells (Hara *et al.*, 2006; Komatsu *et al.*, 2006).

Hypoxic/ischemic (H/I) brain injury at birth causes neurological impairment such as cerebral palsy, mental retardation, and epilepsy. A large number of studies have shown that the hippocampus is the most fragile brain region to ischemic insult (Ness *et al.*, 2006; Sheldon *et al.*, 1998). After H/I injury, pyramidal neurons in the neonatal hippocampus undergo pyknosis during the early stages and the number of pyknotic neurons continues to increase. Approximately 35% of these dying pyramidal neurons are positive for activated caspase-3, indicating that H/I injury-induced pyramidal neuron death occurs, at least in part, via caspase-3-dependent apoptosis, whereas caspase-independent cell death is the other mechanism by which neurons die (Koike *et al.*, 2008).

Our recent studies provide direct evidence that autophagy is involved in the H/I brain injury-induced neuron death (Koike *et al.*, 2008; Uchiyama *et al.*, 2008b). When pyramidal layers of the neonatal hippocampus, damaged as a result of H/I brain injury, are examined by Western blotting and immunohistochemistry for LC3 as a marker protein of autophagy (Kabeya *et al.*, 2000), the amount of membrane-bound LC3-II is significantly increased in the ipsilateral hippocampus, and granular staining of LC3 appears in the ipsilateral pyramidal neurons but not in the contralateral hippocampus. By electron microscopy, damaged neurons in the ipsilateral pyramidal layer are found to contain abundant autophagic vacuoles during the early stages of cell death. Moreover, induction of autophagy after H/I brain injury also occurs in pyramidal neurons of the hippocampus deficient in caspase-3 or CAD (caspase-activated DNase). Surprisingly, the neonatal mice lacking Atg7, which are unable to execute autophagy, are largely resistant to both caspase-dependent and caspase-independent neuron death after H/I brain injury. The data suggest that autophagy in pyramidal neurons of the neonatal hippocampus induced by H/I brain injury may act as an initiator of caspase-dependent and -independent neuron death.

This chapter describes experimental models of neurodegeneration and both morphological and biochemical techniques commonly used in our laboratory for the measurement of neuron death.

2. Experimental Models of Neurodegeneration

2.1. Cell culture model

Since PC12 cells, a rat adrenal pheochromocytoma cell line, which were established by Greene and Tischler in 1976, respond to nerve growth factor and show gene expression profiles similar to that of neurons (Greene and Tischler, 1976; Leonard *et al.*, 1987), they are a good model for neurons and have been used by many researchers. PC12 cells proliferate when cultured in the presence of serum, but they extend processes and differentiate into neuron-like cells when cultured in the absence of serum but in the presence of NGF (Greene and Tischler, 1976). Furthermore, PC12 cells undergo apoptosis when cultured in the absence of serum, and this cell death is prevented by caspase-inhibitors such as acetyl-DEVD-fmk and Z-VAD-fmk (Isahara *et al.*, 1999). However, it is also true that when PC12 cells are cultured in serum-free medium, autophagic vacuoles/autolysosomes appear in the cytoplasm within 3 h of serum deprivation (Ohsawa *et al.*, 1998). The cells then start to die, and cathepsin D (CD) activity increases, while cathepsin B (CB) activity decreases (Isahara *et al.*, 1999; Ohsawa *et al.*, 1998; Shibata *et al.*, 1998). Simultaneously, PC12 cell death following serum deprivation is largely rescued by the addition of 3-methyladenine (3-MA) (Uchiyama, 2001). These results indicate that PC12 cells may die in both caspase-dependent and autophagic/lysosomal proteinase-dependent manners, when cultured under serum-deprived conditions. Therefore, this culture model might be useful for the study of autophagic neuron death.

2.1.1. Serum-deprived cultures of PC12 cells

a. The day prior to initiation of serum-deprived cultures, seed PC12 cells in the presence of high glucose (4.5 g/L) Dulbecco's Modified Eagle Medium (DMEM) supplemented with 10% FBS.

b. Wash culture dishes with serum-free DMEM 3 times.
c. Scrape the cells with a silicon rubber cell scraper in 5 ml of serum-free DMEM.
d. Count the cell number, seed the cells at a density of $1 \times 10^4/cm^2$, and incubate them at 37 °C in 5% CO_2.

2.1.2. Sample preparation for biochemical analysis

a. Remove the adhered cells using a cell scraper. Transfer cells that were attached and floating cells and the conditioned medium to 50 ml of centrifuge tubes.
b. Centrifuge the tubes at 2000*g* for 5 min and collect the cell pellets. Wash the cell pellets 3 times with phosphate-buffered saline (PBS).
c. Lyse the cells with a lysis buffer (1% Triton X-100 in PBS with or without a proteinase inhibitor cocktail (1 m*M* PMSF, 0.8 μM aprotinin, 15 μM E-64, 20 μM leupeptin, 50 μM bestatin, 10 μM pepstatin A) (Nacalai Tesque, Japan)) and centrifuge at 10,500*g* for 10 min to remove cell debris. The resultant supernatant fractions are used for samples for biochemical analyses.

2.1.3. Immunostaining for confocal laser scanning microscopy

a. Culture the cells on cover glasses coated with poly-L-lysine in a multi-well plate.
b. Centrifuge the plate at 2000*g* for 5 min, and do not discard the supernatant fraction.
c. For fixation of the cells, add gently an equal volume of 8% paraformaldehyde (PA) buffered with 0.1 *M* phosphate buffer (PB) (a doubly concentrated fixative buffered with a buffer of normal molarity), pH 7.2, to the culture medium from step b (final concentration 4% PA) for 15 min on ice.
d. Wash the cells with PBS containing Tween 20 (TPBS: 10 m*M* PB, pH 7.2, 0.5 *M* NaCl, 0.1% Tween 20) at room temperature (RT) for 5 min 3 times. The concentration of NaCl used in this step and subsequent steps is dependent on the affinity of the antigen-antibody complex. To avoid nonspecific reactions, it is recommended to use 0.5 *M* NaCl.
e. Block the cells with 10% host serum in TPBS, corresponding to the host species of the 2nd antibody, for 20 min at RT.
f. Incubate the cells with the 1st antibody diluted in TPBS for 1 to 3 days.
g. Wash the cells with TPBS at RT for 5 min 3 times.
h. Incubate the cells with the fluorescence-labeled 2nd antibody for 30 min in the dark.
i. Wash the cells with TPBS at RT for 5 min 3 times in the dark.
j. Mount cover glasses onto glass slides using a mounting medium (VECSTASHIELD (H-1400), Vector Lab., CA).
k. Observe the cells with a confocal laser-scanning microscope.

2.1.4. Electron microscopy

Cultured cells can either be embedded in resin as pellets (Ohsawa *et al.*, 1998) or processed for flat embedding (Yoshida *et al.*, 2005): follow the procedures described in sections 3.1.2 and 3.1.3 for resin and flat embedding, respectively.

2.2. Genetic models

The most common inherited neurodegenerative disease in childhood is neuronal ceroid-lipofuscinosis (NCL or Batten disease), which is categorized by the accumulation of proteolipids, such as subunit c of mitochondrial ATP synthase or sphingolipid activator proteins, in the lysosomes of neurons (Fearnley *et al.*, 1990; Hall *et al.*, 1991; Kominami *et al.*, 1992; Palmer *et al.*, 1989). There are 10 different subtypes (CLN1-CLN10) of NCLs, for which 7 causal genes have been identified (Haltia, 2006; Siintola *et al.*, 2006a). CLN2 and CLN10 are caused by dysfunction/deficiency of the lysosomal proteinases, tripeptidyl peptidase I (TPP-I) and CD, respectively (Siintola *et al.*, 2006b; Sleat *et al.*, 1997; Steinfeld *et al.*, 2006).

CD-deficient ($CD^{-/-}$) mice are phenotypically normal at birth, but die at postnatal day 26 ± 1 because of massive intestinal necrosis, thromboembolia, and lymphopenia (Saftig *et al.*, 1995). CNS neurons in $CD^{-/-}$ mice are affected by a new form of lysosomal accumulation disease with a phenotype that resembles NCL (Koike *et al.*, 2000; Koike *et al.*, 2003; Koike *et al.*, 2005; Nakanishi *et al.*, 2001). More recently, the CD gene has been shown to be responsible for a congenital form of NCL, which is currently denoted as CLN10 (Siintola *et al.*, 2006b; Steinfeld *et al.*, 2006). $CD^{-/-}$ mice manifest neurological phenotypes such as seizures with trembling and stiff tails, and blindness during the terminal stages, because of massive neuron death in various brain regions, particularly in the thalamus, cerebral cortex, hippocampus, and retina (Koike *et al.*, 2000, 2003; Nakanishi *et al.*, 2001). The most striking feature of the $CD^{-/-}$ CNS neurons is the storage of autophagosome/autolysosome-like bodies, granular osmiophilic deposits, and fingerprint profiles; they emit autofluorescence and are immunopositive for both CB and subunit c of mitochondrial ATP synthase, indicating that they contain ceroid lipofuscin (Koike *et al.*, 2000). Neuron death is executed by nitric oxide (NO) that is produced via inducible NO synthase activity in microglial cells (Nakanishi *et al.*, 2001). Moreover, the phenotypes of mice deficient in both CB and cathepsin L (CL), which are lysosomal cysteine proteinases, are similar to those of $CD^{-/-}$ mice: the accumulation of autophagosome/autolysosome-like bodies, granular osmiophilic deposits, and fingerprint profiles (Koike *et al.*, 2005).

Atg7 is an autophagy-related gene that is essential to form autophagosomes and execute autophagy. The autophagic pathway executed by Atg7 has two ubiquitin-like conjugation systems: Atg8 and Atg12 conjugation systems (Ohsumi, 2001), in which Atg7 functions as a key regulatory E1-like enzyme (Tanida *et al.*, 2002). Mice with a conditional knockout of *Atg7* in CNS tissue (*Atg7*$^{flox/flox}$: Nestin-*Cre*) lose cerebral and cerebellar cortical neurons and accumulate ubiquitin aggregates in neuronal perikarya and axons, leading to neurodegeneration, abnormal neurological signs, and cell death (Komatsu *et al.*, 2006). Similar phenotypes are observed in mice with conditional knockout of *Atg5* in CNS tissue (*Atg5*$^{flox/flox}$: Nestin-*Cre*) (Hara *et al.*, 2006). Although autophagy is typically a nonselective degradation pathway induced under fasting conditions to supply nutrients such as amino acids, ubiquitin-containing aggregates accumulate in Atg7-deficient CNS neurons, to which nutrients are constantly supplied under feeding conditions. Moreover, proteasome activity in brains of Atg7-deficient mice is normal. The results indicate that basal autophagy plays an essential role in the elimination of unfavorable proteins; however, the candidates for proteins that are ubiquitinated and form aggregates in the brain remains to be identified.

2.3. Hypoxic ischemic brain injury model

As described earlier, autophagy is involved in neurodegenerative disorders (Chu, 2006; Koike *et al.*, 2005; Nixon, 2006; Zhu *et al.*, 2007). Autophagy is also induced in damaged neurons after ischemic brain injury in rodent models. In fact, autophagy is highly induced in CA1 pyramidal neurons of gerbil hippocampus after brief forebrain ischemia and such damaged pyramidal neurons undergo delayed neuronal death as demonstrated by electron microscopy (Nitatori *et al.*, 1995). Induction of autophagy, assessed using LC3 as a marker protein of autophagy (Kabeya *et al.*, 2000), has also been shown in pyramidal neurons of the hippocampus in various rodent models, as follows: (1) neonatal and adult mouse brains after H/I brain injury (Adhami *et al.*, 2006; Koike *et al.*, 2008; Uchiyama *et al.*, 2008a,b; Zhu *et al.*, 2005, 2006); (2) neonatal rats after H/I injury (Koike *et al.*, unpublished data) and adult rats after focal cerebral ischemia (Rami *et al.*, 2008); and (3) rodent brains after other traumatic injuries (Adhami *et al.*, 2006; Clark *et al.*, 2008; Erlich *et al.*, 2007; Lai *et al.*, 2008; Sadasivan *et al.*, 2008).

Among these experimental brain injury models, the mouse H/I injury model has most often been used to determine if autophagy is neuroprotective or antineuroprotective in the execution of neuron death. We analyzed this H/I injury-induced neuron death using *Atg7*-conditional knockout mice, in which *Atg7* is absent specifically in CNS tissue (Koike *et al.*, 2008; Komatsu *et al.*, 2006). Our data show that H/I injury induces autophagy in neonatal hippocampal pyramidal neurons, whereas this pyramidal neuron death is prevented by *Atg7* deficiency. Moreover, pyramidal neuron death in the adult hippocampus after H/I injury is caspase independent and accompanied by autophagosome formation. In particular, morphological features of the degenerating adult neurons resemble the features of type 2 neuron death, as defined by Peter G. Clarke (1990). Our data provide direct evidence for autophagy-induced neuron death following neonatal H/I brain injury based on studies with mice that cannot execute autophagy specifically in CNS tissue (Koike *et al.*, 2008; Uchiyama *et al.*, 2008a,b).

C57BL/6J mice and Wistar rats are usually used for H/I brain injury on P7 and at 8 weeks of age, respectively, according to the Rice-Vannucci model (Rice *et al.*, 1981) with minor modifications (Zhu *et al.*, 2005). In general, hypoxia alone does not produce brain damage in mice or rats; rather, ischemic alterations occur in the ipsilateral hemisphere when the unilateral common carotid artery is occluded prior to hypoxia (Ditelberg *et al.*, 1996; Ferriero *et al.*, 1996; Levine, 1960; Rice *et al.*, 1981). It is well known that CA1 pyramidal neurons in the hippocampus are the most vulnerable to ischemic insult (Ness *et al.*, 2006; Sheldon *et al.*, 1998). It is also possible to produce damage that is restricted to the hippocampus of C57BL/6J mouse brains after H/I injury by adjusting the duration of hypoxia to 45 min for neonates and 35 min for adults (Koike *et al.*, 2008). H/I injury resulting from 60 min of hypoxia produces alterations mainly in the hippocampal region in neonatal Wistar rats (Koike *et al.*, unpublished data) (Fig. 3.1F–J).

2.3.1. Generation of hypoxic-ischemic brain injury model

a. Warm a plastic chamber in a 37 °C water bath.
b. Anesthetize animals with isoflurane (2%). The duration of anesthesia should be <5 min.
c. Using a stereomicroscope, make a midline neck incision.
d. Use tweezers to remove adipose tissue and locate the left carotid sheath under the deep cervical fascia (Fig. 3.1A).
e. Lift the internal jugular vein and vagal nerve carefully from the (left) common carotid artery (Fig. 3.1B).
f. Ligate the left common carotid artery with silk sutures (6/0) at two points to obtain complete occlusion. Cut the artery between the two ligations (Fig. 3.1C–E). This procedure is very important to obtain stable results that show ischemic alterations in the brains including the hippocampal pyramidal neurons.
g. After closure of the neck incision, allow the animals to recover for approximately 60 min in a glass chamber placed in a 37 °C water bath.
h. Place the animals in a container in which the oxygen concentration is adjusted to 8% (balance, nitrogen) and the temperature is maintained at 37 °C. To induce H/I injury in the ipsilateral hemisphere (mainly in the

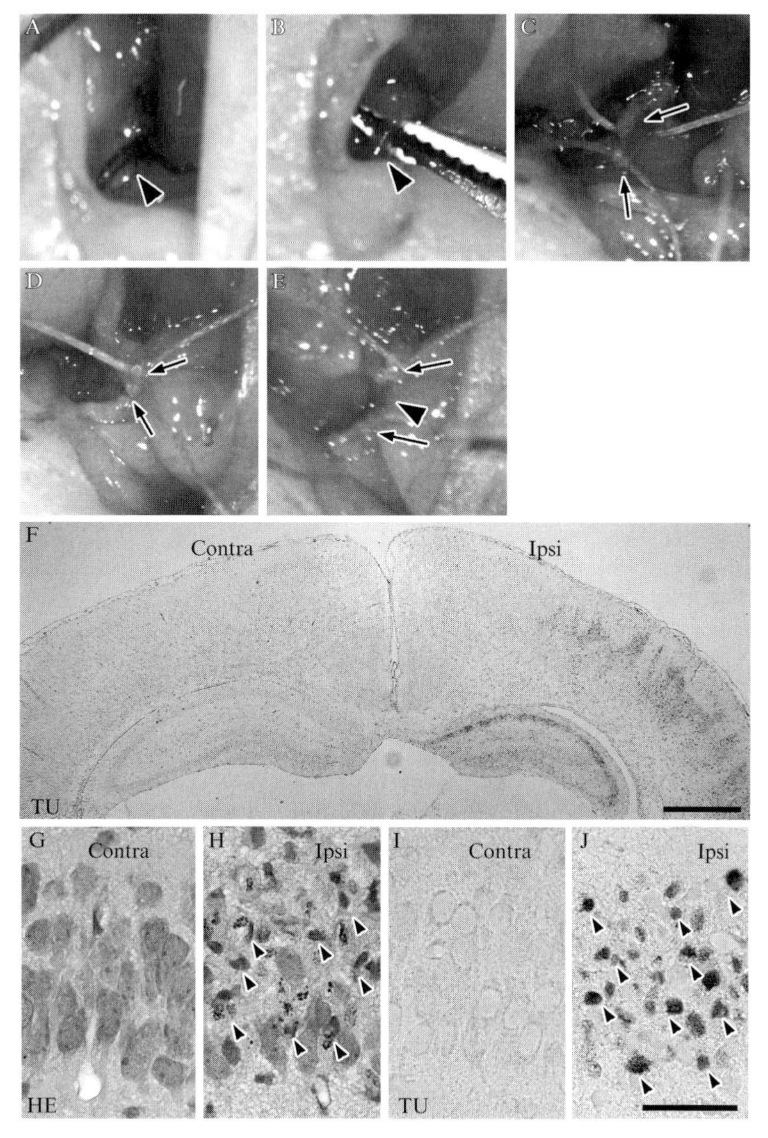

Figure 3.1 Hypoxic-ischemic brain injury. (A–E) Surgical microscopy photographs showing ligation of the left carotid artery of a neonatal Wistar rat at P7. (A) The left carotid sheath (arrowhead) is visible under the deep cervical fascia. (B) After peeling back the internal jugular vein and vagus nerve, a left carotid artery is exposed (arrowhead). (C) Silk sutures (6/0) are placed under the left carotid artery at two positions (arrows). (D) The artery is ligated at two locations (arrows). (E) The artery (arrowhead) between the two ligatures (arrows) is cut. (F) A low-power view showing TUNEL staining (TU) in the hippocampus of a wild-type neonatal rat 3 days after H/I injury. TUNEL-positive cells are abundant in the ipsilateral (Ipsi) hippocampus and cerebral cortex. Contra: contralateral hippocampus and cerebral cortex. Bar: 1000 μm. (G–J)

hippocampus), expose neonatal and adult mice to hypoxic conditions (8% oxygen) for 45 and 35 min, respectively. Neonatal rats require a 60-min exposure to hypoxia.

i. Return the animals to either dams or plastic cages after hypoxic exposure.
j. Remove brains promptly at designated time points after H/I insult (for example; 3, 8, 24, or 72 h, or 7 days), and process them for biochemical and morphological analyses as described subsequently.
k. Expose but do not ligate the unilateral common carotid artery of control littermates that are not subjected to H/I injury.

3. Measurements of Neuron Death

3.1. Morphological analysis

On the basis of morphological criteria using electron microscopy, physiological neuron death is categorized into 3 types that can be detected in CNS tissue during development: apoptotic, autophagic, and nonlysosomal vesiculate (Clarke, 1990). Although LC3 is a well-known marker of autophagosome formation/autophagic vacuoles/autophagosomes (Kabeya *et al.*, 2000), immunohistochemical staining for LC3 alone is not always sufficient to detect autophagy, because LC3 accumulates in lysosomes under certain conditions (Tanida *et al.*, 2005), and it also localizes to organelles other than autophagic vacuoles such as phagosomes (Sanjuan *et al.*, 2007). Therefore, it is important to use electron microscopy for the identification of autophagic vacuoles in addition to LC3 staining. Moreover, either TUNEL or hematoxylin-eosin (HE) staining alone cannot distinguish between nuclei of cells undergoing apoptotic, necrotic, or autophagic cell death. Electron microscopy is required to know morphological characteristics of dying cells and their nuclei, in addition to the presence of autophagic vacuoles (Uchiyama, 1995).

When samples are embedded in conventional plastic resins (e.g., Epon 812), autophagic vacuoles can be identified by morphology (Eskelinen, 2008; Koike *et al.*, 2005, 2008) (also see the chapter by Anttila *et al.*, in volume 452). Furthermore autophagic vacuoles can be classified into 2 types: early or initial autophagic vacuoles (i.e., autophagosomes), which

High power views showing CA1 pyramidal neurons of contralateral (G, I) and ipsilateral (H, J) hippocampi of the rat shown in F. (G, H) Hematoxylin-eosin staining (HE). (I, J) TUNEL staining. There are many CA1 neurons that possess pyknotic nuclei (arrowheads in H). The pyknotic nuclei in such neurons also show positive-TUNEL staining (arrowheads in J). Bar: 40 μm.

contain morphologically intact cytoplasm; and late or degradative autophagic vacuoles (i.e., autolysosomes), which contain partially degraded but identifiable cytoplasmic materials (Koike *et al.*, 2005; Liou *et al.*, 1997; Tanaka *et al.*, 2000).

Fixation of brain tissues should be performed by whole body perfusion via the left ventricle whenever possible (Koike *et al.*, 2000, 2003, 2005, 2008). In the case of cultured cells, cells can be either embedded in resin as pellets (Ohsawa *et al.*, 1998) or processed for flat embedding (Yoshida *et al.*, 2005). Cell suspensions are usually embedded as pellets (Somboonthum *et al.*, 2007). Cells in pellets are often coated in low-melting-point agarose after postfixation with osmium tetroxide, because the sample blocks are easy to prepare and handle.

3.1.1. Resin embedding of rodent brain tissue

a. Anesthetize animals with pentobarbital (25 mg/kg i.p.).
b. Open the chest and insert a needle into the left ventricle of the heart.
c. Fix animals by cardiac perfusion with 2% PA-2% glutaraldehyde (GA) buffered with 0.1 *M* PB.
d. Quickly remove brains from the animals and further immerse them in the same fixative for at least 2 h at 4 °C.
e. Prepare 1-mm thick brain slices using Rodent Brain Matrices (ASI Instruments, Warren, MI).
f. Immerse the brain slices in the same fixative overnight at 4 °C.
g. Wash tissues in 7.5% sucrose in 0.1 *M* PB 3 times at 4 °C.
h. Postfix tissue slices in 1% osmium tetroxide in 0.1 *M* PB containing 7.5% sucrose for 2 h at 4 °C in the dark.
i. Wash tissues in distilled water 3 times at RT.
j. Stain tissues in 2% aqueous uranyl acetate in water at RT for 1 h.
k. Wash tissues in distilled water 3 times at RT.
l. Dehydrate tissues with a graded series of ethanol (from 50% ~ 90%). At each step agitate the tissues well for 5 min at RT.
m. Dehydrate tissues with 100% ethanol 3 times for 5 min at RT.
n. Immerse tissues in a 1:1 mixture of ethanol and QY-1 (n-butyl glycidyl ether) (Nisshin EM, Tokyo, Japan) for 10 min.
o. Immerse tissues in 100% QY-1 for 10 min.
p. Immerse tissues in a 1:1 mixture of QY-1 and Epon 812 (TAAB, Barks, UK) for 1 h.
q. Immerse tissues in a 1:3 mixture of QY-1 and Epon 812 for 1 h.
r. Immerse tissues in pure Epon 812 and place them in a vacuum chamber overnight.
s. Embed tissues in fresh Epon 812 and polymerize the resin at 80 °C for 48 h.

t. Trim tissues with a glass knife. Using an ultramicrotome (Ultracut N, Reichert-Nissei, Tokyo, Japan; or, Leica UC6, Leica, Vienna, Austria), cut semithin sections (~500 nm to 1 μm) with a diamond knife.
u. Stain the semithin sections with 0.05% toluidine blue for 5 min at RT.
v. Further trim the sample blocks with a glass knife, and cut 70- to 80-nm ultrathin sections with a diamond knife.
w. Stain the sections with 4% aqueous uranyl acetate and 0.4% lead citrate for 5 min each.
x. Observe sections with a transmission electron microscope.

3.1.2. Resin embedding of cell pellets

a. Culture adherent cells to semiconfluence in 10-cm Petri dishes. The day prior to fixation, carefully replace the culture medium by adding 5 ml of fresh medium.
b. Add 5 ml of 4% PA-4% GA (doubly concentrated fixative) in 0.1 *M* PB at RT to the medium of a, and mix gently. After 1 or 2 min, change the fixative to 2% PA-2% GA (normally concentrated fixative) in 0.1 *M* PB and further fix the cells in the same fixative for at least 2 h at 4 °C. After fixation, scrape the cells from the dishes with a rubber policeman or cell-scraper and transfer them to microcentrifuge tubes for further processing.

   For cultures of nonadherent cells, mix the suspended cells with an equal volume of 4% PA-4% GA (double-concentrated fixative) in 0.1 *M* PB at RT, to the culture medium of step a. After gentle centrifugation (<2000*g*), remove the supernatant fraction and resuspend the cell pellets in 1–1.5 ml of 2% PA-2% GA in 0.1 *M* PB (freshly prepared) and transfer to microcentrifuge tubes for further processing. Immerse the cells in the same fixative for at least 2 h at 4 °C.
c. Wash the cells in PBS 3 times.
d. Postfix the cells in 1% osmium tetroxide in PBS for 1 h at 4 °C in the dark.
e. Wash the cells in distilled water 3 times.
f. Gently centrifuge the cells at 1500*g* for 5 min to form a loose pellet and remove the distilled water.
g. Using a warm plastic pipette, quickly transfer 300 μl of warm (60 °C) 1% low melting agarose onto the pellet in the tube and gently resuspend the cells in the warm agarose.
h. Using a swing rotor, gently centrifuge the cells at 1500*g* for 5 min at RT and place the tube on ice to solidify the agarose.
i. Trim the agarose into 1- to 2-mm cubes using a sharp razor blade.
j. Wash the cubes in distilled water once.
k. Follow the steps a–x as described in section 3.1.1.

3.1.3. Resin flat embedding of cells grown on glass or plastic coverslips

a. Grow cells on poly-L-lysine-coated glass (Matsunami, Osaka, Japan) or plastic coverslips (Celldesk; Sumitomo Bakelite, Tokyo, Japan). For a 24-well plastic dish, use coverslips that are approximately 12 mm in diameter. Large coverslips are not easy to handle and tend to break during the procedure. However, the diameter of the coverslips should be greater than that of the gelatin capsules, as described subsequently. Keep the cells wet at all times during the procedure.
b. Centrifuge the plate at 2000*g* for 5 min and do not remove the supernatant fraction (medium).
c. For fixation of the cells, add gently an equal volume of 4% PA-4% GA buffered with 0.1 *M* PB, pH 7.2, to the culture medium of step b for 15 min at RT.
d. Gently aspirate the fixative and replace with 2% PA-2% GA in 0.1 *M* PB. Fix the cells at RT for 2 h.
e. Wash the cells in PBS 3 times.
f. Postfix the cells in 1% osmium tetroxide in PBS for 1 h at 4 °C in the dark.
g. Wash the cells in distilled water 3 times.
h. Stain the cells in 2% uranyl acetate in water at RT for 1 h.
i. Wash the cells in distilled water 3 times at RT.
j. Dehydrate the cells with a graded series of ethanol (from 50% ~ 90%). At each step agitate the tissues for 5 min at RT.
k. Dehydrate the cells with absolute ethanol three times for 5 min at RT.
l. Immerse the coverslips in a 1:1 mixture of ethanol and Epon 812 for 1 h.
m. Immerse the coverslips in a 1:3 mixture of ethanol and Epon 812 for 1 h.
n. Immerse the coverslips in pure Epon 812, swirl them to remove ethanol, and place them in a vacuum chamber overnight.
o. Place the coverslips upside down on gelatin capsules that are filled with Epon 812. Make sure that there are no air bubbles under the coverslips.
p. Polymerize the resin at 80 °C for 48 h.
q. Remove the glass coverslips by heating the capsules on a hot plate. Make sure that all glasses are removed. Celldesks are easily removed by tweezers.
r. Cut 70- to 80-nm ultrathin sections with a diamond knife.
s. Stain the sections with 4% aqueous uranyl acetate and 0.4% lead citrate for 5 min each, and observe as described previously.

3.2. Biochemical analysis

3.2.1. Genomic DNA laddering

Many of the morphological changes that occur in cell nuclei as a result of autophagic neuron death are similar to those that occur with apoptosis. Although TUNEL staining has been used for the detection of apoptotic nuclei after neuron death, necrotic nuclei also stain positive for TUNEL. DNA ladder formation by electrophoresis of genomic DNA is a hallmark of apoptosis and also autophagic neuron death. This section describes easy methods for both the purification of genomic DNA (Koike *et al.*, 2003, 2008) and highly sensitive detection using ligation-mediated PCR (LMPCR) (Koike *et al.*, 2008; Staley *et al.*, 1997). LMPCR can detect even naturally occurring (programmed) cell death that appears to occur in a fewer number of cells in the hippocampus of mice around P8.

3.2.1.1. Extraction of genomic DNA

a. Homogenize brain sections in 1 ml of lysis buffer (4 *M* guanidine thiocyanate and 0.1 *M* Tris-HCl, pH 7.0) and incubate them at RT for 1 h.
b. Add 20 μl of QIAEX II suspension (QIAGEN, Hilden, Germany), vortex, and incubate at RT for 1 h.
c. Centrifuge at 12,500*g* for 2 min and discard the supernatant fractions.
d. Wash the pellet fractions and resuspend them in washing buffer (50% Ethanol, 100 m*M* NaCl, 1 m*M* EDTA, 10 m*M* Tris-HCl, pH 7.5).
e. Centrifuge at 12,500*g* for 2 min, discard the supernatant fractions, and repeat steps d and e twice.
f. Resuspend the pellet fractions in TE and incubate them at RT for 10 min.
g. Centrifuge at 12,000*g* for 2 min and recover the supernatant fractions.
h. Measure the concentration of genomic DNA. Apply 1 μg of genomic DNA to a 2.0% agarose gel and separate the DNA fragments via electrophoresis.

3.2.1.2. LMPCR

a. Synthesize two 24-bp oligonucleotides: 5′-AGCACTCTCGAG-CCTCTCACCGCA-3′ and 12-bp: 5′-TGCGGTGAGAGG-3′.
b. Anneal the two oligonucleotides by heating at 55 °C for 10 min in a T4 DNA ligase buffer (commercially available), and cool to 10 °C over a 55-min period.
c. Mix 1 μg genomic DNA, 1 nmol annealed oligonucleotides, and 3 U of T4 DNA ligase in 50 μl T4 DNA ligase buffer.
d. Incubate the mixture at 16 °C overnight.
e. Dilute the mixture with TE to a final concentration of 5 ng/μl.
f. PCR solution: 25 ng of ligated genomic DNA and 62 pmol of 24-bp oligonucleotides, as the primer, and 1.75 U Expand high-fidelity PCR system (Roche, Basel, Switzerland)) in a 50-μl reaction volume.
g. PCR conditions: 94 °C for 20 s and 72 °C for 3 min for 20 cycles (neonate) or 25 cycles (adult), and postcycling at 72 °C for 15 min.
h. Separate amplified DNA by electrophoresis on a 2% agarose gel.

3.2.2. Western blotting

Autophagic neuron death is morphologically and pathologically distinct from apoptotic cell death (Koike *et al.*, 2008). In the neonatal hippocampus, caspase-3 (caspase-7) is in part activated in H/I injury-induced pyramidal neurons, whereas caspase-3 or -7 is not activated in the adult hippocampus after H/I injury. In neonatal mice, therefore, the activated forms of caspases can be detected by Western blotting using specific antibodies to the cleaved forms. The conversion of LC3 from LC3-I to LC3-II, an autophagosome marker, can also be examined by Western blotting.

a. Homogenize brain tissues in 1 ml of lysis buffer (per 100 mg brain tissue) (1% Triton X-100, 1% Nonidet P-40 in PBS containing a proteinase inhibitor cocktail) by pipetting several dozen times.
b. Place the samples on ice for 15 min.
c. Centrifuge the samples at 10,500*g* for 10 min at 4 °C.
d. Recover the supernatant fractions and centrifuge them at 10,500*g* for 10 min at 4 °C.
e. Measure the protein concentrations of the supernatant fractions using BSA as a standard.
f. Separate the proteins using 12.5% SDS-PAGE. Load appropriate amounts of protein (15 μg (LC3 and glyceraldehyde-3-phosphate dehydrogenase (GAPDH) as a loading control) or 40 μg (caspases)) onto the gel.
g. Transfer proteins from PAGE gels to PVDF membranes (at 2 mA/cm^2 for 1 h).
h. Block the membranes with skim milk in TPBS at RT for 20 min. For caspases, Tris-buffered saline with Tween 20 (TBST) is first substituted for TPBS prior to treatment with skim milk.
i. Incubate the membranes with the 1st antibody diluted with TPBS either at RT for 1 h or at 4 °C overnight. The preparation of rabbit antibodies against rat LC3 is described elsewhere (Koike *et al.*, 2005). The following antibodies are commercially available: rabbit polyclonal antibodies against caspase-3 (Cell Signaling, Danvers, MA), and mouse monoclonal antibodies against GAPDH (clone: 6C5) (Ambion, Austin, TX), caspase-7 (clone: 10-1-62) (BD Biosciences, San Jose, CA).
j. Wash the membranes in TPBS 3 times at RT.
k. Incubate the membranes with the HRP-conjugated 2nd antibody diluted with TPBS at RT for 1 h.
l. Wash the membrane in TPBS 3 times at RT.
m. Detect the activated forms of caspases (17 kD, caspase-3; 20 kD, caspase-7) and the conversion of LC3-I to LC3-II using a chemiluminescent substrate (also see the chapter by Kimura *et al.*, in volume 452).

3.2.3. Enzyme activities (caspases and cathepsins)

Proteolytic activity of both the caspases (caspase-3/7) and lysosomal cathepsins changes during autophagic neuron death (Uchiyama, 2001). Measurements of the activity of these enzymes may be useful for understanding of the death modes involved (Barrett and Kirschke, 1981; Koike *et al.*, 2003, 2008; Shibata *et al.*, 2002).

3.2.3.1. Caspases (caspase-3/7) activity

a. Homogenize brain tissues in 100 μl (per 10 mg of brain tissue) of lysis buffer (10 m*M* HEPES, 42 m*M* KCl, 5 m*M* $MgCl_2$, 1 m*M* DTT, 0.5% CHAPS, and protease inhibitor cocktail, pH 7.5).
b. Centrifuge the samples at 10,500*g* for 15 min at 4 °C and recover the supernatant fractions.
c. Incubate 10 μl of lysate with 190 μl of assay buffer (25 m*M* HEPES, 1 m*M* EDTA, 3 m*M* DTT, 0.1% CHAPS, 10% sucrose, proteinase inhibitor cocktail, pH 7.5) containing 30 μM of the substrate for caspase-3/-7, N-acetyl-Asp-Glu-Val-Asp-AMC (3171-v; Peptide, Japan), at 37 °C.
d. Monitor the fluorescence intensity (excitation, 365 nm; emission, 465 nm) for 15 min at 30-s intervals. Make sure that the accumulation of fluorescence is linear over at least a 15-min interval.
e. Determine the initial reaction rate from the initial velocity.

3.2.3.2. Cathepsin B activity

a. Homogenize brain tissues in 100 μl (per 10 mg of brain tissue) of lysis buffer (1% Triton X-100 in PBS).
b. Centrifuge the samples at 10,500*g* for 15 min at 4 °C and recover the supernatant fractions.
c. Dissolve cysteine in H_2O to a final concentration of 0.2 *M* prior to the enzyme assay and store on ice. Prepare the reaction buffer (4 m*M* EDTA, 0.4 M NaAc, pH 5.5, 4 μg/ml pepstatin A) and the substrate (20 μM of Z-Arg-Arg-MCA (3123-v; Peptide Inc., Japan) in H_2O). Prewarm the reaction buffer and the substrate solution to 37 °C.
d. Mix brain-tissue lysates with 8 μl of 0.2 *M* cysteine and 50 μl of reaction buffer, and adjust the final volume of the lysate solution to 100 μl by addition of H_2O.
e. Preincubate the solution for 1 min at 37 °C.
f. Add 100 μl of substrate to the lysate solution.
g. Incubate the mixture at 37 °C.
h. Monitor the fluorescence intensity (excitation, 370 nm; emission, 460 nm) for 15 min at 30-s intervals. Make sure that the accumulation of fluorescence is linear over at least a 15-min interval.
i. Determine the initial reaction rate from the initial velocity.

3.2.3.3. Cathepsin D activity

Follow section 3.2.3.2 with the following modifications:

a. Prepare the reaction buffer (1 *M* NaAc, pH 4.0), 100 μg/ml leupeptin, and the substrate (20 μ*M* of MOCAc-Gly-Lys-Pro-Ile-Leu-Phe-Phe-Arg-Leu-Lys(Dnp)-D-Arg-NH_2) (3200-v; Peptide, Japan).
b. Monitor the fluorescence intensity (excitation, 328 nm; emission, 393 nm) for 15 min at 30-s intervals.

REFERENCES

Adhami, F., Liao, G., Morozov, Y. M., Schloemer, A., Schmithorst, V. J., Lorenz, J. N., Dunn, R. S., Vorhees, C. V., Wills-Karp, M., Degen, J. L., Davis, R. J., Mizushima, N., *et al.* (2006). Cerebral ischemia-hypoxia induces intravascular coagulation and autophagy. *Am. J. Pathol.* **169,** 566–583.

Barrett, A. J., and Kirschke, H. (1981). Cathepsin B, cathepsin H, and cathepsin L. *Methods Enzymol.* **80 Pt C,** 535–561.

Chu, C. T. (2006). Autophagic stress in neuronal injury and disease. *J. Neuropathol. Exp. Neurol.* **65,** 423–432.

Clark, R. S., Bayir, H., Chu, C. T., Alber, S. M., Kochanek, P. M., and Watkins, S. C. (2008). Autophagy is increased in mice after traumatic brain injury and is detectable in human brain after trauma and critical illness. *Autophagy* **4,** 88–90.

Clarke, P. G. (1990). Developmental cell death: morphological diversity and multiple mechanisms. *Anat. Embryol. (Berl.)* **181,** 195–213.

Ditelberg, J. S., Sheldon, R. A., Epstein, C. J., and Ferriero, D. M. (1996). Brain injury after perinatal hypoxia-ischemia is exacerbated in copper/zinc superoxide dismutase transgenic mice. *Pediatr. Res.* **39,** 204–208.

Erlich, S., Alexandrovich, A., Shohami, E., and Pinkas-Kramarski, R. (2007). Rapamycin is a neuroprotective treatment for traumatic brain injury. *Neurobiol. Dis.* **26,** 86–93.

Eskelinen, E.-L. (2008). To be or not to be? Examples of incorrect identification of autophagic compartments in conventional transmission electron microscopy of mammalian cells. *Autophagy* **4,** 257–260.

Fearnley, I. M., Walker, J. E., Martinus, R. D., Jolly, R. D., Kirkland, K. B., Shaw, G. J., and Palmer, D. N. (1990). The sequence of the major protein stored in ovine ceroid lipofuscinosis is identical with that of the dicyclohexylcarbodiimide-reactive proteolipid of mitochondrial ATP synthase. *Biochem. J.* **268,** 751–758.

Ferriero, D. M., Holtzman, D. M., Black, S. M., and Sheldon, R. A. (1996). Neonatal mice lacking neuronal nitric oxide synthase are less vulnerable to hypoxic-ischemic injury. *Neurobiol. Dis.* **3,** 64–71.

Greene, L. A., and Tischler, A. S. (1976). Establishment of a noradrenergic clonal line of rat adrenal pheochromocytoma cells which respond to nerve growth factor. *Proc. Natl. Acad. Sci. USA* **73,** 2424–2428.

Hall, N. A., Lake, B. D., Dewji, N. N., and Patrick, A. D. (1991). Lysosomal storage of subunit c of mitochondrial ATP synthase in Batten's disease (ceroid-lipofuscinosis). *Biochem. J.* **275(Pt 1),** 269–272.

Haltia, M. (2006). The neuronal ceroid-lipofuscinoses: From past to present. *Biochim. Biophys. Acta* **1762,** 850–856.

Hara, T., Nakamura, K., Matsui, M., Yamamoto, A., Nakahara, Y., Suzuki-Migishima, R., Yokoyama, M., Mishima, K., Saito, I., Okano, H., and Mizushima, N. (2006).

Suppression of basal autophagy in neural cells causes neurodegenerative disease in mice. *Nature* **441,** 885–889.

Isahara, K., Ohsawa, Y., Kanamori, S., Shibata, M., Waguri, S., Sato, N., Gotow, T., Watanabe, T., Momoi, T., Urase, K., Kominami, E., and Uchiyama, Y. (1999). Regulation of a novel pathway for cell death by lysosomal aspartic and cysteine proteinases. *Neuroscience* **91,** 233–249.

Kabeya, Y., Mizushima, N., Ueno, T., Yamamoto, A., Kirisako, T., Noda, T., Kominami, E., Ohsumi, Y., and Yoshimori, T. (2000). LC3, a mammalian homologue of yeast Apg8p, is localized in autophagosome membranes after processing. *EMBO J.* **19,** 5720–5728.

Koike, M., Nakanishi, H., Saftig, P., Ezaki, J., Isahara, K., Ohsawa, Y., Schulz-Schaeffer, W., Watanabe, T., Waguri, S., Kametaka, S., Shibata, M., Yamamoto, K., *et al.* (2000). Cathepsin D deficiency induces lysosomal storage with ceroid lipofuscin in mouse CNS neurons. *J. Neurosci.* **20,** 6898–6906.

Koike, M., Shibata, M., Ohsawa, Y., Nakanishi, H., Koga, T., Kametaka, S., Waguri, S., Momoi, T., Kominami, E., Peters, C., Figura, K., Saftig, P., *et al.* (2003). Involvement of two different cell death pathways in retinal atrophy of cathepsin D-deficient mice. *Mol. Cell Neurosci.* **22,** 146–161.

Koike, M., Shibata, M., Tadakoshi, M., Gotoh, K., Komatsu, M., Waguri, S., Kawahara, N., Kuida, K., Nagata, S., Kominami, E., Tanaka, K., and Uchiyama, Y. (2008). Inhibition of autophagy prevents hippocampal pyramidal neuron death after hypoxic-ischemic injury. *Am. J. Pathol.* **172,** 454–469.

Koike, M., Shibata, M., Waguri, S., Yoshimura, K., Tanida, I., Kominami, E., Gotow, T., Peters, C., von Figura, K., Mizushima, N., Saftig, P., and Uchiyama, Y. (2005). Participation of autophagy in storage of lysosomes in neurons from mouse models of neuronal ceroid-lipofuscinoses (Batten disease). *Am. J. Pathol.* **167,** 1713–1728.

Komatsu, M., Waguri, S., Chiba, T., Murata, S., Iwata, J., Tanida, I., Ueno, T., Koike, M., Uchiyama, Y., Kominami, E., and Tanaka, K. (2006). Loss of autophagy in the central nervous system causes neurodegeneration in mice. *Nature* **441,** 880–884.

Kominami, E., Ezaki, J., Muno, D., Ishido, K., Ueno, T., and Wolfe, L. S. (1992). Specific storage of subunit c of mitochondrial ATP synthase in lysosomes of neuronal ceroid lipofuscinosis (Batten's disease). *J. Biochem. (Tokyo)* **111,** 278–282.

Lai, Y., Hickey, R. W., Chen, Y., Bayir, H., Sullivan, M. L., Chu, C. T., Kochanek, P. M., Dixon, C. E., Jenkins, L. W., Graham, S. H., Watkins, S. C., and Clark, R. S. B. (2008). Autophagy is increased after traumatic brain injury in mice and is partially inhibited by the antioxidant g-glutamylcysteinyl ethyl ester. *J. Cereb. Blood. Flow. Metab.* **28,** 540–550.

Leonard, D. G., Ziff, E. B., and Greene, L. A. (1987). Identification and characterization of mRNAs regulated by nerve growth factor in PC12 cells. *Mol. Cell Biol.* **7,** 3156–3167.

Levine, S. (1960). Anoxic-ischemic encephalopathy in rats. *Am. J. Pathol.* **36,** 1–17.

Liou, W., Geuze, H. J., Geelen, M. J., and Slot, J. W. (1997). The autophagic and endocytic pathways converge at the nascent autophagic vacuoles. *J. Cell Biol.* **136,** 61–70.

Mizushima, N. (2007). Autophagy: Process and function. *Genes Dev.* **21,** 2861–2873.

Nakanishi, H., Zhang, J., Koike, M., Nishioku, T., Okamoto, Y., Kominami, E., von Figura, K., Peters, C., Yamamoto, K., Saftig, P., and Uchiyama, Y. (2001). Involvement of nitric oxide released from microglia-macrophages in pathological changes of cathepsin D-deficient mice. *J. Neurosci.* **21,** 7526–7533.

Ness, J. M., Harvey, C. A., Strasser, A., Bouillet, P., Klocke, B. J., and Roth, K. A. (2006). Selective involvement of BH3-only Bcl-2 family members Bim and Bad in neonatal hypoxia-ischemia. *Brain Res.* **1099,** 150–159.

Nitatori, T., Sato, N., Waguri, S., Karasawa, Y., Araki, H., Shibanai, K., Kominami, E., and Uchiyama, Y. (1995). Delayed neuronal death in the CA1 pyramidal cell layer of the gerbil hippocampus following transient ischemia is apoptosis. *J. Neurosci.* **15,** 1001–1011.

Nixon, R. A. (2006). Autophagy in neurodegenerative disease: Friend, foe or turncoat? *Trends Neurosci.* **29,** 528–535.

Ohsawa, Y., Isahara, K., Kanamori, S., Shibata, M., Kametaka, S., Gotow, T., Watanabe, T., Kominami, E., and Uchiyama, Y. (1998). An ultrastructural and immunohistochemical study of PC12 cells during apoptosis induced by serum deprivation with special reference to autophagy and lysosomal cathepsins. *Arch. Histol. Cytol.* **61,** 395–403.

Ohsumi, Y. (2001). Molecular dissection of autophagy: Two ubiquitin-like systems. *Nat. Rev. Mol. Cell Biol.* **2,** 211–216.

Palmer, D. N., Martinus, R. D., Cooper, S. M., Midwinter, G. G., Reid, J. C., and Jolly, R. D. (1989). Ovine ceroid lipofuscinosis. The major lipopigment protein and the lipid-binding subunit of mitochondrial ATP synthase have the same NH2-terminal sequence. *J. Biol. Chem.* **264,** 5736–5740.

Rami, A., Langhagen, A., and Steiger, S. (2008). Focal cerebral ischemia induces upregulation of Beclin 1 and autophagy-like cell death. *Neurobiol. Dis.* **29,** 132–141.

Rice, J. E., 3rd, Vannucci, R. C., and Brierley, J. B. (1981). The influence of immaturity on hypoxic-ischemic brain damage in the rat. *Ann. Neurol.* **9,** 131–141.

Sadasivan, S., Dunn, W. A., Jr., Hayes, R. L., and Wang, K. K. (2008). Changes in autophagy proteins in a rat model of controlled cortical impact induced brain injury. *Biochem. Biophys. Res. Commun.* **374,** 478–481.

Saftig, P., Hetman, M., Schmahl, W., Weber, K., Heine, L., Mossmann, H., Koster, A., Hess, B., Evers, M., von Figura, K., *et al.* (1995). Mice deficient for the lysosomal proteinase cathepsin D exhibit progressive atrophy of the intestinal mucosa and profound destruction of lymphoid cells. *EMBO J.* **14,** 3599–3608.

Sanjuan, M. A., Dillon, C. P., Tait, S. W., Moshiach, S., Dorsey, F., Connell, S., Komatsu, M., Tanaka, K., Cleveland, J. L., Withoff, S., and Green, D. R. (2007). Toll-like receptor signalling in macrophages links the autophagy pathway to phagocytosis. *Nature* **450,** 1253–1257.

Sheldon, R. A., Sedik, C., and Ferriero, D. M. (1998). Strain-related brain injury in neonatal mice subjected to hypoxia-ischemia. *Brain Res.* **810,** 114–122.

Shibata, M., Kanamori, S., Isahara, K., Ohsawa, Y., Konishi, A., Kametaka, S., Watanabe, T., Ebisu, S., Ishido, K., Kominami, E., and Uchiyama, Y. (1998). Participation of cathepsins B and D in apoptosis of PC12 cells following serum deprivation. *Biochem. Biophys. Res. Commun.* **251,** 199–203.

Shibata, M., Koike, M., Waguri, S., Zhang, G., Koga, T., and Uchiyama, Y. (2002). Cathepsin D is specifically inhibited by deoxyribonucleic acids. *FEBS Lett.* **517,** 281–284.

Siintola, E., Lehesjoki, A. E., and Mole, S. E. (2006a). Molecular genetics of the NCLs: Status and perspectives. *Biochim. Biophys. Acta* **1762,** 857–864.

Siintola, E., Partanen, S., Stromme, P., Haapanen, A., Haltia, M., Maehlen, J., Lehesjoki, A. E., and Tyynela, J. (2006b). Cathepsin D deficiency underlies congenital human neuronal ceroid-lipofuscinosis. *Brain* **129,** 1438–1445.

Sleat, D. E., Donnelly, R. J., Lackland, H., Liu, C. G., Sohar, I., Pullarkat, R. K., and Lobel, P. (1997). Association of mutations in a lysosomal protein with classical late-infantile neuronal ceroid lipofuscinosis. *Science* **277,** 1802–1805.

Somboonthum, P., Yoshii, H., Okamoto, S., Koike, M., Gomi, Y., Uchiyama, Y., Takahashi, M., Yamanishi, K., and Mori, Y. (2007). Generation of a recombinant Oka varicella vaccine expressing mumps virus hemagglutinin-neuraminidase protein as a polyvalent live vaccine. *Vaccine* **25,** 8741–8755.

Staley, K., Blaschke, A. J., and Chun, J. (1997). Apoptotic DNA fragmentation is detected by a semi-quantitative ligation-mediated PCR of blunt DNA ends. *Cell Death Differ.* **4,** 66–75.

Steinfeld, R., Reinhardt, K., Schreiber, K., Hillebrand, M., Kraetzner, R., Brück, W., Saftig, P., and Gärtner, J. (2006). Cathepsin D deficiency is associated with a human neurodegenerative disorder. *Am. J. Hum. Genet.* **78,** 988–998.

Tanaka, Y., Guhde, G., Suter, A., Eskelinen, E.-L., Hartmann, D., Lullmann-Rauch, R., Janssen, P. M., Blanz, J., von Figura, K., and Saftig, P. (2000). Accumulation of autophagic vacuoles and cardiomyopathy in LAMP-2-deficient mice. *Nature* **406,** 902–906.

Tanida, I., Minematsu-Ikeguchi, N., Ueno, T., and Kominami, E. (2005). Lysosomal turnover, but not a cellular level, of endogenous LC3 is a marker for autophagy. *Autophagy* **1,** 84–91.

Tanida, I., Tanida-Miyake, E., Komatsu, M., Ueno, T., and Kominami, E. (2002). Human Apg3p/Aut1p homologue is an authentic E2 enzyme for multiple substrates, GATE-16, GABARAP, and MAP-LC3, and facilitates the conjugation of hApg12p to hApg5p. *J. Biol. Chem.* **277,** 13739–13744.

Uchiyama, Y. (1995). Apoptosis: The history and trends of its studies. *Arch Histol. Cytol.* **58,** 127–137.

Uchiyama, Y. (2001). Autophagic cell death and its execution by lysosomal cathepsins. *Arch. Histol. Cytol.* **64,** 233–246.

Uchiyama, Y., Koike, M., and Shibata, M. (2008a). Autophagic neuron death in neonatal brain ischemia/hypoxia. *Autophagy* **4,** 404–408.

Uchiyama, Y., Shibata, M., Koike, M., Yoshimura, K., and Sasaki, M. (2008b). Autophagy-physiology and pathophysiology. *Histochem. Cell. Biol.* **129,** 407–420.

Yoshida, H., Kawane, K., Koike, M., Mori, Y., Uchiyama, Y., and Nagata, S. (2005). Phosphatidylserine-dependent engulfment by macrophages of nuclei from erythroid precursor cells. *Nature* **437,** 754–758.

Zhu, C., Wang, X., Xu, F., Bahr, B. A., Shibata, M., Uchiyama, Y., Hagberg, H., and Blomgren, K. (2005). The influence of age on apoptotic and other mechanisms of cell death after cerebral hypoxia-ischemia. *Cell Death Differ.* **12,** 162–176.

Zhu, C., Xu, F., Wang, X., Shibata, M., Uchiyama, Y., Blomgren, K., and Hagberg, H. (2006). Different apoptotic mechanisms are activated in male and female brains after neonatal hypoxia-ischaemia. *J. Neurochem.* **96,** 1016–1027.

Zhu, J. H., Horbinski, C., Guo, F., Watkins, S., Uchiyama, Y., and Chu, C. T. (2007). Regulation of autophagy by extracellular signal-regulated protein kinases during 1-methyl-4-phenylpyridinium-induced cell death. *Am. J. Pathol.* **170,** 75–86.

CHAPTER FOUR

Assessing Metabolic Stress and Autophagy Status in Epithelial Tumors

Robin Mathew,[*,†] Vassiliki Karantza-Wadsworth,[§,¶] *and* Eileen White[*,†,‡,¶]

Contents

[*] University of Medicine and Dentistry of New Jersey, Robert Wood Johnson Medical School, Piscataway, New Jersey, USA
[†] Center for Advanced Biotechnology and Medicine, Rutgers University, Piscataway, New Jersey, USA
[‡] Department of Molecular Biology and Biochemistry, Rutgers University, Piscataway, New Jersey, USA
[§] Division of Medical Oncology, Department of Internal Medicine, University of Medicine and Dentistry of New Jersey, Robert Wood Johnson Medical School, Piscataway, New Jersey, USA
[¶] Cancer Institute of New Jersey, New Brunswick, New Jersey, USA

Methods in Enzymology, Volume 453
ISSN 0076-6879, DOI: 10.1016/S0076-6879(08)04004-4

Abstract

Autophagy is a survival mechanism activated in response to metabolic stress. In normal tissues autophagy plays a major role in energy homeostasis through catabolic self-digestion of damaged proteins and organelles. Contrary to its survival function, autophagy defects are implicated in tumorigenesis suggesting that autophagy is a tumor suppression mechanism. Although the exact mechanism of this tumor suppressor function is not known, it likely involves mitigation of cellular damage leading to chromosomal instability. The complex role of functional autophagy in tumors calls for model systems that allow the assessment of autophagy status, stress management and the impact on oncogenesis both *in vitro* as well as *in vivo*. We developed model systems that involve generation of genetically defined, isogenic and immortal epithelial cells from different tissue types that are applicable to both wild-type and mutant mice. This permits the study of tissue- as well as gene-specific tumor promoting functions. We successfully employed this strategy to generate isogenic, immortal epithelial cell lines from wild-type and mutant mice deficient in essential autophagy genes such as *beclin 1* (*beclin* $1^{+/-}$) and *atg5* (*atg* $5^{-/-}$). As these cell lines are amenable to further genetic manipulation, they allowed us to generate cell lines with apoptosis defects and stable expression of the autophagy marker EGFP-LC3 that facilitate *in vitro* and *in vivo* assessment of stress-mediated autophagy induction. We applied this model system to directly monitor autophagy in cells and 3D-morphogenesis *in vitro* as well as in tumor allografts *in vivo*. Using this model system we demonstrated that autophagy is a survival response in solid tumors that co-localizes with hypoxic regions, allowing tolerance to metabolic stress. Furthermore, our studies have established that autophagy also protects tumor cells from genome damage and limits cell death and inflammation as possible means to tumor suppression. Additionally these cell lines provide an efficient way to perform biochemical analyses, and high throughput screening for modulators of autophagy for potential use in cancer therapy and prevention.

1. Introduction

Hypoxia is a common occurrence in human solid tumors. Hypoxic regions, largely due to inefficient vasculature and rapid tumor growth, may influence tumor progression and negatively affect clinical outcome, as they are implicated in resistance to therapy (Barlogie *et al.*, 1982; Folkman, 2003). Metabolic stress often triggers apoptotic cell death that results in cancer cell elimination, and hence mutations in the apoptotic pathways are common in human cancers (Nelson *et al.*, 2004). For cancer cells to better survive metabolic stress, this apoptotic resistance must also be accompanied by activation of alternative pathways supporting cell survival under stress. One such mechanism is the up-regulation of the transcription factor hypoxia-inducible factor 1-α (Hif1-α) to promote metabolic adaptation and angiogenesis (Dang *et al.*, 2008; Semenza, 2003). Another mechanism involves activation of the catabolic pathway of autophagy to facilitate cell survival (Mathew *et al.*, 2007a). Elucidating the molecular intricacies of the pathways will hopefully lead to the development of targeted therapeutic strategies, and thus may have profound implications for cancer therapy.

Autophagy (i.e., macroautophagy) is a response to stress and starvation whereby cellular organelles and proteins are sequestered and targeted for lysosomal degradation as an alternate energy source (Levine and Kroemer, 2008). The role of autophagy as a survival mechanism under metabolic stress is well documented (Mizushima, 2007). Immortalized baby mouse kidney epithelial (iBMK) cells rendered autophagy-deficient by allelic loss of *beclin 1* or *atg5* deficiency display increased susceptibility to metabolic stress (Degenhardt *et al.*, 2006; Mathew *et al.*, 2007b). Similarly, autophagy mitigates metabolic stress in immortalized mouse mammary epithelial cells (iMMECs) (Karantza-Wadsworth *et al.*, 2007) and promotes survival of immortalized, nontumorigenic human mammary epithelial cell lines (MCF10A) and primary human mammary cells during extracellular matrix detachment (anoikis) (Debnath, 2008; Fung *et al.*, 2008). In apoptosis-competent cells, autophagy delays apoptotic death under metabolic stress, whereas apoptosis defects unmask autophagy-mediated cell survival (Mathew *et al.*, 2007a). Moreover, *atg5*$^{-/-}$ mouse embryonic fibroblasts (MEFs) show signs of ATP depletion, and *atg5*$^{-/-}$ mice do not survive neonatal starvation, suggesting that autophagy promotes survival under metabolic stress during mammalian development (Kuma *et al.*, 2004; Lum *et al.*, 2005). Importantly, in solid tumors autophagy localizes to regions of metabolic stress, suggesting that it may be exploited by cancer cells for survival (Degenhardt *et al.*, 2006; Karantza-Wadsworth *et al.*, 2007; Mathew *et al.*, 2007b). Thus, the functional status of autophagy is an important determinant of tumor cell response to metabolic stress.

Although autophagy induction under metabolic stress is well established as a survival strategy, interactions between metabolic stress and defective autophagy are more complex. Intuitively contradictory to the survival function of autophagy under metabolic stress, defects in autophagy are associated with increased tumorigenicity in mice as well as humans. Allelic loss of *beclin 1* is frequently observed in human breast, ovarian, and prostate cancers (Aita *et al.*, 1999; Liang *et al.*, 1999), and *beclin* $1^{+/-}$ and *atg4C*$^{-/-}$ mice are tumor-prone, suggesting that autophagy is a tumor suppression mechanism (Marino *et al.*, 2007). Moreover, growth factor– and nutrient-driven oncogenic pathways, such as the PI3-kinase pathway, inhibit autophagy, whereas inhibitors of this pathway, such as the tumor suppressor protein PTEN, activate autophagy (Arico *et al.*, 2001). Although a clear understanding of autophagy-mediated tumor suppression is only beginning to emerge, one of the likely mechanisms by which autophagy inhibits tumorigenesis is suppression of necrotic cell death (Degenhardt *et al.*, 2006). Impairment of autophagy by monoallelic deletion of *beclin 1*, RNAi-mediated knockdown of *beclin 1* or *atg5*, or constitutive activation of Akt, induces necrotic cell death when apoptosis is blocked (Degenhardt *et al.*, 2006). In tumors *in vivo*, this necrosis is associated with inflammation, activation of the cytokine-responsive NF-κB pathway and tumor progression (Degenhardt *et al.*, 2006). Remarkably, autophagy defects in mouse liver cause excessive hepatocyte cell death, steatohepatitis and hepatocellular carcinoma (HCC) suggesting that support of cell survival and suppression of inflammation may be important autophagy functions in normal tissues, as well as tumors (Komatsu *et al.*, 2007).

Another insight into the role of autophagy in tumor suppression came from the discovery that immortalized mouse epithelial cell lines with autophagy defects show signs of genome damage, which is exacerbated under metabolic stress. iBMK and iMMEC cells rendered autophagy-defective by *beclin 1* monoallelic loss or *atg5* deletion display activation of the DNA damage response, gene amplification, and accelerated progression to aneuploidy (Karantza-Wadsworth *et al.*, 2007; Mathew *et al.*, 2007a,b). These phenotypes are accentuated in an apoptosis-defective background, together suggesting that autophagy functions to limit chromosomal instability, preferentially manifested in cells with checkpoint and apoptosis defects. Thus, autophagy-mediated housekeeping and mitigation of genome damage play important roles in the cellular response to metabolic stress and in tumorigenesis.

However, the exact mechanism by which autophagy suppresses tumorigenesis is not known. To further investigate the role of metabolic stress–induced autophagy in tumorigenesis, we developed an *in vitro* metabolic stress assay that combines hypoxia (defined gas mixture composed of 1% O_2, 5% CO_2, and 94% N_2) with glucose deprivation, thus mimicking metabolic stress in the tumor microenvironment *in vivo* (Nelson *et al.*, 2004).

Immortalized epithelial cell model systems, such as iBMK cells (Mathew *et al.*, 2008) and iMMECs (Karantza-Wadsworth and White, 2008), have several advantages over conventional MEF or human cancer cell lines widely used to model human cancer (Mathew *et al.*, 2008). Being epithelial in origin, they provide a superior representation of human tumor cell physiology compared to MEFs, can be generated from different tissues from any mouse that survives to birth (for isolation of kidney, prostate, liver, and lung tissue) and to young adulthood (for mammary gland isolation), are immortalized by well-defined genetic events, are amenable to additional genetic manipulation, and can be used for the generation of tumor allografts. Immortalized epithelial cell lines derived from wild-type and mutant mice extend the utility of mouse models by enabling biochemical and cell biological analysis. Stable expression of fluorescent fusion proteins, fluorescent or luminescent reporter- and cell tracking-constructs, or RNAi-mediated knockdown of specific proteins extend the analyses to the study of the role of compound mutations in tumorigenesis. These cell models together with our *in vitro* metabolic stress assay have been successfully used to characterize epithelial cell response to metabolic stress *in vitro* and *in vivo* (Degenhardt *et al.*, 2006; Karantza-Wadsworth *et al.*, 2007; Karp *et al.*, 2008; Mathew *et al.*, 2007b; Nelson *et al.*, 2004; Shimazu *et al.*, 2007).

2. Mouse Epithelial Cell Models for Studying the Role of Autophagy in Cancer

Autophagy is a highly conserved process tightly regulated by a set of essential genes such as *atg5*, *atg7*, and *beclin 1*, which produce a profound autophagy-defective phenotype when allelically lost (*atg5*$^{-/-}$, *atg7*$^{-/-}$, or *beclin 1*$^{+/-}$). With the genetic landscape of autophagy regulation quickly emerging, several transgenic mice specifically targeting autophagy are currently available for tumorigenicity studies *in vivo*. Primary epithelial cells from wild-type, *atg5*$^{-/-}$, *atg7*$^{-/-}$, and *beclin 1*$^{+/-}$ mice can be immortalized through expression of the adenoviral protein E1A and a dominant negative p53 mutant (p53DD) to generate isogenic epithelial cell lines that are suitable for studying the role of autophagy in cancer (Mathew *et al.*, 2008). We have generated iBMK cells and iMMECs that are wild-type or autophagy-deficient (*beclin 1*$^{+/-}$, *atg5*$^{-/-}$, *atg7*$^{-/-}$) with and without a functional apoptotic pathway (Degenhardt *et al.*, 2006; Karantza-Wadsworth *et al.*, 2007; Mathew *et al.*, 2007b, 2008). These cell lines have been successfully employed to demonstrate that their autophagy-defective phenotype is independent of the mode of autophagy impairment and the tissue of origin (Degenhardt *et al.*, 2006; Karantza-Wadsworth *et al.*, 2007; Mathew *et al.*, 2007b).

2.1. Generation of stable iBMK and iMMEC cell lines to monitor autophagy *in vitro* and *in vivo*

The process of autophagy is characterized by the formation of isolation membranes (phagophores) that mature into double-membrane vesicles called autophagosomes (Levine and Kroemer, 2008). Under conditions of metabolic stress, the product of the essential autophagy gene *LC3/atg8* is proteolytically cleaved, lipidated, and translocated to the forming autophagosomes, as demonstrated by the redistribution of the EGFP-LC3 fusion protein from a diffuse cytoplasmic localization under normal conditions to discrete, perinuclear puncta under metabolic stress (Fig. 4.1) (Klionsky, 2007; Mizushima, 2004). Similar induction of LC3 translocation occurs under growth factor deprivation. This process is impaired by deficiencies in essential autophagy genes, as indicated by the failure to form EGFP-LC3 puncta on monoallelic loss of *beclin 1* (Fig. 4.1) or complete *atg5* deficiency (Mizushima, 2004).

In cells competent for apoptosis and autophagy, the predominant phenotype under conditions of metabolic stress is apoptosis, the defects in which induce prolonged autophagy-supported cell survival. Thus, the assessment of autophagy under metabolic stress is facilitated in an apoptosis-defective background. To monitor autophagy induction under metabolic stress, we generated apoptosis-defective *beclin* $1^{+/+}$ and *beclin* $1^{+/-}$ iBMK and iMMEC cells stably expressing the autophagy marker EGFP-LC3.

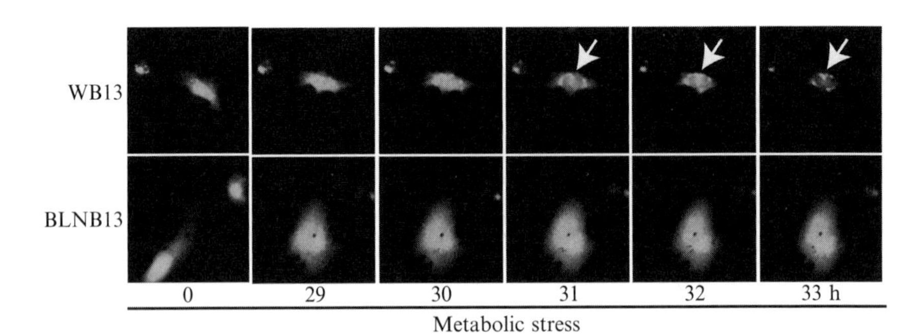

Figure 4.1 Live cell imaging demonstrates the process of autophagy in living cells. The product of the essential autophagy gene *LC3* becomes posttranslationally modified and translocated to autophagosomes during autophagy induction, as indicated by the transition from diffuse cytoplasmic to punctate perinuclear localization. The process is impaired by heterozygosity in the essential autophagy gene *beclin 1*, as indicated by failure of LC3 translocation. To capture the process of autophagy in living cells, apoptosis-defective *beclin* $1^{+/+}$ or *beclin* $1^{+/-}$ iBMK cells stably expressing EGFP-LC3 are subjected to metabolic stress and observed by Computerized Fluorescence Video Time-Lapse Microscopy for 33 h. Autophagy is induced (arrow) in a representative *beclin* $1^{+/+}$ (WB13) cell (top row), but not in a representative *beclin* $1^{+/-}$ (BLNB13) cell (bottom row) under metabolic stress.

These cell lines and their tumor allografts in nude mice allow *in vitro* and *in vivo* visualization and quantification of autophagy under a wide variety of experimental conditions (Karantza-Wadsworth and White, 2008; Mathew *et al.*, 2008; Nelson *et al.*, 2004).

2.1.1. Equipment and reagents

1. pEGFP-C1 (Clontech, 6084-1), pEGFP-C1-LC3 (Mizushima, 2004), pCDNA3.1 zeo (Invitrogen, V86020) mammalian expression vectors
2. Geneticin and zeocin (Invitrogen, 11811 and R-250)
3. iBMK regular growth medium: DMEM, 10% fetal bovine serum (FBS), 1% Pen/Strep (Invitrogen, 15140)
4. iMMEC regular growth medium: F12 (Invitrogen, 11765054), 10% FBS, 5 μg/ml insulin (Sigma, I0516), 1 μg/ml hydrocortisone (Sigma, H0888), 5 ng/ml EGF (Sigma, E4127), 1% Pen/Strep
5. Electroporation system (Bio-Rad Gene Pulser II)
6. Freezing medium: 92% FBS, 8% tissue culture grade DMSO (Sigma, D2650)

2.1.2. Generation of cell lines stably expressing EGFP-LC3

Primary baby mouse kidney epithelial (BMK) cells are isolated from wild-type, *bax*$^{-/-}$/*bak*$^{-/-}$, *beclin 1*$^{+/-}$, and *atg5*$^{-/-}$ mice and immortalized by E1A and a dominant negative p53 mutant (p53DD) to generate iBMK cells. Apoptosis-competent iBMK cells are then engineered to express either vector control (pCEP4) or Bcl-2 (pCEP-Bcl-2) as described previously (Degenhardt *et al.*, 2006; Mathew *et al.*, 2007b). Mouse mammary epithelial cells (MMECs) are isolated from 6–8-week-old female *beclin 1*$^{+/+}$ and *beclin 1*$^{+/-}$ mice, immortalized by E1A and p53DD (to generate iMMECs) and rendered apoptosis-defective by stable Bcl-2 expression, as previously described (Karantza-Wadsworth *et al.*, 2007). The detailed protocols for generating iBMK cells and iMMECs are available in an earlier volume of *Methods in Enzymology* (Karantza-Wadsworth and White, 2008; Mathew *et al.*, 2008).

iBMK cells are further engineered to express EGFP-LC3 as described subsequently:

1. Two 1.5-ml sterile microcentrifuge tubes are labeled and 10 μg of pEGFP-C1 is transferred into one and 10 μg of pEGFP-C1-LC3 into the other tube.
2. 2×10^6 cells are plated per 10-cm plate in duplicate and allowed to grow in normal tissue culture conditions (38.5 °C and 8.5% CO_2). When 80% confluency is reached, cells from each plate are harvested separately by trypsinization using 0.05% trypsin-EDTA (Invitrogen, 25300), resuspended in 10 ml of tissue culture medium, and centrifuged at 1500 rpm (approximately $300\times g$).

3. Each cell pellet is resuspended in 500 μl of phosphate buffered saline (PBS). 250 μl are transferred into a 1.5-ml microcentrifuge tube containing 10 μg of pEGFP-C1 and the other 250 μl are transferred into a microcentrifuge tube containing 10 μg of pEGFP-C1-LC3 plasmid.
4. Each cell suspension is mixed by gentle pipetting, transferred to a 0.4-cm electroporation cuvette (Bio-Rad, 1652088), and pulsed at 0.22 V and 950 μF.
5. Transfected cells are allowed to sit for 5 min, resuspended in 12 ml of regular growth medium, dispensed into 1-, 2-, and 3-ml aliquots and plated in 10-cm plates in duplicate.
6. 48 h later, the medium is replaced by geneticin (G418)-containing regular growth medium, and G418-resistant colonies arise in 7–10 days. G418 is used at 2 mg/ml for $bax^{-/-}/bak^{-/-}$-derived cell lines, and at 1 mg/ml for wild-type and Bcl-2-expressing cell lines. Selection medium is changed every 3 days.
7. Typically, up to 12 independent (well-isolated and from different plates) G418-resistant colonies per genotype are isolated and expanded as described in section 2.1.3.

iMMECs are further engineered to express EGFP-LC3 as described subsequently:

1. Apoptosis-competent iMMECs are engineered to express EGFP-LC3 as described earlier for iBMK cells. The only protocol changes involve use of the iMMEC-specific medium for cell growth and a much lower geneticin concentration (300 μg/ml) for selection.
2. Bcl-2 expressing iMMECs, which are already selected and grown in G418-contaning medium (Karantza-Wadsworth and White, 2008), are cotransfected with pEGFP-C1-LC3 and pcDNA3.1zeo plasmids in a 10:1 ratio, followed by zeocin selection in the presence of geneticin (300 μg/ml). Zeocin is used at 100 μg/ml for 10–14 days. Independent G418- and zeo-double-resistant colonies are isolated and expanded to stable cell lines, as described subsequently.

2.1.3. Cloning, expansion, and preservation of cell lines

1. When colonies reach 0.6 cm to 0.8 cm in diameter, they are examined under an inverted fluorescence microscope, and well-separated green fluorescent colonies are marked on the bottom of the plates for ring cloning.
2. Cells are washed once with PBS, which is then gently aspirated. Sterile cloning rings (6 × 8 mm or 10 × 10 mm; Belco Biotechnology, 2090-00608 or 2090-01010) are placed over the marked colonies and held in place by autoclaved vacuum grease (VWR, 59344-055).
3. Cells are trypsinized with a small amount of 0.05% Trypsin-EDTA placed inside the cloning rings and transferred into 96-well plates.

Individual clones are sequentially expanded into 24- and 12-well plates, then 6-cm and finally 10-cm plates. EGFP and EGFP-LC3 expression is tested by fluorescence microscopy and western blotting using anti-GFP antibody (BD Living Colors, 632381) at 1:6000 dilution.
4. For long-term storage, cells are trypsinized at 70% confluency (approximately 6×10^6 cells/10-cm plate), resuspended in 2 ml of freezing medium per 10-cm plate, stored in a –70 °C freezer wrapped in paper towels overnight, and then cryopreserved in vapor phase nitrogen.

3. Protocols for Monitoring Autophagy in iBMK Cells and iMMECs *In Vitro*

A cell-based system that enables functional autophagy monitoring is important not only to better understand the role of autophagy in metabolic stress management and cancer progression but also to screen for autophagy modulators. Immortalized epithelial cells from wild-type and autophagy-deficient mice stably expressing EGFP-LC3 (Karantza-Wadsworth and White, 2008; Mathew *et al.*, 2008) can be used for real-time observation of autophagy (see subsequently), as well as high-throughput screens for identifying novel autophagy inhibitors and stimulators.

3.1. Autophagy induction under metabolic stress as monitored by *in vitro* EGFP-LC3 assay

3.1.1. Reagents

1. Coverslips: Fisherbrand Coverglass (Fisher Scientific, 12-545-82-12CIR-1D)
2. Metabolic stress assay medium
   a. iBMK cells: Glucose-free DMEM (Invitrogen, 11966), 10% FBS, 1% Pen/Strep
   b. iMMECs: Glucose-free DMEM, 10% FBS, 5 μg/ml insulin, 1 μg/ml hydrocortisone, 5 ng/ml EGF, 1% Pen/Strep
3. 10% buffered formalin solution: (Formaldefresh; Fisher Scientific, SF94)
4. ProLong Gold antifade reagent (Molecular Probes, P36930)

3.1.2. *In vitro* metabolic stress assay

1. Ethanol-sterilized coverslips are aseptically placed in 10-cm plates. One plate per cell line per time point is used.
2. 2×10^6 cells engineered to stably express EGFP-LC3 are plated in each coverslip-containing plate and grown to 80% confluency.
3. At this point (usually 24 h later), regular growth medium is replaced with glucose-free medium (iBMK- or iMMEC-specific) and plates are placed in modular incubator chambers (Billups-Rothenberg; Model MIC-101),

which are flushed with a gas mixture containing 1% O_2, 5% CO_2 and 94% N_2 (GTS-Welco) for 5 min and placed at 37 °C for 1, 2, and 3 days to induce metabolic stress (Nelson *et al.*, 2004). One chamber is used per time point to prevent reoxygenation upon chamber opening for sample collection. For longer than 24-h incubations, chambers are flushed with the same gas mixture for 5 min daily.
4. Chambers are opened at designated time-points and cover slips are collected.

3.1.3. Autophagy quantification by EGFP-LC3 translocation

1. Coverslips are fixed in 10% buffered formalin solution for 15 min at room temperature (RT), washed, and stored in PBS at 4 °C for up to 2 weeks.
2. Coverslips are rinsed with distilled water, mounted on glass slides with the antifade agent Prolong and imaged using a fluorescence microscope at 600X magnification.
3. EGFP-LC3 shows diffuse cytoplasmic distribution under normal conditions. Induction of autophagy is indicated by translocation of the EGFP-LC3 signal to discrete perinuclear puncta (autophagosomes). Cells expressing just GFP show no punctate staining under metabolic stress and are used as negative controls.
4. Autophagy is quantified as percent EGFP-LC3 translocation, determined as the fraction of green fluorescent cells that demonstrate punctate staining. Three independent experiments are performed (with at least 100 cells counted each time) and the mean values with standard deviation are presented.

3.2. Autophagy induction under metabolic stress as monitored by EM

3.2.1. Reagents

1. EM fixative: 4% paraformaldehyde and 2.5% glutaraldehyde in 0.1 *M* cacodylate buffer (Electron Microscopy Sciences, 11650)

3.2.2. Cell preparation

1. Cells are plated at 2 × 10^6 cells per 10-cm plate and 24 h later are exposed to metabolic stress as described previously. At successive time points, cells are harvested by trypsinization, resuspended in 10 ml of culture medium, and centrifuged at 1500 rpm (approximately 300×*g*) for 5 min.
2. The cell pellet is washed, resuspended in 2 ml of PBS, transferred into 2 1.5-ml microcentrifuge tubes (1 ml of cell suspension per tube), and centrifuged at 3000 rpm (approximately 800×*g*) for 5 min.

3. The cell pellet is fixed by gentle addition of EM fixative (1 ml), taking care to avoid disturbing the pellet, and is stored at 4 °C overnight.
4. The cell pellet is processed for EM by standard procedures (e.g., see the chapter by Ylä-Anttila *et al.*, in volume 452). Images are acquired at 4000–5000X magnification.

3.3. Protocols for monitoring autophagy and metabolic stress in three-dimensional morphogenesis

3.3.1. Reagents

1. Growth factor–reduced Matrigel (BD, 356231)
2. 8-well RS glass slides (BD Falcon, 354108)
3. iMMEC regular growth medium (F12, 10% FBS, 5 μg/ml insulin, 1 μg/ml hydrocortisone, 5 ng/ml EGF, 1% Pen/Strep)
4. Hypoxyprobe-1 kit (Chemicon International, HP2-1000)
5. ProLong Gold antifade reagent

3.3.2. Monitoring metabolic stress in 3D-morphogenesis

1. A 3D-culture of EGFP-LC3-expressing iMMECs is performed according to previously published protocols (Debnath *et al.*, 2003; Karantza-Wadsworth and White, 2008).
2. Spatial localization of metabolic stress in mammary acini is performed by hypoxyprobe (pimonidazole) immunohistochemistry and immunofluorescence using a hypoxyprobe kit (Fig. 4.2). Mammary acini generated by *beclin* $1^{+/+}$ and *beclin* $1^{+/-}$ iMMECs expressing Bcl-2 are grown in Matrigel for 12 days and are then incubated with 200 μM hypoxyprobe in regular growth medium at 37 °C for 2 h. For immunohistochemistry, acini are fixed in 10% formalin, scraped from the glass slide with a razor blade, pelleted, embedded in paraffin, and sectioned. For immunofluorescence, acini are fixed and processed as previously described (Debnath *et al.*, 2003; Karantza-Wadsworth and White, 2008).

3.3.3. Monitoring autophagy in 3D-morphogenesis

1. Apoptosis-defective iMMECs stably expressing EGFP-LC3 are grown in Matrigel for 12 days and fixed as previously.
2. Acinar structures are washed with PBS (3 times) and mounted with the antifade agent Prolong.
3. EGFP-LC3 fluorescence is imaged by confocal microscopy.

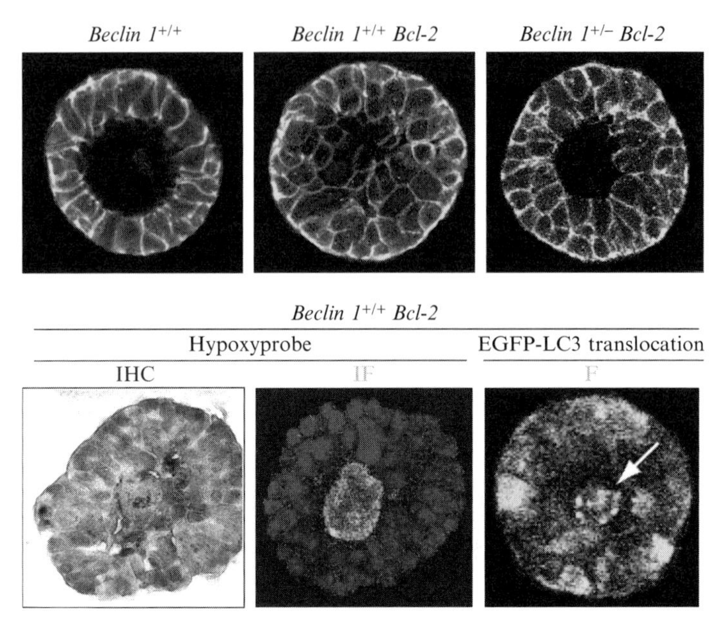

Figure 4.2 In 3D-morphogenesis, allelic loss of *beclin 1* accelerates death of mammary epithelial cells in the acinar center, where metabolic stress and autophagy localize. Top panel: Apoptosis-competent *beclin* $1^{+/+}$ iMMECs form polarized acini that generate lumens via apoptosis (left). Defective apoptosis delays lumen formation (middle), whereas concurrent autophagy defects abrogate the survival advantage that Bcl-2 expression confers to central acinar cells (right). Mammary acini are immunostained with β-catenin (green) and nuclei are counterstained with DAPI (blue). Bottom panel: Hypoxyprobe immunohistochemistry (IHC, left) and immunofluorescence (IF, middle) on acini generated by apoptosis-defective *beclin* $1^{+/+}$ iMMECs. Fluorescence (F) of *beclin* $1^{+/+}$ iMMECs with Bcl-2 and stably expressing EGFP-LC3 (right). (See Color Insert.)

4. Protocols for Monitoring Autophagy in Tumors *In Vivo*

Rapid tumor growth is often associated with metabolic stress, as cellular proliferation outstrips vascular supply and results in hypoxic regions within tumors. Induction of autophagy in tumors *in vivo* can be visualized in the first few days following subcutaneous or orthotopic implantation of apoptosis-defective iBMK cells (Fig. 4.3) and iMMECs stably expressing EGFP-LC3 in nude mice (Degenhardt *et al.*, 2006; Karantza-Wadsworth *et al.*, 2007; Mathew *et al.*, 2007b). Tumor allografts of EGFP-LC3-expressing cells in the abdominal flank (iBMK) or the mammary fat pad (iMMECs) of nude mice provide an excellent system to monitor metabolic stress and autophagy induction during tumorigenesis (Karantza-Wadsworth *et al.*, 2007; Mathew *et al.*, 2007b).

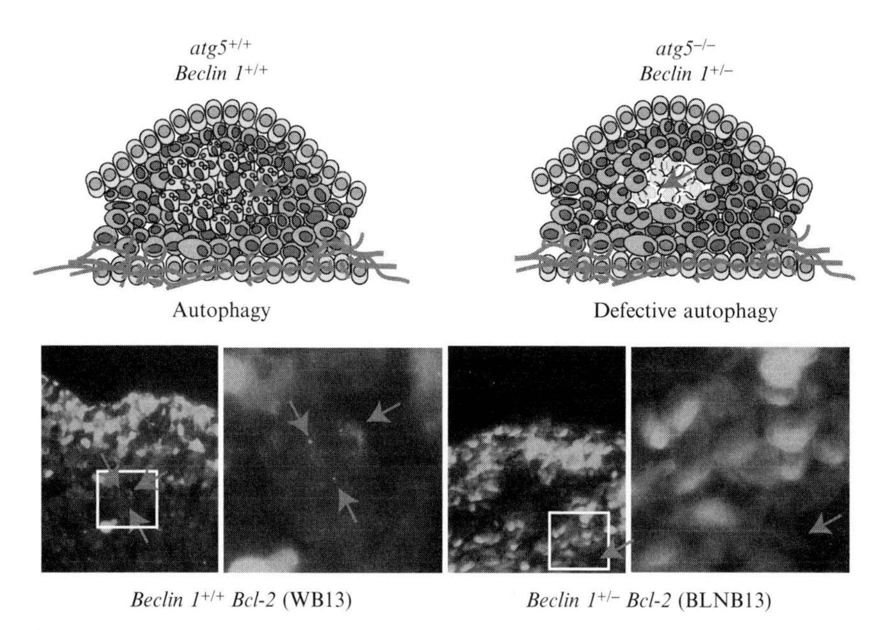

Figure 4.3 Autophagy is a response to metabolic stress in solid tumors. Tumor cells suffer metabolic stress that can either trigger apoptosis to promote tumor suppression or survival by autophagy in tumor cells with apoptosis defects. Autophagy can be visualized in tumors by stable expression of the autophagy marker EGFP-LC3 and generation of subcutaneous tumor allografts. Autophagy-competent (WB3; left panel) or autophagy-defective (BLNB13; right panel) iBMK cells stably expressing EGFP-LC3 shown in figure 1, are implanted subcutaneously to generate allografts for spatial and temporal mapping of autophagy. Tumors are excised at various times post-implantation, fixed and GFP-LC3 fluorescence is imaged by confocal microscopy to indicate autophagy status. Top left panel is a schematic representation of the autophagy-competent (*beclin* $1^{+/+}$) tumor section (on day 3 post-implantation) shown in the bottom left panel: autophagy induction in the center promotes tumor cell survival (red arrows). Top right panel is a schematic representation of the autophagy-deficient (*beclin* $1^{+/-}$) tumor section (on day 3 post-implantation) shown in the bottom right: autophagy defects impair tumor cell survival leading to metabolic catastrophe and necrosis. (See Color Insert.)

Tumors generated by cells stably expressing EGFP-LC3 are excised at various times post-implantation (days 1, 3, 8 and 15) allowing spatial and temporal correlation of functional autophagy status with histological markers (e.g., hypoxia, vascularization, inflammation).

4.1. Monitoring autophagy induction in subcutaneous tumor allografts *in vivo*

4.1.1. Reagents

1. 10% buffered formalin solution
2. 15% sucrose solution

3. 30% sucrose solution
4. Tissue-Tek O.C.T. (VWR, 25608-930)
5. ProLong Gold antifade reagent

4.1.2. Tumor allograft growth

1. Tumor allografts are generated by subcutaneous implantation of iBMK cell lines expressing EGFP or EGFP-LC3 in nude mice (2 animals per time point per genotype) (Mathew *et al.*, 2008).
2. Apoptosis-defective iBMK cells are grown in 15-cm plates (roughly 1 plate per animal injection) to 90% confluency, harvested by trypsinization, centrifuged at 1500 rpm (approximately $300\times g$), and washed twice with PBS. Cells are counted by trypan blue exclusion.
3. Cells are resuspended at a final concentration of 1×10^8 cells/ml in sterile PBS, and 100 μl (10^7 cells) are implanted under the skin in the abdominal flank of nude mice (5 weeks old, male, NCR Nu/Nu; Taconic) per an IACUC-approved protocol.
4. Tumors appear as small subcutaneous bumps as early as 1 day post-implantation. Animals are euthanized on day 1, 3, 8, and 15 after tumor cell implantation and tumors are carefully excised and fixed for frozen sections.

4.1.3. Tumor fixation and processing for frozen sections

1. Tumors are fixed in 10% buffered formalin solution overnight at -20 °C and are then subjected to sequential sucrose dehydration. The dehydration step is critical for preventing ice formation and resultant tissue artifacts upon cryosectioning.
2. Tumors are dehydrated in 15% sucrose at 4 °C for 6 h, and then in 30% sucrose at 4 °C overnight.
3. Tumors are subsequently embedded in Tissue-Tek O.C.T. in a dark chamber at 4 °C for 24 h.
4. Frozen sections (5-μm in thickness) are obtained with a cryomicrotome (Leica Reichert-Jung CryoCut 1800, Leica Microsystems) and collected on poly-L-lysine (1%)-coated glass slides (Lab Scientific, 7799).
5. Frozen sections are mounted with the antifade agent Prolong, stored at 4 °C, and imaged by confocal microscopy using a FITC filter.

4.2. Monitoring autophagy induction in orthotopically implanted mammary tumors *in vivo*

4.2.1. Reagents

1. Ketamine (Ketaset; Fort Dodge, 4402A)
2. Xylazine (Xyla-ject, Phoenix Pharmaceutical, 600077)
3. See reagents in section 4.1.1

4.2.2. Orthotopic tumor growth

1. *beclin 1*$^{+/+}$ and *beclin 1*$^{+/-}$ iMMECs with Bcl-2 and stably expressing EGFP-LC3 are grown in 15-cm plates to 90% confluency, harvested by trypsinization, washed, and resuspended in PBS (10^8 cells/ml).
2. Orthotopic iMMEC implantation is performed using a previously described IACUC-approved protocol (Karantza 2008 MIE). In summary, 5- to 8-week-old nude female mice are anesthetized with ketamine (100 mg/kg intraperitoneally, IP) and xylazine (10 mg/kg IP). A small incision is made to reveal the right third mammary gland, and 10^7 cells (100 μl) are implanted into the mammary fat pad. The incision is closed with surgical clips.

4.2.3. Tumor fixation and processing for frozen sections

1. Mammary tumors are excised at different time points post-implantation, fixed and processed for frozen sections as described in section 4.1.3.

5. Monitoring Chromosomal Instability Due to Autophagy Defects

Centrosomes are cellular organelles that ensure uniform distribution of DNA during mitosis through bipolar division of chromosomes. Centrosomes themselves divide once per cell cycle during S phase, maintaining a tightly regulated centrosome number of 1 (in G_1 phase) or 2 (in G_2 phase) per cell. However, when this regulation is impaired, numerical abnormalities such as supernumerary centrosomes (more than 2 per cell) can result in multipolar spindles and abnormal segregation of chromosomes, leading to aneuploidy. Centrosome abnormalities are common among solid tumors and are indicative of genomic instability (Fukasawa, 2007).

5.1. Protocols for studying centrosome abnormalities

5.1.1. Reagents

1. Coverslips: Fisherbrand Coverglass
2. Methanol:acetone (1:1 v/v), chilled
3. Mouse anti-γ-tubulin antibody (Axell, BYA96861)
4. TRITC-conjugated goat antimouse secondary antibody (Jackson Immuno Research, 115-025-146)
5. Blocking buffer (3% BSA and 0.05% Triton X-100 in PBS)
6. ProLong Gold antifade reagent

5.1.2. γ-tubulin staining and quantification

1. Ethanol-sterilized coverslips are aseptically placed in 10-cm plates.
2. 2×10^6 cells are plated per coverslip-containing plate and grown (for at least 24 h) to 80% confluency.
3. Coverslips are fixed in chilled methanol:acetone for 10 min, rinsed twice with PBS, and incubated with anti-γ-tubulin antibody at 1:100 dilution in blocking buffer in a humidified chamber at 37 °C for 1 h.
4. Coverslips are washed with PBS (3 times) and incubated with TRITC-conjugated goat antimouse secondary antibody at 1:50 dilution in blocking buffer in a humidified chamber at 37 °C for 30 min. Nuclei are counterstained with DAPI (0.5 ng) for 15 min at RT.
5. Coverslips are rinsed in PBS and then distilled water, mounted with the antifade agent Prolong, and imaged at 600X using fluorescence microscopy.
6. The number of centrosomes per cell is recorded for at least 100 cells. Any cell with centrosome number greater than 2 is considered abnormal.

5.2. Protocols for ploidy determination by flow cytometry

Flow cytometry is a simple, but powerful, tool to study ploidy abnormalities by measuring cell DNA content. Cells are fixed and stained with propidium iodide (PI), which stains DNA by intercalating between the bases. PI also binds to RNA, necessitating treatment with ribonuclease (RNase) to minimize interference with DNA staining. Once PI is bound to nucleic acids, the fluorescence excitation and emission maxima are shifted and fluorescence emission is enhanced 20- to 30-fold and is proportional to the total amount of the DNA. Flow cytometry allows the measurement of this fluorescence per cell, thus permitting the quantification of the total amount of DNA per cell. A DNA-PI fluorescence histogram for a normal cell population is typically comprised of a 2-peak profile with fluorescence intensity on an arbitrary scale on the X-axis and frequency on the Y-axis. The first peak represents the diploid population of cells in G_1 phase of the cell cycle (2N DNA content) and the second peak corresponds to cells in G_2 and M phases of the cell cycle (4N DNA content). The valley connecting the two peaks represents cells with intermediate amounts of DNA (2N-4N) corresponding to cells in S phase that are undergoing DNA synthesis at the time of fixation. It may be noted that 2N and 4N notations are merely relative amounts of DNA. In samples where cell death has occurred, there can be a sub-G_1 peak (less than 2N DNA content) and this is often used as a measure of apoptosis.

A major problem in the determination of ploidy abnormalities by flow cytometry is that only a relative, but not the absolute, DNA content is obtained.

This is further complicated by variations in DNA staining intensity, due to cell concentration and instrument parameter variability, which can affect peak positioning and may lead to misinterpretation of the histogram. Extreme consistency in sample preparation is thus critical. These problems are circumvented by using an internal biological DNA standard with a known genome size (C-value), such as Chicken Erythrocyte Nuclei (CEN) Singlets. CEN have a C-value of 1.25 pg (2N chromosome number 18) compared to that of mouse (*Mus musculus*) cells (C-value = 3.25; 2N chromosome number 40) and human (*Homo sapiens*) cells (C-value = 3.5; 2N chromosome number 46). Therefore, when stained and analyzed together with the mouse or human cell lines under investigation, CEN provides a single reference peak to the left of the diploid (2N) peak of these cell lines and the relative position of the other peaks in reference to the CEN singlet provides a satisfactory ploidy measure. Spontaneous ploidy abnormalities due to allelic loss of *beclin 1* in iBMK cells and iMMECs are analyzed as described subsequently.

5.2.1. Reagents

1. Propidium iodide (PI) (Sigma; P4170), 1 mg/ml in dH_2O.
2. RNase A (Sigma, R4875), 1 mg/ml in PBS: 50 mg RNase A are dissolved in 50 ml of PBS containing 0.1% Tween 20 (Sigma, P1379). The solution is placed in a 95 °C water bath for 30 min allowed to cool on ice for 1 h, then filtered through a 0.2-μm filter. The final reagent can be stored for up to 6 months at 4 °C.
3. 70% (v/v) ethanol in dH_2O.
4. Chicken Erythrocyte Nuclei (CEN) Singlets (2×10^4 nuclei/μL) (BioSure, 1013), 1:10 fresh working dilution in PBS.

5.2.2. PI staining and DNA quantification

1. Cells are harvested by trypsinization, centrifuged at 1500 rpm (approximately 300×*g*), resuspended in PBS (1×10^6 cells/ml) and mixed by gentle pipetting to remove cell clumps and prepare a uniform single cell suspension.
2. 1×10^6 cells are transferred to a 15-ml snap-cap polypropylene tube (Falcon, 352059) (polystyrene/carbonate tubes are not suitable as cells may stick to the tube following fixation).
3. Chilled 70% ethanol (5 mL) is added drop wise with simultaneous gentle vortexing (level 3–4) to prevent cell clumping. Cells are fixed on ice for 30 min.
4. Cells are washed with 10 ml of PBS and centrifuged at 2000 rpm (approximately 600×*g*) for 10 min. The supernatant fraction is carefully aspirated, leaving approximately 200 μl in the tube. It is important

that this volume is consistent between samples. Fixed cells are lighter; therefore, care must be taken during supernatant aspiration.
5. PBS (300 μl) is added to the above fixed cell suspension, followed by gentle vortexing (levels 3–4).
6. CEN (1:10 dilution in PBS, 10 μl) is added as an internal standard and the suspension is gently mixed by pipetting.
7. RNase A solution (100 μl) and PI (50 μl) are added to the cell suspension, followed by gentle vortexing and incubation at RT for 30 min.
8. After PI staining, samples are analyzed by flow cytometry (FC500; Beckman Coulter). Cell staining and analysis are repeated at least twice with 10,000 cells per genotype in each experiment. Voltage is adjusted for each sample using CEN as the internal standard peak.
9. Alternatively, PI-stained cells can be sealed airtight in capped tubes (to prevent evaporation) and kept at 4 °C until analysis by flow cytometry.

5.3. Protocols for studying chromosomal abnormalities by metaphase spreads

Ploidy abnormalities are often caused by numerical aberrations in chromosome numbers that are telltale signs of an underlying genomic instability (Rajagopalan and Lengauer, 2004). Giemsa staining allows visualization of gross numerical and structural chromosome abnormalities (Brown and Baltimore, 2000). Prior to Giemsa staining, cells are treated with a microtubule poison, such as nocodazole, for arrest in mitosis (metaphase), where chromatin material is condensed into individual chromosomes. Cells in mitosis are spherical, and therefore loosely attached to the tissue culture plate, so they can be easily removed by shake-off. Isolated mitotic cells are then allowed to swell up by hypotonic treatment followed by methanol: acetic acid fixation before being dropped on a glass slide. A detailed protocol follows.

5.3.1. Reagents

1. Nocodazole (Sigma, M1404) is dissolved in tissue culture grade DMSO, 10 mg/mL stock solution. This is further diluted in regular growth medium to a final concentration of 0.5 mg/ml. The stock solution can be aliquoted and stored at −20 °C.
2. Hypotonic potassium chloride solution (KCl 0.075M).
3. 3:1 (v/v) mix of chilled methanol and glacial acetic acid (methanol:acetic acid fixative), which can be stored at −20 °C for extended periods of time.
4. KaryoMax Giemsa Stain (Gibco, 10092-013): 1X Giemsa staining buffer is prepared by mixing 5 ml of KaryoMax Geimsa stain with 45 ml of Gurr's buffer.

5. Gurr's buffer tablets (Gibco, 10582-013): Gurr's buffer (pH = 6.8) is prepared by dissolving 1 tablet in 1 L of distilled water.
6. CytoSeal mounting medium (Richard-Allan Scientific, 8310-4).

5.3.2. Protocol for Giemsa staining and analysis

1. 2×10^6 cells are plated in a 10-cm plate in regular growth medium and allowed to reach 80% confluency at 38.5 °C and 8.5% CO_2 (usually 24 h).
2. Cells are washed with PBS and treated with nocodazole (0.5 μg/mL) for 2 h. Longer periods of incubation with nocodazole result in overcondensed chromosomes that may make interpretation difficult. Cells are examined by light microscopy at 400–1000X magnification to verify accumulation of mitotic cells.
3. Plates are shaken to remove mitotic cells that are loosely attached compared with cells in interphase. Mitotic cells are collected and transferred to a 50-ml conical tube. Plates are washed 1 more time with PBS to collect the remaining mitotic cells and samples are pooled together.
4. Remaining attached cells are carefully harvested by trypsinization and collected in 10 ml of regular growth medium. Cells are gently mixed by pipetting, and 2 ml of this cell suspension is added into the mitotic cell pool from step 3 (these cells act as carrier cells in subsequent steps to reduce mitotic cell loss).
5. Mitotic cells are pelleted at 1000 rpm (approximately 300×g) at 4 °C for 5 min, gently resuspended in 0.075M KCl (5 mL), and incubated at RT for 20 min.
6. Cells are pelleted again at 1000 rpm (approximately 300×g) at 4 °C for 5 min. The supernatant fraction is carefully aspirated and discarded. Cells are resuspended in methanol:acetic acid fixative (1 ml), transferred to a 1.5-ml microcentrifuge tube, and incubated at 4 °C for 20 min (cells may be stored at 4 °C in the fixative).
7. Glass slides (1 glass slide per mitotic spread) are labeled and chilled at −20 °C.
8. Fixed cells from step 6 are pelleted by centrifugation at 2000 rpm (approximately 600×g) at 4 °C for 2 min and resuspended in 300 μl chilled fixative.
9. When ready to prepare mitotic spreads, slides are taken out of the freezer and moisture is removed by quickly wiping with a clean and dry tissue paper.
10. Fixed cells are gently mixed and dropped from 30–45 cm height onto the prechilled glass slides held in a slanted position. Maintaining cells and slides at low temperatures prevents quick evaporation of the fixative and provides better quality spreads. Slides are allowed to air-dry at RT for 4 min and then cured overnight at 37 °C in an oven (Hybaid).

11. Mitotic spreads are stained with Giemsa staining buffer at RT for 20 min in coplin jars (Fisher, 08-817).
12. Slides are rinsed with Gurr's buffer (2–3 times), air-dried at RT for 30 min, and mounted with Cytoseal-60 mounting medium.
13. Giemsa-stained chromosomes appear purple in color and can be visualized and photographed at 200X magnification using a standard upright microscope. Approximately 150 individual mitotic figures per cell line are photographed and carefully analyzed for numerical and structural chromosome abnormalities. The total number of chromosomes is carefully counted in each spread and depicted as a scatter and X–Y scatter plot. Mean and median chromosome numbers are calculated for each cell line and are used to compute the average of the mean chromosome number in each genotype (Mathew *et al.*, 2007b). The normal mouse chromosome number is 40 and therefore any metaphase spread that contains a chromosome number higher or lower than 40 is considered abnormal.

5.4. Protocols for studying chromosomal gains and losses by array-based comparative genome hybridization (aCGH)

One of the major genomic instability phenotypes associated with autophagy defects is the random gain and losses of chromosomes (Albertson *et al.*, 2003). Such aberrations lead to variations in DNA copy numbers and constitute a major genomic instability phenotype in cancer (Albertson, 2006). As chromosomal gains often also accompany chromosomal losses in the genome, these variations may not necessarily be reflected in the total amount of DNA per cell and therefore cannot be identified by DNA quantification by flow cytometry. Microarray-based aCGH is a powerful technique to identify DNA copy number variations by comparing the hybridization intensities between a normal (reference genome) and a test genome to detect the relative copy number variations signifying chromosomal losses and gains (Kallioniemi *et al.*, 1992).

5.4.1. Reagents

1. Tris-EDTA (TE): 10 m*M* Tris base, pH 7.6, 0.5 m*M* EDTA, pH 8.0
2. Proteinase K (Promega, V3021): 20 mg/mL in water (stock solution); this should be made fresh or kept frozen to prevent self-digestion
3. Sodium dodecyl sulfate (SDS) (Bio-Rad, 161-0302) (20% in water)
4. Saturated sodium chloride (NaCl) solution: 36 g NaCl are dissolved in 100 ml water (solubility of NaCl in water at 20 °C is 36 g/100 mL)

5.4.2. Isolation of genomic DNA from cultured cells

1. Cells are plated in a 10-cm plate, allowed to grow to 80% confluency, harvested by trypsinization and pelleted by centrifugation at 1500 rpm (approximately 300×*g*).
2. The supernatant fraction is carefully aspirated and discarded, and the cells are resuspended in 3 ml of TE and 100 μl 20% SDS.
3. Proteinase K (20 mg/ml, 20 μl) is added to the preceding cell suspension, which is gently mixed by inversion and incubated in a water bath at 55 °C overnight.
4. Saturated NaCl solution (1 ml) is added to the cell digest, and the suspension is mixed gently and uniformly by inversion.
5. 100% ethanol (10 ml) is added, and the tube is gently inverted 3–4 times. A long strand of genomic DNA is visible at this time. Tubes are incubated at RT overnight on a slow rocker.
6. The DNA precipitate is gently transferred to a 15-ml tube containing 10 ml of 70% ethanol and incubated at RT for 6–8 h on a slow rocker.
7. The DNA precipitate is spooled using a slightly bent 22G needle and transferred to the bottom of a 1.5-ml microcentrifuge tube. As much of the ethanol as possible is aspirated. The location of the precipitate is marked outside the tube and the precipitate is allowed to air-dry (once dry, DNA will not be readily visible).
8. Sterile water (100–300 μl) is added toward the tube side, and the DNA pellet is resuspended gently by flicking the tube and allowing it to sit at RT for 1–2 h. Genomic DNA should not be pipetted, as this may cause DNA shearing.
9. Dissolved DNA is further diluted 1:500 in distilled water, and the absorbance is measured at 260 m*M* and 280 m*M* using a UV spectrophotometer (Beckman Coulter). DNA concentration is calculated using the following formula: DNA concentration (μg/μl) $= A_{260} \times CF \times DF/\text{path length}$, where the concentration factor (CF) for double-stranded DNA is 0.05 (1 A_{260} unit = 50 μg/ml) and the dilution factor (DF) is 500.
10. An absorbance ratio ($A_{260/280}$) of 1.8–1.9 indicates that the DNA is pure. $A_{260/280}$ less than 1.8 indicates protein contamination, whereas a ratio higher than 2.0 indicates RNA contamination. In either case, DNA purity can be further improved by an additional phenol/chloroform/isoamyl alcohol extraction as described below.
11. Equal volume of phenol/chloroform/isoamyl alcohol (25:24:1) is added to DNA in a polypropylene tube, and the mixture is vigorously shaken until an emulsion is formed and then centrifuged at 1600×*g* for 3 min. The aqueous phase (containing DNA) is transferred to a fresh 1.5-ml polypropylene microcentrifuge tube. The process is repeated with

chloroform/isoamyl alcohol without phenol, the aqueous phase is transferred into a fresh polypropelene tube and DNA is precipitated by addition of 100% ethanol (2.5 volumes) and 3 *M* sodium acetate (0.1 volume). The solution is mixed by inversion (3-4 times) and centrifuged at $1600\times g$ for 15 min to pellet the DNA, which is finally resuspended in water or TE (resuspension in TE requires 2–3 h of incubation at RT).

5.4.3. Determination of DNA copy number variations

1. Genomic DNA isolated from iBMK cell lines and DNA from a normal mouse kidney are fluorescently labeled (Cy3 and Cy5), hybridized to a BAC array in triplicates and analyzed for chromosome gains and losses using aCGH at the DNA Array Core Facility at the University of California, San Francisco, as described elsewhere (Snijders *et al.*, 2005).
2. Mean $\log_2$ ratios of the total integrated Cy3 and Cy5 intensities of each sample from triplicate spots as estimated by UCSF SPOT software (Jain *et al.*, 2002) and SPROC software (Snijders *et al.*, 2005) are plotted in genome order. An average $\log_2$ ratio of more than ± 0.5 is considered as a loss or gain (Mathew *et al.*, 2007b).

5.5. Protocols for studying gene amplification by PALA assay

Gene amplification results from DNA double-strand breaks (DSBs) due to increased oxidative stress or defects in DNA repair. It is facilitated by inactivation of the p53 DNA damage checkpoint (Lin *et al.*, 2001; Little and Chartrand, 2004; Livingstone *et al.*, 1992; Mondello *et al.*, 2002) and is a major mechanism of oncogene activation (Albertson, 2006; Hennessy *et al.*, 2005; Shen *et al.*, 1986). Therefore, iBMK and iMMEC cell lines with inactivated p53 and pRb pathways are expected to be prone to gene amplification at similar frequencies. Gene amplification is the only known mechanism of resistance to N-phosphonacetyl-l-aspartate (PALA) that prevents *de novo* pyrimidine biosynthesis by inhibiting the aspartate transcarbamylase activity of the carbamoyl-P synthetase/aspartate transcarbamylase/di-hydroorotase (CAD) enzyme complex (Livingstone *et al.*, 1992). Indeed, PALA-resistant cells demonstrate amplification of the CAD gene. Thus, the frequency of clonogenic resistance to PALA is a direct measure of gene amplification, and therefore of the underlying DNA damage and genomic instability. Autophagy suppresses DNA damage and gene amplification, as monoallelic loss of *beclin 1* promotes PALA resistance mediated by gene amplification (Karantza-Wadsworth *et al.*, 2007; Mathew *et al.*, 2007b). This function of autophagy is one of the mechanisms by which autophagy may function as a tumor suppressor pathway (Mathew *et al.*, 2007a).

5.5.1. Reagents

1. N-(phophonoacetyl)-L-aspartate (PALA) (Drug Synthesis and Chemistry Branch, National Cancer Institute): PALA solution (1 m*M*) is prepared fresh by dissolving 175.525 mg of PALA in 500 ml of regular growth medium supplemented with 10% dialyzed, rather than regular, FBS. PALA is light sensitive and therefore exposure to light should be minimized.
2. Dialyzed FBS (Invitrogen, 26400-044): Use of dialyzed FBS eliminates the presence of metabolic precursors that could interfere with PALA toxicity.
3. Trypan blue stain (0.4% solution) (Invitrogen, 15250-061).
4. Methanol (100%).
5. Giemsa Stain (15X) (Sigma; GS1L).

5.5.2. Determination of PALA LD_{50}

1. 1×10^5 cells are plated in 10-cm plates and allowed to adhere for 24 h. Each cell line is plated in triplicate. Medium is then replaced with regular growth medium containing increasing concentrations (0, 10, 20, 30, 40, 50, 60, 70, 80, 90, and 100 μM) of PALA, and cells are incubated for 3 days at 38.5 °C and 8.5% CO_2. Untreated cells are harvested 24 h after plating and counted by trypan blue exclusion to determine the cell number at time 0.
2. PALA-treated cells are rinsed with PBS and the adherent viable cells are harvested by trypsinization and collected in culture medium (5 ml).
3. Viable cells are counted by trypan blue exclusion and the mean viability at each PALA concentration is normalized to cell number at time 0. The data are plotted as a smooth line curve with standard deviations and the concentration of PALA that kills 50% of the cells (LD_{50}) is calculated from the graph. PALA LD_{50} is 20 and 17 μM for Bcl-2-expressing iBMK cells and iMMECs, respectively.

5.5.3. Determination of plating efficiency (PE)

1. 1×10^3 cells are plated per 10-cm plates (in triplicate per cell line) and incubated at 38.5 °C and 8.5% CO_2 until well-separated colonies emerge (7–10 days).
2. Colonies, once visible to the naked eye, are washed with PBS and are fixed in 100% methanol at RT for 15 min.
3. Colonies are stained with Giemsa (15X stock) diluted 1:15 with distilled water for 15–20 min or until colonies become purple in color.
4. Stained plates are rinsed with water several times, inverted, and allowed to dry.
5. Plating efficiency (PE) is calculated as: (number of colonies per plate/ number of cells plated) × 100.

5.5.4. Quantification of PALA-resistance frequency

1. 5×10^5 cells are plated per 10-cm plate (in triplicate per cell line) and allowed to adhere overnight by incubation at 38.5 °C and 8.5% CO_2.
2. The medium is then replaced with regular growth medium containing PALA at 3X LD_{50} and 5X LD_{50} for 10–14 days or until well-separated PALA-resistant colonies emerge. PALA-containing medium is replaced every 3–4 days.
3. PALA-resistant colonies are fixed in methanol, stained with Giemsa (as described in section 5.5.3) and counted.
4. The frequency of PALA-resistance is calculated as the number of PALA-resistant colonies per plate/(PE × number of cells plated).

5.5.5. Polymerase chain reaction (PCR) for *CAD* gene amplification

1. Genomic DNA is isolated from multiple PALA-resistant colonies as described in section 5.4.2.
2. A 756-base-pair fragment flanking two exons of the *Mus musculus CAD* gene (accession no. NM 023525) on chromosome 5 (accession no. AC_109608) is amplified by PCR using the following set of primers:
   Fwd: (5′-GGAGCTGGAGACTCCGACG-3′)
   Rev: (5′-CTAATGAACAGGAAGATCCGGTATC-3′)
3. The PCR is performed for 19, 21, 23, 25, and 27 (N) cycles in the presence of 2.5 units *Taq* DNA polymerase (Invitrogen, 18038-042) and 1.5 m*M* $MgCl_2$ using the program shown below:

Step	*Description*
1	Initial denaturation at 94 °C for 5 min
2	Denaturation at 94 °C for 30 s
3	Annealing at 50 °C for 30 s
4	Extension at 72 °C for 30 s
5	Repeat steps 2–4 for N times
6	Additional extension at 72 °C for 5 min

4. PCR products are separated on a 1% agarose gel and stained with ethidium bromide.

5.6. Computerized video time-lapse (CVTL) microscopy

Genomic instability due to defective autophagy likely plays a major role in tumorigenesis. Conventional tissue culture techniques cannot provide the spatial and temporal information that is crucial for understanding the dynamic cellular processes leading to genomic instability. Long-term observations are also restricted by the demands for physiological growth conditions and by limitations of microscopy instrumentation. CVTL, especially in combination with fluorescence protein tagging, is a powerful tool to monitor cellular processes that occur over extended periods of time. We have developed a model system comprising of a wide array of genetically defined epithelial cells that closely recapitulate *in vivo* tumor cell phenotypes under physiological conditions *in vitro* (Mathew *et al.*, 2008). Combining the CVTL system and a panel of fluorescent probes stably expressed in these cells, we have captured and characterized cellular processes such as apoptosis, autophagy, necrosis, mitosis and cell division, wound healing response, and 3D morphogenesis that play key roles in tumorigenesis (Degenhardt *et al.*, 2006; Karantza-Wadsworth *et al.*, 2007; Karp *et al.*, 2008; Shimazu *et al.*, 2007). The fully automated and environmentally controlled system enables us to culture and film isogenic cell lines of varying genotype under multiple growth conditions within the same experiment. Time-lapse observations typically span up to 12 days or more and can be performed under regular and metabolic stress conditions, including drug treatments (Degenhardt *et al.*, 2006; Karantza-Wadsworth *et al.*, 2007; Karp *et al.*, 2008; Shimazu *et al.*, 2007).

5.6.1. Equipment and reagents

1. Inverted microscope with 10X, 20X, and 40X objectives
2. Temperature, humidity and CO_2-controlled environmental chamber
3. Cooled CCD camera
4. Image analysis software

5.6.2. Protocol for CVTL microscopy

1. Wild-type and autophagy-defective cells are plated as described in steps 2 or 3 subsequently. Alternatively, cells stably expressing fluorescent markers can also be used to track fluorescent markers for monitoring cellular processes such as mitosis (for special settings, see step 6 here).
2. For time-lapse observations under normal growth conditions, cells are plated at a density of 2×10^5 cells per well in a 6-well plate in 2 ml of medium so that all the cell lines and culture conditions are represented on the same 6-well plate. Cells are allowed to attach and grow overnight.
3. For time-lapse observations under metabolic stress conditions that require examination of one cell line at a time, cells are plated in a T25 tissue culture flask (Corning cell culture flask with canted neck, plug seal cap, 430168) at a density of 2×10^5 per flask in 8 ml of regular culture medium and allowed to grow for 16–18 h. Normal growth medium is then replaced with 8 ml of glucose-free medium (glucose-free DMEM supplemented with 10% FBS and 1% Pen/Strep) and the flasks are then sealed air-tight using a rubber stopper (VWR, 59582-585) modified to accommodate tubing that allows flushing of the flasks with a defined

ischemic gas mixture containing 1% oxygen, 5% CO_2 and 94% N_2 (GTS-Welco) (Nelson *et al.*, 2004) for 5 min. After flushing, the flask and the tubes are sealed and the procedure continues with step 4.

4. Approximately 16–18 h after the plating, cells normally growing in 6-well plates (from step 2) or subjected to metabolic stress in T25 flasks (from step 3) are transferred into the time-lapse chamber, equipped with controlled environmental conditions.
5. The time-lapse microscopy system consists of an Olympus IX-71 inverted microscope fitted with temperature, humidity and CO_2-controlled environmental chamber (37 °C and 8.5% CO_2) (Solent Scientific) and a CoolSNAP ES-cooled CCD camera.
6. Phase contrast images (100X) at multiple fields are obtained at 10-min intervals for 3–5 days using ImagePro Plus software (Media Cybernetics). An example of the settings used for bright field images are given here:
   Binning: 1 × 1
   Exposure: 16 ms.
   Interval: 600 s.
   For fluorescence time-lapse experiments images are captured using GFP filters. An example of the settings used for GFP images are given here:
   Binning: 4 × 4
   Expose: 10 ms.
   Interval: 1800 s.
7. Captured images are assembled into time-lapse sequences and movies (.avi). Analysis of still images and sequences are performed using ImagePro Plus software.
8. For examining differences in cell survival, time-lapse sequence files (.seq files) are played using ImagePro software, in parallel. Still images are used to assemble a panel for direct comparison of cell survival.

6. Concluding Remarks and Future Perspectives

Metabolic stress and genomic instability have been implicated in human cancer pathogenesis for a long time; however, the exact functional interaction between these factors has been largely elusive. By developing an *in vitro* assay that faithfully recapitulates metabolic stress in the tumors *in vivo*, we discovered that autophagy plays a major role in linking the two by mitigating the deleterious consequences of metabolic stress and the resultant DNA damage and instability (Karantza-Wadsworth *et al.*, 2007; Mathew *et al.*, 2007b). Therefore, activation of autophagy not only supports survival under stress but also protects the genome, and therefore promoting autophagy may be beneficial as a cancer prevention strategy.

Even more important, cancer cells overtly depend on an uninterrupted nutritional supply for meeting their proliferative needs, and this high demand in conjunction with inadequate supply is exactly what causes hypoxic regions in tumors. On one hand, metabolic stress compromises treatment efficacy, as tumor hypoxia is associated with resistance to radiation and chemotherapy. On the other hand, the greater susceptibility of cancer cells to metabolic stress can be exploited for therapeutic benefit. As a major cellular stress response, autophagy may facilitate tumor cell survival creating dormant tumor cells that cause disease recurrence. A clearer understanding of how tumor cells use autophagy to survive metabolic stress is essential for the successful use of autophagy modulation in cancer therapy. Therefore, screening for small-molecule autophagy inducers and inhibitors may prove extremely important in the development of novel preventative and therapeutic strategies. In the past few years, major strides have been made in understanding the role of autophagy in cancer progression and treatment, and there is great enthusiasm in targeting this important pathway for clinical outcome improvement. The wild-type and autophagy-defective immortalized epithelial cell lines that we have developed—in particular the cells with fluorescent readouts for autophagy monitoring—are powerful tools available to the cancer research community for this new and exciting endeavor.

REFERENCES

Aita, V. M., Liang, X. H., Murty, V. V., Pincus, D. L., Yu, W., Cayanis, E., Kalachikov, S., Gilliam, T. C., and Levine, B. (1999). Cloning and genomic organization of *beclin 1*, a candidate tumor suppressor gene on chromosome 17q21. *Genomics* **59,** 59–65.

Albertson, D. G. (2006). Gene amplification in cancer. *Trends Genet.* **22,** 447–455.

Albertson, D. G., Collins, C., McCormick, F., and Gray, J. W. (2003). Chromosome aberrations in solid tumors. *Nat. Genet.* **34,** 369–376.

Arico, S., Petiot, A., Bauvy, C., Dubbelhuis, P. F., Meijer, A. J., Codogno, P., and Ogier-Denis, E. (2001). The tumor suppressor PTEN positively regulates macroautophagy by inhibiting the phosphatidylinositol 3-kinase/protein kinase B pathway. *J. Biol. Chem.* **276,** 35243–35246.

Barlogie, B., Johnston, D. A., Smallwood, L., Raber, M. N., Maddox, A. M., Latreille, J., Swartzendruber, D. E., and Drewinko, B. (1982). Prognostic implications of ploidy and proliferative activity in human solid tumors. *Cancer Genet. Cytogenet.* **6,** 17–28.

Brown, E. J., and Baltimore, D. (2000). ATR disruption leads to chromosomal fragmentation and early embryonic lethality. *Genes Dev.* **14,** 397–402.

Dang, D. T., Chun, S. Y., Burkitt, K., Abe, M., Chen, S., Havre, P., Mabjeesh, N. J., Heath, E. I., Vogelzang, N. J., Cruz-Correa, M., Blayney, D. W., Ensminger, W. D., *et al.* (2008). Hypoxia-inducible factor-1 target genes as indicators of tumor vessel response to vascular endothelial growth factor inhibition. *Cancer Res.* **68,** 1872–1880.

Debnath, J. (2008). Detachment-induced autophagy during anoikis and lumen formation in epithelial acini. *Autophagy* **4,** 351–353.

Debnath, J., Muthuswami, S. K., and Brugge, J. (2003). Morphogenesis and oncogenesis of MCF-10A mammary epithelial acini grown in three-dimensional basement membrane cultures. *Methods* **30,** 256–268.

Degenhardt, K., Mathew, R., Beaudoin, B., Bray, K., Anderson, D., Chen, G., Mukherjee, C., Shi, Y., Gelinas, C., Fan, Y., Nelson, D. A., Jin, S., *et al.* (2006). Autophagy promotes tumor cell survival and restricts necrosis, inflammation, and tumorigenesis. *Cancer Cell* **10,** 51–64.

Folkman, J. (2003). Angiogenesis and apoptosis. *Semin. Cancer Biol.* **13,** 159–167.

Fukasawa, K. (2007). Oncogenes and tumour suppressors take on centrosomes. *Nat. Rev. Cancer* **7,** 911–924.

Fung, C., Lock, R., Gao, S., Salas, E., and Debnath, J. (2008). Induction of autophagy during extracellular matrix detachment promotes cell survival. *Mol. Biol. Cell* **19,** 797–806.

Hennessy, B. T., Smith, D. L., Ram, P. T., Lu, Y., and Mills, G. B. (2005). Exploiting the PI3K/AKT pathway for cancer drug discovery. *Nat. Rev. Drug Discov.* **4,** 988–1004.

Jain, A. N., Tokuyasu, T. A., Snijders, A. M., Segraves, R., Albertson, D. G., and Pinkel, D. (2002). Fully automatic quantification of microarray image data. *Genome Res.* **12,** 325–332.

Kallioniemi, A., Kallioniemi, O. P., Sudar, D., Rutovitz, D., Gray, J. W., Waldman, F., and Pinkel, D. (1992). Comparative genomic hybridization for molecular cytogenetic analysis of solid tumors. *Science* **258,** 818–821.

Karantza-Wadsworth, V., Patel, S., Kravchuk, O., Chen, G., Mathew, R., Jin, S., and White, E. (2007). Autophagy mitigates metabolic stress and genome damage in mammary tumorigenesis. *Genes Dev.* **21,** 1621–1635.

Karantza-Wadsworth, V., and White, E. (2008). A mouse mammary epithelial cell model to identify molecular mechanisms regulating breast cancer progression. *Methods Enzymol.* **446,** 61–76.

Karp, C. M., Tan, T. T., Mathew, R., Nelson, D., Mukherjee, C., Degenhardt, K., Karantza-Wadsworth, V., and White, E. (2008). Role of the polarity determinant crumbs in suppressing mammalian epithelial tumor progression. *Cancer Res.* **68,** 4105–4115.

Klionsky, D. J. (2007). Autophagy: From phenomenology to molecular understanding in less than a decade. *Nat. Rev. Mol. Cell Biol.* **8,** 931–937.

Komatsu, M., Waguri, S., Koike, M., Sou, Y. S., Ueno, T., Hara, T., Mizushima, N., Iwata, J. I., Ezaki, J., Murata, S., Hamazaki, J., Nishito, Y., *et al.* (2007). Homeostatic levels of p62 control cytoplasmic inclusion body formation in autophagy-deficient mice. *Cell* **131,** 1149–1163.

Kuma, A., Hatano, M., Matsui, M., Yamamoto, A., Nakaya, H., Yoshimori, T., Ohsumi, Y., Tokuhisa, T., and Mizushima, N. (2004). The role of autophagy during the early neonatal starvation period. *Nature* **432,** 1032–1036.

Levine, B., and Kroemer, G. (2008). Autophagy in the pathogenesis of disease. *Cell* **132,** 27–42.

Liang, X. H., Jackson, S., Seaman, M., Brown, K., Kempkes, B., Hibshoosh, H., and Levine, B. (1999). Induction of autophagy and inhibition of tumorigenesis by *beclin 1*. *Nature* **402,** 672–676.

Lin, C. T., Lyu, Y. L., Xiao, H., Lin, W. H., and Whang-Peng, J. (2001). Suppression of gene amplification and chromosomal DNA integration by the DNA mismatch repair system. *Nucleic Acids Res.* **29,** 3304–3310.

Little, K. C., and Chartrand, P. (2004). Genomic DNA is captured and amplified during double-strand break (DSB) repair in human cells. *Oncogene* **23,** 4166–4172.

Livingstone, L. R., White, A., Sprouse, J., Livanos, E., Jacks, T., and Tlsty, T. D. (1992). Altered cell cycle arrest and gene amplification potential accompany loss of wild-type p53. *Cell* **70,** 923–935.

Lum, J. J., Bauer, D. E., Kong, M., Harris, M. H., Li, C., Lindsten, T., and Thompson, C. B. (2005). Growth factor regulation of autophagy and cell survival in the absence of apoptosis. *Cell* **120,** 237–248.

Marino, G., Salvador-Montoliu, N., Fueyo, A., Knecht, E., Mizushima, N., and Lopez-Otin, C. (2007). Tissue-specific autophagy alterations and increased tumorigenesis in mice deficient in Atg4C/autophagin-3. *J. Biol. Chem.* **282,** 18573–18583.

Mathew, R., Degenhardt, K., Haramaty, L., Karp, C. M., and White, E. (2008). Immortalized mouse epithelial cell models to study the role of apoptosis in cancer. *Methods Enzymol.* **446,** 77–106.

Mathew, R., Karantza-Wadsworth, V., and White, E. (2007a). Role of autophagy in cancer. *Nat. Rev. Cancer* **7,** 961–967.

Mathew, R., Kongara, S., Beaudoin, B., Karp, C. M., Bray, K., Degenhardt, K., Chen, G., Jin, S., and White, E. (2007b). Autophagy suppresses tumor progression by limiting chromosomal instability. *Genes Dev.* **21,** 1367–1381.

Mizushima, N. (2004). Methods for monitoring autophagy. *Int. J. Biochem. Cell Biol.* **36,** 2491–2502.

Mizushima, N. (2007). Autophagy: Process and function. *Genes Dev.* **21,** 2861–2873.

Mondello, C., Guasconi, V., Giulotto, E., and Nuzzo, F. (2002). Gamma-ray and hydrogen peroxide induction of gene amplification in hamster cells deficient in DNA double strand break repair. *DNA Repair (Amst.)* **1,** 483–493.

Nelson, D. A., Tan, T. T., Rabson, A. B., Anderson, D., Degenhardt, K., and White, E. (2004). Hypoxia and defective apoptosis drive genomic instability and tumorigenesis. *Genes Dev.* **18,** 2095–2107.

Rajagopalan, H., and Lengauer, C. (2004). Aneuploidy and cancer. *Nature* **432,** 338–341.

Semenza, G. L. (2003). Targeting HIF-1 for cancer therapy. *Nat. Rev. Cancer* **3,** 721–732.

Shen, D. W., Fojo, A., Chin, J. E., Roninson, I. B., Richert, N., Pastan, I., and Gottesman, M. M. (1986). Human multidrug-resistant cell lines: Increased mdr1 expression can precede gene amplification. *Science* **232,** 643–645.

Shimazu, T., Degenhardt, K., Nur, E. K. A., Zhang, J., Yoshida, T., Zhang, Y., Mathew, R., White, E., and Inouye, M. (2007). NBK/BIK antagonizes MCL-1 and BCL-XL and activates BAK-mediated apoptosis in response to protein synthesis inhibition. *Genes Dev.* **21,** 929–941.

Snijders, A. M., Nowak, N. J., Huey, B., Fridlyand, J., Law, S., Conroy, J., Tokuyasu, T., Demir, K., Chiu, R., Mao, J. H., Jain, A. N., Jones, S. J., *et al.* (2005). Mapping segmental and sequence variations among laboratory mice using BAC array CGH. *Genome Res.* **15,** 302–311.

CHAPTER FIVE

Autophagic Clearance of Aggregate-Prone Proteins Associated with Neurodegeneration

Sovan Sarkar,* Brinda Ravikumar,* *and* David C. Rubinsztein*

Contents

Abstract

Autophagy has emerged as a field of rapidly growing interest with implications in several disease conditions, such as cancer, infectious diseases, and neurodegenerative diseases. Autophagy is a major degradation pathway for aggregate-prone, intracytosolic proteins causing neurodegenerative disorders, such as Huntington's disease and forms of Parkinson's disease. Up-regulating autophagy may be a tractable therapeutic intervention for clearing these disease-causing proteins. The identification of autophagy-enhancing compounds would be beneficial not only in neurodegenerative diseases but also

* Department of Medical Genetics, University of Cambridge, Cambridge Institute for Medical Research, Addenbrooke's Hospital, Cambridge, UK

Methods in Enzymology, Volume 453
ISSN 0076-6879, DOI: 10.1016/S0076-6879(08)04005-6

in other conditions where up-regulating autophagy may act as a protective pathway. Furthermore, small molecule modulators of autophagy may also be useful in dissecting pathways governing mammalian autophagy. In this chapter, we highlight assays that can be used for the identification of autophagy regulators, such as measuring the clearance of mutant aggregate-prone proteins or of autophagic flux with bafilomycin A_1. Using these methods, we recently described several mTOR-independent autophagy-enhancing compounds that have protective effects in various models of Huntington's disease.

1. Introduction

Several neurodegenerative diseases, such as Alzheimer's disease, Parkinson's disease (PD), and Huntington's disease (HD) share a common pathological manifestation: aggregation of misfolded mutant proteins in neurons in specific areas of the brain (Ross and Poirier, 2004). In many cases, the disease is caused by genetic mutations, such as point mutations in α-synuclein in some familial forms of PD, or polyglutamine (polyQ) expansion mutations in HD, which make the mutant proteins more aggregate-prone. The mutant protein aggregates and sequesters certain cellular proteins, including some heat-shock proteins. Although it appears that the large inclusions visible by light microscopy are not the most toxic species (Arrasate *et al.*, 2004), several studies report a strong correlation between the accumulation of the mutant protein and the disease severity (Haass and Selkoe, 2007; Ross and Poirier, 2004, 2005; Rubinsztein, 2006). As many of these mutations cause disease by gain-of-function mechanisms (Rubinsztein, 2006), a simple but attractive treatment strategy for these complex and diverse protein conformation diseases may be to enhance the clearance of the mutant protein.

The two major pathways for protein clearance in mammalian cells are the ubiquitin-proteasome system and the autophagy-lysosomal pathway (Rubinsztein, 2006). Although a role for the proteasome has been implicated in many of these diseases, the proteasome is not capable of clearing protein oligomers or complexes. Macroautophagy (hereafter referred to simply as autophagy) is a nonspecific bulk degradation system that is highly conserved from yeast to man (Mizushima *et al.*, 2008). Unlike the proteasome, autophagy is capable of clearing protein oligomers, complexes and even organelles (Rubinsztein, 2006). The formation of isolation membranes called phagophores marks the initiation of the autophagy process. The phagophore membranes then elongate and fuse while engulfing portions of cytosol to form mature autophagosomes. The autophagosomes may initially fuse with endosomes forming hybrid intermediates called amphisomes, which ultimately fuse with lysosomes to form the degradative autolysosomes. The acidic lysosomal proteases degrade the engulfed contents and the nutrients thus obtained are recycled back to the cell (Xie and Klionsky, 2007). Mammalian cells undergo autophagy at a basal level, which can be induced under physiological stress conditions such as starvation. Conditional knockouts of key autophagy genes in the brains of mice result in a neurodegenerative phenotype and the formation of protein aggregates in neurons (Hara *et al.*, 2006; Komatsu *et al.*, 2006). Thus, constitutive autophagy may play a crucial role in the clearance of normally occurring misfolded proteins in the cells. Autophagy is also implicated in several human diseases such as cancer, certain neurodegenerative disorders, myopathies, and infectious diseases (Mizushima *et al.*, 2008).

Genetic studies in yeast have identified several *Atg* (autophagy-related) genes that are categorized into 5 functional groups: two ubiquitin-like systems, a phosphatidylinositol 3-kinase (PI3K) complex, Atg1 and its regulators, and the Atg2-Atg18 complex. Orthologs for many of the yeast *Atg* genes are known to exist in mammalian cells, and a subset of these have been characterzed (Levine and Klionsky, 2004; Meijer and Codogno, 2006; Ohsumi, 2001). The classical autophagy pathway in mammalian cells is similar to yeast and involves the serine/threonine kinase, mammalian target of rapamycin (mTOR). Inhibition of mTOR results in activation of the autophagic pathway. Rapamycin, a macrolide antibiotic, forms a complex with the immunophilin FKBP12, which inhibits mTOR (Noda and Ohsumi, 1998; Rubinsztein *et al.*, 2007). Recent studies have uncovered several key players in a novel mTOR-independent pathway regulating autophagy (Sarkar *et al.*, 2005; Williams *et al.*, 2008). Several kinases are involved in the regulation of mammalian autophagy. Inhibition of class III PI3K by treatment with 3-methyladenine (3-MA) or wortmannin blocks the autophagy pathway at a very early stage, preventing the formation of mature autophagosomes (Petiot *et al.*, 2000). Furthermore, acidification of lysosomes is also important for the final autophagosome-lysosome fusion event. Thus, bafilomycin A_1, which prevents lysosomal acidification, blocks autophagy by inhibiting lysosomal hydrolase activity (Fass *et al.*, 2006) that subsequently causes impairment of autophagosome-lysosome fusion (Yamamoto *et al.*, 1998). This results in the aberrant accumulation of enlarged autophagosomes.

LC3 (microtubule-associated protein 1 light chain 3 or MAP-LC3), the mammalian homolog of yeast Atg8, is the only known protein that specifically associates with autophagosomes and phagophores and not with other vesicular structures (Kabeya *et al.*, 2000). The cytosolic precursor of LC3 is cleaved at its C terminus by Atg4 to form LC3-I. LC3-I is covalently conjugated to phosphatidylethanolamine to form LC3-II. LC3-II is specifically targeted to autophagosome precursors and remains associated with autophagosomes even after fusion with lysosomes; LC3-II on the outer surface of the autophagosome can be delipidated and recycled

(Kabeya *et al.*, 2000; Tanida *et al.*, 2004). Thus, LC3-II levels on Western blots correlate with autophagic vacuole number, which can also be assessed by scoring LC3-positive vesicle numbers (Kabeya *et al.*, 2000; Klionsky *et al.*, 2008; Mizushima, 2004).

2. Aggregate-Prone Intracytoplasmic Proteins Associated with Neurodegenerative Disorders are Autophagy Substrates

Autophagy is a key route for clearance of mutant huntingtin containing expanded polyQ repeats (Rubinsztein, 2006). Chemical inhibition of the autophagy-lysosomal pathway at the autophagosome formation step with 3-MA or at the autophagosome-lysosome fusion or degradation stage with bafilomycin A_1 inhibit the clearance of mutant huntingtin and enhance aggregate formation and toxicity in cell models. Induction of autophagy by treatment with rapamycin enhances the clearance of the mutant protein and decreases aggregate formation and associated toxicity (Ravikumar *et al.*, 2002). The wild-type fragments of huntingtin show minimal accumulation when autophagy is blocked, in contrast to the mutant constructs. This phenomenon is probably because the non-aggregate-prone wild-type species are efficiently cleared by the proteasome, compared to the mutant forms. It is likely that autophagy induction decreases the levels of aggregates by enhancing the clearance of monomeric and/or oligomeric species of mutant huntingtin (the inclusion precursors), rather than the large inclusions themselves. Mutant huntingtin inclusions visible by light microscopy are not membrane-bound (Rubinsztein, 2006). Rapamycin also protects against neurodegeneration in a *Drosophila* HD model and a rapamycin analogue, CCI-779, improves various HD-associated behavioral tasks and decreases aggregate formation in a transgenic HD mouse model (Ravikumar *et al.*, 2004). The protective effect of rapamycin in fly models of diseases similar to HD is autophagy dependent (Berger *et al.*, 2006; Pandey *et al.*, 2007). This pathway appears to be equally important for the clearance of several other intracellular proteins that cause neurodegeneration, including the A53T or A30P mutants of α-synuclein, ataxin 3, and tau (Berger *et al.*, 2006; Cuervo *et al.*, 2004; Rubinsztein, 2006; Webb *et al.*, 2003). Thus, a potential therapeutic intervention for these protein conformation diseases may be to induce autophagy.

Although rapamycin is the most widely used compound to induce autophagy in diverse cell types, inhibition of mTOR can have other effects such as translation inhibition (Sarbassov *et al.*, 2005). This might result in side effects with long-term rapamycin therapy, including poor wound healing and immunosuppression. Thus, it is important to identify other pathways independent of mTOR to induce autophagy. We have now identified several components of a novel mTOR-independent pathway of autophagy regulation (Sarkar *et al.*, 2005; Williams *et al.*, 2008). This is potentially useful, as simultaneous up-regulation of autophagy by mTOR-dependent and mTOR-independent mechanisms results in a greater protection against mutant huntingtin aggregation in cell and *Drosophila* models of HD (Sarkar *et al.*, 2005, 2008). Thus, combination treatment using a low dose of compounds inhibiting the two pathways simultaneously may be effective in patients and allow reduced side effects.

We use the following assays in our studies:

1. Clearance of mutant huntingtin and the A53T mutant form of α-synuclein.
2. Aggregation of mutant huntingtin and associated toxicity.
3. Measurement of autophagic flux with bafilomycin A_1.

3. Assays for The Clearance of Aggregate-Prone Proteins

We have previously shown that autophagy upregulation may be a possible therapeutic strategy for neurodegenerative diseases by enhancing the clearance of aggregate-prone proteins. Most of the autophagy-inducing compounds were initially identified by assessing the clearance of intracytosolic, aggregate-prone proteins, which are autophagy substrates. For clearance studies, we used stable inducible rat pheochromocytoma (PC12; neuronal precursor) cell lines expressing enhanced green fluorescent protein (EGFP)-tagged huntingtin exon 1 fragment with 74 polyQ repeats (EGFP-HDQ74) (Wyttenbach *et al.*, 2001) (Figs. 5.1A and 5.2A), or hemagglutinin (HA)-tagged A53T or A30P mutants of α-synuclein (Webb *et al.*, 2003) (Fig. 5.1B). In these cell lines, the transgene expression is first induced by adding doxycycline and then switched off by removing doxycycline from the medium. If the transgene expression level (amount of mutant proteins) is followed at various times after switching off expression after an initial induction period, one can assess whether specific agents alter the clearance of the transgene product, as the amount of transgene product decays when synthesis is stopped (Figs. 5.1A,B, and 5.2A). In this paradigm, treatment with rapamycin (autophagy inducer) or 3-MA (autophagy inhibitor) enhances or retards the clearance of mutant aggregate-prone proteins, respectively (Ravikumar *et al.*, 2002; Webb *et al.*, 2003).

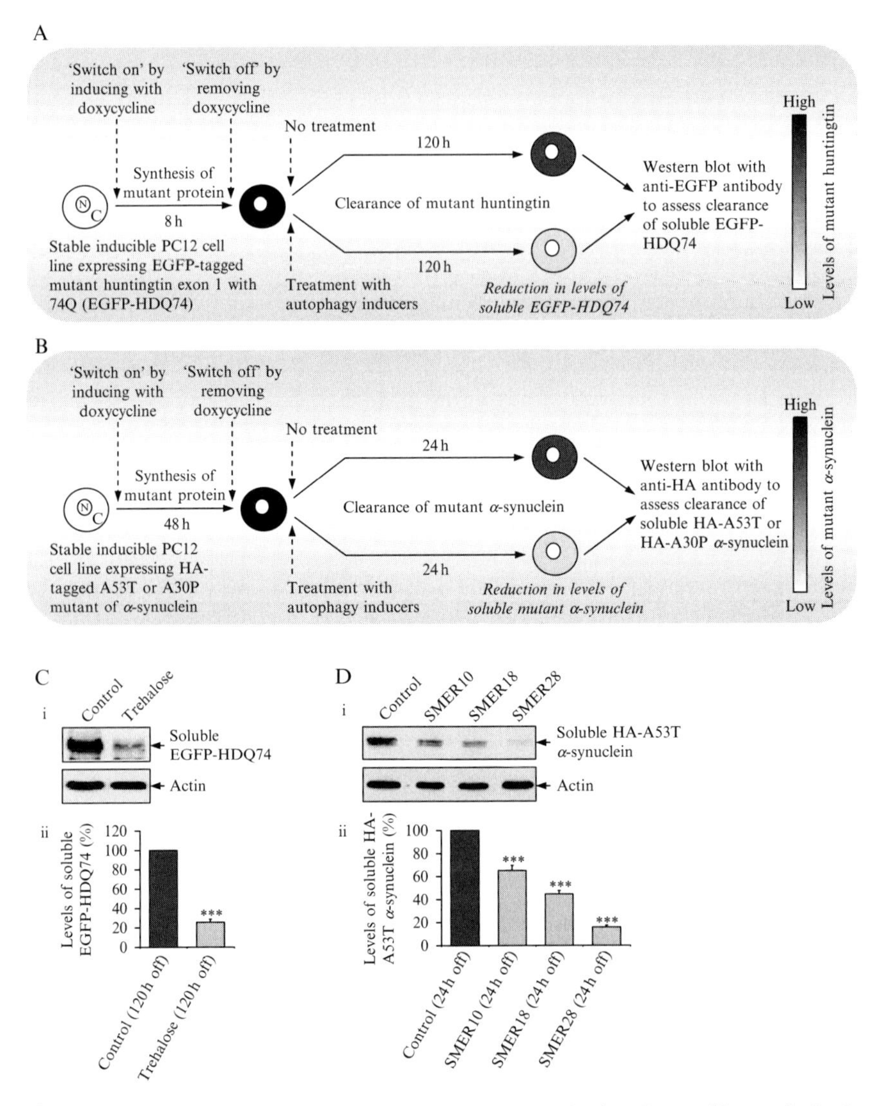

Figure 5.1 Clearance of mutant aggregate-prone proteins by Western blot analysis. A, Stable inducible PC12 cells expressing EGFP-HDQ74 are induced by 1 μg/ml doxycycline for 8 h to switch on the expression of the transgene. Doxycycline is then removed and the cells are washed with culture medium to switch off the expression of the transgene. Cells are treated with or without autophagy inducers for 120 h. Respective vehicles are used as controls. Levels of the transgene product decay when its synthesis is stopped, and subsequent treatment with an autophagy inducer facilitates greater degradation of the transgene product compared to the control condition. The clearance of mutant huntingtin, which is an autophagy substrate, can be assessed by Western blotting with anti-EGFP antibody and densitometric analysis of soluble EGFP-HDQ74 relative to actin (loading control). Cells treated with autophagy enhancers have markedly lower levels of the mutant protein compared to the control condition. N = Nucleus,

3.1. Materials

3.1.1. Cell culture

1. Stable inducible PC12 cell lines expressing EGFP-HDQ74 (Wyttenbach *et al.*, 2001) or HA-A53T or HA-A30P mutants of α-synuclein (Webb *et al.*, 2003); COS-7 cells; wild-type (*Atg5*$^{+/+}$) or *Atg5*-deficient (*Atg5*$^{-/-}$) mouse embryonic fibroblasts (MEFs) (Kuma *et al.*, 2004).
2. Dulbecco's Modified Eagle's Medium (DMEM) (Sigma, D6546).
3. Fetal bovine serum (FBS) (Sigma, F7524).
4. Horse serum (Sigma, H1270).
5. L-Glutamine (Sigma, G7513).
6. Penicillin/Streptomycin solution (Sigma, P0781).
7. 1X Trypsin-EDTA solution (Sigma, T3924).
8. Geneticin/G418 sulphate (Gibco, 11811-031), which is dissolved in sterile water to make a 100 mg/ml stock and stored at −20 °C as frozen aliquots.
9. Hygromycin B (Calbiochem, 400051) solution, which is mixed with DMEM to a 26 mg/ml stock and stored at −20 °C as frozen aliquots.
10. Doxycycline (Sigma, D9891), which is dissolved in sterile water to make a 100 mg/ml stock and stored at −20 °C as frozen aliquots in dark microcentrifuge tubes to protect it from direct light. This stock can

---

C = Cytoplasm. B, Stable inducible PC12 cells expressing HA-A53Tor HA-A30P mutant forms of α–synuclein are induced by 1 μg/ml doxycycline for 48 h. This cell line follows a similar methodological paradigm to that expressing EGFP-HDQ74. After the initial induction period, the cells are treated with or without autophagy inducers for 24 h. Respective vehicles are used as controls. The clearance of mutant α-synucleins, which are autophagy substrates, can be assessed by Western blotting with anti-HA antibody and densitometric analysis of soluble HA-A53Tor HA-A30P α-synuclein mutant relative to actin (loading control). N = Nucleus, C = Cytoplasm. C, Stable inducible PC12 cells expressing EGFP–HDQ74 are induced with 1 μg/ml doxycycline for 8 h, and transgene expression is then switched off (by removing doxycycline) for 120 h, with (trehalose) or without (control) 100 m*M* trehalose. Clearance of soluble EGFP-HDQ74 is analyzed by immunoblotting with anti–EGFP antibody (i) and densitometric analysis relative to actin (ii). The control condition is set to 100% and the error bars denote standard error of mean. Trehalose significantly enhanced the clearance of soluble EGFP-HDQ74 at 120 h ($p < 0.0001$). D, Stable inducible PC12 cells expressing HA–A53T α-synuclein are induced with 1 μg/ml doxycycline for 48 h, and transgene expression is then switched off (by removing doxycycline) for 24 h, with DMSO (control), 47 μ*M* SMER10, 43 μ*M* SMER18, or 47 μ*M* SMER28. Clearance of soluble HA-A53T α-synuclein is analyzed by immunoblotting with anti-HA antibody (i) and densitometric analysis relative to actin (ii). The control condition is set to 100% and the error bars denote standard error of mean. The SMERs significantly enhanced the clearance of soluble HA-A53T α-synuclein at 24 h ($p = 0.0001$ for SMER10; $p = 0.0002$ for SMER18; $p < 0.0001$ for SMER28). ***$p < 0.001$.

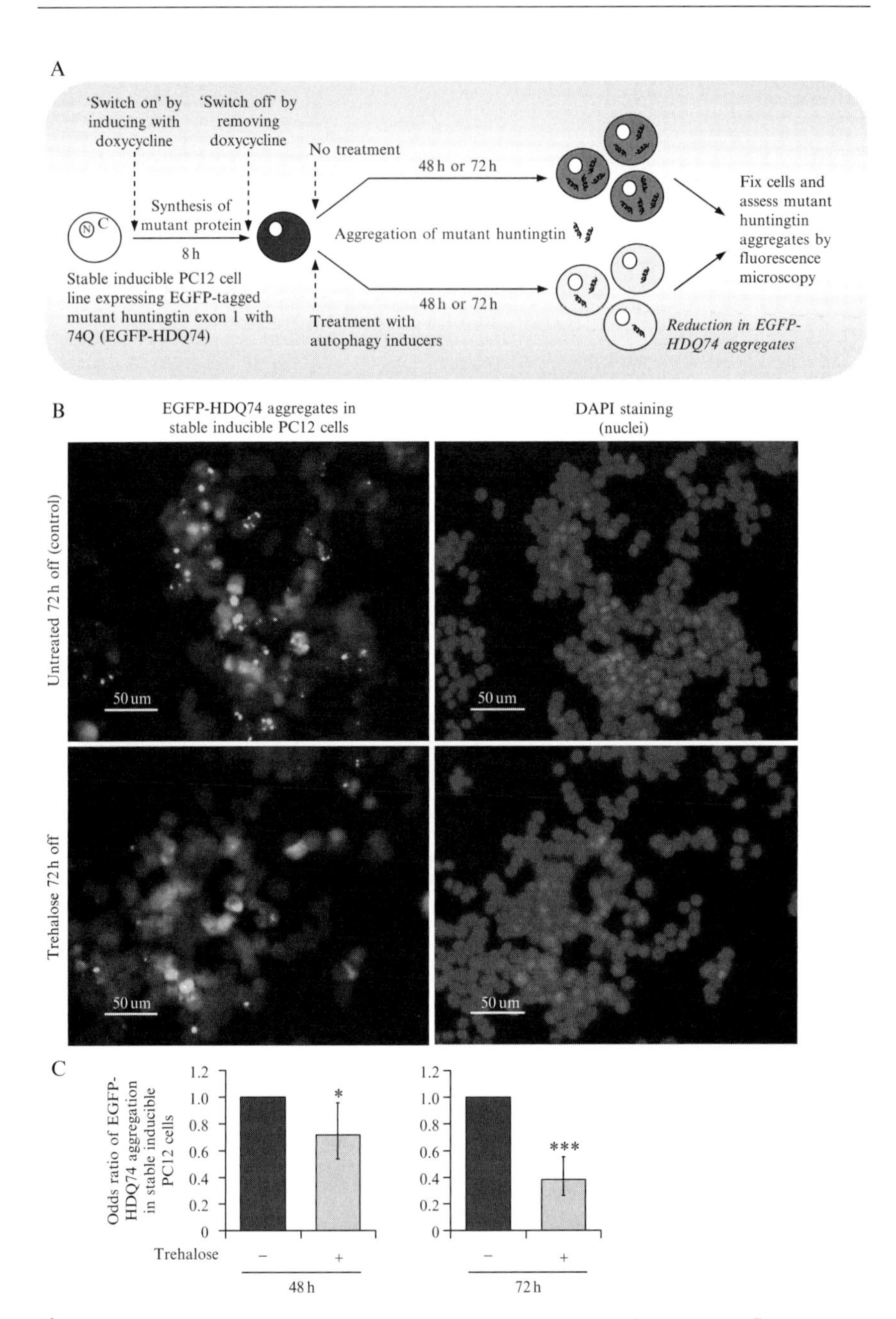

Figure 5.2 Clearance of mutant aggregate–prone proteins by immunofluorescence analysis. (A), Stable inducible PC12 cells expressing EGFP-HDQ74 are induced by doxycycline, following a similar switch on and switch off paradigm as in Fig. 5.1A. Cells are then treated with or without autophagy inducers for 48 h or 72 h. The clearance of mutant aggregate-prone protein can be assessed by fixing the cells for

be diluted in sterile water to a 10 mg/ml working stock solution, which can also be stored at −20 °C and can be used for a few months. The working stock is then diluted again in the cell culture medium to a final concentration of 1 μg/ml.

11. 6-well plates (Greiner; Cellstar, 657160).
12. 22 × 22-mm thickness No. 1 glass coverslips (VWR international).
13. 75-cm^2 flasks (NUNC; Invitrogen).
14. Cell scrapers (Sarstedt, 831830).

3.1.2. Autophagy enhancers

We discuss the methods sections using the following compounds, which we use as examples to show the effects of an autophagy inducer or inhibitor.

1. Trehalose: Trehalose is a disaccharide, which is an mTOR-independent autophagy inducer (Sarkar *et al.*, 2007a). It is dissolved in DMEM by warming in a water bath at 37 °C to make a 1.25 *M* stock that can be stored at 4 °C for a month. During experiments, trehalose is diluted in cell culture medium to make a working concentration of 100 m*M*.
2. Small molecule enhancers of rapamycin (SMER): The SMERs 10, 18, and 28 are mTOR-independent autophagy inducers (Sarkar *et al.*, 2007b). These compounds are dissolved in DMSO to make a 5 mg/ml stock and can be stored at −20 °C as frozen aliquots. During experiments, these SMERs are used at 1:400 dilutions in cell culture medium.
3. Calpastatin: Calpastatin is a calpain inhibitor peptide that induces mTOR-independent autophagy (Goll *et al.*, 2003; Williams *et al.*, 2008). It is dissolved in sterile distilled water to make a stock of 1 m*M* and can be stored at −20 °C as frozen aliquots. Calpastatin is used at 10 μ*M* concentration in cell culture media during experiments.

---

immunofluorescence analysis and quantifying EGFP-HDQ74 aggregation. Cells treated with autophagy enhancers have reduced EGFP-HDQ74 aggregates compared to the control condition. N = Nucleus, C = Cytoplasm. B and C, Stable inducible PC12 cells expressing EGFP–HDQ74 are induced with 1 μg/ml doxycycline for 8 h, and transgene expression is then switched off (by removing doxycycline) for 48 h or 72 h, with (+) or without (−) 100 m*M* trehalose. Cells are fixed and assessed by fluorescence microscopy. Images in the left panel show EGFP–HDQ74 aggregates (bright green aggregated structures) in stable PC12 cells (diffuse green staining) that have been treated with or without trehalose for 72 h in the switch off period, and those in the right panel show DAPI staining for nuclei (B). Bar, 50 μm. EGFP-HDQ74 aggregates appear to be fewer in trehalose–treated cells, compared to the control condition. The proportion of cells with EGFP–HDQ74 aggregates is counted and expressed as odds ratio with 95% confidence interval, where the control condition is fixed at one (C). Trehalose significantly reduced EGFP-HDQ74 aggregates at 72 h ($p < 0.0001$), and to a lesser extent at 48 h ($p = 0.024$). ***$p < 0.001$; *$p < 0.05$. (See Color Insert.)

4. Bafilomycin A_1: Bafilomycin A_1 is dissolved in DMSO to make a stock of 100 μM and stored at $-20\,^{\circ}C$ as frozen aliquots. It is diluted in cell culture media to a working concentration of 400 n*M*, which is saturating for its effect on blocking autophagosome-lysosome fusion (Sarkar *et al.*, 2007a; Yamamoto *et al.*, 1998) or lysosomal hydrolase activity (Fass *et al.*, 2006).

3.1.3. Lysis of mammalian cells

1. Lysis buffer (2X): 20 μM Tris-HCl, pH 6.8, 137 μM NaCl, 1 m*M* EGTA, 1% Triton X-100, 10% glycerol; stored at $4\,^{\circ}C$. 1X lysis buffer is prepared fresh with protease inhibitor (see below) just before harvesting cells for lysis.
2. Complete protease inhibitor cocktail tablet (Roche, 11697498001) is dissolved in 2 ml of sterile water and stored at $-20\,^{\circ}C$. While making 1X lysis buffer, complete protease inhibitor cocktail solution is added at 1:25 dilution.

3.1.4. SDS–polyacrylamide gel electrophoresis (SDS–PAGE)

1. Resolving gel (12%): 30% acrylamide:bis-acrylamide solution, 1.5 *M* Tris-HCl, pH 8.8, 10% sodium dodecyl sulphate (SDS), 10% ammonium persulphate (APS), N,N,N,N′-tetramethylethylenediamine (TEMED) and distilled water. APS must be prepared fresh every time before pouring the gel. TEMED is best stored at room temperature and as its quality may decline after opening, it is better to buy small bottles.
2. Stacking gel (5%): 30% acrylamide:bis-acrylamide solution, 1 *M* Tris-HCl, pH 6.8, 10% SDS, 10% APS, TEMED, and distilled water.
3. Water-saturated isobutanol: Take equal volumes of distilled water and isobutanol in a glass bottle, mix in a stirrer overnight, allow to separate, and store at room temperature. Use the top (butanol) layer.
4. Prestained molecular weight marker: Kaleidoscope marker (Bio-Rad).
5. 3X sample buffer: 187.5 m*M* Tris-HCl, pH 6.8, 6% w/v SDS, 30% glycerol, 150 m*M* DTT, 0.03% w/v bromophenol blue.
6. 10X running buffer: 250 m*M* Tris, 1.92 *M* glycine. While making 1X gel running buffer with distilled water, add 10% SDS at 1:100 dilution.
7. SDS-PAGE apparatus (Bio-Rad).
8. Power pack (Bio-Rad).

3.1.5. Western blotting

1. 10X transfer buffer: 250 m*M* Tris, 1.92 *M* glycine. Make 1X transfer buffer with distilled water.
2. Hybond nitrocellulose membrane (Amersham Bioscience, GE Healthcare, RPN203B).

3. Trans blot SD semidry transfer cell (Bio-Rad).
4. Power pack (Bio-Rad).
5. Extrathick filter paper (Bio-Rad).
6. Blocking buffer: 6% (w/v) nonfat dry milk, 0.1% (v/v) Tween 20 in 1X phosphate buffered saline (PBS).
7. Platform rocker (Bibby Sterilin).
8. Primary antibody: Mouse anti-EGFP (632375, BD Clontech), mouse anti-HA (MMS-101P, Covance). Dilutions (1:2000 for anti-EGFP antibody and 1:1000 for anti-HA antibody) are made in blocking buffer.
9. Secondary antibody: Enhanced chemiluminescent (ECL) antimouse IgG conjugated to horseradish peroxidise (HRP) (NA931, Amersham Biosciences, GE Healthcare). Dilution (1:2000) is made in blocking buffer.
10. ECL Western blotting detection reagent (Amersham Bioscience, GE Healthcare, RPN2134).
11. High-performance chemiluminescence film (Hyperfilm ECL, Amersham Biosciences, GE Healthcare).
12. Hypercassette (Amersham Biosciences, GE Healthcare).
13. RP X-OMAT Processor, Model M6B (Developer, Kodak).

3.1.6. Stripping and reprobing immunoblots

1. Stripping buffer: 10% SDS, 1 *M* Tris-HCl, pH 6.8, 14 *M* β-mercaptoethanol.
2. Hot block (Bibby Sterilin).
3. Primary antibody: Rabbit anti-actin (A 2066, Sigma). Dilution (1:2000) is made in blocking buffer.
4. Secondary antibody: ECL anti-rabbit IgG conjugated to HRP (NA934, Amersham Biosciences, GE Healthcare). Dilution (1:2000) is made in blocking buffer.

3.1.7. Immunofluorescence analysis

1. 1X PBS solution.
2. Paraformaldehyde (PFA) is dissolved in 1X PBS to make a 4% stock. While dissolving, the solution is warmed to 50 °C with constant stirring in a fume hood. If the solution is cloudy, a few drops of 1 *M* NaOH may be added to make the solution clear. It is then cooled at room temperature and stored at −20 °C as frozen aliquots.
3. Antifadent, Citifluor AF1 (Citifluor, AF1-100ml).
4. 4′,6-diamidino-2-phenylindole (DAPI) (Sigma, D9564), which is dissolved in citifluor to make a 3 μg/ml stock and stored at 4 °C.
5. Nikon Eclipse E600 fluorescence microscope with Nikon Digital Camera DXM1200, 40x plan-fluor 40x/0.75 lens and plan-apo 60x/1.4 oil immersion lens (Nikon).

3.1.8. Transient Transfection

1. Lipofectamine 2000 transfection reagent (Invitrogen, 11668-019).
2. Plasmid DNA: EGFP-tagged huntingtin exon 1 fragment with 74 polyQ repeats (EGFP-HDQ74) (Narain *et al.*, 1999).
3. Serum-free DMEM: DMEM containing 2 m*M* L-glutamine only.

3.2. Cell culture

3.2.1. Cell culture medium and growth conditions

1. Stable PC12 cell lines, expressing either EGFP-HDQ74 (Wyttenbach *et al.*, 2001) or the HA-A53T or HA-A30P mutant of α-synuclein (Webb *et al.*, 2003), are maintained in DMEM supplemented with 10% horse serum (Sigma, H1270), 5% FBS (Sigma, F7524), 100 U/ml penicillin/streptomycin, 2 m*M* L-glutamine, 100 g/ml G418 (to maintain tet-on element) and 75 μg/ml hygromycin (to maintain the mutant huntingtin exon 1 or α-synuclein transgenes) at 37 °C, 10% CO_2 with humidity.
2. COS-7 cells, *Atg5*$^{+/+}$ MEFs and *Atg5*$^{-/-}$ MEFs (Kuma *et al.*, 2004) are cultured in DMEM supplemented with 10% FBS, 100 U/ml penicillin/streptomycin, and 2 m*M* L-glutamine at 37 °C, 5% CO_2 with humidity.

3.2.2. Subculture

1. The cells are passaged until they reach 80%–90% confluency. The cell culture medium from the flask is removed and the cells are rinsed once with DMEM (without serum).
2. 2.5 ml of 1X trypsin-EDTA is added to cover the cell layer and the flask is incubated at 37 °C for 3–5 min or until the cells are dissociated from the flask (but no longer than 10–15 min).
3. The enzyme activity is quenched by adding 8 ml of fresh supplemented medium and the cells are dissociated by pipetting the medium up and down a few times. Because cells tend to form clumps, care is taken to dissociate the clumps to a fine suspension before reseeding them.
4. The cells are passaged in 1:5 or 1:10 ratios in fresh supplemented medium and maintained at 37 °C, in a 5% or 10% CO_2 incubator with humidity.

3.3. Clearance of aggregate-prone proteins

3.3.1. Experimental cell culture setup

1. The stable inducible PC12 cells are seeded at 1–2×10^5 per well in 6-well plates either directly (for Western blot analysis in EGFP-HDQ74- or HA–A53T α-synuclein-expressing cells) or on glass coverslips (for immunofluorescence analysis in EGFP-HDQ74-expressing cells) for 24 h prior to induction.

2. The expression of the transgene is switched on with 1 μg/ml doxycycline treatment for 8 h (for EGFP-HDQ74) or 48 h (for HA–A53T α-synuclein).
3. The expression of the transgene is then switched off by removing the medium containing doxycycline followed by a few rinses with fresh supplemented doxycycline-free medium.
4. After switching off the transgene expression, the cells are treated with the mTOR-independent autophagy inducers, such as trehalose (Sarkar *et al.*, 2007a) for 120 h (for clearance analysis of soluble EGFP-HDQ74 by western blotting) (Figs. 5.1A,C), or SMERs 10, 18, and 28 (Sarkar *et al.*, 2007b) for 24 h (for clearance analysis of soluble A53T α-synuclein by western blotting) (Figs. 5.1B,D). Equal amounts of sterile distilled water or DMSO are added as controls, where relevant. For immunofluorescence analysis of EGFP-HDQ74 aggregates, stable inducible PC12 cells expressing EGFP-HDQ74 are treated with trehalose for 48 h or 72 h before fixing (Figs. 5.2A–C). The autophagy enhancers are normally diluted in cell culture media, and 1 ml of the final diluted media is added per well.
5. The cells are collected for western blot analysis (as explained in section 3.3.2), or fixed with PFA and mounted on coverslips for immunofluorescence analysis (as explained in section 3.3.3).

3.3.2. Western blot analysis

3.3.2.1. Preparation of samples

1. The cells from 6-well plates are scraped into the medium and collected in labeled microcentrifuge tubes for each sample.
2. The cells are centrifuged at 5000 rpm at 4 °C for 5 min.
3. The supernatant fraction is discarded and the cell pellets are washed once with 1X PBS.
4. The cells are then lysed with appropriate volumes of lysis buffer for 30 min on ice. Approximately 70–100 μl of lysis buffer is used for cell pellets from each well of a 6-well plate.
5. The lysed cells are centrifuged at 13,000 rpm for 5 min at 4 °C to remove cell debris and any unlysed cells.
6. The supernatant fractions are transferred to new tubes and can be stored at −80 °C until further analysis or processed immediately.
7. Protein assay is performed on the samples with a protein assay kit (Bio-Rad).
8. 30 μg of protein from each sample are mixed with 3X sample buffer (to a final concentration of 1X) and boiled at 100 °C for 5 min on a heating block before loading on the gel for SDS-PAGE.

3.3.2.2. SDS-PAGE

1. SDS-PAGE is performed using a gel apparatus (Bio-Rad).
2. The resolving gel (12%) is first poured in between a gel plate with a 0.75- or 1.5-mm integrated spacer and a glass plate sandwich that are held

together in the gel casting unit. The gel is poured up to 75% of the length of the plates, leaving space for pouring the stacking gel.

3. Approximately 500 μl of water-saturated butanol is added on the surface of the resolving gel, which is later washed with distilled water when the gel has polymerized.
4. The stacking gel is poured over the resolving gel and a 10- or 15-well comb is inserted into it.
5. After polymerization of the stacking gel, the combs are removed and the wells thus created are rinsed immediately with distilled water.
6. The casted gels are fitted into the gel apparatus and 1X gel running buffer is poured into the tank.
7. The boiled samples are then loaded into the wells along with a prestained molecular weight marker.
8. The gels are run at a constant current of 15 mA per gel until the marker bands migrate to the desired position.

3.3.2.3. Western blotting

1. The proteins in the samples that have been separated by SDS-PAGE are transferred electrophoretically onto a nitrocellulose membrane using a semidry transfer apparatus (Bio-Rad).
2. Nitrocellulose membranes are cut according to the size of the gel and equilibrated in 1X transfer buffer along with extra-thick filter papers for 5–10 min. The gel is taken out and is also equilibrated in the transfer buffer.
3. The gel-transfer unit is then assembled by placing a filter paper on the base plate of the transfer apparatus, then the nitrocellulose membrane, followed by the gel on top of the membrane and finally 2 more filter papers on the top. While placing the gel/filter papers on the membrane, the surface is rolled gently with a plastic pipette so as to remove any air bubbles.
4. After placing the compression plate and connecting the lid, the transfer is performed at a constant voltage of 15 V for approximately 1 h.
5. The membranes (immunoblots) are then removed and incubated in 25 ml of blocking buffer with gentle shaking on a rocker at room temperature for 1 h.
6. The blocking buffer is discarded and primary antibody diluted in 10 ml of blocking buffer (anti-EGFP at 1:2000 dilution for detecting EGFP-HDQ74 and anti-HA at 1:1000 dilution for detecting mutant α–synuclein) is added to the immunoblots and incubated overnight on a rocker in a cold room at 4 °C. Blotting can also be carried out by incubating the membrane with anti-EGFP antibody for 1 h at room temperature on a rocker. For best results, it is recommended to incubate the membrane with primary antibody overnight.

7. The nonspecific binding of the primary antibody is then removed by washing the immunoblots with 0.1% Tween 20 in 1X PBS (PBS-T) solution, 3 times for 10 min each on a rocker at room temperature.
8. Secondary antibody (anti-mouse IgG conjugated to HRP at 1:2000 dilution in blocking buffer) is added to the immunoblots and incubated for a minimum of 1 h (not more than 3 h) on a rocker at room temperature.
9. The nonspecific binding of the secondary antibody is then removed by washing the immunoblots in PBS-T solution, 3 times for 10 min each on a rocker at room temperature.
10. After the final wash and discarding PBS-T, the following procedures are performed in a dark room under safe light conditions.
11. 1.5 ml each of the detection reagents 1 and 2 of the ECL Western blotting detection system are mixed and added to the immunoblots for 1 min, while ensuring coverage of the entire surface.
12. The detection solution is discarded and the immunoblots are then wrapped in cling films and placed in hypercassettes along with a hyperfilm with appropriate exposure times. Detection of soluble EGFP-HDQ74 is quick with a very short exposure time (usually 1–5 s), and the band appears at 50 kDa (Fig. 5.1C). Detection of soluble HA-A53T (or HA-A30P) α-synuclein requires relatively short exposure time (usually 1–2 min), with the band appearing at 24 kDa (Fig. 5.1D).
13. The films are developed in an automated Kodak developer.

3.3.2.4. Stripping and reprobing the immunoblots

1. The immunoblots are stripped of the signal for EGFP or HA by immersing them in stripping buffer at 65 °C on a hot block for 10 min.
2. The immunoblots are then washed with 1X PBS twice for 5 min each on a rocker at room temperature, followed by blocking with blocking buffer for 1 h.
3. After discarding the blocking buffer, primary antibody (rabbit anti-actin at 1:2000 dilution in blocking buffer) is added to the immunoblots and incubated either overnight at 4 °C or for 1–2 h at room temperature on a rocker.
4. The immunoblots are washed 3 times with PBS-T as before, incubated with secondary antibody (ECL anti-rabbit IgG conjugated to HRP at 1:2000 dilution in blocking buffer) for a minimum of 1 h on a rocker at room temperature, followed by 3 washes with PBS-T.
5. The signal is detected with the ECL Western blotting detection system as before. Detection of actin is also quick, requiring a very short exposure time; usually 1–5 s if the antibody is applied overnight, or 1–5 min if the antibody is incubated for 1 h. The band appears at 45 kDa (Figs. 5.1C and 5.1D).
6. Densitometric analysis on the immunoblots for assessing the clearance of aggregate-prone proteins is described in section 3.5.1.

3.3.3. Fixing cells for immunofluorescence analysis of aggregate formation

1. The stable PC12 cells expressing EGFP-HDQ74 are rinsed once with 1X PBS and fixed with 4% PFA in 1X PBS for 20 min.
2. 2. The fixed cells on the coverslips are further rinsed twice with 1X PBS and air-dried inside the hood with the lights off.
3. The coverslips are mounted onto glass slides on the antifadent, citifluor, supplemented with 3 μg/ml DAPI (to allow visualization of nuclei).
4. The cells are quantified for the effect of trehalose in reducing EGFP-HDQ74 aggregates with a Nikon Eclipse E600 fluorescence microscope, as described in section 3.4.3. Statistical analysis by odds ratio is described in section 3.5.2 (Figs. 5.2B,C).

3.4. Analysis of aggregate formation and cell death

To test the effect of autophagy modulators on the aggregation and toxicity of mutant huntingtin fragments, COS-7 (nonneuronal) or SH-N-SH (neuronal precursor) cells can be used. COS-7 cells are particularly easy for counting apoptotic nuclei, which are convenient to assess under fluorescence microscopy. To further test whether the autophagy modulators are affecting mutant huntingtin aggregation by influencing autophagic degradation, this assay can be performed in both *Atg5*$^{+/+}$ (autophagy-competent) and *Atg5*$^{-/-}$ (autophagy-deficient) MEFs (Fig. 5.3A). These cells have been established by Mizushima and colleagues (Kuma *et al.*, 2004), and provide a powerful tool to dissect autophagy-dependent processes. We have previously shown that *Atg5*$^{-/-}$ MEFs have massive accumulation of autophagy substrates (as seen by mutant huntingtin aggregates), compared to the *Atg5*$^{+/+}$ MEFs (Figs. 5.3A–C). Furthermore, a variety of mTOR-independent autophagy enhancers fail to reduce mutant huntingtin aggregation in *Atg5*$^{-/-}$ MEFs, whereas their protective effects (reduction of mutant huntingtin-associated aggregation and toxicity) are observed in *Atg5*$^{+/+}$ MEFs (Figs. 5.3A, 5.3C), thereby confirming that their effects are autophagy dependent (Sarkar *et al.*, 2007a,b; Williams *et al.*, 2008).

3.4.1. Transient transfection with EGFP-HDQ74

1. *Atg5*$^{+/+}$ and *Atg5*$^{-/-}$ MEFs are grown on coverslips in 6-well plates for 24 h up to 60% confluency prior to transfection.
2. Transfection is performed using Lipofectamine 2000 reagent (Invitrogen). 5 μl of the reagent prediluted in 100 μl of serum-free DMEM and incubated for at least 5 min at room temperature is mixed with 1.5–2 μg of plasmid DNA (EGFP-HDQ74) also diluted in 100 μl of serum-free DMEM.

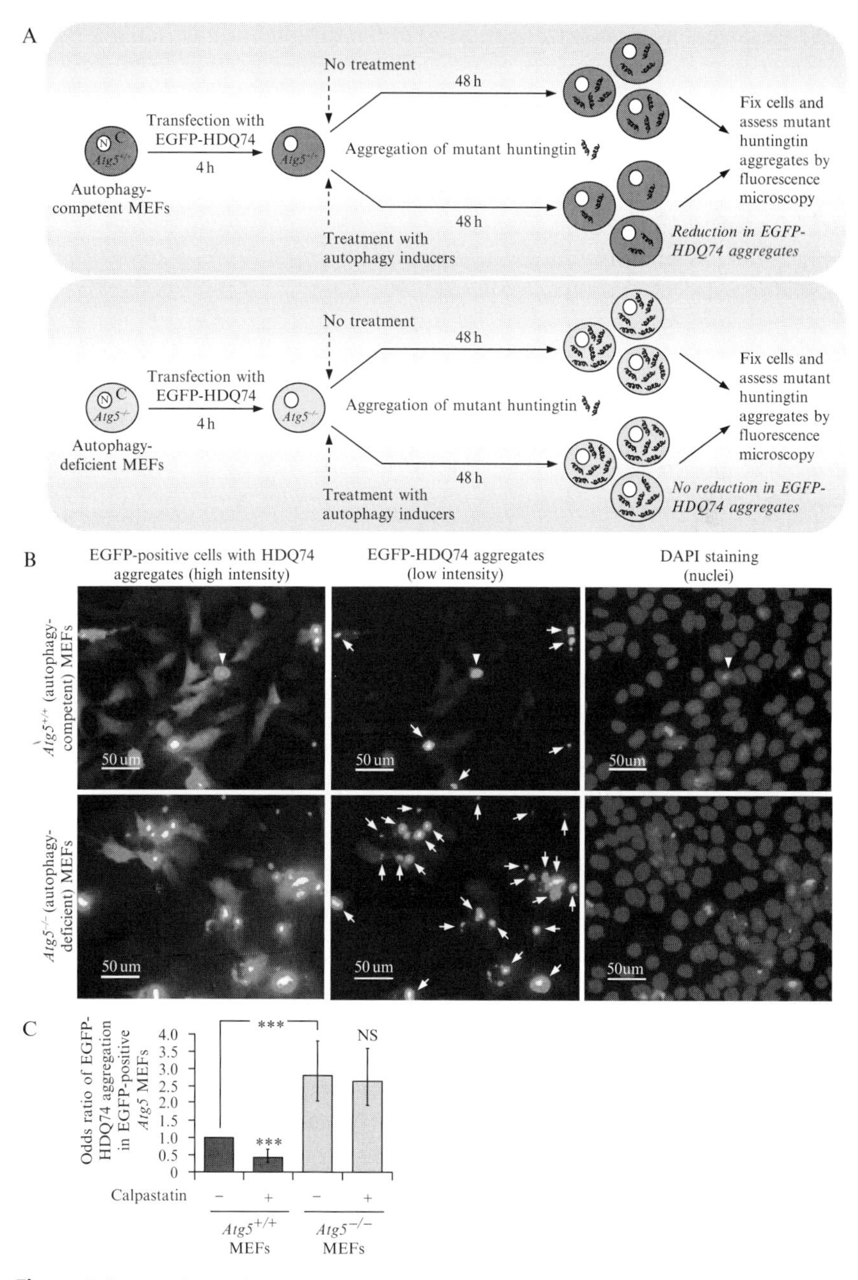

Figure 5.3 Analysis of aggregate formation in autophagy-limiting conditions. A and B, Analysis of mutant huntingtin–associated aggregate formation and cell death is performed in *Atg5*$^{+/+}$ (autophagy-competent) and *Atg5*$^{-/-}$ (autophagy-deficient) MEFs, which specifically indicates the autophagy-dependent effects of an autophagy modulator. These cells are transfected with EGFP-HDQ74 for 4 h, treated with or without

3. The mixture is incubated at room temperature for 20-30 min.
4. Close to the end of this incubation period, the cells in 6-well plates are rinsed once with serum-free DMEM.
5. Serum-free DMEM (800 μl per well) is then added to the Lipofectamine 2000-DNA mixture so that 1 ml of the final volume of the transfection mixture can be added to each well. The cells are incubated for 4–5 h at 37 °C, 5% CO_2.
6. The transfection mixture is replaced with fresh supplemented culture medium after the transfection period. The transfected cells are then treated with an mTOR–independent autophagy enhancer, calpastatin, for 48 h.
7. Cells are fixed for immunofluorescence analysis of EGFP-HDQ74 aggregate formation and cell death (described in section 3.4.3) 48 h post-transfection.

3.4.2. Fixing cells for immunofluorescence analysis

1. Cells are fixed for quantification of aggregate formation and cell death by immunofluorescence analysis, as described in section 3.3.3 (Figs. 5.3B,C).

3.4.3. Quantification of aggregate formation and cell death

1. Aggregate formation and nuclear morphology are assessed using a Nikon eclipse E600W fluorescent microscope with plan-apo 60x/1.4 oil immersion lens at room temperature.

---

autophagy enhancers for 48 h post-transfection, and fixed for analysis by fluorescence microscopy. *Atg5*$^{-/-}$ MEFs have increased EGFP-HDQ74 aggregation compared to *Atg5*$^{-/-}$ MEFs as mutant huntingtin is an autophagy substrate. Treatment with an autophagy inducer can reduce EGFP-HDQ74 aggregation in *Atg5*$^{+/+}$ MEFs, but not in *Atg5*$^{-/-}$ MEFs (A). Images of cells in untreated (control) conditions show a massive accumulation of autophagy substrates (as determined by increased mutant huntingtin aggregates) in *Atg5*$^{-/-}$ MEFs, compared to the wild–type *Atg5*$^{+/+}$ MEFs (B). Bar, 50 μm. Images taken at high intensity (left panel) denote the EGFP-positive transfected cells (diffuse green staining), while those taken at low intensity (middle panel) show the EGFP-HDQ74 aggregates (bright green fluorescent aggregated structures; indicated by arrows). Note that the bright green structure in *Atg5*$^{+/+}$ MEFs (indicated by an arrowhead) is an apoptotic cell with condensed nuclei, and this must not be counted as an aggregate. Nuclei are shown by DAPI staining (right panel). C, The proportions of EGFP-positive *Atg5*$^{+/+}$ and *Atg5*$^{-/-}$ MEFs with aggregates, transfected with EGFP-HDQ74 for 4 h and treated with or without 10 μM calpastatin for 48 h post-transfection, are expressed as odds ratio with 95% confidence interval. The amount of EGFP-HDQ74 aggregates in *Atg5*$^{+/+}$ MEFs is set to one to facilitate comparison with the extent of aggregation in *Atg5*$^{-/-}$ MEFs. EGFP–HDQ74 aggregation is significantly higher in *Atg5*$^{-/-}$ MEFs, compared to *Atg5*$^{+/+}$ MEFs ($p < 0.0001$). Furthermore, calpastatin fails to reduce EGFP-HDQ74 aggregation in *Atg5*$^{-/-}$ MEFs ($p = 0.648$), whereas it significantly reduces EGFP-HDQ74 aggregates in *Atg5*$^{+/+}$ MEFs ($p < 0.0001$), thereby confirming that its effects are autophagy dependent. $^{***}p < 0.001$; NS, nonsignificant. (See Color Insert.)

2. Approximately 200 EGFP-positive cells are randomly selected and the proportion of cells with EGFP-HDQ74 aggregates (bright fluorescent foci) is assessed (Figs. 5.2B and 5.3B). If a cell has no aggregates, a score of zero is given, whereas a cell having one or more aggregates is given a score of one (Ravikumar *et al.*, 2002; Sarkar *et al.*, 2005). The observer is blinded to the identity of the slides and the experiments are performed in triplicate and are repeated twice. Quantification of aggregate formation by odds ratio is described in section 3.5.2. Images are acquired with Nikon Digital Camera DXM1200 using a Nikon Eclipse E600 fluorescence microscope with 40x plan-fluor 40x/0.75 lens at room temperature (Figs. 5.2B and 5.3B). Acquisition software is Nikon ACT-1 version 2.12 and Adobe Photoshop 7.0 (Adobe Systems) is used for subsequent image processing.
3. Cells are considered dead if the DAPI-stained nuclei show apoptotic morphology (fragmentation or pyknosis). Pyknotic nuclei are typically less than 50% the diameter of normal nuclei and show obvious increased DAPI intensity. These criteria are specific for cell death, as they show a very high correlation with propidium iodide staining of live cells (Wyttenbach *et al.*, 2002). Furthermore, these nuclear abnormalities are reversed with caspase inhibitors (Wyttenbach *et al.*, 2001, 2002). Quantification of cell death by odds ratio is described in section 3.5.2.

3.5. Statistical analysis

3.5.1. Densitometric analysis on immunoblots for quantifying the clearance of mutant aggregate-prone proteins

1. Densitometric analysis on immunoblots from 3 independent experiments is performed with ImageJ software. The control condition is set to 100%. The y-axis of the graph denotes the levels of mutant proteins in control and treatment conditions (expressed in percentage), and the error bars denote standard error of the mean. The significance levels (*p* values) are determined by Annova Factorial test using Stat View software (Abacus Concepts).★★★, $p<0.001$; ★★, $p<0.01$; ★, $p<0.05$; NS, nonsignificant.
2. The clearance of the mutant aggregate-prone proteins is determined by the ratio of the intensity of EGFP to actin (loading control) for mutant huntingtin clearance, or the ratio of the intensity of HA to actin (loading control) for mutant α-synuclein clearance.
3. Trehalose significantly enhanced the clearance of soluble EGFP-HDQ74 at 120 h (Fig. 5.1C), and the SMERs significantly increased the clearance of HA-A53T α-synuclein at 24 h (Fig. 5.1D) in respective stable inducible PC12 cell lines expressing these autophagy substrates.

3.5.2. Odds ratio for quantifying mutant huntingtin-associated aggregation and toxicity

1. Pooled estimates for the changes in aggregate formation, resulting from perturbations assessed in multiple experiments, are calculated as odds ratios with 95% confidence intervals. Odds ratio of aggregation = (percentage of cells expressing construct with aggregates in perturbation conditions/ percentage of cells expressing construct without aggregates in perturbation conditions)/(percentage of cells expressing construct with aggregates in control conditions/percentage of cells expressing construct without aggregates in control conditions). Odds ratios were considered to be the most appropriate summary statistic for reporting multiple independent replicate experiments of this type, because the percentage of cells with aggregates under specified conditions can vary between experiments on different days, whereas the relative change in the proportion of cells with aggregates induced by an experimental perturbation is expected to be more consistent. We use this method frequently to allow analysis of data from multiple independent experiments (Carmichael *et al.*, 2002; Sarkar *et al.*, 2005; Williams *et al.*, 2008; Wyttenbach *et al.*, 2001, 2002).
2. Odds ratios and p values are determined by unconditional logistical regression analysis, using the general log-linear analysis option of SPSS 9 software (SPSS, Chicago). The control condition is set to one. ★★★, $p < 0.001$; ★★, $p < 0.01$; ★, $p < 0.05$; NS, nonsignificant.
3. In a stable inducible PC12 cell line expressing EGFP-HDQ74, trehalose significantly reduces EGFP-HDQ74 aggregates at 72 h, but to a lesser extent at 48 h (Fig. 5.2C). This suggests that analysis of aggregate formation in this cell line can be best assessed at 72 h.
4. In *Atg5*$^{+/+}$ MEFs, calpastatin could significantly reduce EGFP-HDQ74 aggregates but not in *Atg5*$^{-/-}$ MEFs, suggesting that calpastatin facilitated degradation of mutant huntingtin by autophagy (Fig. 5.3C). The extent of EGFP-HDQ74 aggregation is markedly increased in *Atg5*$^{-/-}$ MEFs, compared to *Atg5*$^{+/+}$ MEFs, since mutant huntingtin is an autophagy substrate (Figs. 5.3B, 5.3C).

4. Measurement of Autophagic Flux Using Bafilomycin A_1

To specifically determine the role of an autophagy modulator, it is essential to measure autophagic flux that involves the complete flow of the autophagosomes from their formation to fusion with the lysosomes. LC3-II levels as a function of actin/tubulin (loading control) on western blots, and LC3-positive vesicle numbers assessed by immunocytochemistry correlate with autophagosome numbers (Klionsky *et al.*, 2008; Mizushima and Yoshimori, 2007). LC3-II levels can increase due to an increase in autophagosome formation, but can also accumulate if there is impairment in autophagosome-lysosome fusion. The two possibilities can be distinguished by assessing LC3-II levels under conditions where autophagosome degradation is blocked by a saturating concentration of bafilomycin A_1 (Sarkar *et al.*, 2007a), which inhibits lysosomal hydrolase activity (Fass *et al.*, 2006) and results in a subsequent block in autophagosome-lysosome fusion (Yamamoto *et al.*, 1998) (Fig. 5.4A). Further blockage of autophagosome-lysosome fusion via a bafilomycin A_1-independent mechanism, using the dynein inhibitor erythro-9-[3-(2-hydroxynonyl)] adenine (EHNA) (Ekstrom and Kanje, 1984; Ravikumar *et al.*, 2005), along with this dose of bafilomycin A_1, results in no increase in LC3-II compared to bafilomycin A_1 alone (Sarkar *et al.*, 2007a). An autophagy enhancer will further increase LC3-II levels in the presence of bafilomycin A_1, compared to bafilomycin A_1 alone, due to an increase in the synthesis of autophagosomes (Figs. 5.4A-C). In contrast, an autophagy inhibitor that prevents autophagosome formation will decrease LC3-II levels in the presence of bafilomycin A_1, compared to bafilomycin A_1 alone. We have recently characterized various autophagy regulators using this assay (Sarkar *et al.*, 2007a,b, 2008; Williams *et al.*, 2008).

Another study designed an assay to differentiate between these possibilities using lysosomal inhibitors (E64d and pepstatin A) to block lysosomal degradation of LC3-II (Tanida *et al.*, 2005). We modified this assay since E64d is also a calpain inhibitor and may be less than ideal for these purposes, as we have recently shown that calpain inhibition induces autophagy (Williams *et al.*, 2008).

4.1. Materials

1. All materials listed in sections 3.1.1, 3.1.3, 3.1.4, 3.1.5, and 3.1.6, except cell lines and antibodies.
2. Cells: This assay can be performed in various cell lines, such as COS-7, HeLa and SK-N-SH cells (by immunoblotting with anti-LC3 antibody for detection of endogenous LC3-II), and also in a stable HeLa cell line expressing EGFP-LC3 (Bampton *et al.*, 2005) (by immunoblotting with anti-EGFP antibody for detection of EGFP-LC3-II). We have used COS-7 cells to describe this assay.
3. Primary antibody: Rabbit anti-LC3 (NB100-2220, Novus Biologicals). Dilution (1:4000) is made in blocking buffer.
4. Secondary antibody: ECL anti-rabbit IgG conjugated to HRP (NA934, Amersham Biosciences, GE Healthcare). Dilution (1:2000) is made in blocking buffer.
5. Autophagy modulators: SMER10 and bafilomycin A1 (described in section 3.1.2).

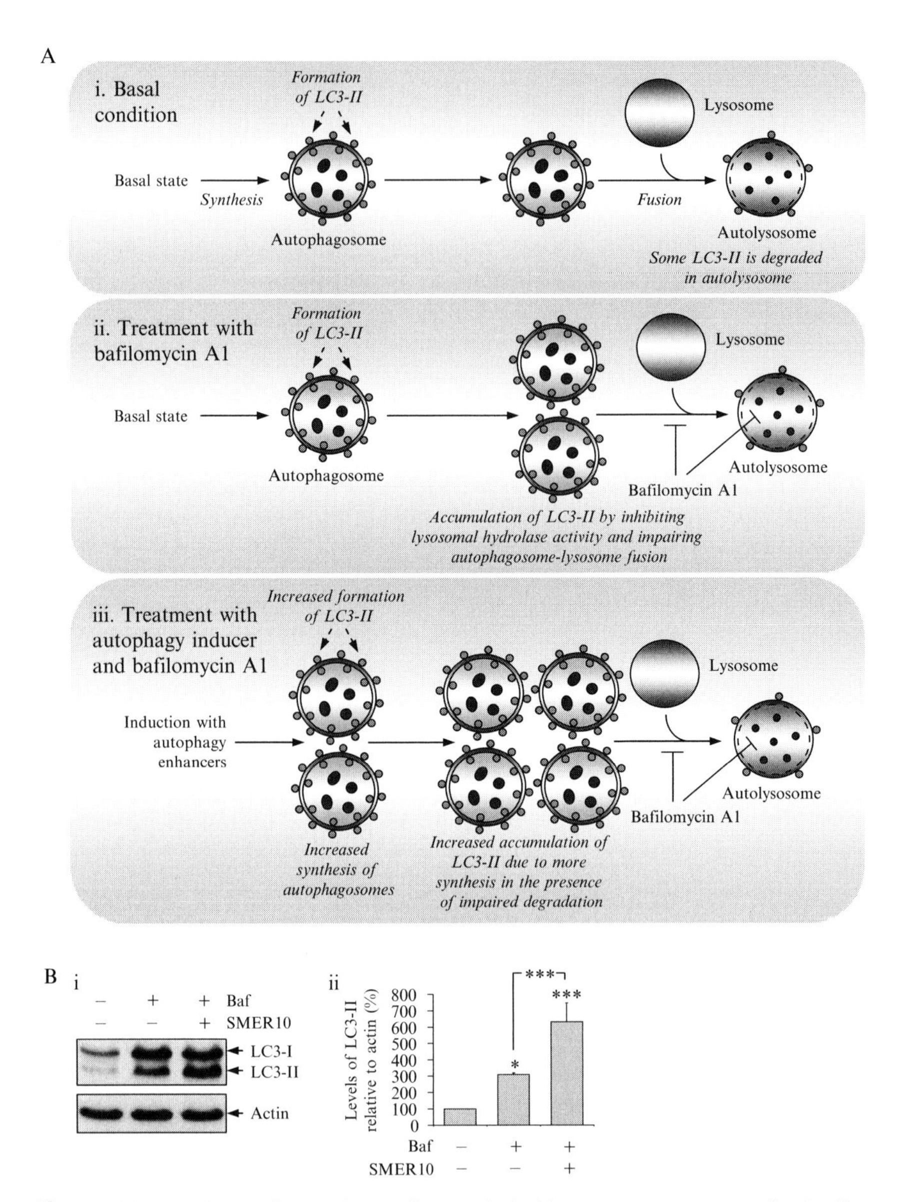

Figure 5.4 Analysis of autophagic flux with bafilomycin A_1. A, Autophagic flux involves the complete flow of the autophagosomes from their formation to fusion with lysosomes. Autophagy is a constitutive process prevalent in all mammalian cells, and under basal conditions, synthesis of autophagosomes and autophagosome-lysosome fusion occur at a normal rate (i). Short-term treatment (4 h) with bafilomycin A_1, which inhibits lysosomal hydrolases and subsequently blocks autophagosome-lysosome fusion, accumulates autophagosomes to some extent and prevents LC3-II degradation. Long-term bafilomycin A_1 treatment (24 h) completely inhibits autophagosome-lysosome fusion. This is reflected by an increase in LC3-II levels, compared to the basal

4.2. Cell culture

1. All cell types are maintained under conditions as described in section 3.2.1.
2. All cell types are subcultured as described in section 3.2.2.

4.3. Assessing autophagic flux using bafilomycin A_1

4.3.1. Experimental cell culture setup

1. COS-7 cells are seeded at 1–2 × 10^5 per well in 6-well plates for 24 h prior to treatment with autophagy modulators.
2. Cells are treated for 24 h with or without 47 μM SMER10, an mTOR-independent autophagy enhancer (Sarkar *et al.*, 2007b), and incubated at 37 °C, 5% CO_2 with humidity.
3. In the last 4 h of the 24-h treatment period, cells (both untreated and SMER10-treated) are incubated with or without 400 n*M* bafilomycin A_1.
4. The cells are collected for western blot analysis to detect endogenous LC3-II, as described in section 4.3.2.

4.3.2. Western blot analysis

1. Western blot analysis is performed as explained in section 3.3.2, except for the following conditions.

---

state (ii). Treatment with autophagy enhancers (for 24 h) prior to adding bafilomycin A_1 (which is added at a saturating concentration in the last 4 h of the 24 h treatment period) leads to a greater accumulation of autophagosomes due to increased autophagosome synthesis upstream of autophagosome-lysosome fusion, compared to bafilomycin A_1 alone. Therefore, an agent that increases LC3-II levels in bafilomycin A_1-treated cells compared to treatment with bafilomycin A_1 alone is a potential inducer of autophagy (iii) (Sarkar *et al.*, 2007a,b, 2008; Williams *et al.*, 2008). In contrast, an agent that decreases LC3-II levels in bafilomycin A_1–treated cells compared to treatment with bafilomycin A_1 alone is a potential inhibitor of autophagy by preventing autophagosome synthesis (Sarkar *et al.*, 2008). Furthermore, an agent that increases LC3-II levels by itself, but does not change LC3-II levels in bafilomycin A_1–treated cells compared to treatment with bafilomycin A_1 alone, is also a potential inhibitor of autophagy, and is likely to act at the level of autophagosome-lysosome fusion (Sarkar *et al.*, 2007a; Williams *et al.*, 2008). Other autophagosome-lysosome fusion blockers cannot cause a further increase in LC3-II levels in the presence of bafilomycin A_1 because the concentration of bafilomycin A_1 used in this assay is saturating (400 n*M*). *B*, COS-7 cells are left untreated, or treated with DMSO (control) or 47 μM SMER10, for 24 h. In the last 4 h of the 24-h treatment period, 400 n*M* bafilomycin A_1 (Baf) is added to the untreated and SMER10-treated cells. Autophagic flux is determined by immunoblotting with anti-LC3 antibody and densitometric analysis of LC3-II levels relative to actin (ii). The control condition is set to 100% and the error bars denote standard error of mean. SMER10 significantly increased LC3-II levels in bafilomycin A_1–treated cells compared to treatment with bafilomycin A_1 alone ($p < 0.0001$), suggesting that SMER10 increased autophagosome synthesis. $^{***}p < 0.001$; $^{*}p < 0.05$.

2. Approximately 10 μg of protein per sample is loaded on a 12% gel. Loading more protein will give rise to strong signals that may obscure the interpretation of results.
3. The primary antibody is rabbit anti-LC3 at a dilution of 1:4000 in blocking buffer, which is incubated with the immunoblot overnight at 4 °C on a rocker platform in the cold room.
4. The secondary antibody is ECL anti-rabbit IgG conjugated to HRP at a dilution of 1:4000 in blocking buffer, which is incubated with the immunoblot for 1 h at room temperature on a rocker platform.
5. Detection of endogenous LC3-I and LC3-II proteins requires a very short exposure time (1–5 s). The bands for LC3-I and LC3-II appear at 18 and 16 kDa, respectively (Fig. 5.4B). Because the LC3-II band is readily detectable by western blot analysis, a short exposure immunoblot is shown to assess the effect of SMER10 [a small molecule inducer of autophagy (Sarkar *et al.*, 2007b)] in the presence of bafilomycin A_1, compared to bafilomycin A_1-treated samples alone. A longer exposure time will intensify the LC3-II bands and will make the interpretation of results difficult.
6. Densitometric analysis on the immunoblots for assessing autophagic flux is described in section 4.4.

4.4. Statistical analysis

1. Densitometric analysis on immunoblots from 3 independent experiments is performed with ImageJ software. The control condition is set to 100%. The y-axis of the graph denotes endogenous LC3-II levels (expressed in percentage), and the error bars denote standard error of the mean. The significance levels (*p* values) are determined by Annova Factorial test using Stat View software (Abacus Concepts). $^{***}p < 0.001$; $^{**}p < 0.01$; $^{*}p < 0.05$; NS, nonsignificant.
2. The extent of autophagosome formation is determined by the ratio of the intensity of LC3-II relative to actin (loading control). Note that the analysis of LC3-I to LC3-II conversion is incorrect for a number of reasons: some LC3-II can be converted back to LC3-I, LC3-II detection is more sensitive by immunoblotting than LC3-I, and the levels of LC3-I may vary between tissues and cell lines and may obscure the LC3-II band on a gel. The levels of LC3-II relative to actin correlate with autophagosome numbers, and this has been endorsed by several autophagy experts in the guidelines for autophagy assays (Klionsky *et al.*, 2008).
3. Bafilomycin A_1 increased LC3-II levels by blocking autophagosome-lysosome fusion. In the presence of bafilomycin A_1, SMER10 increased LC3-II levels to a greater extent as compared to bafilomycin A_1 alone,

suggesting that SMER10 induced autophagosome synthesis upstream of autophagosome-lysosome fusion (Fig. 5.4B).

5. Concluding Remarks

Autophagy has been recently implicated in diverse disease conditions, including neurodegenerative diseases. Enhancing autophagy may be beneficial in the treatment of these disorders by clearing disease-associated, intracytosolic, aggregate-prone proteins. Identification of small molecule autophagy enhancers may not only provide a tractable therapeutic intervention for neurodegenerative diseases and other conditions where upregulating autophagy acts as a protective pathway, but also for dissecting signaling pathways regulating mammalian autophagy. The assays described in this chapter are specifically designed to identify autophagy regulators, which are relevant to various aspects in the field of autophagy, from fundamental biology to therapeutic targets.

ACKNOWLEDGMENTS

We are grateful to the MRC, Wellcome Trust (Senior Fellowship to D.C.R.), EU Framework VI (EUROSCA) and the National Institute for Health Research Biomedical Research Centre at Addenbrooke's Hospital for funding work that we have discussed in this chapter.

REFERENCES

Arrasate, M., Mitra, S., Schweitzer, E. S., Segal, M. R., and Finkbeiner, S. (2004). Inclusion body formation reduces levels of mutant huntingtin and the risk of neuronal death. *Nature* **431,** 805–810.

Bampton, E. T. W., Goemans, C. G., Niranjan, D., Mizushima, N., and Tolkovsky, A. M. (2005). The dynamics of autophagy visualised in live cells: From autophagosome formation to fusion with endo/lysosomes. *Autophagy* **1,** 23–36.

Berger, Z., Ravikumar, B., Menzies, F. M., Oroz, L. G., Underwood, B. R., Pangalos, M. N., Schmitt, I., Wullner, U., Evert, B. O., O'Kane, C. J., and Rubinsztein, D. C. (2006). Rapamycin alleviates toxicity of different aggregate-prone proteins. *Hum. Mol. Genet.* **15,** 433–442.

Carmichael, J., Sugars, K. L., Bao, Y. P., and Rubinsztein, D. C. (2002). Glycogen synthase kinase-3β inhibitors prevent cellular polyglutamine toxicity caused by the Huntington's disease mutation. *J. Biol. Chem.* **277,** 33791–33798.

Cuervo, A. M., Stefanis, L., Fredenburg, R., Lansbury, P. T., and Sulzer, D. (2004). Impaired degradation of mutant α-synuclein by chaperone-mediated autophagy. *Science* **305,** 1292–1295.

Ekstrom, P., and Kanje, M. (1984). Inhibition of fast axonal transport by erythro-9-[3-(2-hydroxynonyl)]adenine. *J. Neurochem.* **43,** 1342–1345.

Fass, E., Shvets, E., Degani, I., Hirschberg, K., and Elazar, Z. (2006). Microtubules support production of starvation-induced autophagosomes but not their targeting and fusion with lysosomes. *J. Biol. Chem.* **281,** 36303–36316.

Goll, D. E., Thompson, V. F., Li, H., Wei, W., and Cong, J. (2003). The calpain system. *Physiol. Rev.* **83,** 731–801.

Haass, C., and Selkoe, D. J. (2007). Soluble protein oligomers in neurodegeneration: Lessons from the Alzheimer's amyloid beta-peptide. *Nat. Rev. Mol. Cell Biol.* **8,** 101–112.

Hara, T., Nakamura, K., Matsui, M., Yamamoto, A., Nakahara, Y., Suzuki-Migishima, R., Yokoyama, M., Mishima, K., Saito, I., Okano, H., and Mizushima, N. (2006). Suppression of basal autophagy in neural cells causes neurodegenerative disease in mice. *Nature* **441,** 885–889.

Kabeya, Y., Mizushima, N., Ueno, T., Yamamoto, A., Kirisako, T., Noda, T., Kominami, E., Ohsumi, Y., and Yoshimori, T. (2000). LC3, a mammalian homologue of yeast Apg8p, is localized in autophagosome membranes after processing. *EMBO J.* **19,** 5720–5728.

Klionsky, D. J., Abeliovich, H., Agostinis, P., Agrawal, D. K., Aliev, G., Askew, D. S., Baba, M., Baehrecke, E. H., Bahr, B. A., Ballabio, A., *et al.* (2008). Guidelines for the use and interpretation of assays for monitoring autophagy in higher eukaryotes. *Autophagy* **4,** 151–175.

Komatsu, M., Waguri, S., Chiba, T., Murata, S., Iwata, J. I., Tanida, I., Ueno, T., Koike, M., Uchiyama, Y., Kominami, E., and Tanaka, K. (2006). Loss of autophagy in the central nervous system causes neurodegeneration in mice. *Nature* **441,** 880–884.

Kuma, A., Hatano, M., Matsui, M., Yamamoto, A., Nakaya, H., Yoshimori, T., Ohsumi, Y., Tokuhisa, T., and Mizushima, N. (2004). The role of autophagy during the early neonatal starvation period. *Nature* **432,** 1032–1036.

Levine, B., and Klionsky, D. J. (2004). Development by self-digestion: Molecular mechanisms and biological functions of autophagy. *Dev. Cell* **6,** 463–477.

Meijer, A. J., and Codogno, P. (2006). Signalling and autophagy regulation in health, aging and disease. *Mol. Aspects Med* **27,** 411–425.

Mizushima, N. (2004). Methods for monitoring autophagy. *Int. J. Biochem. Cell Biol* **36,** 2491–2502.

Mizushima, N., Levine, B., Cuervo, A. M., and Klionsky, D. J. (2008). Autophagy fights disease through cellular self-digestion. *Nature* **451,** 1069–1075.

Mizushima, N., and Yoshimori, T. (2007). How to interpret LC3 immunoblotting. *Autophagy* **3,** 542–545.

Narain, Y., Wyttenbach, A., Rankin, J., Furlong, R. A., and Rubinsztein, D. C. (1999). A molecular investigation of true dominance in Huntington's disease. *J. Med. Genet.* **36,** 739–746.

Noda, T., and Ohsumi, Y. (1998). Tor, a phosphatidylinositol kinase homologue, controls autophagy in yeast. *J. Biol. Chem.* **273,** 3963–3966.

Ohsumi, Y. (2001). Molecular dissection of autophagy: Two ubiquitin-like systems. *Nat. Rev. Mol. Cell Biol.* **2,** 211–216.

Pandey, U. B., Nie, Z., Batlevi, Y., McCray, B. A., Ritson, G. P., Nedelsky, N. B., Schwartz, S. L., DiProspero, N. A., Knight, M. A., Schuldiner, O., Padmanabhan, R., Hild, M., *et al.* (2007). HDAC6 rescues neurodegeneration and provides an essential link between autophagy and the UPS. *Nature.* **447,** 859–863.

Petiot, A., Ogier-Denis, E., Blommaart, E. F., Meijer, A. J., and Codogno, P. (2000). Distinct classes of phosphatidylinositol 3′-kinases are involved in signaling pathways that control macroautophagy in HT-29 cells. *J. Biol. Chem.* **275,** 992–998.

Ravikumar, B., Acevedo-Arozena, A., Imarisio, S., Berger, Z., Vacher, C., O'Kane, C. J., Brown, S. D., and Rubinsztein, D. C. (2005). Dynein mutations impair autophagic clearance of aggregate-prone proteins. *Nat. Genet.* **37,** 771–776.

Ravikumar, B., Duden, R., and Rubinsztein, D. C. (2002). Aggregate-prone proteins with polyglutamine and polyalanine expansions are degraded by autophagy. *Hum. Mol. Genet.* **11,** 1107–1117.

Ravikumar, B., Vacher, C., Berger, Z., Davies, J. E., Luo, S., Oroz, L. G., Scaravilli, F., Easton, D. F., Duden, R., O'Kane, C. J., and Rubinsztein, D. C. (2004). Inhibition of mTOR induces autophagy and reduces toxicity of polyglutamine expansions in fly and mouse models of Huntington disease. *Nat. Genet.* **36,** 585–595.

Ross, C. A., and Poirier, M. A. (2004). Protein aggregation and neurodegenerative disease. *Nat. Med.* **10 Suppl,** S10–S17.

Ross, C. A., and Poirier, M. A. (2005). What is the role of protein aggregation in neurodegeneration? *Nat. Rev. Mol. Cell Biol.* **6,** 891–898.

Rubinsztein, D. C. (2006). The roles of intracellular protein-degradation pathways in neurodegeneration. *Nature.* **443,** 780–786.

Rubinsztein, D. C., Gestwicki, J. E., Murphy, L. O., and Klionsky, D. J. (2007). Potential therapeutic applications of autophagy. *Nat. Rev. Drug Discov.* **6,** 304–312.

Sarbassov, D. D., Ali, S. M., and Sabatini, D. M. (2005). Growing roles for the mTOR pathway. *Curr. Opin. Cell Biol.* **17,** 596–603.

Sarkar, S., Davies, J. E., Huang, Z., Tunnacliffe, A., and Rubinsztein, D. C. (2007a). Trehalose, a novel mTOR-independent autophagy enhancer, accelerates the clearance of mutant huntingtin and α-synuclein. *J. Biol. Chem.* **282,** 5641–5652.

Sarkar, S., Floto, R. A., Berger, Z., Imarisio, S., Cordenier, A., Pasco, M., Cook, L. J., and Rubinsztein, D. C. (2005). Lithium induces autophagy by inhibiting inositol monophosphatase. *J. Cell Biol.* **170,** 1101–1111.

Sarkar, S., Krishna, G., Imarisio, S., Saiki, S., O'Kane, C. J., and Rubinsztein, D. C. (2008). A rational mechanism for combination treatment of Huntington's disease using lithium and rapamycin. *Hum. Mol. Genet.* **17,** 170–178.

Sarkar, S., Perlstein, E. O., Imarisio, S., Pineau, S., Cordenier, A., Maglathlin, R. L., Webster, J. A., Lewis, T. A., O'Kane, C. J., Schreiber, S. L., and Rubinsztein, D. C. (2007b). Small molecules enhance autophagy and reduce toxicity in Huntington's disease models. *Nat. Chem. Biol.* **3,** 331–338.

Tanida, I., Minematsu-Ikeguchi, N., Ueno, T., and Kominami, E. (2005). Lysosomal turnover, but not a cellular level, of endogenous LC3 is a marker for autophagy. *Autophagy.* **1,** 84–91.

Tanida, I., Sou, Y. S., Ezaki, J., Minematsu-Ikeguchi, N., Ueno, T., and Kominami, E. (2004). HsAtg4B/HsApg4B/autophagin-1 cleaves the carboxyl termini of three human Atg8 homologues and delipidates microtubule-associated protein light chain 3- and GABAA receptor-associated protein-phospholipid conjugates. *J. Biol. Chem.* **279,** 36268–36276.

Webb, J. L., Ravikumar, B., Atkins, J., Skepper, J. N., and Rubinsztein, D. C. (2003). α-Synuclein is degraded by both autophagy and the proteasome. *J. Biol. Chem.* **278,** 25009–25013.

Williams, A., Sarkar, S., Cuddon, P., Ttofi, E. K., Saiki, S., Siddiqi, F. H., Jahreiss, L., Fleming, A., Pask, D., Goldsmith, P., O'Kane, C. J., Floto, R. A., and Rubinsztein, D. C. (2008). Novel targets for Huntington's disease in an mTOR-independent autophagy pathway. *Nat. Chem. Biol.* **4,** 295–305.

Wyttenbach, A., Sauvageot, O., Carmichael, J., Diaz-Latoud, C., Arrigo, A. P., and Rubinsztein, D. C. (2002). Heat shock protein 27 prevents cellular polyglutamine toxicity and suppresses the increase of reactive oxygen species caused by huntingtin. *Hum. Mol. Genet.* **11,** 1137–1151.

Wyttenbach, A., Swartz, J., Kita, H., Thykjaer, T., Carmichael, J., Bradley, J., Brown, R., Maxwell, M., Schapira, A., Orntoft, T. F., Kato, K., and Rubinsztein, D. C. (2001). Polyglutamine expansions cause decreased CRE-mediated transcription and early gene expression changes prior to cell death in an inducible cell model of Huntington's disease. *Hum. Mol. Genet.* **10,** 1829–1845.

Xie, Z., and Klionsky, D. J. (2007). Autophagosome formation: core machinery and adaptations. *Nat. Cell Biol.* **9,** 1102–1109.

Yamamoto, A., Tagawa, Y., Yoshimori, T., Moriyama, Y., Masaki, R., and Tashiro, Y. (1998). Bafilomycin A1 prevents maturation of autophagic vacuoles by inhibiting fusion between autophagosomes and lysosomes in rat hepatoma cell line, H-4-II-E cells. *Cell Struct. Funct.* **23,** 33–42.

CHAPTER SIX

Monitoring Autophagy in Alzheimer's Disease and Related Neurodegenerative Diseases

Dun-Sheng Yang,[*,†] Ju-Hyun Lee,[*,†] *and* Ralph A. Nixon[*,‡,1]

Contents

* Center for Dementia Research, Nathan S. Kline Institute, Orangeburg, New York, USA
† Department of Psychiatry, New York University School of Medicine, New York, USA
‡ Department of Cell Biology, New York University School of Medicine, New York, USA
1 Corresponding author: Ralph A. Nixon, Email: nixon@nki.rfmh.org

Methods in Enzymology, Volume 453
ISSN 0076-6879, DOI: 10.1016/S0076-6879(08)04006-8

Abstract

This chapter describes detailed methods to monitor autophagy in neurodegenerative disorders, especially in Alzheimer's disease. Strategies to assess the competence of autophagy-related mechanisms in disease states ideally incorporate analyses of human disease and control tissues, which may include brain, fibroblasts, or other peripheral cells, in addition to animal and cell models of the neurodegenerative disease pathology and pathobiology. Cross-validation of pathophysiological mechanisms in the diseased tissues is always critical. Because of the cellular heterogeneity of the brain and the differential vulnerability of the neural cells in a given disease state, analyses focus on regional comparisons of affected and unaffected regions or cell populations within a particular brain region and include ultrastructural, immunological, and cell and molecular biological approaches.

1. Introduction

Autophagy in mammalian cells refers to at least three processes by which intracellular constituents enter lysosomes for degradation: chaperone-mediated autophagy (CMA), microautophagy, and macroautophagy (Cuervo, 2004). In CMA, cytosolic proteins containing a KFERQ motif are selectively targeted to the lysosomal lumen for degradation. In microautophagy, small quantities of cytoplasm nonselectively enter lysosomes when the lysosomal membrane invaginates and pinches off small vesicles for digestion within the lumen. Finally, macroautophagy, the subject of this chapter, mediates large-scale degradation of cytoplasmic constituents. This process and its implications in diseases are extensively detailed in recent reviews (Levine and Kroemer, 2008; Mizushima *et al.*, 2008) and is only briefly described herein.

During macroautophagy, an elongated isolation membrane, created from a phagophore assembly site or preautophagosomal structure (PAS), sequesters a region of cytoplasm to form a double-membrane-limited autophagosome. Autophagosomes receive hydrolases by fusing with either lysosomes to form autolysosomes or late endosomes/multivesicular bodies to form amphisomes. Efficient digestion of substrates within these compartments yields lysosomes containing mainly acid hydrolases with little residual cytoplasm. The term *autophagic vacuole* (AV) refers to any of the aforementioned compartments of the autophagic pathway, except lysosomes.

The induction of macroautophagy, through its principal route, involves the mTOR kinase (mammalian target of rapamycin), which is regulated by growth factors (especially insulin) and nutrient levels. A mTOR-independent induction of macroautophagy involving Beclin 1 and the class III phosphatidylinositol 3-kinase hVps34 can also occur (Yamamoto *et al.*, 2006). The clearance of autophagosomes, including their sequestered material, requires competent mechanisms of vesicular fusion and intrinsic lysosomal functions including acidification of the lysosomal environment and delivery and activation of lysosomal hydrolases.

Hereafter, macroautophagy is referred to by the general term *autophagy* unless otherwise indicated. Dysfunction of autophagy may occur at any stages of the autophagic-lysosomal pathway and pathological alterations that either increase or reduce induction or impair clearance of AVs could have different implications in the design of disease therapies. The characterization of autophagy dysfunction in disease states, therefore, should aim to discriminate the stage(s) along the pathway at which the impairment occurs, which can be achieved by employing appropriate methods to detect stage-related molecules/event(s).

2. General Approaches to Investigations of Human Neurodegeneration

2.1. Postmortem brain tissue

A general strategy for using human diseased tissue to understand underlying pathophysiological mechanisms is to assess the suspected etiologically relevant process, in this case autophagy, in brain regions that are pathologically affected and those that are not detectably affected. In addition, normal control brains are compared to those from individuals with the disease, ideally taken from cases ranging from the earliest stage of the disease to progressively more advanced stages. In later stages of the disease, effects that are secondary, tertiary, or even more indirectly related to the primary insult may complicate detection or interpretation of more primary events. Pathophysiology recognized before neural cells have begun to degenerate is most easily related to primary disease factors. Finally, phenomena that are observed in the disease of interest are optimally interpreted in light of the specificity of their occurrence in a given disease condition, which necessitates analyses of other related diseases to distinguish disease-specific pathobiology from general responses to injury or general effects of the neurodegenerative process itself. In the case of autophagy, the presence of increased numbers of autophagic vacuoles in neurons is a frequent observation in many neurodegenerative diseases. Only with a much more detailed assessment, however, is it possible to understand the basis and pathogenic significance of a specific pattern of autophagic pathology in a particular disease (Boland *et al.*, 2008). For example, pathological AV accumulation in cells could reflect either increased induction or decreased clearance of AVs, which imply two very different underlying mechanisms.

In line with the foregoing investigative strategy, studies on Alzheimer's disease (AD) brains should ideally include at least three groups of cases: AD individuals with clinical and neuropathological diagnoses, individuals who do not have AD but have other neurodegenerative diseases such as Parkinson's disease or Huntington's disease, and age-matched neurologically normal individuals (Cataldo and Nixon, 1990; Cataldo *et al.*, 1991, 2004a). In addition to matching age and sex of these cases, the postmortem interval (PMI), the interval between time of death and freezing or fixing the brain, should also be matched among the groups, and brains with shorter PMI (less than 12 h if possible), should be used.

Several criteria are available for neuropathological assessment of AD. The Khachaturian (1985) criteria rely on numerical cutoffs in the quantification of neocortical plaques of both diffuse and neuritic types per field corrected for age. The Consortium to Establish a Registry for Alzheimer's Disease (CERAD) (Mirra *et al.*, 1991, 1993) emphasizes semiquantitative neuritic plaque counts with corrections for age, along with clinical history to establish the level of likelihood of AD. The more recent NIA-Reagan criteria, proposed by the NIA-Reagan Institute Consensus conference (Hyman and Trojanowski, 1997), which assign cases to high-, intermediate-, or low-likelihood categories, involve combining semiquantitation of neuritic plaques using the CERAD approach with Braak and Braak's (1991) topographic staging of neurofibrillary changes. The latter staging procedure is based primarily on the distribution pattern of neurofibrillary tangles and neuropil threads, resulting in the differentiation of six stages of progressive pathology (Braak Stage I–VI). Frozen brain tissue (for biochemical analyses) and fixed brain tissue (for morphological analyses) can be obtained from any of a wide network of national brain banks, many supported by the NIH and often affiliated with the Alzheimer's Disease Research Center. Examples include the Bronx Veteran's Administration Medical Center and Mt. Sinai Medical Centers, the Harvard Brain Tissue Resource Center at McLean Hospital (Belmont, MA), and the Neuropathology Core Facility of the Massachusetts Alzheimer's Disease Resource Center (Massachusetts General Hospital, Boston, MA).

Because AD neuropathology begins in the entorhinal cortex/hippocampus and later spreads to neocortical areas, the following regions, as suggested by the NIA-Reagan criteria, should be sampled: (1) hippocampal formation including entorhinal cortex at the level of the uncus; (2) hippocampal formation at the level of the lateral geniculate nucleus; (3) neocortical areas including superior temporal gyrus, inferior parietal lobe, mid-frontal cortex, and occipital cortex; (4) substantia nigra and locus coeruleus.

As mentioned earlier, brain regions that are pathologically affected should be studied at the same time with regions that are minimally affected. In AD, the cerebellum is often used as a minimally affected region even at late disease stages. An additional level of analysis exploits the temporal-spatial progression of AD and could be modified to apply to other diseases. Tissue from the precentral gyrus, which is almost devoid of neurofibrillary changes in most layers even at Stage VI (Braak and Braak, 1991), appears to be a very late affected cortical area and can therefore serve as another type of control for assessing other regions affected at early or intermediate stages (e.g., Braak Stage I–IV). At the earliest stages of AD, areas of brain cortex other than the limbic regions, which are vulnerable to develop disease as it advances, can also serve as control tissues.

Given the heterogeneity of cell types in the brain (i.e., neurons, glial cells, and their subtypes), biochemical analyses such as Western blotting on dissected areas have limitations in distinguishing changes associated with certain cell types. This underscores the need to combine biochemical analyses with cell-specific morphological and genetic approaches, such as standard histochemical staining, immunocytochemistry, *in situ* hybridization, and DNA microarray.

2.2. Transgenic and gene deletion mouse models

A variety of genetically modified mice with mutations in autophagy genes have been generated in the past several years, including those exhibiting neurodegenerative phenotypes (e.g., mice deficient in the gene(s) for either cathepsin D, cathepsin B/L, atg5, or atg7; for review, see Levine and Kroemer, 2008).

As for AD, a large number of mouse models has been available, most of which target one of the four major AD-related genes encoding either amyloid precursor protein (APP), presenilins (PS), tau protein, or apolipoprotein E (APOE); comprehensive lists of genetically engineered AD mouse models can be found from the Alzheimer Research Forum at http://www.alzforum.org/res/com/tra/default.asp, which list more than 100 singly transgenic or gene-deletion mouse models plus numerous models harboring double or multiple engineered AD-related genes. Specific features of AD neuropathology include loss of synapses and neurons, amyloid plaques, neurofibrillary tangles, and dystrophic neurites. However, none of the AD mouse models are true, complete models that reproduce all these AD pathological features, and most of the mouse lines model a specific aspect of AD pathology (Duyckaerts *et al.*, 2008; Radde *et al.*, 2008). Mouse models carrying one or more APP mutations, especially double or triple transgenic mice harboring mutant forms of the presenilin 1 and APP genes, usually exhibit amyloid pathology along with extensive autophagic-lysosomal pathology such as dystrophic neurites that are filled with large numbers of AVs containing undigested material. Mouse models under investigation in our laboratory, which model amyloid and/or endosomal-autophagic-lysosomal pathologies in the brain, are listed in Table 6.1. In assessing autophagy pathology in these models, use of multiple models with different origins may be instructive, but it is always important to relate findings back to the pathology and mechanisms seen in the human disease itself.

To begin structural analyses in these models, we recommend the following protocol:

1. The mice are anesthetized with a mixture (0.01 ml/g body weight, i.p.) of ketamine (10 mg/ml) and xylazine (1 mg/ml).
2. For immunocytochemical studies, mice are usually fixed by cardiac perfusion using 4% paraformaldehyde (PFA) in 0.1 *M* sodium cacodylate buffer [pH 7.4, Electron Microscopy Sciences (EMS), 15713, 11652, respectively].
3. Following perfusion fixation, the brains are immersion-fixed in the same fixative overnight at 4 °C. For electron microscopy (EM) study, glutaraldehyde is included as specified subsequently.
4. For biochemical analyses, brains may be quick frozen on dry ice and stored at −70 °C or anesthetized mice may be briefly perfused with saline to wash out the blood, and the brains are then removed and stored at −70 °C. The latter procedure is used when both morphological and biochemical analyses are planned on the same mouse. In this case, the brain is removed after perfusion, and one hemisphere is immersion fixed in 4% PFA for 3 days at 4 °C while the other half is frozen at −70 °C. The brains can then be processed for different biological analyses including Western blotting (see subsequently), isolation of membrane proteins, and ELISA as described in detail previously (Schmidt *et al.*, 2005).

2.3. Culture and treatment of cell lines and primary cortical neurons

Cell models have been crucial to the analysis of autophagy in neurodegenerative disease states because of the ability to manipulate the cells genetically to induce a pathogenic condition, manipulate the autophagy pathway pharmacologically or genetically to define specific impairments, or introduce probes that enable specific aspects of autophagy to be monitored more easily. In general, these cell models have different attributes that contribute complementary information to the assessment. Immortalized cell lines are easily produced and grown in bulk for transfection or other manipulation followed by biochemical analysis that may be more difficult to do from primary cell cultures. Immortalized blastocysts or fibroblasts derived from genetic animal models have advantages in enabling complementary information relevant to the live model to be obtained more easily.

Table 6.1 Mice modeling AD-related amyloid and/or endosomal-autophagic-lysosomal pathologies

Models	Genetically modified gene(s)	Modeling	Ref
Tg2576	APP Swedish mutation	Amyloidosis, dystrophic neurites	Hsiao, Chapman *et al.* 1996
PS/APP	APP Swedish mutation and PS1M146L	Amyloidosis, autophagy/lysosomal pathology, neuronal cell death	Cataldo, Peterhoff *et al.* 2004; Yu, Cuervo *et al.*, 2005; Yang *et al.* 2008
GFP-LC3 Tg	LC3 overexpression	Autophagy	Mizushima, Yamamoto *et al.* 2004
TgCRND8	APP Swedish and Indiana mutations	Amyloidosis, autophagy/lysosomal pathology	Chishti, Yang *et al.* 2001
GFP-LC3 Tg x TgCRND8	LC3 overexpression, APP Swedish and Indiana mutations	Amyloidosis, autophagy/lysosomal pathology	
TgCRND8 x CstB-KO	APP Swedish and Indiana mutations, Cystatin B-KO	Amyloidosis, autophagy/lysosomal pathology	
Ts65Dn	Segmental trisomy 16	Endosomal pathology	Cataldo, Petanceska *et al.* 2003
CatD Tg	Cathepsin D overexpression	Autophagy/lysosomal pathology	
CatD Tg x Tg2576	Cathepsin D overexpression, APP Swedish mutation	Amyloidosis, autophagy/lysosomal pathology	

Primary fibroblasts from patients with an inherited form of the neurodegenerative disease are an invaluable and often underutilized model that authentically reproduces the cellular pathobiology of the disease and frequently provides a valuable direct window on the mechanisms underlying the neuropathology seen in the disease. Recent studies of fibroblasts from individuals with Down syndrome (DS) have shown remarkable reproduction of AD-related endosomal-lysosomal pathology seen in neurons of DS brains (Cataldo *et al.*, 2008). There are, however, limitations of to how faithfully the same cellular defect in fibroblast may reproduce the downstream disease manifestation in neurons, which have many distinctive morphological and biochemical features. Primary cortical neurons are increasingly used to evaluate uniquely neuronal aspects of autophagy, such as the transport of autophagy-related organelles and constituents back and forth along the long neuritic processes. These are usually derived from genetic animal models of the disease or wild-type controls, which are then manipulated *in vitro* to impose additional disease-related factors or to probe autophagy processes in specific ways. Although in subsequent descriptions we will mostly use examples of models related to AD, similar cell and tissue resources are available for many other neurological diseases.

1. Presenilin gene-ablated murine blastocysts (A. Bernstein, Mount Sinai Hospital, Toronto, Canada) (Donoviel *et al.*, 1999; Lai *et al.*, 2003) are grown in 35-mm dishes, precoated with poly-d-lysine (50 μg/ml, Sigma, P7280), in DMEM (Invitrogen, 10567-014) supplemented with penicillin/streptomycin (Invitrogen, 11995-073), 15% fetal bovine serum (FBS; Hyclone, SH30070.03), and β-mercaptoethanol (Sigma, M6250) at 37 °C and 5% CO_2. Cells are grown to 70%–80% confluency.

   Familial Alzheimer's disease (FAD)-human fibroblasts lines, acquired from the Coriell Institute (Camden NJ), are maintained in MEM (Invitrogen, 11095-098) with penicillin/streptomycin, 15% FBS at 37 °C and 5% CO_2. The fibroblasts are grown in 150-mm dishes for biochemical analyses or 35-mm dishes for EM and metabolic studies and grown to confluence prior to the experiment.

2. Cells are extensively washed with Hanks Balanced Salt Solution (HBSS; Invitrogen, 24020-117) prior to adding MEM in the presence or absence of serum and maintained for 12 h prior to harvesting. All nonneuronal cells are used before passage number 20, because extensive passaging may alter growth and metabolic characteristics significantly.
3. Primary cortical neurons are derived from E17 pups of C57/B6 mice (Charles River, C57/B6 strain code 027) or genetic mouse models of the disease. Pup brains are harvested and placed in ice-cold Hibernate-E medium (BrainBits LLC, HE-Ca 500 ml) where the meninges are removed and the cerebral cortices are dissected.

4. Cortices are minced using a scalpel and dissociated by incubating the tissue in Hibernate-E medium containing 15 units/ml of papain (Worthington Biochemical's, LK003178) for 30 min at 37 °C before triturating in Neurobasal medium (Invitrogen, 12348-017) containing 20% FBS, DNAse (0.2 μg/ml; Sigma, AMPD1), and 0.1 M $MgSO_4$.
5. Undissociated brain tissue is removed by passing the cell suspension through a 40-μm cell strainer (Fisher Scientific, 08-771-23).
6. Dissociated neurons are centrifuged at 200×*g* for 3 min at room temperature (RT) and the pellet is resuspended in Neurobasal medium supplemented with B27 (2%, Invitrogen, 17504-044), penicillin (100 U/ml), streptomycin (100 U/ml), and glutamine (0.5 m*M*, Invitrogen, 21051024).
7. Viable neurons are plated at a density of 100,000 cells per 13-mm circular cover glass and 250,000 cells per well in 6-well tissue culture dishes, precoated with poly-D-lysine (50 μg/ml), and incubated in a humidified atmosphere containing 5% CO_2:95% atmosphere at 37 °C.
8. Half of the plating medium is replaced with fresh medium after 3 days. Serum-free, B27 supplemented Neurobasal medium ensures minimal growth of glial cells (<5%) after 5 days in culture.

3. Characterization of Autophagic Vacuoles, Evaluation of Autophagosome and Autolysosome Formation, and Autolysosomal Clearance

3.1. Standard histochemical stains for neuropathology

The methods described in this first section are not autophagy specific. However, information revealed by the methods such as gross cell loss and neuritic changes is useful for pathological evaluation of the status/stages of the diseased samples being used for autophagy studies.

3.1.1. Cresyl violet stain for Nissl substance and nuclei

Neurons contain Nissl substance, composed primarily of rough endoplasmic reticulum. Nissl substance is lost after cell injury. Cresyl violet staining is a frequently used method for staining nuclei of all types of cells in the nervous system and Nissl substance in neuronal cell bodies.

a. Paraffin sections are deparaffinized in xylene (twice, 3 min each), passed through descending ethanol solutions (100%, 90%, 70%, 50%, 3 min each) and hydrated in distilled water (dH_2O). Vibratome sections are mounted, air-dried, and passed through 100% ethanol, xylene, descending ethanol solutions, dH_2O (3 min each).

b. Hydrated sections are then stained in cresyl violet working solution [6 ml of 0.1 % cresyl violet (Fisher Scientific, C-581) + 94 ml of 0.1 *M* acetic acid + 6 ml of 0.1 M sodium acetate] for 20 min.
c. After rinsing in 3 changes of dH_2O, sections are differentiated in 1% acetic acid water for 3 min, washed in dH_2O twice, then in 95% ethanol for 3 min.
d. Sections are further dehydrated in 100% ethanol twice, 3 min each, cleared in xylene twice, 3 min each and mounted with Permount (Fisher Scientific, SP15-100). The sections should be almost grossly transparent. Microscopically, Nissl substance and nuclei (the nucleolus, scattered chromatin and nuclear membrane) are stained as blue or dark purple while other cellular parts are clear.

3.1.2. Silver-staining methods

Silver-staining methods for tissue sections are widely used for histopathological identification of pathological deposits such as neuritic plaques and neurofibrillary tangles of Alzheimer's type. AV accumulation within the focal swellings of dystrophic neurites is common in neurodegenerative conditions especially in the AD brain. The rapid detection of dystrophic neurites by silver staining can therefore facilitate the assessment of the degree and distribution of this autophagy-related pathological change in addition to being a necessary tool for AD pathological assessment and diagnosis. Major silver-staining methods for neuropathological diagnosis include Bielschowsky, Bodian, Gallyas (Fig. 6.1G) and Campbell-Switzer, and their modifications (Uchihara, 2007). The Bielschowsky method, which is frequently used for examining sections from human autopsy brains with degenerative diseases, is described here (Yamamoto and Hirano, 1986).

a. Paraffin-embedded brain sections are dewaxed, and hydrated in dH_2O as described earlier.
b. The hydrated sections are placed in 20% (w/v) silver nitrate (Sigma, S6506) for 20 min and then held in dH_2O.
c. A 20% ammoniacal silver solution is freshly prepared by adding evaporated ammonia (prepared by evaporating 28% ammonium hydroxide) to the 20% silver nitrate solution, drop by drop with vigorous stirring until the precipitates turn clear. Sections are then placed into this solution for 15 min in the dark and rinsed in a jar of dH_2O containing 3 drops of evaporated ammonia.
d. Next, sections are transferred into a jar containing the 20% ammoniacal silver solution with 3 drops of developer (20 ml of formalin, 100 ml of dH_2O, 1 drop concentrated nitric acid, and 0.5 g of citric acid) and allowed to remain in this solution until the fibers are black with a tan background, which is monitored with a microscope.

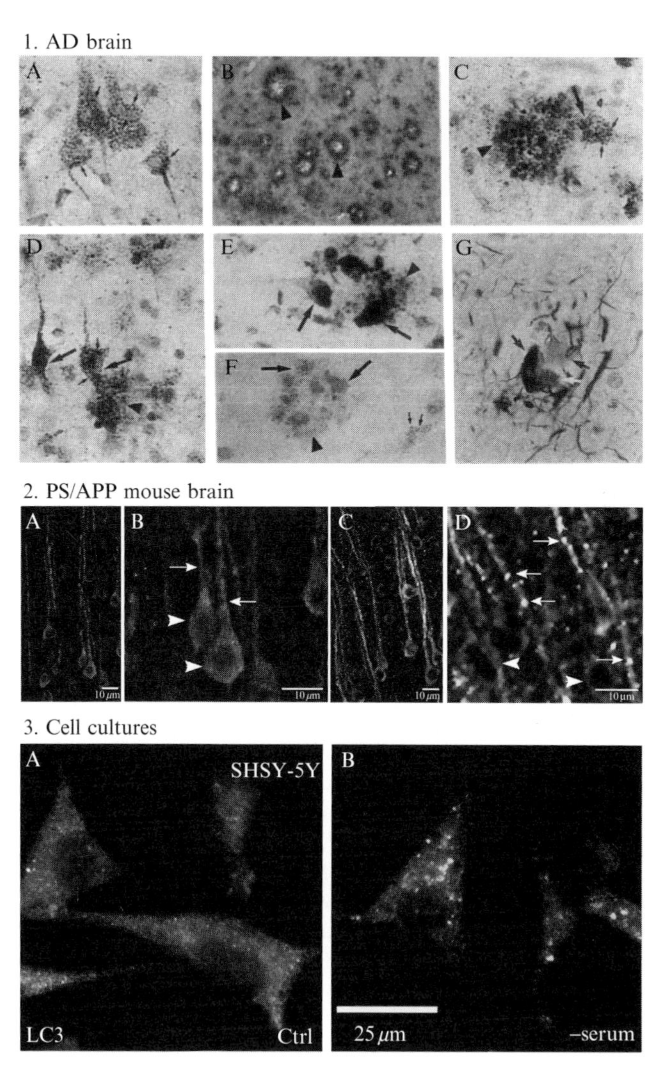

Figure 6.1 1: Staining of neuronal perikarya and senile plaques in the prefrontal cortex of Alzheimer brain. Antiserum directed against cathepsin D invariably stained lysosomes within neuronal perikarya (A, C, D, and F; small arrows) and extracellular lesions (B–F; arrowheads) identified by thioflavin S histofluorescence as senile plaques. Cathepsin-immunoreactive plaques were most numerous within cortical laminae III and V and exhibited thioflavin S-positive cores (B). Neuronal perikarya undergoing degeneration (C–E; large arrows) were often observed in plaques and, after Gallyas staining (G), displayed large amounts of lipid-positive material (white arrow). Note probable degenerated neurons (arrows) within the plaque, one of which contains argyrophilic material (G). Reprinted from Cataldo and Nixon, 1990, with permission. 2: LC3 immunofluorescence in 9-month-old mouse brains. LC3 immunostaining is weak, diffuse and uniform in neurons of nontransgenic mice (A and B) but is strong, predominantly vesicular and distributed more to the dendrites (arrows) than the cell soma

e. Sections are washed in dH_2O, dehydrated in ascending ethanol solutions (50%, 70%, 90%, 100%, 3 min each), cleared in xylene (twice, 3 min each) and mounted.

3.2. Immunocytochemistry: Immunofluorescence, double-immunofluorescence, colorimetric method

With the growing availability of antibodies directed against molecules in the autophagic-lysosomal pathway, immunocytochemistry is an essential tool for autophagy research, which enables the detection of changes in the numbers, types and distribution of vacuoles, and expression levels for individual autophagy molecules. Examples of the worthwhile information about autophagic-lysosomal changes gathered from analyses of AD brain and AD models using immunocytochemistry are growing (Boland *et al.*, 2008; Moreira *et al.*, 2007; Pickford *et al.*, 2008; Yang *et al.*, 2008; Yu *et al.*, 2005) (Fig. 6.1).

3.2.1. Brain sections

a. Immunocytochemistry using the avidin-biotin complex (ABC) method (Hsu *et al.*, 1981) is frequently used to detect endocytic-autophagic-lysosomal pathologies in human (Fig. 6.1.1) and mouse brain sections (Cataldo *et al.*, 1990, 1996, 1997, 2000, 2004b; Shiurba *et al.*, 1998; Yu *et al.*, 2005).
   i. Brain sections are treated with methanolic hydrogen peroxide [1%–3% H_2O_2 (Fisher Scientific, H325-500) in Tris-buffered saline (TBS) containing 10% methanol] for 30 min to block nonspecific endogenous peroxidase activity.
   ii. The sections are then rinsed in a diluting buffer [TBS containing 0.4% Triton X-100, 1% normal goat or rabbit serum (Rockland Immunochemicals for Research, D204-00 and D209-00, respectively) and 2% bovine serum albumin (BSA, Fisher Scientific, BP1605-100), pH 7.4] and blocked with 20% normal serum from the species donating the secondary antibody (e.g., goat or rabbit) to reduce nonspecific background staining.
   iii. Sections are then incubated in appropriate concentrations of primary antibodies diluted in the diluting buffer overnight at 4 °C.

(arrowheads) in the cerebral cortex of the PS1/APP mouse model of AD (C and D). Reprinted from Yu *et al.*, 2005 by permission of Rockefeller University Press. 3: Immunofluorescent labeling by LC3 antibody detects large vesicles in SH-SY5Y cells after macroautophagy induction by serum deprivation (B), which are much less abundant in cells grown in complete medium (A). Reprinted from Yu *et al.*, 2005 by permission of Rockefeller University Press. (See Color Insert.)

iv. After washing with the diluting buffer for 10 min 3 times, the tissue is incubated first in a biotinylated-secondary antibody (1:200 dilution, Vector Laboratories, BA-1000 for antirabbit, BA-2000 for antimouse) for 30 min at RT, rinsed, and subsequently in preformed ABC according to the manufacturer's instruction (Vector Standard ABC, Vector Laboratories, PK-4000) for 1 h at RT.

v. The final reaction is achieved by treating the sections with hydrogen peroxide and 3,3′-diaminobenzidine tetrahydrochloride (DAB) (Vector Substrate Kit, Vector Laboratories, SK-4100) for 5 to 10 min.

vi. Paraffin or frozen sections are then dehydrated in a series of ethanol to xylene as described previously and coverslipped with Permount. Vibratome sections are mounted on gelatin-coated (Fisher Scientific, 74489) slides and air-dried, dehydrated in a series of ethanol to xylene, and coverslipped with Permount.

Immunocytochemical controls consist of either incubating tissue in nonimmune sera or omitting incubation in primary antisera.

b. For immunofluorescent labeling, we use the following protocol:

i. Brain sections (either free-floating vibratome, frozen, or paraffin sections) are incubated in primary antisera overnight at 4 °C and subsequently with either an Alexa Fluor 488-conjugated or Alexa Fluor 568-conjugated secondary antibody of the appropriate species (Invitrogen, e.g., A11034, A11036, A11029, A11031) for 4 h at RT (dilutions of secondary antibodies range from 1:500 to 1:1000).

ii. The sections are then mounted on gelatin-coated slides and coverslipped in an aqueous antifade mounting medium such as Gelmount (Biomeda, M01) or VectaMount AQ (Vector Laboratories, H-5501).

iii. Sections are visualized using a Zeiss Axiovert microscope with dual filters to detect Alexa Fluor 488 and 568, or with a laser scanning confocal microscope, LSM 510 META, with LSM software v3.5 (Carl Zeiss MicroImaging, Thornwood, NY, USA) (Fig. 6.1.2).

3.2.2. Cells in culture

a. Cells (Fig. 6.1.3) grown to 70%–80% confluency on glass coverslips are fixed by removing culture medium, washing (3 times) in prewarmed PBS and adding 4% PFA with 5% sucrose in PBS (pH 7.4) for 20 min at RT or 100% ice-cold methanol for 5 min.

b. Following a wash with PBS, cells are blocked with a blocking buffer (5% normal goat serum, 0.3% Triton X-100 in PBS, PBST) for 15 min, 2 times.

c. After incubation with a primary antibody overnight at 4 °C, cells are washed with PBST for 10 min, 3 times.

d. Secondary antibodies used are from Invitrogen: goat anti-mouse/rabbit/rat Alexa Fluor 488 (A11029, A11034, A11006, respectively) and goat anti-mouse/rabbit/rat Alexa Fluor 568 (A11031, A11036, A11077, respectively) (1:500–1000 in blocking buffer).
e. After incubation with the secondary antibody for 2 h at RT, cells are washed in PBS and the cell-containing coverslips are mounted onto microscope slides with antifade Gelmount or VectaMount AQ. All incubations are carried out in a humidified chamber.
f. Cells are imaged using a plan-Apochromat 40x or 100x/1.4 oil DIC objective lens on the laser scanning confocal microscope, LSM 510 META, with LSM software v3.5. Images were analyzed using ImageJ program (NIH) (Boland *et al.*, 2008).

The same procedure can be used for double-immunofluorescence experiments, except that two primary antibodies (e.g., a polyclonal and a monoclonal) are included in the primary antibody step and accordingly, two secondary antibodies raised from the same species (e.g., a goat-antirabbit and a goat-antimouse) are incubated simultaneously.

3.3. Electron microscopy

3.3.1. Conventional transmission electron microscopy

3.3.1.1. General method Transmission electron microscopy (TEM) is a crucial technique for assessing autophagy morphology at the ultrastructural level for both brain tissue and cell cultures (Boland *et al.*, 2008; Nixon *et al.*, 2005; Yu *et al.*, 2005) (Fig. 6.2A–F) (see also the chapter by Ylä-Anttila in volume 452).

i. Human brain tissue is fixed in 3% glutaraldehyde, 0.1 *M* phosphate buffer, pH 7.4. Mouse brains are fixed by cardiac perfusion using 4% PFA, 2% glutaraldehyde in 0.1 *M* sodium cacodylate buffer. Cell lines and primary cell cultures are fixed by removing the culture medium, washing (3 times) in 37 °C supplement-free medium [e.g., MEM (Invitrogen, 11095-098) for nonneuronal cells; Neurobasal medium (Invitrogen, 12348-017) for primary neurons] and adding 4% PFA, 2% glutaraldehyde and 4% sucrose in 0.1 *M* sodium cacodylate buffer for 3 h at RT.
ii. Following fixation, vibratome brain sections or cell cultures are washed (3 times) in cacodylate buffer.
iii. The sections or cell cultures are postfixed in 1% osmium tetroxide in Sorensen's phosphate buffer (EMS, 19170) for 30 min.
iv. The sections or cell cultures are then progressively dehydrated in a graded series of ethanol as described previously and embedded in Epon (EMS, 14120).

Note: The embedding for cultured cells is usually performed by filling a plastic EM embedding capsule, which is then placed upside down onto the

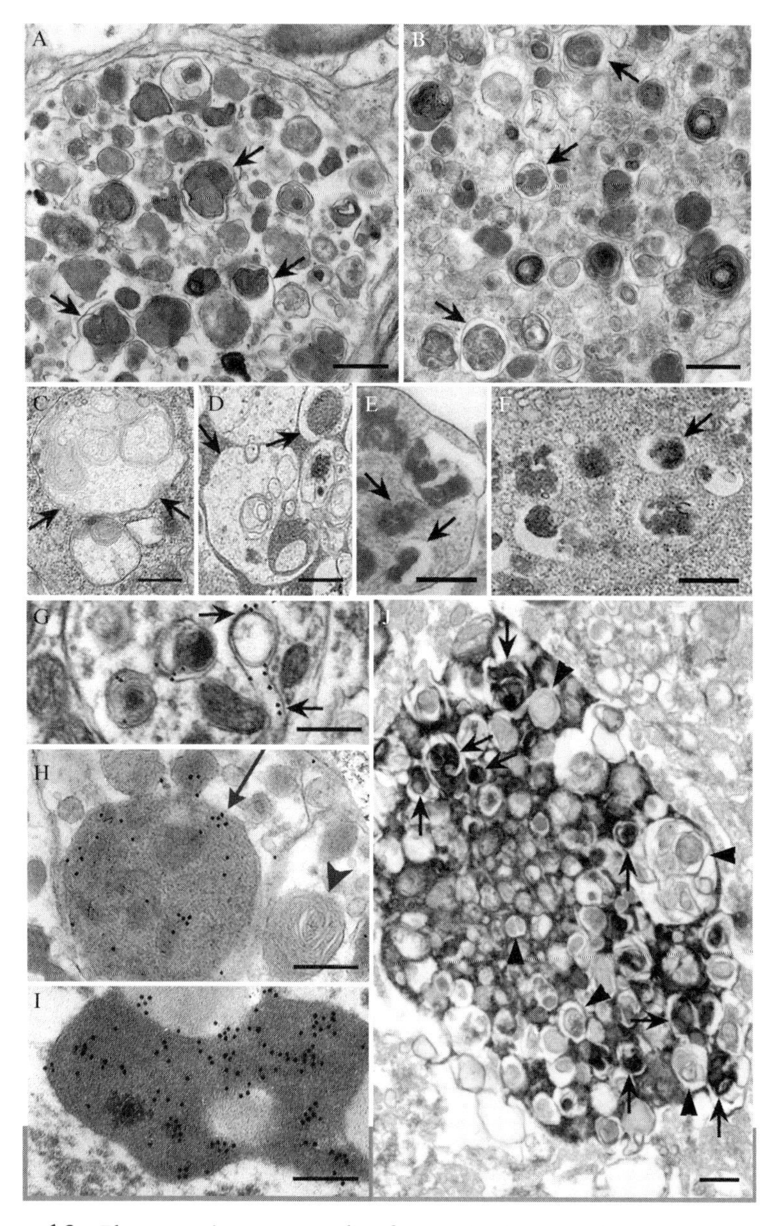

Figure 6.2 Electron microscopy studies for monitoring autophagy in AD and AD models. In AD brain (A) or the brain of PS/APP mice (B), dystrophic neurites in cerebral cortex contain robust accumulations of single- and double-membrane limited AVs with electron-dense content. C–F represent the internal morphology of primary neurons during different autophagy modulating conditions: inducing autophagy with rapamycin (C); inhibiting autophagosome/lysosome fusion with vinblastine (D); inhibiting lysosomal proteolysis with leupeptin (E) and a combination of inducing autophagy by serum deprivation together with inhibiting lysosomal proteolysis with pepstatin (F).

cell culture dish/well. After polymerization, the capsule is removed from the culture dish, which contains the cells on the top of the embedded block.

v. Thin sections of 1 μm are cut from the polymer for light microscopy analysis and followed by ultrathin sections (70–80 nm) using a Recheirt Ultracut S microtome and placed on copper grids for structural analysis. Grids are briefly stained with uranyl acetate (6.6% saturated solution) and lead citrate (0.4%) before being examined with a Philips electron microscope (Model CM 10). Images are captured with a digital camera (Hamamatsu, model C4742-95) using Advantage CCD Camera System software (Advanced Microscopy Techniques Corporation, Danvers, MA).

3.3.1.2. Ultrastructural analysis of autophagy When autophagy is initiated, cytoplasmic constituents are sequestrated by a phagophore resulting in the formation of a double-membrane-limited autophagosome. Autophagosomes then fuse with endosomes/lysosomes or other autophagosomes. In the fusion step, the outer membrane of the autophagosome fuses with the endo/lysosome membrane and its contents are released. The resulting structure is called an autolysosome (Suzuki *et al.*, 2001).

Autophagic vacuoles (AVs) are double- or single-membrane-limited vacuoles with diameters varying between 300 nm and several micrometers (Eskelinen, 2008). AVs are morphologically further classified as initial autophagic vacuoles (AVi or autophagosomes) or late/degradative autophagic vacuoles (AVd or autolysosomes). Generally AVi contain largely intact mitochondria, endoplasmic reticulum membranes, and cytoplasm with ribosomes. Clear double membranes often with an electron lucent space between the two membranes are visible depending on the sample preparation and plane of section. Nevertheless, AVd is usually limited by single membranes or sometimes by double membranes early after autophagosome-lysosome fusion. The partially degraded, amorphous

Note that the treatments used in E and F induce AVs morphologically resembling those in AD brains. Arrows indicate different AV morphologies seen in each treatment. Scale bars, 500 nm. Reprinted from Boland *et al.*, 2008, by permission of the Society of Neuroscience. G–I are immunogold electron microscopy of brain AVs. Highly selective immunogold labeling with anticalnexin antibodies of the double-limiting membranes of AVs in abnormal neuritis in an AD brain (G). A rare ER membrane profile in the process of autophagosome formation can also be seen (arrows). Mitochondria (arrowheads) are indicated and other cytoplasmic structures are unlabeled. Antibodies directed against cathepsin D mainly label electron-dense intralumenal contents of structures with single-limiting membranes (H, arrows), corresponding to late AVs or lysosomes, and strongly label proteinaceous components of lipofuscin (I), but rarely label multilamellar bodies (H, arrowhead). Scale bars, 250 nm. Reprinted from Nixon *et al.*, 2005, by permission of Lippincott, Williams and Wilkins. J: Enzyme histochemistry on brain sections of 16-month-old PS/APP mice identifies acid phosphatase activity in a dystrophic neurite. Some AVs contain hydrolase reaction product (black color, arrows) while others are apparently devoid of the product (arrowheads). Scale bar, 500 nm.

electron-dense appearance of the intralumenal contents serves as a good indicator for late/degradative AVs. Optimally, EM identification of AVs and AV subtypes should be further validated by demonstrating the presence of specific markers of these subtypes by IEM or enzyme cytochemistry (see following). In AD, AVs accumulate profusely within large swellings along dystrophic and degenerating neurites (Nixon *et al.*, 2005). For cell models, the number of AVs per cell body is counted using EM images at direct magnification of 10,500x for the various treatment conditions. Electron micrographs (25–100 images per treatment condition) are examined and values are expressed as AVs per field. For AV quantitative studies in the brain, AVs are counted from electron micrographs at 8000x magnification from a defined neuronal population (e.g., cortical lamina III). Every fourth neuron in the cross-section is photographed within each sector of the entire EM grid. Glial cells are excluded on the basis of their morphology and chromatin patterns.

3.3.2. Immuno-electron microscopy (IEM)

3.3.2.1. Post-embedding IEM

i. Brain tissue from human and mice is fixed and embedded as described above for conventional TEM. Neuronal cultures are fixed with 4% PFA, 2% glutaraldehyde and 4% sucrose in 0.1 M sodium cacodylate buffer in Neurobasal medium for 30 min at 37 °C followed by incubation in fixative without Neurobasal medium at 4 °C for 24 h.
ii. Following fixation, neurons are washed, dehydrated, and embedded in Lowicryl K4M (Polysciences, 15923-1) and polymerized under UV light (360 nm) at −35 °C.
iii. Ultrathin sections are cut from embedded brain tissue or neuronal cultures, mounted on nickel grids, and air-dried.
iv. Ultrathin sections are etched briefly (5 min) with 1% sodium metaperiodate in phosphate-buffered saline (PBS).
v. The sections are then washed in filtered double-distilled deionized water and incubated in 1% BSA in PBS for 2 h.
vi. After incubation with primary antibodies in a humidified chamber overnight at 4 °C, sections are washed several times in PBS, and incubated with 5- to 20-nm gold-conjugated antimouse or antirabbit secondary antibodies (1:50 in PBS) for 2 h at RT. These antibody reagents can be obtained from EMS (for catalog numbers, see http://www.emsdiasum.com/microscopy/products/immunogold/reagent.aspx?mm=3). Grids are washed and briefly stained with uranyl acetate and lead citrate by standard methods. In negative control experiments, the primary antibody is substituted with normal rabbit serum or normal mouse serum depending on the primary antibody used (polyclonal or monoclonal) (Nixon *et al.*, 2005; Yu *et al.*, 2005) (Fig. 6.2G–I).

3.3.2.2. Preembedding IEM Postembedding IEM is usually efficient when the antigen of interest is abundant and/or distributes widely in brain tissue or culture cells and when the antibody used has sufficiently high affinity for the antigen. In case a given antigen of interest exists in only a small number of structures, preembedding IEM can be employed, which allows localization of the immunostained structures prior to EM ultrasectioning.

i. Mouse brains are perfusion fixed with 4% PFA, 0.1% glutaraldehyde in 0.1 *M* sodium cacodylate buffer.
ii. The brains are removed and immersion fixed in the same fixative for 4 h at 4 °C and subsequently transferred to 4% PFA overnight at 4 °C.
iii. Vibratome sections (50-μm thick) are cut into PBS.
iv. Neuronal cultures are fixed with 3% PFA, 0.1% glutaraldehyde and 4% sucrose in 0.1 *M* sodium cacodylate buffer in Neurobasal medium for 30 min at 37 °C followed by incubation in fixative without Neurobasal medium at 4°C for 24h.
v. Brain sections or cell cultures are incubated with 0.1% sodium borohydride in PBS for 15 min, washed 4 times, 10 min each, permeabilized with 0.05% Triton-X100 in PBS for 30 min, washed in PBS 4 times, 10 min each, and blocked with 5% BSA and 5% normal goat serum in PBS for 1 h at 4 °C.
vi. Sections or cells are incubated with a primary antibody diluted in the blocking buffer overnight at 4 °C followed by washing and incubation with ultrasmall gold-conjugated goat-antirabbit or goat-antimouse secondary antibody (1:100 in blocking buffer) for 2 h at RT. Ultrasmall gold is silver-enhanced using a Custom ULTRASMALL Kit (EMS, 25550-06) according to the manufacturer's guidelines (Boland *et al.*, 2008).
vii. Following silver enhancement, sections or cells are postfixed with osmium tetroxide and processed for EM as described previously in 3.3.1.1.iii.

Another preembedding IEM procedure, modified from previous reports (Talbot *et al.*, 2006; Teclemariam-Mesbah *et al.*, 1997) and verified in this laboratory (Yang *et al.*, 2008), also demonstrates high specificity of immunolabeling at the EM level:

i. Vibratome brain sections are cut and treated with the following solutions alternating with PBS washes: 50% ethanol in PBS for 20 min, 0.05% Triton X-100 in PBS for 20 min, freshly made 1% sodium borohydride in PBS for 10 min and 3% H_2O_2 for 10 min.
ii. After blocking in 10% normal goat serum for 1 h at RT, sections are briefly rinsed and incubated in a primary antibody (diluted in 1% BSA in PBS) overnight at 4 °C.
iii. After washing in PBS, sections are incubated in a biotinylated goat antirabbit secondary antibody (Vector Laboratories, BA-1000 for

antirabbit, BA-2000 for antimouse, diluted 1:250 in 1% BSA in PBS) for 1 h, then, incubated in a Vector Standard ABC solution (Vector Laboratories) for 2 h, and reacted with 0.025% DAB (EMS, 13082) in PBS in the presence of 0.006% H_2O_2 for 10 min.

iv. The DAB reaction product is then intensified with silver–gold treatment following the protocol from Teclemariam-Mesbah *et al* (1997).

v. The sections are then postfixed in 1% osmium tetroxide, dehydrated and embedded as described previously.

3.3.3. Special ultrastructural methods

3.3.3.1. Acid phosphatase histochemistry Autophagic-lysosomal compartments contain numerous acid hydrolases. Analysis using enzyme histochemistry for acid phosphatase at the ultrastructural level can provide information about the trafficking and distribution of acid hydrolases in neuronal processes and perikarya and within specific subcellular compartments. We usually use a lead capture technique with cytidine 5′-monophosphate (CMP) as the substrate (Cataldo *et al.*, 1991).

i. Mice are transcardially perfused with 4% PFA, 1% glutaraldehyde in 0.1 *M* sodium cacodylate buffer containing 0.025% calcium chloride, 5% sucrose and 0.075% CMP.

ii. The brains are removed and further immersion-fixed in 4% PFA for 4 h at 4 °C.

iii. Vibratome sections (50 μm) are cut, rinsed in 0.1 *M* sodium cacodylate buffer containing 5% sucrose, then rinsed in 0.05 *M* Tris-maleate buffer containing 5% sucrose.

iv. Sections are incubated in the reaction medium (25 mg of CMP, 7 ml of distilled water, 10 ml of 0.05 *M* Tris-maleate buffer with 5% sucrose, 5 ml of 0.025 *M* manganese chloride, 3 ml of 1% lead nitrate, pH 5.0, filtered with #50 paper) for 1 h at 37 °C.

v. After washing in Tris-maleate buffer and then sodium cacodylate buffer containing 5% sucrose, sections are briefly treated with 1% sodium sulfide in sodium cacodylate buffer containing 5% sucrose, and rinsed well in sodium cacodylate buffer containing 5% sucrose. The sections are then postfixed in 1% osmium tetroxide and processed for EM embedding (Fig. 6.2J).

3.3.3.2. Horseradish peroxidase uptake/trafficking Uptake of horseradish peroxidase (HRP) into neuronal cells occurs via bulk endocytosis, especially at axonal terminals. Endocytosed HRP is transported in a retrograde direction and delivered into autophagic-lysosomal vacuoles in the cell body and, therefore, can serve as a marker for studying endocytic trafficking to autophagic-lysosomal compartments, especially when compared with enzyme histochemistry for acid hydrolases (Broadwell *et al.*, 1980).

i. Mice are anesthetized and placed in a stereotaxic apparatus with a mouse adapter (BenchMark Stereotaxic model from myNeuroLab, 477001).
ii. The scalp is shaved and a midline incision made starting slightly behind the eyes, exposing the skull area. A hole is drilled in the skull and the coordinates for HRP perfusion into the lateral ventricle are AP –0.22 mm to Bregma, ML 1.0 mm to Bregma, and DV 2.5 mm to cranium, according to the Mouse Brain, in *Stereotaxic Coordinates*, by Keith B. J. Franklin and George Paxinos, Academic Press.
iii. 20 μl of a 5% solution of HRP (Type VI-A, Sigma, P6782) in PBS are delivered at a rate of 1 μl/min.
iv. 24 h after the perfusion, mice are fixed by cardiac perfusion using 2% PFA, 2% glutaraldehyde containing 0.025% $CaCl_2$ in 0.1 *M* sodium cacodylate buffer, pH 7.2.
v. Brain sections (50 μm thick) are cut on a Vibratome and rinsed in cacodylate buffer 5 times to wash out glutaraldehyde.
vi. Sections are then incubated in a DAB solution, made following the instruction from the manufacturer (DAB Substrate Kit, SK-4100, Vector Laboratories) and filtered with a 0.22-μm syringe filter, for 5–10 min at room temperature.
vii. Sections are washed several times in cacodylate buffer, then either mounted onto gelatin-coated slides for light microscopy analysis, or post-fixed in 1% osmium tetroxide and 1.5% potassium ferrocyanide, dehydrated and embedded for EM as described previously. The tissues are not stained *en bloc*, nor are ultrathin sections poststained with heavy metals (Broadwell and Brightman, 1983).

3.3.3.3. DAMP IEM Sequestered material within an autophagosome is degraded when it fuses with lysosomes. Lysosomes have numerous enzymes that are activated under acidic pH conditions generated by an ATP-dependent proton pump. For this reason, examining acidification status within lysosomal compartments is one way to investigate the origins of an impairment in autophagic/lysosomal clearance. In cell culture systems, LysoTracker is widely used for this purpose (see section 3.4.). Due to technical difficulties with this probe, *in vivo* acidification analysis in brain tissues can be accomplished using a DAMP [3-(2,4-dinitroanilino)-3′-amino-N-methyldipropylamine] IEM method (Dunn, 1990). The primary amino group of DAMP is protonated and positively charged at acidic pH. This positively charged DAMP can covalently link to proteins fixed with aldehydes, which provides the basis for using DAMP as a marker of organelle acidification (Anderson *et al.*, 1984).

i. Mice are injected with DAMP (20 μl of 30 m*M* stock solution prepared in PBS, Invitrogen, D-1552) by intracerebroventricular administration as described for HRP injection.

ii. After 4 h, animals are anesthetized and perfused with a fixative containing 0.1% glutaraldehyde and 4% PFA in sodium cacodylate buffer.
iii. Brains are dissected and immersed in the same fixative for 4 h and then 40-μm sagittal sections are made using a Vibratome.
iv. The sections are processed routinely for EM and embedded in LR White resin.
v. Ultrathin sections are mounted on nickel grids.
vi. Sections are treated with 4% normal serum for 2 h at RT.
vii. The sections are incubated with a mouse monoclonal antibody to DNP (Invitrogen, LO-DNP-2, 1:50) and a rabbit polyclonal antibody to cathepsin D (1:50, DAKO, A0561) or another marker of lysosomal compartments, overnight.
viii. Grids are subsequently washed with PBS and incubated for 2 h at RT with 10-nm gold-conjugated goat-antimouse IgG and 6-nm gold-conjugated goat-antirabbit IgG (1:50, EMS, 25104, 25129 respectively).
ix. Sections are washed with PBS and stained briefly with uranyl acetate and lead citrate.
x. Sections are examined and photographed with a Philips CM10 electron microscope.

3.4. Florescence probes: LysoTracker, tandem-tagged probes, protease affinity reagents to analyze compartment acidification and proteolysis

3.4.1. Transient DsRed-LC3, mRFP-GFP-LC3 and GFP-Endo transfection

GFP-tagged LC3 (microtubule-associated protein 1 light chain 3) has been used to monitor autophagy in living cells by fluorescence microscopy by measuring levels of punctate LC3 profiles (mainly a marker of autophagosomes) (Kabeya *et al.*, 2000). However, in an acidic environment such as autophagosomes fused with lysosomes, GFP fluorescence is quenched. Recently, LC3 has been tagged with DsRed or mRFP/mCherry fluorescence in combination with GFP fluorescence and used to study both autophagy induction and autophagic/lysosomal protein turnover (Kimura *et al.*, 2007; Pankiv *et al.*, 2007) (see the chapter by Kimura *et al.*, in this volume). The advantage of this doubly tagged LC3 is that detection of the colocalization of both signals indicates that the autophagosome has not fused with lysosomes, whereas the detection of only the mRFP/mCherry signal reflects LC3-positive autophagosome fusion with lysosomes.

i. Polymerase chain reaction (PCR)-amplified coding sequences of LC3 in GFP-LC3 (provided by Noboru Mizushima, Tokyo Medical and Dental University) is subcloned into a pDsRed-Monomer-C1 vector (Clontech, 632466) to create DsRed-LC3, and verified by sequencing.

The pEGFP-Endo reporter vector, which encodes a fusion protein containing the human RhoB GTPase, is obtained from BD Biosciences (632365).

ii. Nonneuronal cell lines and primary cortical neurons after 3–4 days *in vitro* plated in 35-mm glass-bottom dishes are transfected with DsRed-LC3 using Lipofectamine 2000 (Invitrogen) according to the manufacturer's suggested conditions. Briefly, 2 ml of conditioned (pre-transfection) medium is replaced with transfection medium consisting of 1 μg of DNA, 5 μl of Lipofectamine 2000, 500 μl of Opti-Mem (Invitrogen, 22600134), and 1.5 ml of Neurobasal medium without B27. Cells are incubated with transfection medium for 30 min at 37 °C, followed by washing 3 times with fresh Neurobasal medium.

iii. Conditioned medium is readded to the transfected neurons, which are maintained in the incubator for least 24 h before treatments.

3.4.2. LysoTracker and Bodipy-FL-pepstatin A

LysoTracker is an acidic pH indication dye taken up by endocytosis and delivered to lysosomes, which strongly fluoresces in compartments that have an acidic pH below 5 (Via *et al.*, 1998). Live microscopy studies after transfection with fluorescence tagged LC3 and in the presence of Lyso-Tracker can provide information about the fusion between autophagosomes and acidic endo/lysosomal compartments. Bodipy-FL-pepstatin A is a fluorescence-tagged cathepsin D inhibitor that binds to the active site of cathepsin D when its active site is open under acid pH conditions. For that reason, Bodipy-FL-pepstatin A can be used as an indirect indicator of Cathepsin D enzyme activity and acidic pH (Chen *et al.*, 2000). One benefit of this dye is that it can be used under both live and fixed cell conditions.

i. Cells are seeded onto 24-well plates with cover glasses or onto 35-mm glass-bottom dishes for fixing or live imaging, respectively, and grown to 70%–80% confluency.

ii. A final concentration of 50 n*M* LysoTracker red (Invitrogen, L7528) (Köchl *et al.*, 2006) or 1 μg/ml of Bodipy-FL-pepstatin A (Invitrogen, P12271) (Chen *et al.*, 2000) is loaded for 30 min or 1 h, respectively at RT, followed by immunocytochemical analysis as described previously.

4. Metabolic Analyses of Autophagy in Neuronal and Nonneuronal Cell Models

4.1. Analysis of protein degradation

Although many of the foregoing methods provide information about individual compartments in the autophagy pathway and their properties at a single point in time, metabolic labeling analyses of protein turnover by autophagy can be used to assess the overall rate of autophagy and the flux of substrates through the pathway.

Total protein degradation in cultured cells is measured in pulse-chase experiments (Auteri *et al.*, 1983; Cuervo *et al.*, 2004) (also see the chapter by Bauvy *et al.*, in volume 452).

i. Confluent nonneuronal cells are labeled with [^{3}H]leucine (2 μCi/ml) for 48 h at 37 °C to preferentially label long-lived proteins.
ii. Following labeling, cells are extensively washed and maintained in complete medium (DMEM + 10% FBS), under which conditions autophagy is suppressed, or in serum-deprived medium, where autophagy is induced.
iii. Under both conditions, after washing the cells, the medium is supplemented with unlabeled 2.8 m*M* leucine to prevent [^{3}H]-leucine reincorporation into newly synthesized proteins.
iv. Aliquots of the medium taken at different time points are precipitated with 10% TCA, filtered using a 0.22 μm-pore membrane and radioactivity in the flow-through is measured.
v. Proteolysis is expressed as the percentage of the initial acid-precipitable radioactivity (protein) converted to acid-soluble radioactivity (amino acids and small peptides) over time.

The same procedure is used for primary cortical neurons, except that neurons are plated at a density of 1–2$\times$10^5 cells/cm^2 on 24-well plates, and maintained and washed in Neurobasal medium supplemented with B27 (2%), penicillin (100 U/ml), streptomycin (100 U/ml) and glutamine (0.5 m*M*).

To inhibit autophagy, 20 m*M* NH_4Cl or 10 m*M* 3-methyladenine (3MA) is added immediately after the labeling period and maintained at that concentration throughout the chase period. NH_4Cl blocks lysosomal degradation, so this procedure permits one to estimate the relative contributions of lysosomal and non-lysosomal pathways to overall protein degradation (Ahlberg *et al.*, 1982). 3MA blocks formation and fusion of autophagosomes with lysosomes and is used to block macroautophagic contributions to proteolysis (Seglen and Gordon, 1984).

Degradation rates of short-half-life proteins are determined by the same procedure but after a labeling period of 30 min at 37 °C. Nearly all turnover of short-lived proteins is due to proteasomal activity (Ding *et al.*, 2003). Protein synthesis is determined as the incorporation of [^{3}H] leucine into acid-insoluble material in the presence of an excess (2.8 m*M*) unlabeled leucine in the medium. Under those conditions incorporation of radioactivity into protein accurately reflects rates of proteins synthesis (Gulve and Dice, 1989) and minimizes differences due to alteration of amino acid transport and/or intracellular amino acid pool sizes.

[^{3}H]-leucine incorporation and proteolysis of short-lived proteins (reflecting mainly ubiquitin-proteasome-dependent degradation) generally remain unchanged under conditions where autophagy is impaired. When autophagy/lysosomal degradation is induced by removing serum from the medium, rates of long-lived protein turnover are normally increased over the rates in serum-supplemented condition. The proteolysis measured under these conditions is lysosomal and completely inhibited by NH_4Cl (final concentration 20 m*M*, Sigma, A9434). Depending on conditions and cell types, a proportion of this activity is due to macroautophagy, which can be demonstrated by the inhibitory effect of 3MA (final concentration 10 m*M*, Sigma, M9281), a selective inhibitor of macroautophagy. The residual NH_4Cl-sensitive activity after 3MA is attributable to chaperone-mediated autophagy (Massey *et al.*, 2006). A reduced release of amino acids from protein under serum-starvation conditions can reflect defects in induction, autophagosome formation, fusion with lysosomes, and/or proteolytic clearance of autophagic substrates.

4.2. Lysosomal enzyme activity assay

Sequestered materials within autophagosomes are degraded after fusing with lysosomes. Lysosomal proteolysis is mainly mediated by cysteine proteases (e.g., cathepsins B, L, H, S) and aspartyl proteases (e.g., cathepsin D). These lysosomal protease are fully activated in an acidic environment (Kokkonen *et al.*, 2004). Delayed proteolytic clearance of autophagic substrates and their accumulation in autophagic vacuoles may result from the down-regulation or inactivation of lysosomal enzymes. To assess the state of lysosomal enzyme activation, *in vitro* enzyme activity measurements can be used together with Bodipy-FL-pepstatin A immunocytochemistry and an IEM approach, especially for cathepsin D activity as described previously.

1. Cathepsin D activity is assayed using [^{14}C]methemoglobin (Sigma, H2625) as previously described by Dottavio-Martin and Ravel (1978).
   a. A sample (50 μl) is incubated at 37 °C for 1 h with 100 μl of 0.4 *M* acetate buffer, pH 3.2, containing 10 mg/ml [^{14}C]methemoglobin in the absence or presence of 1 n*M* pepstatin A (Sigma, P5318).
   b. The enzyme reaction is stopped by adding 0.5 ml of 10% trichloroacetic acid (TCA) then centrifuged for 5 min at 500×*g*.
   c. The proteolytic activity is calculated from the TCA-soluble radioactivity per microgram of protein extracted from the corresponding sample. Cathepsin D activity is expressed in terms of methemoglobin degrading activity; one unit of enzyme activity is defined as the capacity to degrade 1 nmol of methemoglobin per min. Specificity of hydrolysis by cathepsin D is monitored by inhibition with 1 n*M* pepstatin A; enzymatic activity is expressed as pepstatin A inhibitable

hydrolytic activity obtained by subtracting the cpm released in the presence of pepstatin A from the total cpm in the acid soluble fraction.

2. Cathepsin B and L activities are assayed as described previously (Marks and Berg, 1987).
   a. Samples are preincubated in 0.1 *M* sodium acetate, pH 5.5, containing 1 m*M* EDTA and 2 m*M* cysteine-HCl for 5 min for activation.
   b. Following addition of 5 μM Z-Phe-Arg-amc (AnaSpec 24096) in 0.1% Brij35 (Sigma, P1254), reaction mixtures are further incubated for 10–30 min at 37 °C.
   c. Reactions are stopped by addition of 200 μl of 0.1 *M* sodium monochloroacetate in 0.1 *M* sodium acetate, pH 4.3, then samples are read in a Wallac Victor-2 spectrofluorimetric plate reader (PerkinElmer Life and Analytical Science, Wellesley, MA, USA) with a filter set optimized for detection of 4-methyl-7-aminocoumarin (-amc) with excitation set at 365 nm and emission at 440 nm. For specific assay of cathepsin B, the buffer system is changed to 0.1 *M* MES, pH 6.0, containing 1 m*M* EDTA and 2 m*M* cysteine-HCl. The substrate used in this case is Z-Arg-Arg-amc (AnaSpec, custom synthesis). In this buffer system, we can also assay cathepsin H using Arg-amc (AnaSpec, custom synthesis). Specificity of hydrolysis by these cathepsins is monitored by inhibition with 1 m*M* leupeptin. Enzymatic activity is expressed as leupeptin inhibitable hydrolytic activity obtained by subtracting the fluorescent amc units released in the presence of leupeptin from the total fluorescent amc units in the acid soluble fraction.

It should be noted that lysosomal proteases are ubiquitous and present in all cell types. Moreover, proteases coexist with their endogenous inhibitors and other regulatory factors within the cells. Thus, activities of cathepsins measured in disrupted brain tissue by *in vitro* enzymatic activity assays will not accurately represent actual *in vivo* activities within the lysosomal compartments, especially of a particular cell population.

5. Isolation and Characterization of Autophagic Vacuoles and Lysosomes from Cell Cultures and Brain Tissue

Subcellular fractionation is a procedure for separation of organelles, which can be performed by differential centrifugation yielding fractions containing individual populations of organelles. Separation of organelle types is based on their physical properties such as size, density, and other parameters (Pasquali *et al.*, 1999). The isolated organelles can be used for studying distributions of specific organelle-associated molecules, degradation of substrate within isolated autophagosomal/lysosomal compartments, or organelle trafficking and fusion phenomena.

A. For cell lines, 500 million cells are serum-deprived overnight to induce autophagic activity before AV preparation.
B. For preparation of brain AV, cerebral cortices and hippocampi of 5 mouse brains are pooled.
C. Using a protocol modified from Marzella *et al.* (Marzella *et al.*, 1982), the samples are collected, disrupted by nitrogen cavitation then homogenized in 3 volumes of 0.25 *M* sucrose in a glass homogenizer with Teflon pestle for 10 strokes.
D. The homogenate is filtered through double gauze and then spun at 2000×*g* for 5 min to yield a supernatant fraction and a pellet of unbroken cells and nuclei.
E. The supernatant fraction is centrifuged at 17,000×*g* for 12 min to yield a pellet fraction and a supernatant fraction, which is centrifuged again at 100,000×*g* for 1 h to yield a pellet containing ER and a supernatant fraction containing cytosol.
F. The pellet from the 17,000×*g* centrifugation is resuspended in the same volume of 0.25 *M* sucrose and centrifuged again at 17,000×*g* for 12 min.
G. The pellet is resuspended in 1.9 ml of 0.25 *M* sucrose and 2.8 ml of metrizamide (Mtz, AKSci 69696).
H. This mixture (2.4-ml volume) is loaded on top of a 26% (4 ml), 24% (2 ml), 20% (2 ml) and 15% (2 ml) Mtz step gradient matrix.
I. The sample on the Mtz gradient is centrifuged at 247,000 rpm for 3 h in an ultracentrifuge using an SW41 rotor.
J. Each gradient interface is collected and diluted with same volume of 0.25 *M* sucrose.
K. The samples are then centrifuged at 24,000×*g* for 10 min.
L. Light AVs (AV10) are present in the 15%–20% fraction, heavy AV (AV20) are present in the 20-24% fraction, lysosomes are in the 24%–26% interface, and mitochondria are located in the 26% Mtz area.
M. Fractions are pelleted and immersed in a cacodylate fixation buffer for EM analysis or analyzed directly by Western blot or enzyme assay.

6. Western Blot Analysis of Autophagy Components and Substrates

6.1. General methods

1. Cells used for western blot analysis are lysed in a buffer containing 50 m*M* Tris, pH 7.4, 150 m*M* NaCl, 1 m*M* EDTA, 1 m*M* EGTA, 1%

Triton X-100 and 0.5% Tween 20 with protease inhibitors (Roche, 11836170001) and phosphatase inhibitor cocktail containing microcystin LR, cantharidin, and (–)-p-bromotetramisole (Sigma, P2850).
2. Brains are homogenized in a buffer containing 50 m*M* Tris, pH 7.4, 150 m*M* NaCl, 1 m*M* EDTA with protease and phosphatase inhibitors.
3. Following electrophoresis, proteins are transferred onto 0.2 um nitrocellulose membrane (Whatman, BA83) at 100 mA for 8–12 h depending on the target protein.
4. Membranes are blocked by incubation in 5% milk (Bio-Rad, 170-6404) in Tris-buffered saline with 0.1% Tween 20 (TBST) for 1 h.
5. The membranes are incubated with a primary antibody (1:2000 in 5% milk/TBST) for 2 h at RT or overnight at 4 °C, washed 3 times, 15 min each in TBST, and then incubated 2 h with a goat-antirabbit or goat-antimouse secondary antibody (Jackson Immunoresearch, 115-035-062, 111-035-045 respectively; 1:5,000 in 5% milk/TBST).
6. The blot is exposed onto film following incubation in enhanced chemiluminescent fluid (ECL, GE Healthcare) and developed.

6.2. Analysis of autophagy induction signaling

Autophagy occurs at low basal levels in all cells to maintain protein and organelle homeostasis; however, rapid upregulation occurs when cells need to generate energy under starvation or growth factor withdrawal conditions, or to defend against oxidative stress, infection, and accumulation of toxic protein aggregates (Levine and Kroemer, 2008). Autophagosome formation is mediated by a family of autophagy-related (Atg) genes regulated by the PI3K/AKT signaling pathway that modulates mTOR kinase (Ohsumi, 2001). The mTOR kinase is a negative regulator of autophagy, so inhibition of mTOR with rapamycin leads to autophagy induction. This kinase activity can be examined by measuring the phosphorylation state of its target protein, p70S6K (see the chapter by Ikenoue *et al.*, in volume 452). The p70S6K-Thr389 is directly phosphorylated by mTOR and this event is rapamycin sensitive (Aoki *et al.*, 2001; Sarbassov *et al.*, 2005). Therefore, a decrease in phosphorylation of p70S6K-Thr389 is indicative of mTOR suppression and likely correlates with the signaling of autophagy induction.

1. To assess activation or suppression of the autophagy induction signaling pathway in cultured nonneuronal cells (60%–70% confluence) or primary neurons (5 days *in vitro*), cells are treated with inducing compounds for 1 h, 6 h, and 24 h before harvesting. Treatment conditions and compounds that may be used include rapamycin (10 n*M*, L.C. Laboratories, AY-22989), serum-deprived medium [e.g., DMEM (Invitrogen, 10567-014), MEM (Invitrogen, 11095-098)] or Earle's Balanced Salt Solution (EBSS; Invitrogen, E3024).

2. After treatments, samples are collected and analyzed by Western blot as described earlier.
   a. Briefly, 20 μg of protein with 2x SDS sample buffer are loaded onto 4%–20% Tris-glycine gels (Invitrogen, EC6028BOX) and detected with antibodies directed against phospho-p70S6k (Cell Signaling, polyclonal antibody 9205 or rabbit monoclonal antibody 108D2, 1/1,000) and total p70S6K (Cell Signaling, 9292, 1/1000).
   b. 5% BSA is used for blocking in Western blotting of phospho-p70S6K instead of 5% milk to avoid phosphatase activity in milk.

In neuronal culture, rapamycin, serum starvation, and vinblastine reduce mTOR-mediated p70S6K phosphorylation; however, lysosomal proteolysis inhibitors such as leupeptin and pepstatin, which block later digestive phases of autophagy, have no effect on induction (Boland *et al.*, 2008).

6.3. Autophagosome formation

The initial form of LC3 is termed pro-LC3, which is proteolytically converted to LC3-I and further modified at its C-terminus to become phosphatidylethanolamine conjugated LC3-II. When autophagy is induced, cytosolic LC3-I is converted to membrane-bound LC3-II, which is associated with autophagosomes or possibly other AV subtypes under some conditions (see subsequently). For that reason, the level of LC3-II relative to a control such as actin or tubulin is a marker for autophagy (Mizushima and Yoshimori, 2007) (also see the chapter by Kimura *et al.*, in volume 452).

In cultured human cells, this conversion can be modified by various chemical agents applied for 1 h, 6 h, and 24 h before harvesting. Treatment conditions and compounds used include inducing autophagy with rapamycin or EBSS, impeding LC3-II degradation by blocking autophagosome fusion with lysosomes using vinblastine (10 μ*M*, Sigma, V1377), or inhibiting proteolysis with leupeptin (20 μ*M*, Sigma, L-2884), pepstatin A (20 μ*M*, Sigma, P5318), or ammonium chloride (NH_4Cl, final concentration 20 m*M*). Extracts (20 μg of protein) in 2x SDS sample buffer are loaded onto 16% Tris-glycine gels (Invitrogen, EC6498BOX) and detected with anti-LC3 rabbit polyclonal antibody (Yu *et al.*, 2005) or anti-LC3 mouse monoclonal antibody (1:250, Nanotools USA, 0261-100) (Boland *et al.*, 2008) (numerous commercial suppliers, also see the Autophagy Reagent Forum at http://www.landesbioscience.com/reagent_blog/). After chemiluminescence development, the levels of LC3-II and the loading control are calculated by densitometry.

The measurements are widely, but incorrectly, used to monitor the extent of autophagy induction; however, autophagosome numbers may accumulate either after autophagy induction or when the proteolysis of autophagosome substrates is impaired. In the case of neurons, incomplete proteolysis, lysosome acidification or slowed autophagosome-lysosome fusion can be quite common in disease states. The heterogeneity of cell populations in brain may also obscure LC3-II changes in specific cell types in brain tissue (Yang *et al.*, 2007). In neurons, the ratio of LC3-I to LC3-II is exceptionally high and levels of LC3-II may be difficult to detect because LC3-II positive autophagosomes, once formed, are typically degraded, unless the degradation process is impaired. In this case, LC3-II-positive compartments may include large numbers of amphisomes and autophagosomes, which can be assessed by morphological analysis using EM.

6.4. Autophagic/lysosomal protein degradation

Because autophagic substrate degradation occurs after autophagosomes fuse with lysosomes, turnover of autophagic compartments or autophagic/lysosomal substrates can be estimated by inhibiting lysosomal proteolysis. Lysosomal proteolysis is mainly mediated by cysteine (e.g., cathepsin B) and aspartyl protease (e.g., cathepsin D), which are inhibited by leupeptin and pepstatin, respectively. Measurements of cathepsin activities have been described previously.

To monitor autophagic protein degradation, several experimental approaches can be used: Cultured cells are treated with rapamycin (10 n*M*) for 3 h to induce autophagy. The cells are then further incubated with normal culture medium for an additional 3 h. LC3-II levels are then measured by Western blot analysis or LC3-positive puncta quantified from immunostained specimens. If autophagic/lysosomal degradation is impaired, the rate of recovery reflected by the restoration of LC3-II levels to their basal level is slowed compared to that in control cells.

A second way to distinguish a change in rates of autophagy induction/ autophagosome formation from an impaired clearance/degradation of autophagosomes involves treatment with lysosomal enzyme inhibitors. Cultured cells are treated with cathepsin inhibitors [i.e., leupeptin (20 μM) and pepstatin A (20 μM), and E64d (10 μM, Sigma, E8640)] for 6 h prior to harvesting the cells. A rise in the LC3-II level or LC3-positive puncta is expected due to the impaired clearance caused by protease inhibition, especially in neurons where clearance of constitutively formed autophagosomes is normally very efficient. Autophagy induction (e.g., following rapamycin treatment) in the presence of protease inhibitors will further increase LC3-II levels. However, if AV clearance is impaired in the untreated cells, the rise in LC3 levels under these conditions will be small or reduced relative to that in control cells.

After the preceding experimental treatments, the LC3-II level is monitored by Western blot analysis. Cells are harvested and cell extracts (20 μg of protein) in 2x SDS sample buffer are loaded onto 16% Tris-glycine gels and detected with anti-LC3 rabbit polyclonal antibody (in house, 1:2,000)

(Yu *et al.*, 2005) or anti-LC3 mouse monoclonal antibody (1:250) (Boland *et al.*, 2008). After chemiluminescent development, LC3-II levels are analyzed by Quantity One (Bio-Rad) image analysis software.

As an alternative to assessing LC3-II turnover, p62/SQSTM1 degradation may also be used (Bjorkoy *et al.*, 2005) to evaluate impairments of autophagic protein degradation (see also the chapter by Bjørkøy *et al.*, in this volume). p62 is a ubiquitin binding protein and a common component of protein aggregates found in various neurodegeneration conditions (Kuusisto *et al.*, 2002; Nagaoka *et al.*, 2004; Zatloukal *et al.*, 2002). When autophagy is genetically ablated, neurons accumulate ubiquitinated protein aggregates (Hara *et al.*, 2006; Komatsu *et al.*, 2006). An important aspect of p62 is that it plays a role in the formation of inclusion bodies that are degraded by autophagy. For this reason, p62 increase suggests inhibition of autophagic/lysosomal degradation (Bjorkoy *et al.*, 2005). The procedures used in this analysis are the same as those used with LC3. In this case, p62 is detected with anti-p62 rabbit pAb (for human, 1/2000, BD Transduction Laboratories, 610832) or anti-p62 guinea pig pAb (for mouse, 1/1000, Progen Biotechnic, GP62). It should be noted that p62 is also degraded by the proteasome, so it is not specific to autophagy.

REFERENCES

Ahlberg, J., Henell, F., and Glaumann, H. (1982). Proteolysis in isolated autophagic vacuoles from rat liver. Effect of pH and of proteolytic inhibitors. *Exp. Cell Res.* **142,** 373–383.

Anderson, R. G. W., Falck, J. R., Goldstein, J. L., and Brown, M. S. (1984). Visualization of acidic organelles in intact cells by electron microscopy. *Proc. Natl. Acad. Sci. USA* **81,** 4838–4842.

Aoki, M., Blazek, E., and Vogt, P. K. (2001). A role of the kinase mTOR in cellular transformation induced by the oncoproteins P3k and Akt. *Proc. Natl. Acad. Sci. USA* **98,** 136–141.

Auteri, J. S., Okada, A., Bochaki, V., and Dice, J. F. (1983). Regulation of intracellular protein degradation in IMR-90 human diploid fibroblasts. *J. Cell Physiol.* **115,** 167–174.

Bjørkøy, G., Lamark, T., Brech, A., Outzen, H., Perander, M., Øvervatn, A., Stenmark, H., and Johansen, T. (2005). p62/SQSTM1 forms protein aggregates degraded by autophagy and has a protective effect on huntingtin-induced cell death. *J. Cell Biol.* **171,** 603–614.

Boland, B., Kumar, A., Lee, S., Platt, F. M., Wegiel, J., Yu, W. H., and Nixon, R. A. (2008). Autophagy induction and autophagosome clearance in neurons: Relationship to autophagic pathology in Alzheimer's disease. *J. Neurosci.* **28,** 6926–6937.

Braak, H., and Braak, E. (1991). Neuropathological stageing of Alzheimer-related changes. *Acta. Neuropathol.* **82,** 239–259.

Broadwell, R. D., and Brightman, M. W. (1983). Horseradish peroxidase: A tool for study of the neuroendocrine cell and other peptide-secreting cells. *Methods Enzymol.* **103,** 187–218.

Broadwell, R. D., Oliver, C., and Brightman, M. W. (1980). Neuronal transport of acid hydrolases and peroxidase within the lysosomal system or organelles: Involvement of agranular reticulum-like cisterns. *J. Comp. Neurol.* **190,** 519–532.

Cataldo, A. M., Barnett, J. L., Pieroni, C., and Nixon, R. A. (1997). Increased neuronal endocytosis and protease delivery to early endosomes in sporadic Alzheimer's disease: Neuropathologic evidence for a mechanism of increased β-amyloidogenesis. *J. Neurosci.* **17,** 6142–6151.
Cataldo, A. M., Hamilton, D. J., Barnett, J. L., Paskevich, P. A., and Nixon, R. A. (1996). Properties of the endosomal-lysosomal system in the human central nervous system: Disturbances mark most neurons in populations at risk to degenerate in Alzheimer's disease. *J. Neurosci.* **16,** 186–199.
Cataldo, A. M., Mathews, P. M., Boiteau, A. B., Hassinger, L. C., Peterhoff, C. M., Jiang, Y., Mullaney, K., Neve, R. L., Gruenberg, J., and Nixon, R. A. (2008). Down syndrome fibroblast model of Alzheimer-related endosome pathology. Accelerated endocytosis promotes late endocytic defects. *Am. J. Pathol.* **173,** 370–384.
Cataldo, A. M., and Nixon, R. A. (1990). Enzymatically active lysosomal proteases are associated with amyloid deposits in Alzheimer brain. *Proc. Natl. Acad. Sci. USA* **87,** 3861–3865.
Cataldo, A. M., Paskevich, P. A., Kominami, E., and Nixon, R. A. (1991). Lysosomal hydrolases of different classes are abnormally distributed in brains of patients with Alzheimer disease. *Proc. Natl. Acad. Sci. USA* **88,** 10998–11002.
Cataldo, A. M., Petanceska, S., Peterhoff, C. M., Terio, N. B., Epstein, C. J., Villar, A., Carlson, E. J., Staufenbiel, M., and Nixon, R. A. (2003). App gene dosage modulates endosomal abnormalities of Alzheimer's disease in a segmental trisomy 16 mouse model of down syndrome. *J. Neurosci.* **23,** 6788–6792.
Cataldo, A. M., Petanceska, S., Terio, N. B., Peterhoff, C. M., Durham, R., Mercken, M., Mehta, P. D., Buxbaum, J., Haroutunian, V., and Nixon, R. A. (2004a). Aβ localization in abnormal endosomes: Association with earliest Abeta elevations in AD and Down syndrome. *Neurobiol. Aging* **25,** 1263–1272.
Cataldo, A. M., Peterhoff, C. M., Schmidt, S. D., Terio, N. B., Duff, K., Beard, M., Mathews, P. M., and Nixon, R. A. (2004b). Presenilin mutations in familial Alzheimer disease and transgenic mouse models accelerate neuronal lysosomal pathology. *J. Neuropathol. Exp. Neurol.* **63,** 821–830.
Cataldo, A. M., Peterhoff, C. M., Troncoso, J. C., Gomez-Isla, T., Hyman, B. T., and Nixon, R. A. (2000). Endocytic pathway abnormalities precede amyloid β deposition in sporadic Alzheimer's disease and Down syndrome: Differential effects of APOE genotype and presenilin mutations. *Am. J. Pathol.* **157,** 277–286.
Cataldo, A. M., Thayer, C. Y., Bird, E. D., Wheelock, T. R., and Nixon, R. A. (1990). Lysosomal proteinase antigens are prominently localized within senile plaques of Alzheimer's disease: evidence for a neuronal origin. *Brain. Res.* **513,** 181–192.
Chen, C. S., Chen, W. N., Zhou, M., Arttamangkul, S., and Haugland, R. P. (2000). Probing the cathepsin D using a BODIPY FL-pepstatin A: Applications in fluorescence polarization and microscopy. *J. Biochem. Biophys. Methods* **42,** 137–151.
Chishti, M. A., Yang, D. S., Janus, C., Phinney, A. L., Horne, P., Pearson, J., Strome, R., Zuker, N., Loukides, J., French, J., Turner, S., Lozza, G., *et al.* (2001). Early-onset amyloid deposition and cognitive deficits in transgenic mice expressing a double mutant form of amyloid precursor protein 695. *J. Biol. Chem.* **276,** 21562–21570.
Cuervo, A.M (2004). Autophagy: Many paths to the same end. *Mol. Cell Biochem.* **263,** 55–72.
Cuervo, A. M., Stefanis, L., Fredenburg, R., Lansbury, P. T., and Sulzer, D. (2004). Impaired degradation of mutant α-synuclein by chaperone-mediated autophagy. *Science* **305,** 1292–1295.
Ding, Q., Dimayuga, E., Martin, S., Bruce-Keller, A. J., Nukala, V., Cuervo, A. M., and Keller, J. N. (2003). Characterization of chronic low-level proteasome inhibition on neural homeostasis. *J. Neurochem.* **86,** 489–497.

Donoviel, D. B., Hadjantonakis, A.-K., Ikeda, M., Zheng, H., Hyslop, P. S. G., and Bernstein, A. (1999). Mice lacking both presenilin genes exhibitearlyembryonic patterningdefects. *Genes Dev.* **13,** 2801–2810.

Dottavio-Martin, D., and Ravel, J. (1978). Radiolabeling of proteins by reductive alkylation with [^{14}C]formaldehyde and sodium cyanoborohydride. *Anal Biochem.* **87,** 562–565.

Dunn, W. A., Jr. (1990). Studies on the mechanisms of autophagy: Maturation of the autophagic vacuole. *J. Cell Biol.* **110,** 1935–1945.

Duyckaerts, C., Potier, M. C., and Delatour, B. (2008). Alzheimer disease models and human neuropathology: Similarities and differences. *Acta Neuropathol.* **115,** 5–38.

Eskelinen, E.L (2008). Fine structure of the autophagosome. *Methods Mol. Biol.* **445,** 11–28.

Gulve, E. A., and Dice, J. F. (1989). Regulation of protein synthesis and degradation in L8 myotubes. Effects of serum, insulin and insulin-like growth factors. *Biochem. J.* **260,** 377–387.

Hara, T., Nakamura, K., Matsui, M., Yamamoto, A., Nakahara, Y., Suzuki-Migishima, R., Yokoyama, M., Mishima, K., Saito, I., Okano, H., and Mizushima, N. (2006). Suppression of basal autophagy in neural cells causes neurodegenerative disease in mice. *Nature* **441,** 885–889.

Hsiao, K., Chapman, P., Nilsen, S., Eckman, C., Harigaya, Y., Younkin, S., Yang, F., and Cole, G. (1996). Correlative memory deficits, Aβ elevation, and amyloid plaques in transgenic mice. *Science* **274,** 99–102.

Hsu, S. M., Raine, L., and Fanger, H. (1981). Use of avidin-biotin-peroxidase complex (ABC) in immunoperoxidase techniques: A comparison between ABC and unlabeled antibody (PAP) procedures. *J. Histochem. Cytochem.* **29,** 577–580.

Hyman, B. T., and Trojanowski, J. Q. (1997). Consensus recommendations for the post-mortem diagnosis of Alzheimer disease from the National Institute on Aging and the Reagan Institute Working Group on diagnostic criteria for the neuropathological assessment of Alzheimer disease. *J. Neuropathol. Exp. Neurol.* **56,** 1095–1097.

Kabeya, Y., Mizushima, N., Ueno, T., Yamamoto, A., Kirisako, T., Noda, T., Kominami, E., Ohsumi, Y., and Yoshimori, T. (2000). LC3, a mammalian homologue of yeast Apg8p, is localized in autophagosome membranes after processing. *EMBO J.* **19,** 5720–5728.

Khachaturian, Z.S (1985). Diagnosis of Alzheimer's disease. *Arch. Neurol* **42,** 1097–1105.

Kimura, S., Noda, T., and Yoshimori, T. (2007). Dissection of the autophagosome maturation process by a novel reporter protein, tandem fluorescent-tagged LC3. *Autophagy* **3,** 452–460.

Köchl, R., Hu, X. W., Chan, E. Y. W., and Tooze, S. A. (2006). Microtubules facilitate autophagosome formation and fusion of autophagosomes with endosomes. *Traffic.* **7,** 129–145.

Kokkonen, N., Rivinoja, A., Kauppila, A., Suokas, M., Kellokumpu, I., and Kellokumpu, S. (2004). Defective acidification of intracellular organelles results in aberrant secretion of cathepsin D in cancer cells. *J. Biol. Chem.* **279,** 39982–39988.

Komatsu, M., Waguri, S., Chiba, T., Murata, S., Iwata, J., Tanida, I., Ueno, T., Koike, M., Uchiyama, Y., Kominami, E., and Tanaka, K. (2006). Loss of autophagy in the central nervous system causes neurodegeneration in mice. *Nature* **441,** 880–884.

Kuusisto, E., Salminen, A., and Alafuzoff, I. (2002). Early accumulation of p62 in neurofibrillary tangles in Alzheimer's disease: Possible role in tangle formation. *Neuropathol. Appl. Neurobiol.* **28,** 228–237.

Lai, M. T., Chen, E., Crouthamel, M.-C., DiMuzio-Mower, J., Xu, M., Huang, Q., Price, E., Register, R. B., Shi, X.-P., Donoviel, D. B., Bernstein, A., Hazuda, D., *et al.* (2003). Presenilin-1 and presenilin-2 exhibit distinct yet overlapping γ-secretase activities. *J. Biol. Chem.* **278,** 22475–22481.

Levine, B., and Kroemer, G. (2008). Autophagy in the pathogenesis of disease. *Cell* **132,** 27–42.

Marks, N., and Berg, M. J. (1987). Rat brain cathepsin L: Characterization and differentiation from cathepsin B utilizing opioid peptides. *Arch. Biochem. Biophys.* **259,** 131–143.

Marzella, L., Ahlberg, J., and Glaumann, H. (1982). Isolation of autophagic vacuoles from rat liver: Morphological and biochemical characterization. *J. Cell Biol.* **93,** 144–154.

Massey, A. C., Kaushik, S., Sovak, G., Kiffin, R., and Cuervo, A. M. (2006). Consequences of the selective blockage of chaperone-mediated autophagy. *Proc. Natl. Acad. Sci. USA* **103,** 5805–5810.

Mirra, S. S., Hart, M. N., and Terry, R. D. (1993). Making the diagnosis of Alzheimer's disease. A primer for practicing pathologists. *Arch. Pathol. Lab. Med.* **117,** 132–144.

Mirra, S. S., Heyman, A., McKeel, D., Sumi, S. M., Crain, B. J., Brownlee, L. M., Vogel, F. S., Hughes, J. P., van Belle, G., and Berg, L. (1991). The Consortium to Establish a Registry for Alzheimer's Disease (CERAD). Part II. Standardization of the neuropathologic assessment of Alzheimer's disease. *Neurology* **41,** 479–486.

Mizushima, N., Yamamoto, A., Matsui, M., Yoshimori, T., and Ohsumi, Y. (2004). *In vivo* analysis of autophagy in response to nutrient starvation using transgenic mice expressing a fluorescent autophagosome marker. *Mol. Biol. Cell* **15,** 1101–1111.

Mizushima, N., Levine, B., Cuervo, A. M., and Klionsky, D. J. (2008). Autophagy fights disease through cellular self-digestion. *Nature* **451,** 1069–1075.

Mizushima, N., and Yoshimori, T. (2007). How to interpret LC3 immunoblotting. *Autophagy* **3,** 542–545.

Moreira, P. I., Siedlak, S. L., Wang, X., Santos, M. S., Oliveira, C. R., Tabaton, M., Nunomura, A., Szweda, L. I., Aliev, G., Smith, M. A., Zhu, X., and Perry, G. (2007). Autophagocytosis of mitochondria is prominent in Alzheimer disease. *J. Neuropathol. Exp. Neurol.* **66,** 525–532.

Nagaoka, U., Kim, K., Jana, N. R., Doi, H., Maruyama, M., Mitsui, K., Oyama, F., and Nukina, N. (2004). Increased expression of p62 in expanded polyglutamine-expressing cells and its association with polyglutamine inclusions. *J. Neurochem.* **91,** 57–68.

Nixon, R. A., Wegiel, J., Kumar, A., Yu, W. H., Peterhoff, C., Cataldo, A., and Cuervo, A. M. (2005). Extensive involvement of autophagy in Alzheimer disease: An immuno-electron microscopy study. *J. Neuropathol. Exp. Neurol.* **64,** 113–122.

Ohsumi, Y (2001). Molecular dissection of autophagy: Two ubiquitin-like systems. *Nat. Rev. Mol. Cell Biol.* **2,** 211–216.

Pankiv, S., Clausen, T. H., Lamark, T., Brech, A., Bruun, J.-A., Outzen, H., Øvervatn, A., Bjørkøy, G., and Johansen, T. (2007). p62/SQSTM1 binds directly to Atg8/LC3 to facilitate degradation of ubiquitinated protein aggregates by autophagy. *J. Biol. Chem.* **282,** 24131–24145.

Pasquali, C., Fialka, I., and Huber, L. A. (1999). Subcellular fractionation, electromigration analysis and mapping of organelles. *J. Chromatogr. B. Biomed. Sci. Appl.* **722,** 89–102.

Pickford, F., Masliah, E., Britschgi, M., Lucin, K., Narasimhan, R., Jaeger, P. A., Small, S., Spencer, B., Rockenstein, E., Levine, B., and Wyss-Coray, T. (2008). The autophagy-related protein beclin 1 shows reduced expression in early Alzheimer disease and regulates amyloid β accumulation in mice. *J. Clin. Invest.* **118,** 2190–2199.

Radde, R., Duma, C., Goedert, M., and Jucker, M. (2008). The value of incomplete mouse models of Alzheimer's disease. *Eur. J. Nucl. Med. Mol. Imaging* **35**(Suppl 1), S70–S74.

Sarbassov, D. D., Ali, S. M., and Sabatini, D. M. (2005). Growing roles for the mTOR pathway. *Curr. Opin. Cell Biol.* **17,** 596–603.

Schmidt, S. D., Jiang, Y., Nixon, R. A., and Mathews, P. M. (2005). Tissue processing prior to protein analysis and amyloid-β quantitation. *Methods Mol. Biol.* **299,** 267–278.

Seglen, P. O., and Gordon, P. B. (1984). Amino acid control of autophagic sequestration and protein degradation in isolated rat hepatocytes. *J. Cell Biol.* **99,** 435–444.

Shiurba, R. A., Spooner, E. T., Ishiguro, K., Takahashi, M., Yoshida, R., Wheelock, T. R., Imahori, K., Cataldo, A. M., and Nixon, R. A. (1998). Immunocytochemistry of formalin-fixed human brain tissues: Microwave irradiation of free-floating sections. *Brain Res. Brain Res. Protoc.* **2,** 109–119.

Suzuki, K., Kirisako, T., Kamada, Y., Mizushima, N., Noda, T., and Ohsumi, Y. (2001). The preautophagosomal structure organized by concerted functions of *APG* genes is essential for autophagosome formation. *EMBO J.* **20,** 5971–5981.

Talbot, K., Cho, D. S., Ong, W. Y., Benson, M. A., Han, L. Y., Kazi, H. A., Kamins, J., Hahn, C. G., Blake, D. J., and Arnold, S. E. (2006). Dysbindin-1 is a synaptic and microtubular protein that binds brain snapin. *Hum. Mol. Genet.* **15,** 3041–3054.

Teclemariam-Mesbah, R., Wortel, J., Romijn, H. J., and Buijs, R. M. (1997). A simple silver-gold intensification procedure for double DAB labeling studies in electron microscopy. *J. Histochem. Cytochem.* **45,** 619–621.

Uchihara, T (2007). Silver diagnosis in neuropathology0000 principles, practice and revised interpretation. *Acta Neuropathol.* **113,** 483–499.

Via, L. E., Fratti, R. A., McFalone, M., Pagan-Ramos, E., Deretic, D., and Deretic, V. (1998). Effects of cytokines on mycobacterial phagosome maturation. *J. Cell Sci.* **111,** 897–905.

Yamamoto, T., and Hirano, A. (1986). A comparative study of modified Bielschowsky, Bodian and thioflavin S stains on Alzheimer's neurofibrillary tangles. *Neuropathol. Appl. Neurobiol.* **12,** 3–9.

Yamamoto, A., Cremona, M. L., and Rothman, J. E. (2006). Autophagy-mediated clearance of huntingtin aggregates triggered by the insulin-signaling pathway. *J. Cell Biol.* **172,** 719–731.

Yang, D. S., Kumar, A., Stavrides, P., Peterson, J., Peterhoff, C. M., Pawlik, M., Levy, E., Cataldo, A. M., and Nixon, R. A. (2008). Neuronal apoptosis and autophagy crosstalk in aging PS/APP mice, a model of Alzheimer's disease. *Am. J. Pathol.* **173,** 665–681.

Yang, Y., Fukui, K., Koike, T., and Zheng, X. (2007). Induction of autophagy in neurite degeneration of mouse superior cervical ganglion neurons. *Eur. J. Neurosci.* **26,** 2979–2988.

Yu, W. H., Cuervo, A. M., Kumar, A., Peterhoff, C. M., Schmidt, S. D., Lee, J.-H., Mohan, P. S., Mercken, M., Farmery, M. R., Tjernberg, L. O., Jiang, Y., Duff, K., *et al.* (2005). Macroautophagy: a novel β-amyloid peptide-generating pathway activated in Alzheimer's disease. *J. Cell Biol.* **171,** 87–98.

Zatloukal, K., Stumptner, C., Fuchsbichler, A., Heid, H., Schnoelzer, M., Kenner, L., Kleinert, R., Prinz, M., Aguzzi, A., and Denk, H. (2002). p62 Is a common component of cytoplasmic inclusions in protein aggregation diseases. *Am. J. Pathol.* **160,** 255–263.

CHAPTER SEVEN

Live-Cell Imaging of Autophagy Induction and Autophagosome-Lysosome Fusion in Primary Cultured Neurons

Mona Bains* *and* Kim A. Heidenreich†

Contents

Abstract

The discovery that impaired autophagy is linked to a wide variety of prominent diseases including cancer and neurodegeneration has led to an explosion of research in this area. Methodologies that allow investigators to observe and quantify the autophagic process will clearly advance knowledge of how this process contributes to the pathophysiology of many clinical disorders. The recent identification of essential autophagy genes in higher eukaryotes has made it possible to analyze autophagy in mammalian cells that express autophagy proteins tagged with fluorescent markers. This chapter describes such

* Department of Pharmacology, University of Colorado at Denver, Anchutz Medical Campus, Aurora, Colorado, USA

† VA Research Service, Eastern Colorado Health Care System, Denver, Colorado, USA

Methods in Enzymology, Volume 453

ISSN 0076-6879, DOI: 10.1016/S0076-6879(08)04007-X

methods using primary cultured neurons that undergo up-regulation of autophagy when trophic factors are removed from their medium. The prolonged up-regulated autophagy, in turn, contributes to the death of these neurons, thus providing a model to examine the relationship between enhanced autophagy and cell death.

Neurons are isolated from the cerebellum of postnatal day 7 rat pups and cultured in the presence of trophic factors and depolarizing concentrations of potassium. Once established, the neurons are transfected with an adeno-viral vector expressing MAP1-LC3 with red fluorescent protein (RFP). MAP1-LC3 is the mammalian homolog of the yeast autophagosomal marker Atg8 and when tagged to GFP or RFP, it is the most widely used marker for autophagosomes. Once expression is stable, autophagy is induced by removing trophic factors. At various time points after inducing autophagy, the neurons are stained with LysoSensor Green (a pH-dependent lysosome marker) and Hoechst (a DNA marker) and subjected to live-cell imaging. In some cases, time-lapse imaging is used to examine the stepwise process of autophagy in live neurons.

1. Introduction

Autophagy is a regulated, catabolic pathway for the turnover of long-lived proteins, macromolecular aggregates, and damaged organelles by lysosomal degradation (Levine and Klionsky, 2004). It also plays a role in clearing the cell of invading bacteria and viruses (Levine and Deretic, 2007; Lee and Iwasaki, 2008). In mammalian cells, the lysosomal pathway of intracellular degradation is divided into three distinct pathways: macroautophagy, microautophagy, and chaperone-mediated autophagy (Cuervo, 2004). Macroautophagy (Mizushima, 2007; Xie and Klionsky, 2007) begins with the formation of a unique double-membrane vesicle (autophagosome) that engulfs cytoplasmic constituents such as proteins, lipids, and damaged organelles, including mitochondria. The outer membrane of the autophagosome then docks and fuses with the lysosome to deliver the sequestered cargo. The inner membrane of the fused vesicle (autolysosome), along with the interior contents of the autophagosome are degraded by lysosomal hydrolases, a process that generates nucleotides, amino acids, and free fatty acids that are recycled to provide raw materials and energy to the cell. Microautophagy (Klionsky *et al.*, 2007) circumvents the autophagosome sequestration step and begins with the direct uptake of cytosolic components by invaginations and pinching off of the lysosomal membrane. As in macroautophagy, the internalized cytosolic components are digested by lysosomal enzymes and the macromolecules are released when the vacuolar membrane disintegrates. In chaperone-mediated autophagy (Majeski and Dice, 2004; Massey *et al.*, 2006), specific chaperone proteins bind to targeted proteins containing a KFERQ sequence and direct these proteins to the surface of the lysosome.

These proteins bind to LAMP-2A on lysosomal membranes and are then transported across the membrane with the assistance of chaperone proteins where they are degraded by the lysosomal proteases.

In eukaryotic cells, macroautophagy (hereon referred to as autophagy) occurs constitutively at low levels in all cells to perform housekeeping functions such as degradation of proteins and destruction of dysfunctional organelles. Dramatic up-regulation of autophagy occurs in the presence of external stressors such as starvation, hormonal imbalance, oxidation, extreme temperature, and infection, as well as internal needs such as cellular remodeling and removal of protein aggregates (Levine and Kroemer, 2008). Mediators of phosphoinositide-3 (PI3) kinase (Type III) signaling pathways and trimeric G proteins play major roles in regulating the formation of autophagosomes (Backer, 2008). In addition, the Type I PI3K/Akt signaling pathway modulate autophagy in response to insulin and other growth factor signals (Meijer and Codogno, 2004). Target of rapamycin (TOR) kinase, a highly conserved environmental sensor protein downstream of PI3K, is the predominant regulator of autophagy (Meijer and Codogno, 2004). The eukaryotic initiation factor 2 kinase, Gcn2, and its downstream target, Gcn4, a transcriptional transactivator of autophagy genes, induces autophagy under conditions of cellular starvation (Natarajan *et al.*, 2001; Tallóczy *et al.*, 2002). In the nervous system, autophagy is unresponsive to starvation, but is induced by removal of trophic factors and the presence of aggregate-prone cytosolic proteins, damaged organelles, or prions, factors that contribute to neurodegenerative disease (Levine and Yuan, 2005). Although autophagy can be induced as a protective mechanism for neurons to clear unwanted proteins and damaged organelles, depending on the extent and duration of the insult, it can also lead to autophagic cell death. The regulatory events that switch autophagy from a protective event to a cell death mechanism are poorly defined but important to understand if autophagic mechanisms are to be manipulated for therapeutic gain.

Recent studies in our laboratory have shown that trophic factor withdrawal (TFW) leads to the death of both granule neurons and Purkinje neurons in postnatal cerebellar cultures. Granule neurons (95%–97% of total cells) die by a classic intrinsic apoptotic pathway, whereas Purkinje neurons (3%–5% of total cells) die by an autophagy associated cell death pathway (Florez-McClure *et al.*, 2004). Thus, under the same culture conditions and the same neurotoxic stimulus, the mechanisms mediating cell death in the two types of neurons are very distinct. Why Purkinje neurons choose an autophagic mechanism to die under these experimental conditions is not clear, although this type of cell demise mimics the death of Purkinje neurons in the Lurcher mouse carrying a point mutation in the delta2 glutamate receptor (Zuo *et al.*, 1997). The Lurcher mouse demonstrates extensive Purkinje neuron degeneration followed by secondary loss of granule neurons in the cerebellum. Lurcher Purkinje neurons demonstrate ultrastructural features of autophagy at a time when they begin to degenerate and more recent biochemical studies have linked autophagy as the cause of death (Selimi *et al.*, 2003; Yue *et al.*, 2002). In this report, we describe our model system for studying autophagy in cultured Purkinje neurons and the methods that we have developed for observing and quantifying autophagy via live-cell fluorescence imaging.

2. Cultured Cerebellar Purkinje Neurons as a Model to Study Neuronal Autophagy

2.1. Primary culture

1. Primary rat cerebellar neurons are isolated from 7-day-old Sprague Dawley rat pups as described previously (D'Mello *et al.*, 1993). The day before plating, coverslips in a 24-well plate are treated with poly-D-lysine (40 μg/ml; Sigma-Aldrich, St. Louis, MO, catalog number P0899) and placed in a 37 °C (5% CO_2) incubator for 1 h. Poly-D-lysine is then aspirated and coverslips are left to dry overnight in the tissue culture hood. The following day, laminin (1 μg/mL; BD Biosciences Discovery Labware, Bedford, MA, catalog number 354232) is added to the coverslips for at least 1 h at room temperature. Prior to plating cells, the laminin is aspirated and coverslips are rinsed twice with plating media. Neurons are plated at a density of 2.0×10^6 cells/ml in basal modified Eagle's medium (BME) containing 10% fetal bovine serum, 25 m*M* KCl, 2 m*M* L-glutamine, and penicillin (100 U/ml)-streptomycin (100 μg/ml; Invitrogen, Gaithersburg, MD, catalog number 15140-122). The cultures are grown in a 5% CO_2 incubator at 37 °C.
2. Cytosine arabinoside (10 μM, Sigma-Aldrich, St. Louis, MO, catalog number C6645) is added to the culture medium 24 h after plating to limit the growth of nonneuronal cells.
3. After 5 days in culture, neurons are infected with adeno-viral vector expressing MAP1-LC3 labeled with red fluorescent protein (RFP-LC3, a kind gift from Aviva Tolkovsky, Cambridge University) at a multiplicity of infection of 100 for 24 h.
4. Autophagic cell death is induced by removing the plating medium and replacing it with TFW medium (serum-free BME containing 5 m*M* KCl).

2.2. Immunostaining

Purkinje neurons in cerebellar cultures are identified by immunostaining with a polyclonal antibody to calbindin D-28k (Chemicon, Temecula, CA, catalog number AB1778). For immunocytochemical staining experiments, we use the following protocol:

1. Neuronal cultures are plated on poly-D-lysine/laminin-coated glass coverslips at 2.0×10^6 cells/ml.
2. Following incubation in control medium or TFW media for 24–48 h, the neurons are fixed with 4% paraformaldehyde in phosphate-buffered saline (PBS) for 30 min at room temperature.
3. The neurons are then permeabilized and blocked with PBS, pH 7.4, containing 0.2% Triton X-100 and 5% BSA for 1 h at room temperature.
4. Cells are incubated with primary antibody (Calbindin D-28K, 1:250; Chemicon, Temecula, CA, catalog number AB1778) overnight at 4 °C diluted in PBS containing 0.1% Triton X-100 and 2% BSA.
5. Primary antibodies are then removed and the cells are washed at least 6 times with PBS at room temperature for at least 1 h.
6. The neurons are then incubated with the appropriate Cy3-conjugated secondary antibody (diluted 1:500) and DAPI (1 μg/ml) for 1 h at room temperature.
7. The cells are washed 6 more times with PBS at room temperature, and coverslips are adhered to glass slides with mounting medium (0.1% p-phenylenediamine in 75% glycerol in PBS). Slides are sealed with clear nail polish and stored at −20 °C prior to imaging.

2.3. Fluorescence imaging and cell counts of cultured Purkinje neurons

Imaging is performed using a Zeiss (Thornwood, NY) Axioplan 2 microscope equipped with a Cooke Sensicam deep-cooled CCD camera. Images are captured on the CY3 and DAPI channels and analyzed with the Slidebook (4.2) software program (Intelligent Imaging Innovations, Denver, CO). The number of calbindin-positive cells are counted in 152 fields using a 63X oil objective that are randomly selected by following a fixed grid pattern over the coverslip. The total area counted per coverslip is 14.6mm^2 or about 13% of the coverslip. Cells counted as Purkinje neurons have large rounded cell bodies, elaborate processes and larger, oval nuclei (in comparison to smaller, rounder nuclei of granule neurons). At least 3 coverslips are counted per experimental condition. The numbers are averaged and expressed as a percentage of the average number counted in the appropriate controls. This is repeated for at least three independent experiments. Results shown in Fig. 7.1 demonstrate that TFW induces the death of Purkinje neurons by a mechanism distinct from granule neurons that does not involve nuclear condensation but is associated with extensive cytoplasmic vacuolation. The cytoplasmic vacuoles are identified as autophagosomes and autolysosomes by monodansylcadaverine staining and electron transmission microscopy (Florez-McClure *et al.*, 2004) (see also the chapter by Vázquez and Colombo in volume 452).

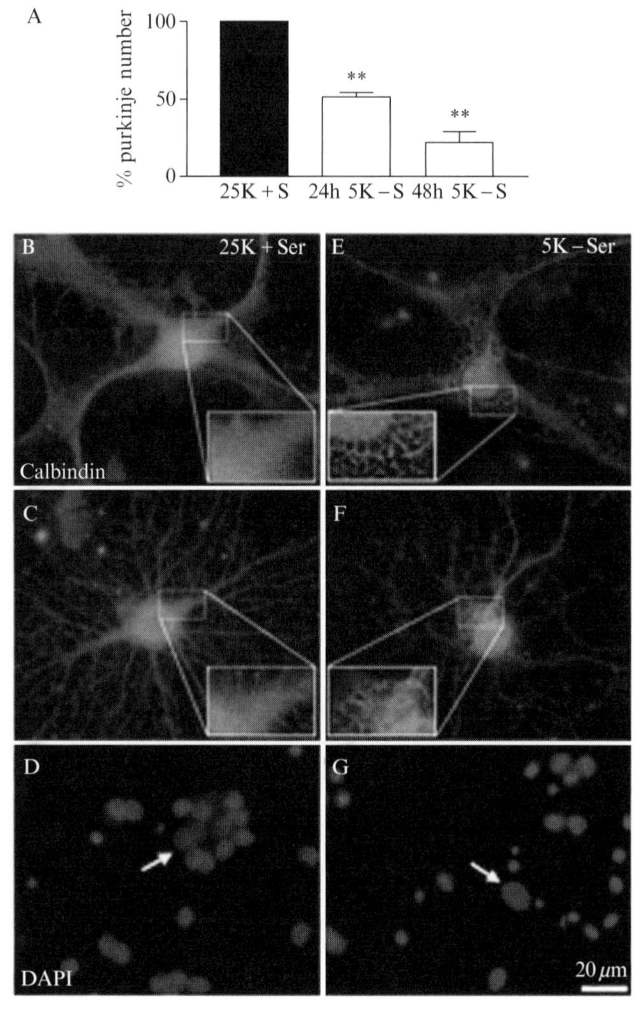

Figure 7.1 Trophic factor withdrawal induces a non-apoptotic death of Purkinje neurons. (A) Calbindin-positive Purkinje cells plotted as a percentage of control. Following 24 h or 48 h of TFW, 51.7 ± 2.2% and 22.3 ± 7.0% of Purkinje neurons survive, respectively, as compared to controls (** indicates significant difference from 25K+S control at $p < 0.01$). (B–G) Cells fixed and stained with polyclonal antibodies against Calbindin D-28K (in red) and the nuclear dye, DAPI (in blue). The remaining Purkinje neurons showed markedly different morphology than control cells, characterized by extensive cytoplasmic vacuolation (compare control cells B, C to trophic factor-deprived cells E, F). In contrast to granule neurons, which demonstrated substantial nuclear condensation and fragmentation characteristic of apoptosis (compare D to G), Purkinje neurons showed no obvious signs of nuclear condensation or fragmentation (compare nuclei indicated by arrows in D and G). This figure was previously published in Florez-McClure *et al.*, (2004), and is reprinted with permission from the Society for Neuroscience, *J. Neuroscience.* (See Color Insert.)

3. Characterization of Autophagic Vacuole Size and Number in Purkinje Neurons

To visualize vacuole size and number, we use the following protocol:

1. Cerebellar neurons are cultured and infected with RFP-LC3 as described earlier.
2. 24 h later, autophagic cell death is induced by removing the plating medium and replacing it with either control medium or TFW medium for various time points (0–24 h).
3. After the appropriate treatment times, cells are stained at 37 °C with Hoechst (20 ng/ml) for 15 min to visualize cellular nuclei, and LysoSensor Green (2 μM) for 5 min to visualize lysosomes.
4. The cells are then washed three times for 5 min with 37 °C phenol red-free control medium (Dulbecco's Modified Eagle Medium (D-MEM) #31053 containing 10% fetal bovine serum, 25m*M* KCl, 2 m*M* L-glutamine, and penicillin (100 U/ml)-streptomycin (100 μg/ml; Invitrogen, Gaithersburg, MD) or TFW medium (serum free D-MEM containing 5 m*M* KCl) to remove nonspecifically bound dyes.
5. The individually timed coverslips to be imaged are attached to a 35-mm dish using Vaseline petroleum jelly and submerged in 3 mL of the appropriate phenol red–free medium.
6. The dish is then fitted into a prewarmed 37 °C heated stage, which maintains the temperature at 37 °C. Imaging is performed as described previously. Images of LysoSensor Green, RFP-LC3, and Hoechst fluorescence are captured on the FITC, Cy3 and DAPI channels, respectively, using the 63X water immersion objective. The length of exposure on the DAPI channel is calibrated by the Slidebook program (Slidebook 4.2, Intelligent Imaging Innovations, Denver, CO). Collected images are used to measure the diameters of all the visible RFP-positive and LysoSensor Green–positive vacuoles in at least seven Purkinje neurons per condition using the Slidebook ruler tool as described in further detail in Fig. 7.2.

Data collected from vacuole size measurements are used to create an autophagic vacuole size distribution profile (from at least 3 independent experiments run in duplicate). The autophagosome size distribution profile of Purkinje neurons exposed to 0, 6, and 24 h of TFW is shown in Fig. 7.3A. In healthy control Purkinje neurons, approximately 90% of RFP-LC3-positive vacuoles are smaller than 0.75 μm in diameter. Following 6 h of TFW, there is a significant shift in the size distribution resulting in a decrease of smaller vacuoles (<0.75 μm) and increase of medium-size autophagosomes (0.75–1.5 μm) (Fig. 7.3, $p^{**} < 0.01$ and $p^{\#\#\#} < 0.001$

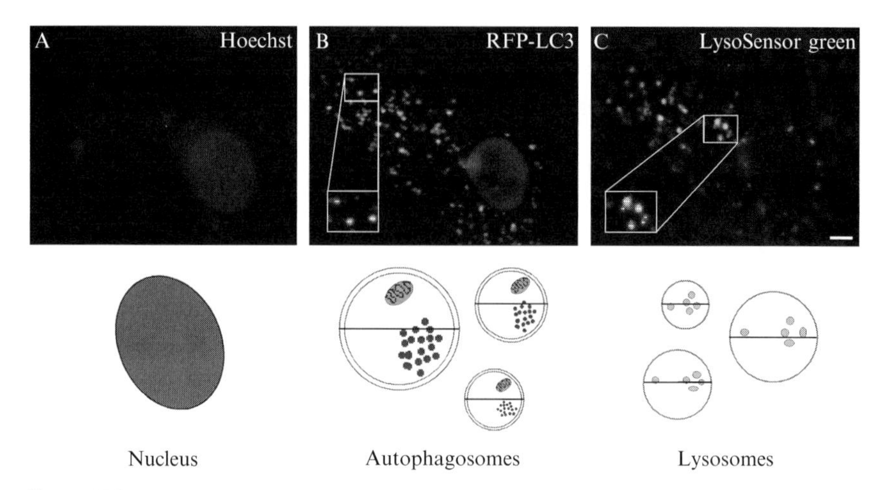

Figure 7.2 Schematic for measuring the size of autophagy-positive vacuoles. Autophagosome and lysosome vacuole size are quantified by measuring the diameters of all visible RFP-LC3- and LysoSensor Green–positive vacuoles. At least 7 images are collected via live cell microscopy per treatment. Using the Slidebook magnification tool, images are increased in size so that the diameter (μm) of each vacuole can be easily measured using the ruler tool. The diameters of all visible vacuoles in 7–12 Purkinje neurons per treatment are measured, which reflects diameter measurements of approximately 200 vacuoles per treatment. The total number of vacuoles are then counted and categorized as small ($<$0.75), medium (0.75–1.5), large (1.5–2.25), or extra large ($>$2.25) based on their diameter, and the size distribution is graphed as percent of total vacuoles within the indicated size ranges. Images shown represent a 6-h treatment of cerebellar cultures in TFW medium. Scale bar represents 5 μm. (See Color Insert.)

compared to their respective controls). This shift in autophagosome size during TFW was also observed in lysosomes as determined with lysosomal diameter measurements (Fig. 7.3B). The increase in the size of autophagic and lysosomal vacuoles present in dying Purkinje neurons is negatively correlated with survival (Florez-McClure *et al.*, 2004). Furthermore, addition of 3-methyladenine at the time of TFW blocks the change in vacuolization size and decreases cell death (Florez-McClure *et al.*, 2004). The addition of rapamycin, the specific inhibitor for mTOR, induces the formation of large autophagosomes (data not shown).

Measuring the vesicle size profile is a good indicator of autophagy induction. Activation of autophagy by TFW results in the formation of vesicles greater than 0.75 μm with significant increases in vesicles between 0.75μm and 1.5 μm. It is important to keep in mind that the measurements are not limited to autophagosomes, as RFP-LC3-positive vacuoles may also reflect autolysosomes that have not yet degraded LC3. However, this is not an accurate representation of the total autophagosome population as these

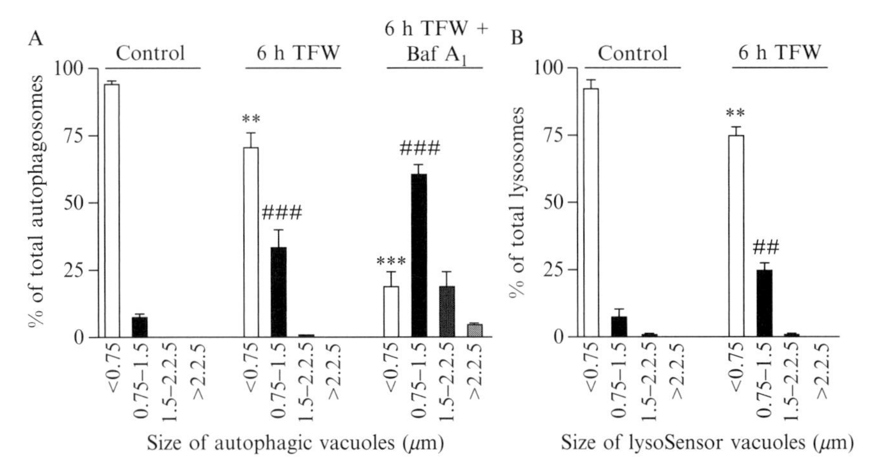

Figure 7.3 Accumulation of autophagic vacuoles in Purkinje neurons. Purkinje neurons were maintained in control medium or TFW medium for 24 hr in the absence and presence of bafilomycin A_1. Autophagosome and lysosome vacuole size were quantified by measuring the diameters of all RFP-LC3-positive vacuoles (A) and LysoSensor Green-positive vacuoles (B) in 7–12 Purkinje neurons per treatment. The size distribution was graphed as percent of total vacuoles within the indicated size ranges. For bafilomycin A_1–treated conditions, the total number of vacuoles per Purkinje neuron was also determined. $^{**}p < 0.01$ and $^{***}p < 0.001$ compared to <0.75 μm control, ##$p < 0.01$ and ###$p < 0.001$ compared to 0.75–1.5 μm control. one-way ANOVA, Tukey's *post hoc* test.

vesicles are continuously being degraded. Total autophagosome numbers can be obtained in the presence of lysosomal protease inhibitors or drugs that block autophagosome degradation such as, bafilomycin A_1, a vacuolar ATPase inhibitor that disrupts lysosomal acidification (Fass *et al.*, 2006). Comparing the vesicle size distribution in the absence and presence of bafilomycin A_1 reveals a further increase in larger vesicles greater than 1.5 μm, which under normal autophagy conditions are rapidly degraded (Fig. 7.3A; see 6-h TFW +Baf A1). The inclusion of bafilomycin A_1 is therefore necessary when quantifying the total number of autophagic vacuoles formed per treatment condition and when making any inferences on autophagic flux.

Changes in autophagic flux between two treatments can also be determined by measuring the vesicle size distribution profile in the absence and presence of bafilomycin A_1. For example, if the total numbers of autophagic vacuoles between two treatment conditions (in the presence of bafilomycin A_1) are similar, but (in the absence of bafilomycin A_1) the large vesicles (>1.5 μm) in one condition are significantly reduced, this could indicate an increase in autophagic turnover as the larger autolysosomes are being more rapidly degraded. However, if the total number of vesicles between each treatment condition is significantly different in the presence of bafilomycin A_1, this would indicate a difference in the formation rate of autophagic vacuoles between the two treatments.

4. Using Colocalization of Fluorescent Tags to Measure Autophagosome-Lysosome Fusion

A distal step in the autophagic pathway is the fusion of the autophagosome with the lysosome and the degradation of both the bulk cytoplasm within the autophagosome and the autophagosome itself. In the described Purkinje neuron autophagy model, fusion can be visualized via live cell imaging as the colocalization of RFP-LC3 with LysoSensor Green.

1. Cerebellar cultures are infected with RFP-LC3 (multiplicity of infection 100, 24 h) and subjected to a trophic factor withdrawal time course (0–24 h).
2. Cells are stained with Hoechst and LysoSensor Green and imaged via live cell imaging as described previously.
3. Quantitation of fluorescence signals from Purkinje neurons is assessed using Slidebook (4.2).

The degree of colocalization between RFP-LC3-positive vacuoles and LysoSensor Green–positive vacuoles is quantified using the Pearson's correlation analysis tool in Slidebook 4.2, which is described in Fig. 7.4 A–D. Pearson's correlation calculates the correlation between the intensity distributions of Cy3 (RFP-LC3) and FITC (LysoSensor Green) and expresses the correlation as a *r*-value between the range of −1 and +1, indicating no colocalization to perfect colocalization, respectively (Manders *et al.*, 1993). The correlation of signal intensities between two channels (Cy3 and FITC) is calculated using the equation below where R_i is the intensity in channel 1 for pixel I, G_i is the intensity in channel 2 for the same pixel, and R_{av} and G_{av} are the average or mean intensity values over all pixels (Slidebook 4.2).

$$r = \frac{\sum_i (R_i - R_{av})\cdot(G_i - G_{av})}{\left\{\sum_i (R_i - R_{av})^2 \cdot \sum_i i(G_i - G_{av})^2\right\}^{\frac{1}{2}}}$$

As shown in Fig. 7.4E, trophic factor withdrawal results in a time-dependent increase in autophagosome-to-lysosome fusion, indicated by the increase in the Pearson's coefficient (r) from 0.15 to 0.75 over a 24-h period.

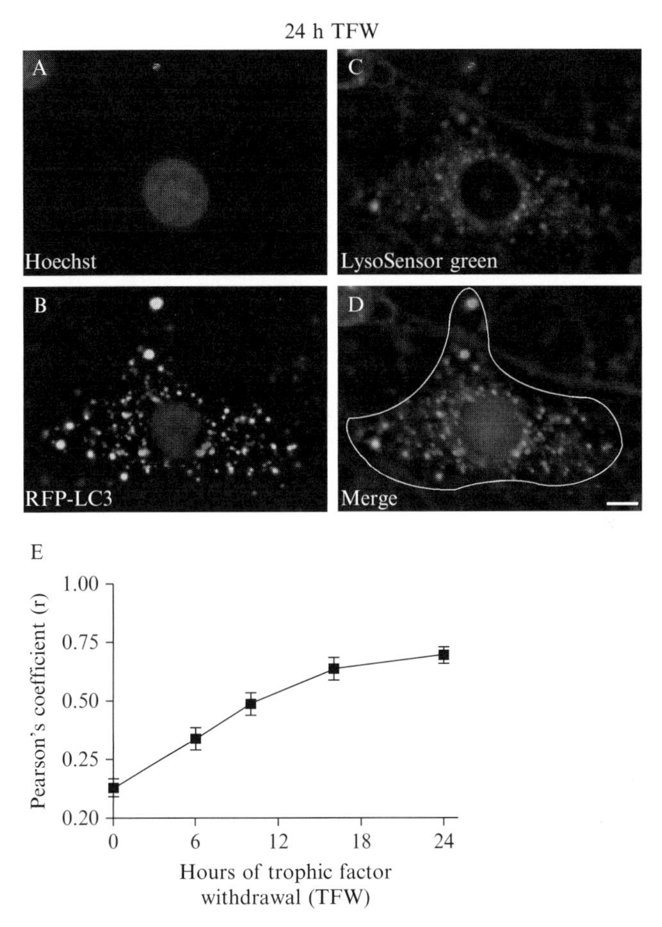

Figure 7.4 (A–D) Schematic for measuring autophagosome-to-lysosome fusion. At least 7–12 live-cell images are captured per treatment condition (from at least 3 independent experiments run in duplicate). Fusion is measured as the degree of colocalization between the 2 fluorescent flours, Cy3 (RFP-LC3) and FITC (LysoSensor Green), which represents the fused vesicles or autolysosome. For each captured Purkinje neuron, a mask is created in Slidebook by outlining the perimeter of the neuron as shown in D. The Pearson's correlation coefficient (*r*) is then determined between the FITC and Cy3 channels by choosing cross channel statistics from the Statistics option under the Mask menu. The degree of colocalized vesicles can be visualized in image D as yellow punctate staining. Images shown represent a 24-h treatment of cerebellar cultures in TFW medium. Scale bar represents 5 μm. (E) Purkinje neuron autophagy fusion rate. RFP-LC3-infected cerebellar cultures were subjected to a time course of TFW (0, 6, 10, 16, 24 h). Coverslips were stained with Hoechst to visualize cellular nuclei and LysoSensor Green to visualize lysosomes, and images were captured via live-cell imaging. Colocalization between RFP-LC3-positive vacuoles and LysoSensor Green–positive vacuoles was quantified using Pearson's correlation coefficient analysis in Slidebook 4.2. The degree of RFP-LC3 and LysoSensor Green colocalization, expressed in Pearson's correlation coefficient (r) increased with time and at 24 h was 0.70, indicative of increased fusion. This time-dependent increase in fusion rate correlates with an increased accumulation of large autophagic vesicles at 24-h TFW (data not shown). (See Color Insert.)

5. Concluding Remarks

The methodologies described herein are directed toward live-cell imaging of autophagy in primary cultured neurons using RFP-tagged LC3 and LysoSensor Green. Examining autophagic vacuole size distribution profiles in combination with autophagosome-lysosome fusion measurements provides useful information regarding autophagic flux. More important, the analyses allow for the investigation of the effects of neuroprotective agents on the autophagic pathway and whether such compounds can alter the formation and/or fusion of autophagic vesicles. However, as with any method, there are caveats to consider when selecting methods to measure autophagy. The advantages and disadvantages of using RFP-LC3 and LysoSensor Green to measure autophagy via live-cell imaging are highlighted subsequently.

Immunofluoresence microscopy is imperative when using primary neurons as a cellular model to study autophagy. Primary neuronal cultures derived from different areas of the brain most often comprise a heterogeneous population of cells including both neurons and nonneuronal cell types (glia). Although glia can be easily removed with the addition of drugs that inhibit cellular proliferation (cytosine arabinoside or 5′-DFUR), cultures that are rich in one neuronal cell type are difficult to produce. Therefore, measuring autophagy induction via LC3 immunoblotting in primary cultures raises the concern of neuronal specificity and caution should be used in data interpretation. Live-cell imaging of autophagy induction using a fluorescent-tagged LC3 as described in this chapter allows one to selectively measure neurons of interest in a mixed population of cells. The difficulty of studying primary neurons, however, is the lack of efficient methods to introduce exogenous genes into neurons. Primary neurons, which are highly sensitive to culture conditions, are in general extremely difficult and labor-intensive to transfect. Adenovirus (Adv)-mediated gene delivery is a commonly used approach that we have found to be effective in cerebellar neurons. Interestingly, the adenoviral RFP-LC3 used in our studies selectively infects Purkinje neurons over granule neurons providing a unique model system to easily identify and examine Purkinje neuron autophagy.

As emphasized by Bampton *et al.* (2005), fluorescent LC3 is a more optimal marker of autophagosomes in comparison to monodansylcadaverine, which was shown to label only late acidic autophagosomes or autolysosomes after fusion with the lysosome. More recently, RFP-LC3 has been described as being more stable in comparison to GFP-LC3, which is sensitive to the acidity of lysosomes resulting in loss of the GFP signal (Kimura *et al.*, 2007; Shaner *et al.*, 2005) during autophagosome-to-lysosome fusion. Therefore, RFP-LC3 represents a marker of early to late autophagosomes including autolysosomes, which may not be labeled consistently using GFP-LC3. Combined with a lysosomal marker such as LysoSensor Green, RFP-LC3 allows for the complete visualization of autophagosome maturation into autolysosomes via time-lapse imaging and/or time-course analyses.

To date, fusion measurements have been quantified by counting the number of autophagosomes (labeled by GFP-LC3 or mCherry-LC3) that colocalize with lysosomal markers, which is then expressed as a percentage of the total number of autophagosomes. We describe a more rapid method of fusion quantification over an autophagy induction time course using Pearson's correlation in Slidebook (4.2) software. Creating masks in Slidebook as described here is a convenient method to define a region of interest for further statistical analysis. Rather than counting the total number of colocalized RFP-LC3/LysoSensor Green vesicles within each neuron, a mask can be created around the area of interest, and the correlation coefficients between the red and green flours are calculated. As a cautionary note, this method is reliable when the captured images are of high quality (without saturation) and have been properly prepared via deconvolution to reduce or eliminate background.

When combined together, autophagic vacuole distribution profiles and autophagosome-lysosome fusion measurements provide detailed information on autophagy induction and flux. Moreover, usage of RFP-LC3 and LysoSensor Green allows for the morphological observation and quantification of autophagosome maturation and fusion with the lysosome. These methods can be used to characterize the process of autophagy in specific neuronal subtypes as well as to characterize drugs that regulate autophagy, which may prove useful in certain neurodegenerative conditions.

ACKNOWLEDGMENTS

The authors would like to thank Aviva Tolkovsky for the kind gift of adeno-RFP-LC3. This research was supported by a Merit Award from the Veteran Affairs.

REFERENCES

Backer, J. M. (2008). The regulation and function of Class III PI3Ks: Novel roles for Vps34. *Biochem. J.* **410,** 1–17.

Bampton, E. T., Goemans, C. G., Niranjan, D., Mizushima, N., and Tolkovsky, A. M. (2005). The dynamics of autophagy visualized in live cells: From autophagosome formation to fusion with endo/lysosomes. *Autophagy* **1,** 23–36.

Cuervo, A. M. (2004). Autophagy: In sickness and in health. *Trends in Cell Biology* **14,** 70–77.

D'Mello, S. R., Galli, C., Ciotti, T., and Calissano, P. (1993). Induction of apoptosis in cerebellar granule neurons by low potassium: Inhibition of death by insulin-like growth factor I and cAMP. *Proc. Natl. Acad. Sci. USA* **90,** 10989–10993.

Fass, E., Shvets, E., Degani, I., Hirschberg, K., and Elazar, Z. (2006). Microtubules support production of starvation-induced autophagosomes but not their targeting and fusion with lysosomes. *J. Biol. Chem* **281,** 36303–36316.

Florez-McClure, M. L., Linseman, D. A., Chu, C. T., Barker, P. A., Bouchard, R. J., Le, S. S., Laessig, T. A., and Heidenreich, K. A. (2004). The p75 neurotrophin receptor can induce autophagy and death of cerebellar Purkinje neurons. *J. Neurosci.* **24,** 4498–4509.

Kimura, A., Noda, T., and Yoshimori, T. (2007). Dissection of the autophagosome maturation process by a novel reporter protein, tandem fluorescent-tagged LC3. *Autophagy* **3,** 452–460.

Klionsky, D. J., Cuervo, A. M., and Seglen, P. O. (2007). Methods for monitoring autophagy from yeast to human. *Autophagy* **3,** 181–206.

Lee, H. K., and Iwasaki, A. (2008). Autophagy and antiviral immunity. *Curr. Opin. Immunol.* **20,** 23–29.

Levine, B., and Deretic, V. (2007). Unveiling the roles of autophagy in innate and adaptive immunity. *Nat. Rev. Immunol.* **7,** 767–777.

Levine, B., and Klionsky, D. J. (2004). Development by self-digestion: Molecular mechanisms and biological functions of autophagy. *Dev. Cell* **6,** 463–477.

Levine, B., and Kroemer, G. (2008). Autophagy in the pathogenesis of disease. *Cell* **132,** 27–42.

Levine, B., and Yuan, J. (2005). Autophagy in cell death: An innocent convict? *J. Clin. Invest.* **115,** 2679–2688.

Majeski, A. E., and Dice, J. F. (2004). Mechanisms of chaperone-mediated autophagy. *Int. J. Biochem. Cell Biol.* **36,** 2435–2444.

Manders, E. M. M., Verbeek, F. J., and Aten, J. A. (1993). Measurement of co-localisation of objects in dual colour confocal images. *J. Microscopy* **169,** 375–382.

Massey, A. C., Zhang, C., and Cuervo, A. M. (2006). Chaperone-mediated autophagy in aging and disease. *Current Topics in Developmental Biology* **73,** 205–235.

Meijer, A. J., and Codogno, P. (2004). Regulation and role of autophagy in mammalian cells. *Int. J. Biochem. Cell Biol.* **36,** 2445–2462.

Mizushima, N. (2007). Autophagy: process and function. *Genes Dev.* **21,** 2861–2873.

Natarajan, K., Meyer, M. R., Jackson, B. M., Slade, D., Roberts, C., Hinnebusch, A. G., and Marton, M. J. (2001). Transcriptional profiling shows that Gcn4p is a master regulator of gene expression during amino acid starvation in yeast. *Mol. Cell. Biol* **21,** 4347–4368.

Selimi, F., Lohof, A. M., Heitz, S., Lalouette, A., Jarvis, C. I., Bailly, Y., and Mariani, J. (2003). Lurcher GRID2-induced death and depolarization can be dissociated in cerebellar Purkinje cells. *Neuron* **37,** 813–819.

Shaner, N. C., Steinbach, P. A., and Tsien, R. Y. (2005). A guide to choosing fluorescent proteins. *Nat. Methods* **2,** 905–909Review.

Tallóczy, Z., Jiang, W., Virgin, H. W. IV,, Leib, D. A., Scheuner, D., Kaufman, R. J., Eskelinen, E.-L., and Levine, B. (2002). Regulation of starvation- and virus-induced autophagy by the eIF2α kinase signaling pathway. *Proc. Natl. Acad. Sci. USA* **99,** 190–195.

Xie, Z., and Klionsky, D. J. (2007). Autophagosome formation: Core machinery and adaptations. *Nat. Cell Biol.* **9,** 1102–1109.

Yue, Z., Horton, A., Bravin, M., DeJager, P. L., Selimi, F., and Heintz, N. (2002). A novel protein complex linking the delta 2 glutamate receptor and autophagy: Implications for neurodegeneration in lurcher mice. *Neuron* **35,** 921–933.

Zuo, J., De Jager, P. L., Takahashi, K. A., Jiang, W., Linden, D. J., and Heintz, N. (1997). Neurodegeneration in Lurcher mice caused by mutation in delta2 glutamate receptor gene. *Nature* **388,** 769–773.

CHAPTER EIGHT

Using Genetic Mouse Models to Study the Biology and Pathology of Autophagy in the Central Nervous System

Zhenyu Yue,* Gay R. Holstein,* Brian T. Chait,† *and* Qing Jun Wang*,†

Contents

* Departments of Neurology and Neuroscience, Mount Sinai School of Medicine, New York, New York, USA
† Laboratory of Mass Spectrometry and Gaseous Ion Chemistry, Rockefeller University, New York, New York, USA

Methods in Enzymology, Volume 453
ISSN 0076-6879, DOI: 10.1016/S0076-6879(08)04008-1

Abstract

Autophagy is a cellular self-eating process that plays an important role in neuroprotection as well as neuronal injury and death. The detailed pathway of autophagy in these two opposing functions remains to be elucidated. Neurons are highly specialized, postmitotic cells that are typically composed of a soma (cell body), a dendritic tree, and an axon. Here, we describe methods for studying autophagy in the central nervous system (CNS). The first involves the use of recently developed transgenic mice expressing the fluorescent autophagosome marker, GFP-LC3. Although CNS neurons show little evidence for the presence of GFP-LC3-containing puncta under normal conditions, under pathological conditions such neurons exhibit many GFP-LC3 puncta. The onset and density of GFP-LC3 puncta have been found to vary significantly in the subcompartments of the affected neurons. These studies suggest that autophagy is distinctly regulated in CNS neurons and that neuronal autophagy can be highly compartmentalized. While transgenic mice expressing GFP-LC3 are a valuable tool for assessing autophagic activity in the CNS, caution needs to be taken when interpreting results solely based on the presence of GFP-LC3 puncta. Therefore, traditional ultrastructural analysis using electron microscopy remains an important tool for studying autophagosomes *in vivo*. Additional reporters of autophagy are constantly being sought. For example, recently a selective substrate of autophagy p62/SQSTM1 has been shown to be specifically regulated by autophagic activity. Therefore, p62/SQSTM1 protein levels can be used as an additional reporter for autophagic activity.

1. Introduction

1.1. Morphological evidence of autophagy in human neurological diseases

There are several excellent reviews covering the topic of autophagy in neurons and human neurological diseases (Chu, 2006; Martinez-Vicente *et al.*, 2005; Nixon, 2006; Rubinszstein *et al.*, 2005). Autophagy was first noticed in injured neurons following axotomy or excitotoxicity (Dixon, 1967;

Matthews and Raisman, 1972). These early studies show that autophagy is frequently associated with axonal stumps (swellings) at lesions, or structures highly related to axon swellings. Recently, autophagic activity (as evidenced by the accumulation of autophagosomes) was found in dysfunctional or degenerating neurons of human brains suffering from neurological diseases, such as Alzheimer's disease (AD), Parkinson's disease (PD), Huntington's disease (HD), and Creutzfeldt-Jackob disease (Anglade *et al.*, 1997; Cataldo *et al.*, 1996; Roizin *et al.*, 1974; Sikorska *et al.*, 2004). For example, in neocortical biopsies from AD brain, large numbers of autophagosomes are observed in perikarya and dystrophic neuritic processes of affected neurons (Nixon *et al.*, 2005). Consistent with the involvement of autophagy in neurodegenerative conditions, genetic disruption of cathepsin D and double deletion for cathepsin B and L in mice induce formation of autophagosomes/autolysosomes in many types of neurons, a scenario closely resembling Batten disease/lysosomal storage disorders (LSDs) (Koike *et al.*, 2005). Neuronal death associated with Niemann-Pick type C disease, a LSD, was recently linked to autophagic action (Ko *et al.*, 2005; Pacheco *et al.*, 2007). Hence, historically autophagy is thought to be expressed excessively in many neurological disorders and suspected to be destructive under those conditions. However, the molecular mechanism whereby autophagy contributes to the neuropathogenesis remains to be shown. For example, it is unclear whether hyperactive or impaired autophagy is involved in the neuropathogenesis. Although many neurological diseases are associated with the accumulation of autophagosomes/autolysosomes, the result can be explained by at least two distinct possibilities: up-regulation of autophagosome biogenesis and blockade in autophagosome maturation/disposal of autophagosomes. More important, a causative role of autophagy in neuropathogenesis remains to be established. To address these questions, it is imperative to develop reliable autophagy assays in neurons by which we can assess how autophagy participates in the progression of neurological disease.

1.2. Basal level of autophagy in the CNS

Autophagy is a key lysosomal pathway, which is responsible for the degradation of long-lived proteins and the turnover of cellular organelles in virtually all cell types. The autophagic process is characterized by the formation, maturation and degradation/recycling of double-membrane bound vacuoles called autophagosomes. Autophagy is generally viewed as a stress-induced process for cells to cope with nutrient and energy crises and to promote cell survival, and until recently the role of autophagy under normal conditions remained unrecognized and hence was considered trivial or was ignored. GFP-LC3 transgenic mice exhibit GFP-containing puncta (an indication of autophagosomes) in many different tissues even under normal conditions, revealing the presence of constitutive (basal) autophagy

(Mizushima *et al.*, 2004). However, the CNS appears to be an exception; GFP-LC3 is observed to be mostly diffuse with no GFP-LC3 puncta even in fasting mice (Mizushima *et al.*, 2004). Indeed, double-membraned autophagosomes are rarely seen in healthy neurons (Nixon *et al.*, 2005). While this evidence seems to suggest that autophagy remains under tight control in CNS neurons (by maintaining the process at a very low level), the most recent studies in knock-out mice with targeted deletion of *Atg5* or *Atg7* unequivocally demonstrate the importance of a basal level of autophagy in CNS neurons (Hara *et al.*, 2006; Komatsu *et al.*, 2006). Our study in mutant mice with Purkinje cell-specific deletion of *Atg7* further reveals that autophagy is essential for prevention of axonal dystrophy and degeneration (Komatsu *et al.*, 2007b). Taken together, these studies clearly establish that autophagy is constitutively active in CNS neurons and plays an important role in the maintenance of protein and membrane homeostasis in CNS neurons. However, it seems paradoxical that autophagy is constitutively active despite the absence of GFP-LC3 puncta, suggesting that the lack of GFP-LC3 puncta does not necessarily reflect the absence of autophagic activity in CNS neurons. This observation raises an important question as to how autophagic activity can be correctly measured at the endogenous level, especially under normal conditions when GFP-LC3 puncta are rarely observed. Moreover, the degree of vulnerability of neurons to dystrophy and degeneration upon the loss of autophagy varies significantly among different neuronal types, suggesting a cell type–specific cellular response to autophagy deficiency (Hara *et al.*, 2006; Komatsu *et al.*, 2006). Therefore, it is also critical to investigate cell type-specific autophagy in neuron populations that are specifically affected in certain human diseases.

1.3. Neuropathology associated with the accumulation of autophagosomes: Impaired versus induced autophagy

Increasing evidence indicates that an elevated number of GFP-LC3 puncta (or alternatively the level of the lipidated form of LC3 (LC3-II)) do not always correlate with up-regulated autophagic activity. As discussed in other chapters, in addition to mechanisms that stimulate the formation of autophagosomes, many events that block autophagic degradation may give rise to an increase in the steady-state number of autophagosomes and elevated levels of LC3-II (see also the chapter by Kimura *et al.*, in volume 452). A block of autophagy may occur at many steps including autophagosome maturation, autophagosome-lysosome fusion or lysosomal degradation. For example, in genetic animal models for juvenile neuronal ceroid lipofuscinosis (Cao *et al.*, 2006) and LSDs (Settembre *et al.*, 2008) where the neuropathologies are associated with the accumulation of autophagosomes, autophagy deficiency in neurons is likely due to impaired autophagosome maturation or lysosomal fusion (Settembre *et al.*, 2008). Two recent studies show that mutations of the ESCRT-III subunit CHMP2B, which are associated with frontotemporal dementia and amyotrophic lateral sclerosis (ALS), can lead to a block of autophagy and the subsequent accumulation of autophagosomes, presumably by interfering with the fusion between multivesicular bodies/late endosomes and autophagosomes (Filimonenko *et al.*, 2007; Lee *et al.*, 2007). The preceding evidence argues strongly that assaying only the number of autophagosomes (based on GFP-LC3-labeled puncta) cannot allow the correct interpretation of autophagic activity under neuropathological conditions.

Moreover, although autophagy is traditionally thought to be a mechanism for nonselective degradation, emerging evidence has pointed to a role of selective degradation for autophagy under certain circumstances. For example, recent studies show that autophagy regulates static levels of p62/SQSTM1 protein, which binds to both ubiquitin and LC3 and plays an important role in the formation of ubiquitinated protein inclusions (Bjørkøy *et al.*, 2005; Komatsu *et al.*, 2007a). We find that steady-state levels of p62/SQSTM1 protein are inversely correlated with autophagic activity (Wang *et al.*, 2006) (see also the chapter by Bjørkøy *et al.*, in this volume). Using p62/SQSTM1 as a readout, we have shown that the excitotoxic *Lurcher* mutation induces, rather than blocks, autophagy because *Lurcher* Purkinje cells accumulate a large number of autophagosomes without an increase in p62/SQSTM1 protein levels (Wang *et al.*, 2006). In addition, a recent study shows that hypoxia-ischemic injury in neonatal mouse brain causes extensive neuronal death accompanied by increased autophagosome formation. This increased number of autophagosomes is likely associated with autophagy activation, because a block of autophagy largely prevents neurodegeneration induced by hypoxia-ischemia (Koike *et al.*, 2008).

These data show that tight control of autophagy is necessary for maintaining the health of CNS neurons; deregulation of neuronal autophagy (either too much or too little) is deleterious. One technical challenge facing us today is the development of robust *in vivo* assays for autophagy, which are essential for the understanding of the exact role of autophagy in neuropathogenesis.

1.4. Essential role of autophagy in axons: Maintenance of homeostasis at the axon terminals

To elucidate the physiological role of neuronal autophagy, we generated mutant mice containing a neural cell type-specific deletion of *Atg7*, an essential gene for autophagy. Characterization of these mutant mice reveals a cell-autonomous function of autophagy in cerebellar Purkinje cells. Our results demonstrate the indispensability of autophagy in the maintenance of axonal homeostasis and the prevention of axonal dystrophy and degeneration under normal conditions (Komatsu *et al.*, 2007b). These results raise interesting questions as to how autophagy proceeds in the axon, which is a highly differentiated neuronal compartment that performs many functions independent of the soma. Since autophagy induction that occurs rapidly in *Lurcher* Purkinje cell axons involves the formation of a large number of autophagosomes in the dystrophic axon terminals (Wang *et al.*, 2006), we hypothesize that autophagosomes may be synthesized locally in axon terminals (especially under pathological conditions) and may undergo axonal transport to the soma where lysosomes are normally present for degradation. To test this hypothesis, we have attempted to set up an *in vitro* system to investigate autophagosome biogenesis and transport in the axons by using a dissociated neuronal culture expressing GFP-LC3, and by live imaging of the behavior of GFP-LC3 puncta (Yue, 2007).

2. Methods

2.1. Transgenic reporter mice GFP-LC3

GFP-LC3 mice (in the background of C57BL/6j) were originally generated by Mizushima *et al.* (2004) and are now distributed by the RIKEN Bio-Resource Center in Japan (http://www.brc.riken.jp/lab/animal/en/dist.shtml). GFP-LC3 mice have now been widely used for monitoring autophagosomes in tissues including the CNS, and provide researchers with an important tool to study autophagic activity *in vivo*.

2.2. Genotyping GFP-LC3 mice

1. Clip the mouse tail (0.3–0.5 cm) and incubate it in a microcentrifuge tube containing 500 *μ*l of tail DNA extraction buffer (100 m*M* Tris-HCl, pH 8.0, 200 m*M* NaCl, 5 m*M* EDTA, 0.2% SDS) supplemented with 10 *μ*l of proteinase K (10 *μ*g/*μ*l stock) at 55 °C overnight. Note that proteinase K should be made fresh or kept frozen to avoid self-digestion.
2. Mix the digested sample by inverting the tube 5 times (avoid vortexing), then centrifuge at 13,000*g* for 2 min. Transfer the supernatant fraction to a new tube and discard the pellet fraction.
3. Precipitate the DNA by mixing the supernatant fraction with an equal volume of isopropanol, and centrifuge at 13,000*g* for 1 min.
4. Aspirate off the supernatant Wash the DNA pellet with 1 ml of 70% ethanol.
5. Discard the ethanol and air-dry the DNA pellet for 10 min at room temperature.
6. Resuspend the DNA pellet with 200 *μ*l of TE buffer (10 m*M* Tris-HCl, pH 7.5, 1 m*M* EDTA) and incubate the sample at 42 °C for 1 h or until

the DNA pellet is fully dissolved. The DNA concentration is measured and adjusted to 100 ng/μl in TE buffer. The DNA prepared using this method can be used for the polymerase chain reaction (PCR) and Southern blot analysis.

7. Take 1 μl of DNA as the template for the PCR. The following primers are designed to amplify a fragment of GFP. The forward primer sequence is 5′-CCT ACG GCG TGC AGT GCT TCA GC-3′. The reverse primer sequence is 5′-CGG CGA GCT GCA CGC TGC GTC CTC-3′. And the PCR is performed using the QIAGEN Tag DNA Polymerase Kit. Each reaction mixture (20 μl total volume) contains 10 μl of double-distilled H_2O, 4 μl of Q solution, 2 μl of 10x buffer, 0.8 μl of 25 m*M* $MgCl_2$, 1.6 μl of 2.5 m*M* dNTP mixture, 0.2 μl of 10 μM forward primer, 0.2 μl of 10 μM reverse primer, 0.2 μl of Tag DNA Polymerase, and 1 μL of DNA template DNA. The PCR conditions are as follows:

   Step 1. 94 °C for 3 min
   Step 2. 94 °C for 30 s
   Step 3. 60 °C for 30 s
   Step 4. 72 °C for 90 s
   Step 5. Repeat steps 2–4, 32 times
   Step 6. 72 °C for 5 min

8. Analyze the PCR-amplified DNA product on a 1% agarose gel.
   Alternative: We have successfully genotyped GFP-LC3 in the neonatal mice (< P10) by visualizing directly GFP fluorescence of the clipped tails. Briefly, place the freshly clipped tail on a glass slide and observe green fluorescence under an inverted fluorescence microscope. The GFP-LC3 transgenic tail normally contains intense and uniform fluorescence and can be easily distinguished from that of a wild-type littermate. Note that this simple typing method is highly reliable with neonates, but not with adult mice.

2.3. Maintaining GFP-LC3 mice

To avoid uncertain effects of transgene insertion in homozygous mice, we recommend that GFP-LC3 mice are maintained as hemizygous (one allele of the transgene). Thus, only one breeder (either male or female) from each breeding pair should contain the GFP-LC3 transgene for genetic crossing, whereas the other breeder should be wild-type (C57BL/6j). However, under some circumstance, GFP-LC3 breeders can be maintained as homozygous to increase the breeding efficiency for having a large number of GFP-LC3-positive offspring in one generation. Recently, the integration site of the GFP-LC3 transgene in the widely used GFP-LC3 line (#53) was identified, allowing the design of specific primers and employing PCR to identify the homozygous GFP-LC3 mice (Kuma and Mizushima, 2008).

2.4. Breeding GFP-LC3 mice to other mouse models

The following will describe a general breeding strategy to produce GFP-LC3 in three different types of genetic mouse models: (1) transgene (e.g., with a disease-related mutation); (2) deletion of the gene of interest with conventional knockout; (3) deletion of the gene of interest with conditional knockout.

2.4.1. Transgenic (Tg) mice

Tg X/+ × GFP-LC3/+

↓

Tg X/+; GFP-LC3/+
+/+; GFP-LC3/+
TgX/+; +/+
+/+; +/+

This is a simple breeding strategy that results in the generation of mice Tg X/+; GFP-LC3/+, which coexpress transgene X and the reporter GFP-LC3. Littermate +/+; GFP-LC3/+ can be used as a control for evaluating the specific effect of transgene X on GFP-LC3 distribution. This type of breeding was used in our previous study of autophagy in Lurcher mice (Wang *et al.*, 2006).

2.4.2. Conventional knockout mice

Gene $X^{+/-}$ × GFP-LC3/+

↓

Gene $X^{+/-}$; GFP-LC3/+ × Gene $X^{+/-}$

↓

Gene $X^{-/-}$; GFP-LC3/+
Gene $X^{+/-}$; GFP-LC3/+
Gene $X^{+/+}$; GFP-LC3/+
Gene $X^{-/-}$; +/+
Gene $X^{+/-}$; +/+
Gene $X^{+/+}$; +/+

To express GFP-LC3 in mutant mice with gene X deletion, mouse breeding should be performed across two generations. The first breeding is responsible for generating Gene $X^{+/-}$; GFP-LC3/+ mice. The second breeding involves a back-cross of Gene $X^{+/-}$; GFP-LC3/+ to Gene $X^{+/-}$. This backcross generates mice Gene $X^{-/-}$;GFP-LC3/+, which allows for the study of GFP-LC3 distribution in a gene X null background.

Littermate Gene $X^{+/-}$; GFP-LC3/+ or Gene $X^{+/+}$; GFP-LC3/+ is used as a control for the specific effect of gene X loss. This type of breeding was previously used in the study of autophagy in cathepsin D knockout mice (Koike *et al.*, 2005).

2.4.3. Conditional knockout mice

Gene $X^{flox/flox}$ × GFP-LC3/+
↓
Gene $X^{flox/+}$; GFP-LC3/+ × Gene $X^{flox/flox}$
↓
Gene $X^{flox/flox}$; GFP-LC3/+ × Gene $X^{flox/+}$; Cre/+
↓
Gene $X^{flox/flox}$; GFP-LC3/+; Cre/+
Gene $X^{flox/+}$; GFP-LC3/+; Cre/+
Gene $X^{flox/flox}$; GFP-LC3/+; +/+
Gene $X^{flox/+}$; GFP-LC3/+; +/+
Gene $X^{flox/flox}$; +/+; Cre/+
Gene $X^{flox/+}$; +/+; Cre/+
Gene $X^{flox/flox}$; +/+; +/+
Gene $X^{flox/+}$; +/+; +/+

To express GFP-LC3 in mutant mice with a tissue/cell type-specific gene X deletion, the breeding will normally involve mice harboring a floxed gene X and mice expressing tissue/cell type-specific Cre, and the entire procedure will span over three generations. As shown in the above flow chart, the first two breedings will generate mice with homozygous floxed gene X and transgene GFP-LC3 (Gene $X^{flox/flox}$; GFP-LC3/+). In the third breeding, Gene $X^{flox/flox}$; GFP-LC3/+ mice will be crossed to Gene $X^{flox/+}$; Cre/+ mice to generate Gene $X^{flox/flox}$; GFP-LC3/+; Cre/+ mice in which GFP-LC3 is produced and gene X is deleted in the specific tissue/cell-type. We performed this type of breeding scheme to express GFP-LC3 in mutant mice with Purkinje cell-specific deletion of the autophagy gene *Atg7* ($Atg7^{flox/flox}$; Pcp2-Cre/+) (Komatsu *et al.*, 2007b).

3. Analysis of GFP-LC3 Expression and Subcellular Localization in the CNS

Perhaps all autophagy-related (ATG) genes including Atg8/LC3 are expressed ubiquitously and their expression levels are highly regulated. However, since the expression of the transgenic GFP-LC3 is under the control of a nonspecific promoter (chicken β-actin promoter) (Mizushima *et al.*, 2004), GFP-LC3 levels may be artificially present in a given tissue/cell type and at a certain time. Therefore, the application of GFP-LC3 transgenic mice is limited to the availability of the GFP-LC3 protein in certain tissues/cell types under study. A prior knowledge of GFP-LC3 expression in the tissue/cell type of interest in GFP-LC3 mice is helpful in designing experiments and interpreting data. The following describes the preparation of brain slices for the imaging of the GFP-LC3 puncta in CNS neurons. All protocols for handling mice should be approved by the appropriate institutional animal care and use committee or the equivalent committee in research institutes.

1. Anesthetize GFP-LC3 mice with ketamine at 100 mg/kg and xylazine at 10 mg/kg through the intraperitoneal (IP) route.
2. Perfuse the anesthetized mice transcardially with PBS (pH 7.4) followed by 4% paraformaldehyde in PBS using a peristaltic pump (Rainin).
3. Carefully remove the brain through dissection, and post fix the brain in 4% paraformaldehyde at 4 °C overnight.
4. Embed the freshly-fixed brain in 4%–5% low-melting-point agarose gel (Cambrex Bio Science Rockland, Rockland, ME, Cat. No. 50110).
5. After the gel block becomes solid, trim the agarose block containing the embedded brain into a cube with minimal size, properly orient and glue it on the stage with Krazy glue.
6. Section the brain tissue at 40–60 μm using a Vibratome (Tissue Sectioning and Bath Refrigeration Systems (Vibratome, St. Louis, MO)). Keep the slice in PBS at 4 °C.

(Optional: after step 3, the fixed brain can be cryoprotected by immersing the brain in 30% sucrose-PBS overnight. The complete penetration of sucrose is achieved when the brain tissue sinks to the bottom of a tube filled with 30% sucrose-PBS. Then the brain can be embedded in Tissue-Tek O. C.T. (Sakura Finetek, Cat. No. 4583) compound and sectioned through the use of a cryostat or sliding microtome on a frozen stage for thinner sections.)

7. For immunostaining, block the brain slices free floating with blocking buffer (PBS supplemented with 0.05% Triton X-100 and 10% goat serum (Invitrogen, Carlsbad, CA, Cat No. 16210-072) for 30 min and then incubate these slices with primary antibody (e.g., anti-GFP antibody) in blocking buffer at 4 °C overnight.
8. Wash the brain slices with PBS 3 times and then incubate these slices with the desired fluorophore-conjugated secondary antibody in blocking buffer in the dark for 45 min at room temperature.
9. Wash the brain slices with PBS 4 times (for the last time, keep the brain slices in PBS for 10 min in the dark).
10. Mount the slices with ProLong Gold antifade reagent (Invitrogen, Carlsbad, CA, Cat. No. P36930 or P36931 (with DAPI)) and 1.5-mm-thick coverslips.

(Optional: It is possible to image direct fluorescence of GFP-LC3 without the immunostaining procedure in certain areas of the CNS where GFP-LC3 levels are relatively high (e.g., cerebellar Purkinje cells, Fig. 8.1A,B); however, staining with anti-GFP antibody was shown to improve significantly the weak GFP-LC3 fluorescent signal in neurons presumably expressing low levels of GFP-LC3, Fig. 8.1C and 8.2).

11. Examine the fluorescent staining of these brain slices using a confocal laser-scanning microscope.

Notes:

1. Transgenic GFP-LC3 mice express high levels of GFP-LC3 in adult cerebellar Purkinje cells. As revealed by imaging direct fluorescence of

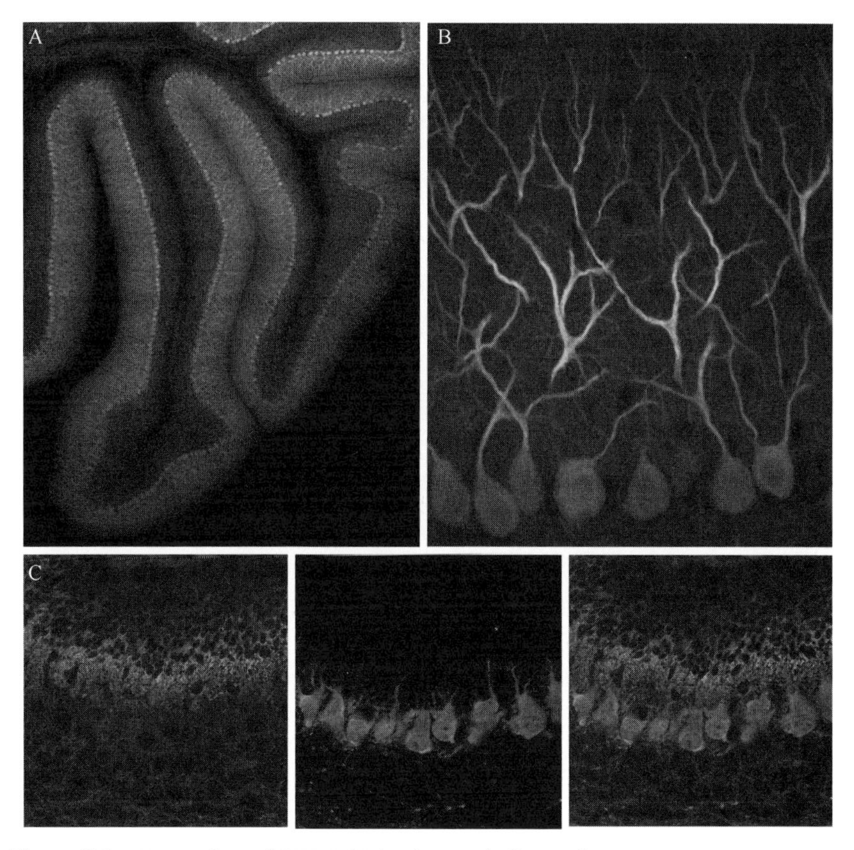

Figure 8.1 Expression of GFP-LC3 in the cerebellum of transgenic GFP-LC3 mice imaged with a confocal laser scanning microscope. (A) and (B) Direct imaging of fluorescence of GFP-LC3 produced in the Purkinje cell layer of an adult mouse at low (A) and high (B) magnifications, respectively. (C) Immunofluorescent staining of the P7 cerebellum with anti-GFP (green, left panel) and anti-calbindin (red, middle panel) antibodies. The merged image is shown in the right panel. GFP-LC3 expression is barely detected in Purkinje cells at P7, in contrast to adult cerebellum. (See Color Insert.)

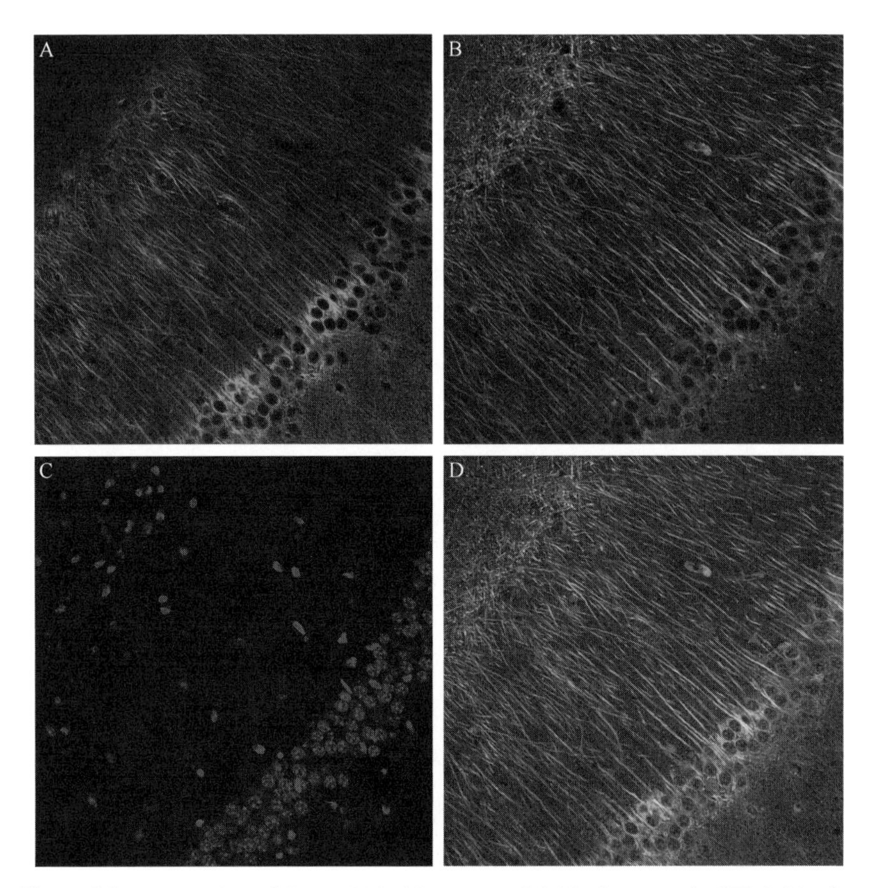

Figure 8.2 Expression of GFP-LC3 in hippocampal CA1 of transgenic GFP-LC3 mice imaged with a confocal laser scanning microscope. Hippocampal slices of GFP-LC3 transgenic mice were stained with anti-MAP2 antibody (A), anti-GFP antibody (B) and DAPI (for nuclei) (C). The merged image is shown in (D). (See Color Insert.)

GFP-LC3 in the cerebellum, intense GFP-LC3 fluorescence is present in the entire Purkinje cell layer (Fig. 8.1A). Furthermore, GFP-LC3 is distributed in the Purkinje cell soma and primary dendrites in a largely diffuse pattern (Fig. 8.1B).

2. Little GFP-LC3 is expressed in Purkinje cells of GFP-LC3 mice at early postnatal days (e.g., P7) (Fig. 8.1C). Cerebellar slices of GFP-LC3 mice at the age of P7 are stained with anti-GFP antibody (affinity purified polyclonal rabbit IgG generated in house (Cristea *et al.*, 2005)) and anti-calbindin antibody (for labeling Purkinje cells in cerebella, monoclonal mouse IgG D-28K, 1:1000; Swant, Bellinzona, Switzerland, Cat. No. 300), followed by labeling with Alexa Fluor 488 goat antirabbit IgG and Alexa Fluor 594 goat antimouse IgG secondary antibodies (1:500, Invitrogen, Carlsbad, CA, Cat. Nos. A11034 and A11032, respectively).

In contrast to adult Purkinje cells, P7 Purkinje cells (in red) express very little GFP-LC3 (in green), whereas cerebellar granule neurons express high levels of GFP-LC3 in the background (Fig. 8.1C).

3. Although imaging direct fluorescence of GFP-LC3 shows that adult hippocampal neurons (CA1) express weak levels of GFP-LC3 (data not shown), staining with anti-GFP antibody and Alexa 488-conjugated secondary antibody yields enhanced GFP-LC3 fluorescence in CA1 neuron at soma and dendrites (Fig. 8.2). The slices are counter-stained with anti-MAP2 antibody (for detection of dendrites), followed by Alexa-594 conjugated secondary antibody (in red). Nuclei of neurons are labeled with DAPI (in blue).

4. Analysis of p62/SQSTM1 and Ubiquitinated Protein Inclusions in the CNS

Targeted deletion of essential autophagy genes in mouse brain provides an opportunity to understand the physiological function of autophagy in the CNS (Hara *et al.*, 2006; Komatsu *et al.*, 2006). In these autophagy-deficient brains, levels of ubiquitinated proteins are markedly increased and a large number of ubiquitinated protein inclusions are formed inside neurons. In parallel, p62/SQSTM1 levels are significantly enhanced and p62/SQSTM1 is accumulated in protein inclusion bodies (Komatsu *et al.*, 2007a). Therefore, increased levels of ubiquitinated proteins and p62/SQSTM1 are important traits of autophagy deficiency in CNS neurons. Assaying p62/SQSTM1 protein levels has increasingly been recognized as a readout for autophagy activity since first demonstrated in our study (Wang *et al.*, 2006).

4.1. Immunohistochemistry staining of p62/SQSTM1

1. Block the brain slices free floating with blocking buffer (PBS supplemented with 0.05% Triton X-100 and 10% goat serum) for 30 min and then incubate slides with guinea pig anti-p62 (1:500, American Research Products, Belmont, MA, Cat. No. 03-GP62-C) in blocking buffer at 4 °C overnight.
2. Wash the brain slices with PBS 3 times and then incubate with Alexa Fluor 488 goat anti-guinea pig secondary antibody (1:1000, Invitrogen, Carlsbad, CA, Cat No. A11073) in blocking buffer in the dark at room temperature for 45 min.
3. Wash the brain slices with PBS 4 times (for the last time, keep the brain slices in PBS for 10 min in the dark).

4. Mount the slice on glass slides with ProLong Gold antifade reagent and 1.5-mm-thick coverslips.
5. Examine the fluorescent staining of these brain slices using a confocal laser-scanning microscope.

Notes:

Immunofluorescent staining of p62 in brain tissues has assisted us in assessing the change in autophagic activity in *Lurcher* Purkinje cells. Previously, we showed that, in both *Lurcher* and *Atg7*$^{flox/flox}$; Pcp2-*Cre* mice, axon terminals of Purkinje cell undergo dystrophic swelling and degeneration (Komatsu *et al.*, 2007; Wang *et al.*, 2006). As expected, *Atg7* deletion-mediated autophagy deficiency causes an increase in p62 protein levels at swollen axon terminals of *Atg7*$^{flox/flox}$; Pcp2-*Cre* Purkinje cells, whereas little change in p62 protein levels is seen at axon terminals of *Lurcher* Purkinje cells (Fig. 8.3). This result suggests that autophagy is induced rather than blocked in axon terminals of *Lurcher* Purkinje cells (Wang *et al.*, 2006).

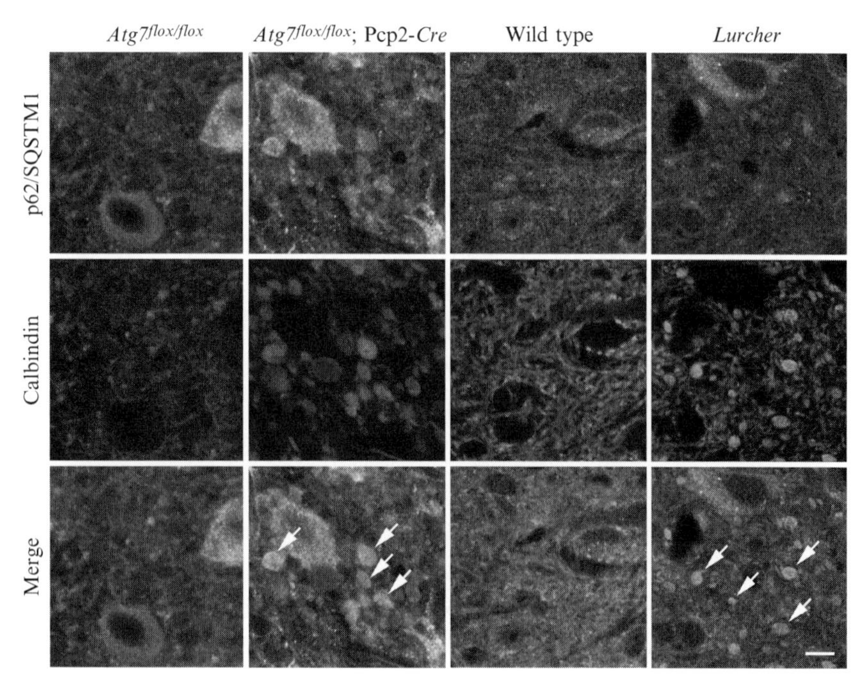

Figure 8.3 Immunostaining of p62/SQSTM1 in dystrophic swellings of Purkinje cell axon terminals. Cerebellar slices at deep cerebellar nuclei from *Atg7*$^{flox/flox}$; Pcp2-*Cre* mice and *Lurcher* mice were stained with anti-p62/SQSTM1 (in green) and anti-calbindin (in red) antibodies. The p62/SQSTM1 is accumulated in the dystrophic swellings of axon terminals (white arrows) from *Atg7*$^{flox/flox}$; Pcp2-*Cre* Purkinje cells but not *Lurcher* Purkinje cells. Scale bar: 10 μm. This figure is modified from (Komatsu *et al.*, 2007b) with permission. (See Color Insert.)

5. Transmission Electron Microscopy (TEM) Analysis of Autophagosomes

5.1. Tissue fixation and processing for ultrastructural studies

1. Perfuse the anesthetized mouse transcardially with 100 mL of room-temperature 1% formaldehyde/2% glutaraldehyde fixative in 0.1 *M* PBS, pH 7.4, after flushing the vasculature with warm (37 °C) heparinized (2 U/mL; American Pharmaceutical Partners, Cat. No. 401586A) 0.01*M* PBS, pH 7.4. The formaldehyde solution should be prepared from EM grade purified paraformaldehyde powder (Electron Microscopy Sciences, Cat. No. 19210). The glutaraldehyde should be specified as EM Grade Distillation Purified and obtained from a reliable source for EM products such as Electron Microscopy Sciences (www.emsdiasum.com), Ernest Fullam (www.fullam.com), or Ted Pella (www.tedpella.com). Although the formaldehyde solution may be prepared several days in advance, the glutaraldehyde should not be added to the fixative solution until shortly before use.
2. After the fixative has been delivered, leave the mouse in place for approximately 60 min.
3. Harvest the brain and immediately immerse it in a vial of cold (4 °C) perfusion fixative. Leave overnight at 4 °C.
4. Cut 50- to 100-μm-thick Vibratome sections through the region(s) of interest, keeping the tissue block and sections cold by using the Tissue Sectioning and Bath Refrigeration System, or by placing 4–5 cubes of frozen 0.1 *M* phosphate buffer, pH 7.4 (PB), in the Vibratome well filled with cold PB. Place the sections in wells of a 24-well plastic plate containing PB, on ice.
5. Further dissect the sections in a large Petri dish of PB so that the sections contain only the region(s) of interest. The sections should be no larger than 5 mm on a side. Place the dissected sections in glass vials containing PB, fitted with screw caps having aluminum (not paper) inner linings, on ice. Sections from the same region/animal/condition may be pooled in a single vial.

Note: All subsequent steps should be performed in a fume hood, with protective eyewear and gloves.

6. Once all the sections are dissected, rinse them twice in ice-cold PB, and then osmicate (1% OsO_4 in PB with 7% dextrose) for 1 h in the covered glass vials, on ice.
7. Dehydrate the specimens in ascending concentrations of ice-cold ethanol diluted in distilled water, on ice, as follows:
   1 rinse in 50% ethanol, 10 min

1 rinse in 70% ethanol, 10 min
2 rinses in 95% ethanol, 10 min each
2 rinses in 100% ethanol, 10 min each
2 rinses in propylene oxide, 10 min each

8. Infiltrate the specimens with resin by immersing them for 30 min in a 1:1 mixture of epoxy resin (Epon, Araldite or Epon-Araldite; Electron Microscopy Sciences Cat. Nos. RT 14120, RT 13920 and RT 13940, respectively) and propylene oxide (Electron Microscopy Sciences, Cat. No. 20410). Replace the solution with 100% resin, and leave overnight on a vertical rotator at 4 °C.
9. Embed the sections in capsules, molds or between Aclar sheets (Electron Microscopy Sciences, Cat. Nos. 70000-B, 70905-01 and 50425-10, respectively) or plastic coverslips (Fisher Scientific, Cat. No. 12-547). Cure in an oven at 60 °C for 3 days.
10. Trim the embedded tissue to contain the area of interest and then thin-section using an ultramicrotome onto mesh or Formvar-coated slot grids (Electron Microscopy Sciences, Cat. Nos. 0200-CU, RT-15830 and G7530-Cu, respectively).
11. Sections can be contrasted using 5% aqueous uranyl acetate (15–30 min) and 0.15% lead acetate or citrate (10 to 30 min). Alternatively, the sections can be stained with uranyl deposits during the dehydration step (step 7 above) using uranyl acetate diluted in 70% ethanol. Insert this step between two 70% ethanol rinses, prior to the 95% ethanol rinse. (One resource for general information about TEM tissue processing is the Practical Methods in Electron Microscopy series (Elsevier) edited by A. M. Glauert. See also the chapter by Ylä-Anttila *et al.*, in this volume.)

5.2. Tissue fixation and processing for pre-embedding immuno-electron microscopy

1. Perfuse the anesthetized mouse transcardially with 100 mL of room temperature 4% formaldehyde/0.25% glutaraldehyde fixative in 0.1 *M* PBS, pH 7.4, after flushing the vasculature with warm (37 °C) heparinized (2 U/mL) 0.01 *M* PBS, pH 7.4. Prepare the fixative as described above.
2. Harvest the brain, block the region(s) of interest, and immerse the block in a beaker containing cold (4 °C) glutaraldehyde-free fixative. Leave for 90 min at 4 °C.
3. Place the blocks in a series of ascending concentrations of sucrose (7%, 10%, 20%) diluted in PB, 30 min each, and leave overnight at 4 °C in fresh 20% sucrose in PB.
4. In a fume hood, place a wide-mouthed beaker containing isopentane within a larger basin of acetone cooled with dry ice. Transfer the tissue

block to a small (e.g., 10 mL) Pyrex beaker, and add just enough fresh 20% sucrose to cover the block. Place the 10 mL beaker in the bath using forceps, and leave until the sucrose appears frozen. Gently remove the beaker using forceps, and allow it to thaw completely at room temperature, undisturbed. Repeat this procedure once. This process aids antibody penetration.

5. Cut 50- to 100-μm-thick Vibratome sections as described previously (*Tissue fixation and processing for ultrastructural studies*, Step 4).
6. Process the sections for peroxidase immunocytochemistry. After quenching endogenous peroxidase by immersing the sections in 0.3% H_2O_2 for 30 min at room temperature (RT), rinse the sections thoroughly with PBS (6 rinses, 10 min each) and then place them in blocking buffer (5% normal serum of the donor species of the secondary antibody in PBS) overnight at 4 °C. Bring the sections to RT, add primary antisera or control sera, and then incubate overnight at 4 °C. Sections are then rinsed thoroughly with PBS (6 rinses, 10 min each) incubated with biotinylated secondary antibodies for 2–3 hrs at RT, rinsed again, and then incubated with streptavidin-HRP (Zymed Laboratories, San Francisco, CA) for 1 h at RT.
7. To visualize peroxidase conjugates, incubate the sections in 0.05% diaminobenzidine (DAB; Sigma Cat No. D4293) with 0.01% H_2O_2 in 0.1 *M* Tris buffer, pH 7.6 (TB; 3.31g Tris-HCl and 0.49 g Tris Base in 500 ml distilled water) for 5–10 min at RT. After the tissue has been incubated in DAB, rinse once with TB, twice with PB, and then process for electron microscopy as described in *Tissue fixation and processing for ultrastructural studies*, starting at step 5. The DAB reaction product is visible by TEM as an electron-dense reaction product, and is readily discernable from the punctate gold particles of varying size used for postembedding labeling.

5.3. Tissue fixation and processing for post-embedding immunogold TEM localization

1. Follow Steps 1 and 2 of the procedure "Tissue fixation and processing for preembedding immuno-electron microscopy."
2. Cut 100-μm to 1-mm thick Vibratome sections through the region(s) of interest, keeping the tissue block and sections cold by using the Tissue Sectioning and Bath Refrigeration System, or by placing 4–5 cubes of frozen 0.1 *M* PB in the Vibratome well filled with cold PB. Place the sections in wells containing PB, on ice.
3. Further dissect the sections in a large Petri dish of PB so the sections contain only the region(s) of interest. The sections should be no larger than 2 mm on a side. Place the dissected sections in glass or plastic vials containing PB, fitted with screw caps.

4. Cryoprotect the tissue at room temperature using increasing concentrations of glycerol in PB (10%, 20%, 30%, 1 h each), then store overnight in 30% glycerol at 4 °C.
5. Freeze-plunging and subsequent tissue embedment require specialized equipment, available from multiple vendors including the electron microscopy sources identified previously. Examples are the Reichert KF 80 and the Leica EM FSP freeze-plunger units and the Leica AFS embedding unit (Leica Microsystems). The sections are glued to aluminum flat-headed pins and then plunge-frozen using the appropriate equipment. The sections are then resin-embedded and UV-polymerized using the automatic reagent dilution and handling systems of the embedment units. In most processing/embedding units, 20–24 specimens can be processed in one run.
6. Trim the embedded tissue to contain the region(s) of interest and then thin-section using an ultramicrotome and transfer onto nickel or gold mesh or Formvar-coated slot grids (Electron Microscopy Sciences, Cat. Nos. G100-Ni, G200-Au, G2010-Ni, and GG205-Ni). Mount the grids in Hiraoka support plates (Electron Microscopy Sciences, Cat. No. 71560-10).
7. The grids are rinsed with TB-saline (4.5 g NaCl in 500 ml TB) containing 0.1% Triton X-100 (TBST), and then with 0.1% $NaBH_4$, 50 m*M* glycine in TBST, blocked in 10% normal serum (NS) in TBST, incubated in primary antibody (e.g., anti-GFP antibody) diluted in 1% NS in TBST (2 h at room temperature), rinsed and blocked again, then incubated for 1 h at room temperature with gold-conjugated secondary antibodies (Electron Microscopy Sciences) diluted 1:20 in 1% NS in TBST with 0.5% polyethylene glycol, and rinsed again. For double labeling using gold particles of different sizes, expose the grids to paraformaldehyde vapors for 1 h, then rinse thoroughly and repeat the labeling procedure beginning with incubation in the primary antibody if the double labeling has both primary antibodies made in the same donor species. At the conclusion of all labeling, contrast the sections with 1% aqueous uranyl acetate (15–30 min) and 0.3% lead citrate diluted in water (10–30 min). Visualization of smaller (1–10 nm) gold particles is often aided by use of silver enhancement (Electron Microscopy Sciences Cat. No. 25521). This process assists localization of the gold particles, but sacrifices the punctate signal provided by nonenhanced gold particles.

5.4. TEM identification of double-membraned vacuoles

Autophagosomes are recognizable in the cytoplasm of neurons as double-membrane bound vacuoles, with cross-sectional diameters of approximately 0.1–0.5 μm. Our studies suggest that autophagosomes may be synthesized locally in axon terminals and may undergo axonal transport to the soma where lysosomes are normally present for final degradation (Komatsu *et al.*, 2007b). Many of the double-membraned vacuoles found in axon terminals appear to be derived from the invagination of neighboring oligodendrocytes, a process which is poorly understood but may be related to endocytosis via the axolemma and cytoplasmic membranes of oligodendrocytes (Eddleman *et al.*, 1998; Zhang *et al.*, 2005). Additionally, we have observed the formation of double-membraned vacuoles that do not appear to be derived from the invasion of oligodendrocytes (Komatsu *et al.*, 2007b). Speculatively, as previously reported, the vacuoles may originate from axonal subsurface cisternae (Li *et al.*, 2005) or smooth ER (Broadwell and Cataldo, 1984).

5.5. Quantification of double-membraned vacuoles

Quantification of double-membraned vacuoles in TEM material can be accomplished using a point-counting method. Construct a grid using Photoshop (or similar software used to analyze the TEM images). Overlay each TEM image with the grid image. Count the number of vacuoles (for profile identification, see *TEM identification of double-membraned vacuoles* section) overlayed by at least one cross-point of the grid and the total number of cross points within the vacuoles (Fig. 8.4). Calculate the total area of each micrograph to determine the density of autophagosomes per unit area. Counting should be performed double-blind, and equal areas of tissue should be analyzed for all conditions. Volumetric densities can be obtained by factoring in the section thickness.

6. Conclusion

Recent studies have begun to explore the potential of autophagy as a drug target in treating neurodegenerative disease. Thus, delineating the neuronal autophagy process and the mechanism by which autophagy is involved in various pathological conditions will be crucial for a better understanding of neurodegeneration and the design of therapeutic drugs. Although GFP-LC3 transgenic mice have been successfully used as a reporter to study autophagic activity in the CNS, limitations in their application as an autophagy reporter have also been recognized. It is of particular importance to dissect the basal level of autophagic activity in healthy neurons where GFP-LC3 may serve as a poor marker. Recent studies indicate that p62/SQSTM1, a selective substrate of autophagy, can also be used as an autophagic marker for neuronal autophagic activity.

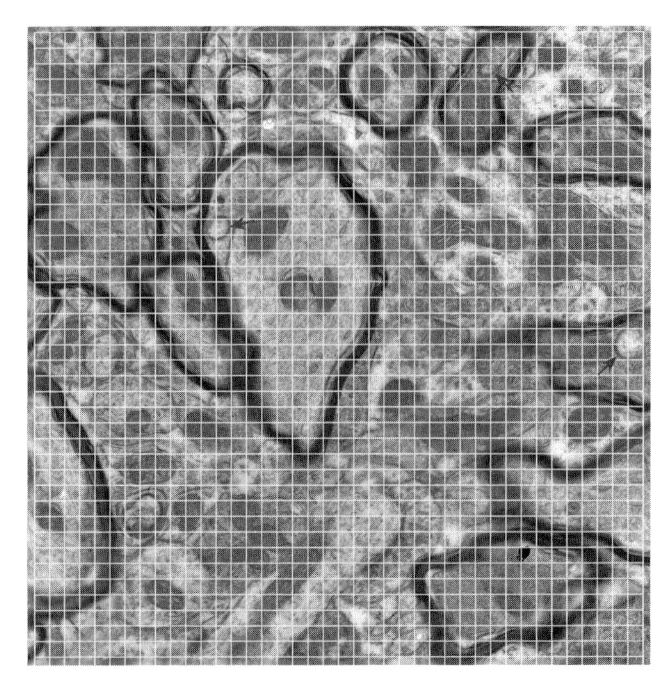

Figure 8.4 An example of using the point-counting method to quantify double-membraned vacuole-like structures in the deep cerebellar nuclei of *Atg7*$^{flox/flox}$; Pcp2-*Cre* cerebellum. The electron micrograph (with an area of 50 μm^2) is overlayed with a Photoshop-generated grid (41 × 41). The mesh size of the grid is chosen depending on the frequency and size of the structures. In this example, the number of these structures containing at least one intersection of the grid was determined to be 3 (arrows). The volume of the structures was counted as the number of intersections within the structures and was determined in this example to be 8.

With our growing knowledge of the neuronal autophagy process, we anticipate identifying ever more specific autophagy markers to assist in assaying autophagic activity in the CNS.

ACKNOWLEDGMENTS

This study was supported by the NIH to Z.Y (NS060123-02), GRH (DC008846-01) and B.T.C (RR00862 and RR022220).

REFERENCES

Anglade, P., Vyas, S., Javoy-Agid, F., Herrero, M. T., Michel, P. P., Marquez, J., Mouatt-Prigent, A., Ruberg, M., Hirsch, E. C., and Agid, Y. (1997). Apoptosis and autophagy in nigral neurons of patients with Parkinson's disease. *Histol. Histopathol.* **12,** 25–31.

Bjørkøy, G., Lamark, T., Brech, A., Outzen, H., Perander, M., Overvatn, A., Stenmark, H., and Johansen, T. (2005). p62/SQSTM1 forms protein aggregates degraded by autophagy and has a protective effect on huntingtin-induced cell death. *J. Cell Biol.* **171,** 603–614.

Broadwell, R. D., and Cataldo, A. M. (1984). The neuronal endoplasmic reticulum: Its cytochemistry and contribution to the endomembrane system. II. Axons and terminals. *J. Comp. Neurol.* **230,** 231–248.

Cao, Y., Espinola, J. A., Fossale, E., Massey, A. C., Cuervo, A. M., MacDonald, M. E., and Cotman, S. L. (2006). Autophagy is disrupted in a knock-in mouse model of juvenile neuronal ceroid lipofuscinosis. *J. Biol. Chem.* **281,** 20483–20493.

Cataldo, A. M., Hamilton, D. J., Barnett, J. L., Paskevich, P. A., and Nixon, R. A. (1996). Properties of the endosomal-lysosomal system in the human central nervous system: Disturbances mark most neurons in populations at risk to degenerate in Alzheimer's disease. *J. Neurosci.* **16,** 186–199.

Cristea, I. M., Williams, R., Chait, B. T., and Rout, M. P. (2005). Fluorescent proteins as proteomic probes. *Mol. Cell. Proteomics* **4,** 1933–1941.

Chu, C. T. (2006). Autophagic stress in neuronal injury and disease. *J. Neuropathol. Exp. Neurol.* **65,** 423–432.

Dixon, J. S. (1967). "Phagocytic" lysosomes in chromatolytic neurones. *Nature* **215,** 657–658.

Eddleman, C. S., Ballinger, M. L., Smyers, M. E., Fishman, H. M., and Bittner, G. D. (1998). Endocytotic formation of vesicles and other membranous structures induced by Ca^{2+} and axolemmal injury. *J. Neurosci.* **18,** 4029–4041.

Filimonenko, M., Stuffers, S., Raiborg, C., Yamamoto, A., Malerod, L., Fisher, E. M., Isaacs, A., Brech, A., Stenmark, H., and Simonsen, A. (2007). Functional multivesicular bodies are required for autophagic clearance of protein aggregates associated with neurodegenerative disease. *J. Cell Biol.* **179,** 485–500.

Hara, T., Nakamura, K., Matsui, M., Yamamoto, A., Nakahara, Y., Suzuki-Migishima, R., Yokoyama, M., Mishima, K., Saito, I., Okano, H., and Mizushima, N. (2006). Suppression of basal autophagy in neural cells causes neurodegenerative disease in mice. *Nature* **441,** 885–889.

Ko, D. C., Milenkovic, L., Beier, S. M., Manuel, H., Buchanan, J., and Scott, M. P. (2005). Cell-autonomous death of cerebellar purkinje neurons with autophagy in Niemann-Pick type C disease. *PLoS Genet.* **1,** 81–95.

Koike, M., Shibata, M., Tadakoshi, M., Gotoh, K., Komatsu, M., Waguri, S., Kawahara, N., Kuida, K., Nagata, S., Kominami, E., Tanaka, K., and Uchiyama, Y. (2008). Inhibition of autophagy prevents hippocampal pyramidal neuron death after hypoxic-ischemic injury. *Am. J. Pathol.* **172,** 454–469.

Koike, M., Shibata, M., Waguri, S., Yoshimura, K., Tanida, I., Kominami, E., Gotow, T., Peters, C., von Figura, K., Mizushima, N., Saftig, P., and Uchiyama, Y. (2005). Participation of autophagy in storage of lysosomes in neurons from mouse models of neuronal ceroid-lipofuscinoses (Batten disease). *Am. J. Pathol.* **167,** 1713–1728.

Komatsu, M., Waguri, S., Chiba, T., Murata, S., Iwata, J., Tanida, I., Ueno, T., Koike, M., Uchiyama, Y., Kominami, E., and Tanaka, K. (2006). Loss of autophagy in the central nervous system causes neurodegeneration in mice. *Nature* **441,** 880–884.

Komatsu, M., Waguri, S., Koike, M., Sou, Y. S., Ueno, T., Hara, T., Mizushima, N., Iwata, J., Ezaki, J., Murata, S., Hamazaki, J., Nishito, Y., *et al.* (2007a). Homeostatic levels of p62 control cytoplasmic inclusion body formation in autophagy-deficient mice. *Cell* **131,** 1149–1163.

Komatsu, M., Wang, Q. J., Holstein, G. R., Friedrich, V. L. Jr., Iwata, J., Kominami, E., Chait, B. T., Tanaka, K., and Yue, Z. (2007b). Essential role for autophagy protein Atg7 in the maintenance of axonal homeostasis and the prevention of axonal degeneration. *Proc. Natl. Acad. Sci. USA* **104,** 14489–14494.

Kuma, A., and Mizushima, N. (2008). Chromosomal mapping of the GFP-LC3 transgene in GFP-LC3 mice. *Autophagy* **4,** 61–62.

Lee, J. A., Beigneux, A., Ahmad, S. T., Young, S. G., and Gao, F. B. (2007). ESCRT-III dysfunction causes autophagosome accumulation and neurodegeneration. *Curr. Biol.* **17,** 1561–1567.

Li, Y. C., Li, Y. N., Cheng, C. X., Sakamoto, H., Kawate, T., Shimada, O., and Atsumi, S. (2005). Subsurface cisterna-lined axonal invaginations and double-walled vesicles at the axonal-myelin sheath interface. *Neurosci. Res.* **53,** 298–303.

Martinez-Vicente, M., Sovak, G., and Cuervo, A. M. (2005). Protein degradation and aging. *Exp. Gerontol.* **40,** 622–633.

Matthews, M. R., and Raisman, G. (1972). A light and electron microscopic study of the cellular response to axonal injury in the superior cervical ganglion of the rat. *Proc. R. Soc. Lond. B. Biol. Sci.* **181,** 43–79.

Mizushima, N., Yamamoto, A., Matsui, M., Yoshimori, T., and Ohsumi, Y. (2004). *In vivo* analysis of autophagy in response to nutrient starvation using transgenic mice expressing a fluorescent autophagosome marker. *Mol. Biol. Cell* **15,** 1101–1111.

Nixon, R. A. (2006). Autophagy in neurodegenerative disease: Friend, foe or turncoat? *Trends Neurosci.* **29,** 528–535.

Nixon, R. A., Wegiel, J., Kumar, A., Yu, W. H., Peterhoff, C., Cataldo, A., and Cuervo, A. M. (2005). Extensive involvement of autophagy in Alzheimer disease: An immuno-electron microscopy study. *J. Neuropathol. Exp. Neurol.* **64,** 113–122.

Pacheco, C. D., Kunkel, R., and Lieberman, A. P. (2007). Autophagy in Niemann-Pick C disease is dependent upon Beclin-1 and responsive to lipid trafficking defects. *Hum. Mol. Genet.* **16,** 1495–1503.

Roizin, L., Stellar, S., Willson, N., Whittier, J., and Liu, J. C. (1974). Electron microscope and enzyme studies in cerebral biopsies of Huntington's chorea. *Trans. Am. Neurol. Assoc.* **99,** 240–243.

Rubinszstein, D. C., DiFiglia, M., Heintz, N., Nixon, R. A., Qin, Z.-H., Ravikumar, B., Stefanis, L, and Tolkovsky, A (2005). Autophagy and its possible roles in nervous system diseases, damage and repair. *Autophagy* **1,** 11–22.

Settembre, C., Fraldi, A., Jahreiss, L., Spampanato, C., Venturi, C., Medina, D., de Pablo, R., Tacchetti, C., Rubinsztein, D. C., and Ballabio, A. (2008). A block of autophagy in lysosomal storage disorders. *Hum. Mol. Genet.* **17,** 119–129.

Sikorska, B., Liberski, P. P., Giraud, P., Kopp, N., and Brown, P. (2004). Autophagy is a part of ultrastructural synaptic pathology in Creutzfeldt-Jakob disease: A brain biopsy study. *Int. J. Biochem. Cell Biol.* **36,** 2563–2573.

Wang, Q. J., Ding, Y., Kohtz, S., Mizushima, N., Cristea, I. M., Rout, M. P., Chait, B. T., Zhong, Y., Heintz, N., and Yue, Z. (2006). Induction of autophagy in axonal dystrophy and degeneration. *J. Neurosci.* **26,** 8057–8068.

Yue, Z. (2007). Regulation of neuronal autophagy in axon: Implication of autophagy in axonal function and dysfunction/degeneration. *Autophagy* **3,** 139–141.

Zhang, P., Land, W., Lee, S., Juliani, J., Lefman, J., Smith, S. R., Germain, D., Kessel, M., Leapman, R., Rouault, T. A., and Subramaniam, S. (2005). Electron tomography of degenerating neurons in mice with abnormal regulation of iron metabolism. *J. Struct. Biol.* **150,** 144–153.

CHAPTER NINE

Biochemical and Morphological Detection of Inclusion Bodies in Autophagy-Deficient Mice

Satoshi Waguri* *and* Masaaki Komatsu[†]

Contents

* Department of Anatomy and Histology, Fukushima Medical University School of Medicine, Fukushima, Japan
[†] Laboratory of Frontier Science, Tokyo Metropolitan Institute of Medical Science, Tokyo, Japan, and PRESTO, Japan Science and Technology Corporation, Kawaguchi, Japan

Methods in Enzymology, Volume 453
ISSN 0076-6879, DOI: 10.1016/S0076-6879(08)04009-3

Abstract

Autophagy-deficient mice exhibit the formation of ubiquitin-inclusions in the liver and brain, which is not attributed to the dysfunction of the ubiquitin-proteasome system. Moreover, it is also clear that a multifunctional protein p62/A170/SQSTM1 (hereafter referred to as p62) links autophagy and inclusion formation, being one of the key components of the ubiquitin inclusions. The ubiquitin/p62 inclusions can be detected in the detergent-insoluble fraction by western blot analysis, while morphological information can be obtained by immunohistochemistry at both the light and electron microscopy levels. Importantly, p62 has become a reliable marker, with which we can identify inclusions and estimate autophagic activity in diseased tissues or cells. In this chapter, we describe the methods used for biochemical and morphological detection of ubiquitin/p62-inclusions in autophagy-suppressed *Atg7*-deficient mice. These methods are suitable for examination of cells and tissues with conditions associated with reduced autophagy (e.g., aging and mice models of intractable diseases such as Alzheimer's disease), and their applications should enhance our understanding of the pathophysiological mechanisms involved in the formation of intracellular inclusions.

1. Introduction

There is ample evidence that dysfunction of the ubiquitin-proteasome system leads to the formation of ubiquitin-positive inclusions in various neurodegenerative diseases, and the detection of such inclusions is thus useful for pathological diagnosis and understanding of the pathogenic mechanisms of these diseases (Goldberg, 2003; Lowe *et al.*, 2005). On the other hand, recent mouse genetic analyses reveal that another degradation system, the autophagy-lysosomal pathway, is also associated with the formation of ubiquitin inclusions. For example, accumulation of numerous ubiquitin-inclusions is seen in hepatocytes or neurons in autophagy-deficient mice with disruption of the autophagy-related (*Atg*) genes *Atg7* or *Atg5*, in spite of the presence of a normal proteasome system (Hara *et al.*, 2006; Komatsu *et al.*, 2005, 2006). Furthermore, p62/A170/SQSTM1 (hereafter referred to as p62), a multifunctional protein known to interact with several signaling molecules (Wooten *et al.*, 2006), is present in the ubiquitin-inclusions. This molecule has a unique feature; an N-terminal Phox and Bem1p (PB1) domain, which retains the ability of self-oligomerization, and a C-terminal ubiquitin-associated (UBA) domain capable of interaction with ubiquitinated proteins. These properties imply the involvement of p62 in inclusion formation. In fact, in autophagy-deficient mouse (Komatsu *et al.*, 2007) and fly (Nezis *et al.*, 2008), additional loss of p62 is associated with marked reduction in the formation of ubiquitin-inclusions. Furthermore, p62 was identified as one of the specific substrates that are degraded through the autophagy-lysosomal pathway (Bjorkoy *et al.*, 2005). This degradation is mediated by interaction with microtubule-associated protein 1 light chain 3 (LC3), a mammalian homolog of Atg8 (Komatsu *et al.*, 2007; Pankiv *et al.*, 2007; Ichimura *et al.*, 2008), which is recruited to the phagophore membrane and remains associated with the completed autophagosome. Therefore, it is possible that intracellular levels of p62 and/or the tendency for ubiquitin-aggregate formation, are regulated by the activity of constitutive autophagy.

The ubiquitin-binding nature of p62 has also attracted many pathologists, because it could be an additional diagnostic marker for diseases characterized by ubiquitin-inclusions. In fact, immunohistochemical studies show a clear-cut signal for p62 in Mallory bodies found in various chronic liver disorders, such as alcoholic hepatitis, nonalcoholic steatohepatitis, and intracytoplasmic hyaline bodies found in hepatocellular carcinoma (Stumptner *et al.*, 2002; Zatloukal *et al.*, 2002). Similar p62-positive cytoplasmic inclusions are found also in several neurodegenerative diseases, including Alzheimer's disease, Parkinson's disease, and amyotrophic lateral sclerosis (Nakano *et al.*, 2004; Kuusisto *et al.*, 2008). Therefore, p62-inclusions are currently considered a common hallmark of conformational diseases, as well as a key molecule for the investigation of pathogenic mechanisms associated with the dysfunction of autophagy. In this section, we describe in detail the methods used for biochemical and morphological detection of ubiquitin- and p62-positive inclusions in autophagy-deficient mice (also see the chapter by Bjørkøy *et al.*, in volume 452).

2. Detection of Ubiquitinated Proteins and p62 in Autophagy-Deficient Mice by Western Blot Analysis

Biochemical analyses of the ubiquitin-inclusions require the detection of ubiquitinated proteins (or polyubiquitinated proteins) in Western blot, which usually appear as a smear with variable molecular weight on the blot. On the other hand, p62 is usually seen as a band (or doublet in the case of mouse liver and brain), which makes this protein more convenient for the analysis of inclusions. More important, the variation of the p62 signal is considered to reflect autophagic activity of tissues or cells (Wang *et al.*, 2006; Komatsu *et al.*, 2007; Nakai *et al.*, 2007; Itoh *et al.*, 2008; Nezis *et al.*, 2008). It should be noted that to correlate the findings of Western blot with inclusions detected by morphological analyses, it is important to prepare detergent-insoluble and detergent-soluble fractions from samples for analysis of both p62 and ubiquitinated substrates.

2.1. Preparation of tissue lysates from the brain and liver

1. Prepare a homogenate buffer consisting of 0.25 *M* sucrose, 10 m*M* 2-[4-(2-hydroxyethyl)-1-piperazinyl] ethanesulfonic acid (HEPES), pH 7.4, and 1 m*M* dithiothreitol (DTT) just before sample preparation.
2. Homogenate freshly excised brain and liver tissues in a 10-fold volume (w/v) of ice-cold homogenate buffer using a Potter-Elvehjem homogenizer (ASONE, Osaka, Japan) with 10 strokes up and down at 1000 rpm.
3. Determine the protein concentration by bicinchoninic acid (BCA) method and prepare an aliquot with 4 μg/μl protein concentration by adding the homogenate buffer. This homogenate is referred to as "total."
4. Add homogenate buffer containing 1% Triton X-100 to the aliquot (100 μl) of the total homogenates so that the final concentration of Triton X-100 becomes 0.5%, and mix vigorously.
5. Centrifuge the mixture at 15,000 rpm for 10 min at 4 °C. The resulting supernatant fraction is used as a detergent-soluble fraction.
6. Dissolve the resulting pellet fraction in 200 μl of homogenate buffer containing 1% sodium dodecyl sulfate (SDS). This is used as a detergent-insoluble fraction.

2.2. Western blot procedure

Basically, a conventional method can be applied for protein separation and membrane transfer procedures. We routinely use the NuPage system (Invitrogen, San Diego, CA) with 12% bis-tris gels and MOPS-SDS buffer and polyvinylidene difluoride (PVDF) membrane as a transfer membrane. Immunodetection of ubiquitin and p62 is carried out as follows:

1. Block the membrane with BlockAce (Dainippon Pharmaceutical, Osaka, Japan, #UK-B80) for 30 min at room temperature (RT).
2. Incubate with primary antibodies for 16 h at 4 °C. The antibodies and dilution to be used are summarized in Table 9.1.
3. Wash extensively in 0.01 *M* Tris-buffered saline (150 m*M* NaCl) containing 0.025% Tween 20 (TBST).
4. Incubate with horseradish peroxidase (HRP)–conjugated secondary antibodies (Jackson ImmunoResearch Laboratories, West Grove, Pennsylvania, #315-035-048 for anti-mouse IgG, #111-035-144 for antirabbit IgG, and #106-035-063 for anti–guinea pig IgG) for 30 min at RT.
5. Visualize the protein bands using enhanced chemiluminescence (Western Lightning, Perkin Elmer).

2.3. Comparison of antibodies in Western blot analysis

Here we show the reactivity of some anti-ubiquitin and anti-p62 antibodies listed in Table 9.1. In this experiment, Cre expression was induced in *Atg7*$^{F/F}$:Mx1 mice by peritoneal injection of polyinosinic

Table 9.1 Antibodies used in this study

	Name (supplier, catalog number) & brief description	Dilution in WB	Dilution in IHC
Anti-ubiquitin			
	Mouse mAb IgM, Fk1 (MBL, D071-3)	1:500	—
	Mouse mAb IgG_1, Fk2 (MBL, D058-3)	1:500	—
	Mouse mAb IgG_1,1B3 (MBL, MK-11-3)	1:500	—
	Rabbit pAb (DAKO, Z0458)	1:500	1:400–800
Anti-p62			
	Rabbit pAb, anti-p62/A170 (Ishii *et al.*, 1996)	1:200	1:400–800
	Guinea pig pAb (Progen, GP62-C), Ag: C-terminal domain (20 amino acids)	1:500	1:200-400
	Rabbit pAb, anti-p62/ SQSTM1 (MBL, PM045), Ag: amino acids 120-440	1:500	1:400–800

Supplier information: MBL, Medical & Biological Laboratories (Nagoya, Japan); DAKO, DakoCytomation (Glostrup, Denmark), anti-p62/A170 (kindly provided by Dr. Tetsuro Ishii, [Ishii *et al.*, 1996]); Progen (Heidelberg, Germany). mAb, monoclonal antibody; pAb, polyclonal antibody, Ag, antigen; WB, western blotting; IHC, immunohistochemistry

acid-polycytidylic acid (pIpC, Sigma Chemical, St. Louis, MO, #P1530), which causes *Atg*7-deficiency in the liver (Komatsu *et al.*, 2005). Then, the amounts of ubiquitinated proteins and p62 in the liver were followed, so that accumulation of these proteins could be monitored in both detergent-soluble and detergent-insoluble fractions (please refer to Komatsu *et al.*, [2007] for more detail). As reported previously, immunoblot analysis with the Fk2 antibody revealed that *Atg*7-deficient liver (8 days after pIpC) contained abundant ubiquitinated proteins in both detergent-soluble and detergent-insoluble fractions (Komatsu *et al.*, 2007). A similar accumulation pattern was also observed when antibody Z0458 was applied, whereas 1B3 appeared to recognize ubiquitinated proteins only poorly. It should be noted that Z0458 and 1B3 detected free ubiquitin whose levels were not affected by the autophagy-deficiency. Intriguingly, a blot with Fk1 antibody hardly showed accumulation of ubiquitinated proteins in the insoluble fraction at 8 days, although it did in the soluble one. Because Fk2 recognizes both poly- and monoubiquitinated proteins, whereas Fk1 detects only polyubiquitinated proteins (Haglund *et al.*, 2003), it is possible that the insoluble fraction consists mainly of monoubiquitinated proteins. Further analysis is needed to determine what proteins are ubiquitinated using which manner of ubiquitination (e.g., mono-, poly-, or multiple-monoubiquitination, and the lysine-position on ubiquitin used for the conjugation) in autophagy-deficient mice.

In the preceding mice, p62 was detected as two bands (51- and 45-kDa) by using all three antibodies, and importantly they were absent in *p62*-deficient liver. The lower band represents either a splicing variant of p62, which is found in the mouse protein database, or a partially cleaved product. These doublets seen in three separate blots in Fig. 9.1 were accumulated at 8 days after disruption of the *Atg7* gene, and their accumulation and insolubility patterns were similar to those observed for ubiquitin detected

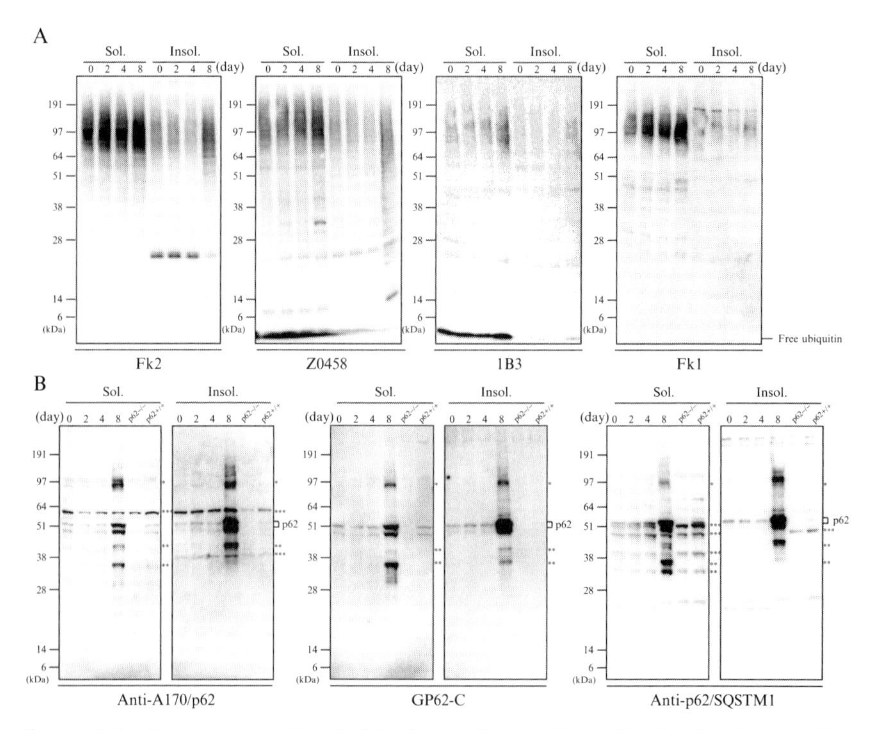

Figure 9.1 Comparison of anti-ubiquitin and anti-p62 antibodies for immunoblot analysis in autophagy-deficient liver. Immunoblot analyses of ubiquitinated proteins (A) and p62 (B) in $Atg7^{F/F}$:Mx1 mice liver at various time points after pIpC injection. Liver homogenates were separated into detergent (0.5% Triton X-100)-soluble (Sol.) and insoluble (Insol.) fractions, and examined by western blotting using the indicated antibodies. Liver lysate from *p62*-deficient mice (Komatsu *et al.*, 2007) was used as negative control for p62 detection. Data shown are representative of three separate experiments. Note: * and ** indicate specific high and low molecular mass bands detected by anti-p62 antibodies, respectively; *** indicates nonspecific bands.

by Fk2 and Z0458. However, blots with anti-p62/SQSTM1 and anti-A170/p62 (Ishii *et al.*, 1996) antibodies showed some nonspecific bands (triple asterisks, Fig. 9.1B) that appeared even in lysates from *p62*-deficient mouse liver, thus care should be taken when using these antibodies. Interestingly, additional bands were also commonly recognized by the three anti-p62 antibodies in autophagy-deficient mice. The high molecular mass bands (single asterisk, Fig. 9.1B) might be tightly aggregated p62 (Aono *et al.*, 2003), while some of the lower bands (double asterisks, Fig. 9.1B) might correspond to the cleaved forms, although no direct evidence for this could be established. The ability of p62 to become insoluble and to segregate and form inclusions is dependent on the PB1 domain (Bjørkøy *et al.*, 2005; Ichimura *et al.*, 2008). Therefore, we speculate that the lower bands that are absent in the insoluble fraction lack the PB1 domain, and conversely, those in both the soluble and insoluble fractions might be cleaved forms including this domain. It is not clear at this stage whether these unknown but specific bands are pathophysiologically important in autophagy-deficient mice.

3. Detection of Ubiquitinated Proteins and p62 in Cultured Hepatocytes Derived from Autophagy-Deficient Mice

Researchers may also want to detect ubiquitin/p62-inclusions in primary cell cultures. We have successfully detected numerous inclusions in isolated hepatocytes, but only rarely in mouse embryonic fibroblasts from *Atg7*-deficient mice (our unpublished data). It is thought that in dividing cells such as embryonic fibroblasts, the concentration of p62 is diluted by repetitive mitosis, which suppresses the formation of inclusions, or that another degradation pathway(s) may compensate for the deficiency in autophagy. Here, immunofluorescence detection of the inclusions in isolated autophagy-deficient hepatocytes is described. For the methods of hepatocyte isolation from mice liver, please refer to Ueno *et al.* (1990).

3.1. Immunofluorescence microscopy in isolated hepatocytes

1. Hepatocytes were plated at 2×10^5 cells in Glass Bottom Culture Dish (MatTek Corporation, Ashland, MA, #P35GC-0-10-C) coated with 0.03% CELLGEN (KOKEN, Tokyo, Japan, #IAC-15), and cultured with 5% CO_2 at 37 °C for 24 h in Williams's Medium E (Invitrogen, #12551) supplemented with 10 % FCS. The hepatocytes should be exposed to the desired experimental regimen, such as amino acid deprivation. In this case, remove the medium by aspiration, wash the

cell with Hanks's Buffered Salt Solution (HBSS) 2 times, followed by the incubation with HBSS including 10 m*M* HEPES for 2 h.

2. Fix in 4% paraformaldehyde (PFA) in 0.1 *M* phosphate buffer, pH 4 (PB) for 15 min.
3. Permeabilize with 50 μg/ml digitonin (Wako, Osaka, Japan, #040-02123) in PB for 5 min.
4. Quench the free aldehyde in 50 m*M* ammonium chloride for 10 min.
5. Block with 0.1% (vol/vol) gelatin (Sigma Chemical, #G-9391) in PB for 30 min.
6. Incubate with anti-ubiquitin (Fk2) and anti-p62/A170 for 1 h at 37 °C.
7. Wash with 0.1% (vol/vol) gelatin in PB.
8. Incubate with Alexa-488 anti-mouse (Molecular Probes, Eugene, OR, #A-11059) and Alexa-647 anti-rabbit antibodies (Molecular Probes, #A-21245) for 30 min.
9. Wash with 0.1% (vol/vol) gelatin in PB and mount on slides with a medium with SlowFade Gold antifade reagent (Invitrogen, #S36936).
10. Examine by epifluorescence microscopy or confocal laser-scanning microscopy.

4. Detection of Ubiquitin- and p62-Positive Inclusions at the Light Microscopy Level

With good antibodies for ubiquitin or p62, any established laboratory method of immunohistochemistry can be applied to detect ubiquitin/p62-inclusions. In principle, this method detects cytoplasmic aggregates. Thus, we can describe the structures according to their shape, size, and number. These morphological features and their alterations might be correlated with the changes in ubiquitin/p62 in the insoluble-fraction examined in Western blot analysis. On the other hand, a diffuse cytoplasmic signal for ubiquitin or p62 might correspond to the soluble form. Evaluation of this type of signal, however, is often difficult, because its intensity varies according to the procedure used for tissue fixation, embedding, immunoreaction and other conditions, and even the observed area. The application of ubiquitin- and p62-immunohistochemistry using paraffin-embedded human tissues was described in recent publications (Lowe *et al.*, 2005; Kuusisto *et al.*, 2008), which are recommended for reading if necessary. Here, we focus on the methods using paraformaldehyde-fixed cryosections from autophagy-deficient mice.

4.1. Fixation and sample preparation

1. *Atg7*-deficient mice are deeply anesthetized with pentobarbital (25 mg/kg intraperitoneally) and perfused via the heart, first with

approximately 30 ml of Lactic injection (Otsuka Pharmaceutical, Tokyo, Japan, #4239) and then with 50 ml of 4% PFA buffered with 0.1 *M* PB, pH 7.4, containing 4% sucrose. A stock solution of 8 or 10% PFA should be freshly prepared and used within 2–3 days.

2. Brain and liver tissues are quickly removed from the mice. The brain is cut in 3- to 5-mm-thick coronal (e.g., the sagittal plane may be cut if necessary). The liver is cut into small pieces.
3. The brain and liver samples are then immersed in the same fixative for 2 h at 4 °C, and washed 3 times with the same buffer containing 7.5% sucrose.
4. Cryoprotection is carried out by successively infusing with 15% and 30% sucrose solutions. The tissue blocks are briefly (approximately 5 min) immersed in optimal cutting temperature (OCT)/sucrose solution prepared by mixing 20% sucrose and OCT-compound (Sakura Finetek Japan, Tokyo, Japan, #4583) at a ratio of 2:1, and then frozen in isopentane with dry ice.
5. Cryosections are cut into 10-μm-thick sections with a cryostat (CM3050S, Leica, Nussloch, Germany) and mounted on silane-coated glass slides (DakoCytomation, Glostrup, Denmark, #S3003).
6. The sections are dried for at least 30 min and stored at −30 °C until use.

4.2. Immunohistochemistry on cryosections

Conventional methods can be applied, except for when mouse antibodies are to be applied to mouse tissues. It is usually recommended not to use mouse antibodies because the secondary anti-mouse antibody recognizes endogenous IgG in the mouse tissue sections, which results in a high level of background staining. To overcome this problem, immunohistochemistry kits with improved blocking reagents are available from Nichirei (Tokyo, Japan, #414321) and Invitrogen (#85-9541). Nevertheless, we sometimes encounter relatively high background staining especially in autophagy-deficient liver. Thus, well-designed control experiments are necessary in this case.

1. Immerse cryosections in a washing solution consisting of 0.01 *M* PB, pH 7.2, 0.5 *M* NaCl, and 0.1% Tween 20 (TPBS) for 5–10 min.
2. Incubate in methanol containing 0.3% H_2O_2 for 30 min for inactivation of endogenous peroxidase.
3. Incubate with a blocking/dilution solution consisting of 0.01 *M* PB, pH 7.2, 0.15 *M* NaCl, 0.05% Tween 20, and 1% normal goat serum (NGS, Vector Laboratories, Burlingame, CA, #S-1000) for 20 min at RT.
4. Dilute the rabbit anti-ubiquitin (Z0458) or goat anti-p62 (Progen) antibody with the above blocking/dilution solution and incubate for 2–3 days at 4 °C or 1 h at RT in a humidified chamber.
5. Rinse with TPBS for 5–10 min, 3 times.

6. For rabbit antibodies, incubate with biotinylated goat anti-rabbit IgG (Vectastain ABC Rabbit IgG Kit, Vector Laboratories, #PK-4001) for 1 h at RT, rinse with TPBS, and then incubate with avidin-biotinylated peroxidase complex (Vectastain ABC Rabbit IgG Kit) for 30 min at RT according to the instructions provided by the manufacturer. For guinea pig antibody, incubate with donkey anti-guinea pig IgG conjugated with HRP (Chemicon, Temecula, CA, #AP193P) for 1 h at RT.
7. Rinse with TPBS for 5–10 min, 3 times
8. Incubate in 0.05 *M* Tris-HCl buffer, pH 7.6, containing 0.0125% diaminobenzidine (DAB, Dojindo Laboratories, Kumamoto, Japan, #349-0093) and 0.002% H_2O_2 for 5–10 min.
9. Immerse successively in 70, 80, 90, 95, 100, and 100% alcohol for 5–10 min each for dehydration, and finally in xylene twice for 5–10 min each, and mount with Canada balsam (Wako, #034-01042).
10. View with a Nomarski differential interference contrast microscope (Olympus, Tokyo).

4.3. Comparison of anti-p62 antibodies in immunohistochemistry

Using the preceding method, we compared immunoreactivities of three anti-p62 antibodies in *Atg7*-deficient liver and brain. In this experiment, we included *Atg7/p62* double-knockout tissues as control tissues, in which p62 should be absent, and as a consequence, the number of inclusion bodies is greatly reduced (Komatsu *et al.*, 2007). The cryosections were immunolabeled with anti-p62 antibodies listed in Table 9.1. In *Atg7*-deficient liver, anti-A170/p62, GP62-C, and anti-p62/SQSTM1 showed similar punctate signals in hepatocytes, which were absent in *Atg7/p62* double-knockout liver (Fig. 9.2A–F). It is noteworthy that when GP62-C and anti-p62/SQSTM1 were used, *Atg7*-deficient hepatocytes often showed a weak diffuse signal in the cytoplasmic region, which was also detected in the *Atg7/p62* double-knockout liver (Fig. 9.2E–F). Therefore, the immunoreactivities probably represent nonspecific signals. In the case of anti-p62/SQSTM1, this background signal may correspond to the bands observed in the soluble fraction in Western blotting. On the other hand, in the *Atg7*-deficient brain (motor cortex), the three antibodies stained the inclusion bodies very well (Fig. 9.2G–I), which were apparently absent in the *Atg7/p62* double-knockout brain (Fig. 9.2J–L). The results suggest that although all three antibodies could be used for the detection of inclusions in conventional immunohistochemistry, caution should be taken when the aim of the study is evaluation of cytoplasmic diffuse staining, which could reflect staining of molecules other than p62.

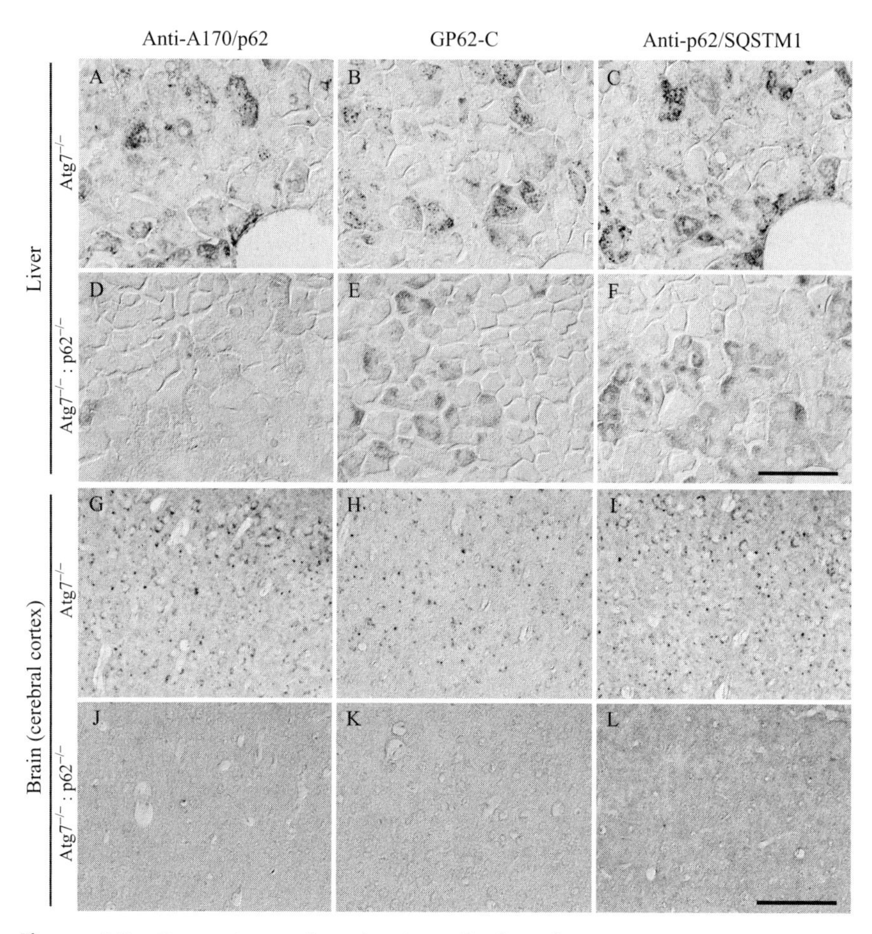

Figure 9.2 Comparison of anti-p62 antibodies for immunohistochemistry in autophagy-deficient tissues. The liver (A–F) and brain (G–L) from *Atg7*$^{-/-}$ (A–C and G-I) or *Atg7*$^{+/-}$*:p62*$^{-/-}$ (D–F and J–L) mice were subjected to immunohistochemistry using anti-A170/p62 (A, D, G, and J), GP62-C (B, E, H, and K), and anti-p62/ SQSTM1 (C, F, I, and L) as indicated. Bars: 100 μm.

4.4. Immunofluorescence microscopy on cryosections

This technique is useful when double-labeling has to be applied. Preparation of the cryosections and immunofluorescence reactions are described previously. Although inactivation of intrinsic peroxidase is unnecessary because a secondary antibody conjugated with an appropriate fluorescent dye is used, it should be noted that removing the step of treatment with methanol/H_2O_2 could either improve or delete the signal for some proteins probably due to changes in their molecular conformation.

5. Detection of Ubiquitin- and p62-Positive Inclusions at the Electron Microscopy Level

Because ubiquitin and p62 appear to be highly concentrated in the inclusions, it is rather easy to detect them in ultrathin cryosections. However, sectioning, staining, and final embedding techniques require experience. Moreover, at the step of observation, defining the aggregate structures in a small specific tissue region or cell (especially in the brain) also requires expert histological knowledge. Therefore, it is strongly recommended that the immunohistochemical analyses described above and conventional electron microscopy (EM) should be mastered before performing immuno-EM.

5.1. Morphological detection of inclusions by conventional EM

Understanding the morphological aspects of ubiquitin/p62-inclusions in conventional EM is important before applying immuno-EM, because the morphology is usually damaged in the latter technique, making it difficult to identify the object of interest. Moreover, accumulation of aberrant structures other than inclusions is also common in autophagy-deficient liver and brain (Komatsu *et al.*, 2005, 2006), and some cells show several features of cell damage or cell death, some of which may look like aggregated structures in immuno-EM. Furthermore, the morphology of the ubiquitin/p62-inclusions themselves is different depending on the cell type. For example, the inclusions in autophagy-deficient liver contain lipid dropletlike structures, membranous structures, and amorphous substances, and are usually surrounded by clusters of vesicular structures (Fig. 9.3A; Komatsu *et al.*, 2005), whereas those in autophagy-deficient neurons consist mainly of fibrillar elements (Fig. 9.3B; Komatsu *et al.*, 2007). We also found other types of aggregated structures in some neurons (our unpublished data). Here, we briefly describe our routine procedure for conventional EM. Other established methods used in individual institutes can also be adopted (see also the chapter by Ylä-Anttila *et al.*, in volume 452).

1. The mice are fixed with 2% PFA-2% glutaraldehyde (GA) buffered with 0.1 *M* PB, pH 7.4.
2. The brain is cut into 0.2- to 0.5-mm-thick slices using a microslicer so that the specific areas of interest can be excised in later steps. The liver is cut into small pieces (1 × 1 mm).
3. The brain and liver samples are further immersed in the same fixative for 24 h at 4 °C.
4. After washing 3 times with the same buffer containing 7.5% sucrose, the samples are postfixed with 1% OsO_4 in the same buffer containing 7.5%

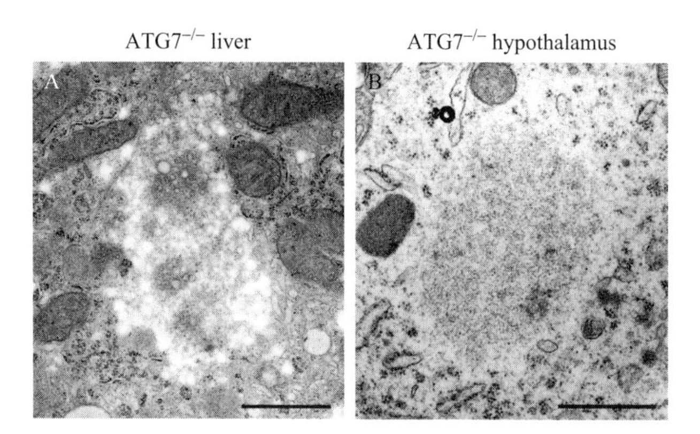

Figure 9.3 Fine structures of inclusions in autophagy-deficient tissues. The liver and brain from *Atg7*$^{-/-}$ mice were subjected to conventional microscopy as described in the text. Inclusions in hepatocytes (A) and hypothalamic neurons (B) are shown. Note the different morphology of the inclusions among tissues. Bars: 1 μm.

sucrose at 4 °C for 2 h, dehydrated using graded alcohol solutions as described previously, and embedded in Epon 812.

5. Ultrathin sections (approximately 60 nm thickness) are cut using a microtome, stained for 5 min with uranyl acetate (Wako, #219-00692) that is prepared by diluting saturated uranyl acetate solution with acetone at 1:3, and then with 0.25%–0.3% lead citrate (TAAB, Berkshire, UK, #L018) for 5 min. They were observed under an electron microscope (JEM1200EX, JEOL).

5.2. Immuno-EM on ultrathin-cryosections

We usually follow the method previously established by Tokuyasu (1986).

1. Briefly, mice are fixed by cardiac perfusion with 4% PFA buffered with 0.1 *M* PB, pH 7.4, containing 4% sucrose. Although including 0.01%–0.1% GA is known to greatly improve morphology, it also reduces the antigenicity. The optimal fixation condition should be determined case by case to obtain both a good signal and morphological preservation.
2. Tissues are cut into small pieces, washed, and infused with 1.84 *M* sucrose and 20% polyvinylpyrrolidone (PVP; Sigma, #PVP40). To prepare the sucrose/PVP solution, mix 80 ml of 2.3 *M* sucrose, 0.1 *M* PB, pH 7.4, 4 ml of 0.25 *M* $NaCO_3$, and 20 g of PVP.
3. After rapidly freezing in liquid nitrogen, ultrathin cryosections are cut using a microtome (Ultracut UCT, Leica) equipped with a cryochamber.
4. The sections are picked up with a drop of solution containing 1.15 *M* sucrose and 1% methylcellulose and placed on Formvar/carbon-coated nickel grids.

5.3. Immunoreactions

1. Incubate with 0.01 *M* glycine in PBS for 5 – 10 min at RT to quench free aldehyde and to remove sucrose.
2. Incubate with 1% bovine serum albumin (BSA, Sigma, #A4503) in PBS for 20 min at RT.
3. Incubate with anti-ubiquitin (1B3) and/or anti-p62/A170 antibody for 1 h at RT or 4 °C overnight.
4. Rinse in 0.1% BSA/PBS for 1–5 min, 3–5 times.
5. Incubate with the secondary antibody conjugated with colloidal gold (5 nm or 10 nm in diameter; GE Healthcare [Uppsala, Sweden, #RPN420/421 for anti-rabbit IgG and #RPN424/425 for anti-mouse IgG]) for 1 h at RT.
6. Rinse in 0.1% BSA/PBS for 1–5 min, 3–5 times.
7. Fix with 2% GA/PBS for 5 min.
8. Rinse in MilliQ-water for 1–5 min, 3–5 times.
9. Stain with 2% uranyl acetate for 5 min.
10. Rinse briefly in MilliQ-water.
11. Embed in 0.33% polyvinyl alcohol (PVA; Sigma, #P1763) and 0.17% uranyl acetate. To prepare this solution, mix 2% PVA and 0.2% uranyl acetate at a ratio of 1:5.
12. Dry for 30 min.
13. Observe with an electron microscope (JEM1200EX, JEOL).

6. Conclusion

We described the methods used for detection of ubiquitin-inclusions in *Atg7*-deficient mice, which mainly include biochemical and morphological techniques. Both sets of methods are complementary to each other and should be used in combination to understand the process of inclusion formation.

Investigation of the new marker, p62, in addition to the well-known ubiquitin, should enhance our understanding of conformation diseases associated with dysfunctional autophagy machinery. Importantly, $Atg7^{F/F}$ mice should be used for establishing various mice models with disrupted autophagy in specific tissues or cells. These mice could be used for investigating yet unsolved important pathological mechanisms. Recent studies show reduced autophagic activity in rat liver with aging, a finding that conversely correlates with increased accumulation of oxidized proteins (Bergamini, 2006). Thus, age-dependent onset of human disorders most likely correlates with the age-dependent decline of autophagic activity. Moreover, two recent studies report reduced autophagic activity in a model mouse of Alzheimer's disease and in affected brain regions of Alzheimer's patients (Boland *et al.*, 2008; Pickford *et al.*, 2008). The methods described here could be adopted to examine various mice models of autophagic dysfunction (e.g., autophagy deficiency, aging, and a mouse model of Alzheimer's disease), which would enhance our understanding of the pathophysiological mechanisms involved in the formation of intracellular inclusions.

ACKNOWLEDGMENTS

We thank A. Yabashi, K. Kanno, and T. Kouno for the excellent technical assistance, and Dr. Ishii for providing anti-p62/A170 antibody. This work was supported in part by grants from the Ministry of Education, Science and Culture of Japan (S.W. and M.K) and the Japan Science and Technology Agency (M.K.)

REFERENCES

Aono, J., Yanagawa, T., Itoh, K., Li, B., Yoshida, H., Kumagai, Y., Yamamoto, M., and Ishii, T. (2003). Activation of Nrf2 and accumulation of ubiquitinated A170 by arsenic in osteoblasts. *Biochem. Biophys. Res. Commun.* **305,** 271–277.

Bergamini, E. (2006). Autophagy: A cell repair mechanism that retards ageing and age-associated diseases and can be intensified pharmacologically. *Mol. Aspects. Med.* **27,** 403–410.

Bjørkøy, G., Lamark, T., Brech, A., Outzen, H., Perander, M., Øvervatn, A., Stenmark, H., and Johansen, T. (2005). p62/SQSTM1 forms protein aggregates degraded by autophagy and has a protective effect on huntingtin-induced cell death. *J. Cell. Biol.* **171,** 603–614.

Boland, B., Kumar, A., Lee, S., Platt, F. M., Wegiel, J., Yu, W. H., and Nixon, R. A. (2008). Autophagy induction and autophagosome clearance in neurons: Relationship to autophagic pathology in Alzheimer's disease. *J. Neurosci.* **28,** 6926–6937.

Goldberg, A. L. (2003). Protein degradation and protection against misfolded or damaged proteins. *Nature* **426,** 895–899.

Haglund, K., Sigismund, S., Polo, S., Szymkiewicz, I., Di Fiore, P. P., and Dikic, I. (2003). Multiple monoubiquitination of RTKs is sufficient for their endocytosis and degradation. *Nat. Cell Biol.* **5,** 461–466.

Hara, T., Nakamura, K., Matsui, M., Yamamoto, A., Nakahara, Y., Suzuki-Migishima, R., Yokoyama, M., Mishima, K., Saito, I., Okano, H., and Mizushima, N. (2006). Suppression of basal autophagy in neural cells causes neurodegenerative disease in mice. *Nature.* **441,** 885–889.

Ichimura, Y., Kumanomidou, T., Sou, Y. S., Mizushima, T., Ezaki, J., Ueno, T., Kominami, E., Yamane, T., Tanaka, K., and Komatsu, M. (2008). Structural basis for sorting mechanism of p62 in selective autophagy. *J. Biol. Chem.* **283,** 22847–22857.

Ishii, T., Yanagawa, T., Kawane, T., Yuki, K., Seita, J., Yoshida, H., and Bannai, S. (1996). Murine peritoneal macrophages induce a novel 60-kDa protein with structural similarity to a tyrosine kinase p56lck-associated protein in response to oxidative stress. *Biochem. Biophys. Res. Commun.* **226,** 456–460.

Itoh, T., Fujita, N., Kanno, E., Yamamoto, A., Yoshimori, T., and Fukuda, M. (2008). Golgi-resident small GTPase Rab33B interacts with Atg16L and modulates autophagosome formation. *Mol. Biol. Cell.* **19,** 2916–2925.

Komatsu, M., Waguri, S., Chiba, T., Murata, S., Iwata, J., Tanida, I., Ueno, T., Koike, M., Uchiyama, Y., Kominami, E., and Tanaka, K. (2006). Loss of autophagy in the central nervous system causes neurodegeneration in mice. *Nature* **441,** 880–884.

Komatsu, M., Waguri, S., Koike, M., Sou, Y. S., Ueno, T., Hara, T., Mizushima, N., Iwata, J., Ezaki, J., Murata, S., Hamazaki, J., Nishito, Y., *et al.* (2007). Homeostatic levels of p62 control cytoplasmic inclusion body formation in autophagy-deficient mice. *Cell* **131,** 1149–1163.

Komatsu, M., Waguri, S., Ueno, T., Iwata, J., Murata, S., Tanida, I., Ezaki, J., Mizushima, N., Ohsumi, Y., Uchiyama, Y., Kominami, E., Tanaka, K., *et al.* (2005). Impairment of starvation-induced and constitutive autophagy in Atg7-deficient mice. *J. Cell Biol.* **169,** 425–434.

Kuusisto, E., Kauppinen, T., and Alafuzoff, I. (2008). Use of p62/SQSTM1 antibodies for neuropathological diagnosis. *Neuropathol. Appl. Neurobiol.* **34,** 169–180.

Lowe, J., Hand, N., and Mayer, R. J. (2005). Application of ubiquitin immunohistochemistry to the diagnosis of disease. *Methods. Enzymol.* **399,** 86–119.

Nakai, A., Yamaguchi, O., Takeda, T., Higuchi, Y., Hikoso, S., Taniike, M., Omiya, S., Mizote, I., Matsumura, Y., Asahi, M., Nishida, K., Hori, M., *et al.* (2007). The role of autophagy in cardiomyocytes in the basal state and in response to hemodynamic stress. *Nat. Med.* **13,** 619–624.

Nakano, T., Nakaso, K., Nakashima, K., and Ohama, E. (2004). Expression of ubiquitin-binding protein p62 in ubiquitin-immunoreactive intraneuronal inclusions in amyotrophic lateral sclerosis with dementia: Analysis of five autopsy cases with broad clinicopathological spectrum. *Acta Neuropathol.* **107,** 359–364.

Nezis, I. P., Simonsen, A., Sagona, A. P., Finley, K., Gaumer, S., Contamine, D., Rusten, T. E., Stenmark, H., and Brech, A. (2008). Ref(2)P, the *Drosophila melanogaster* homologue of mammalian p62, is required for the formation of protein aggregates in adult brain. *J. Cell. Biol.* **180,** 1065–1071.

Pankiv, S., Clausen, T. H., Lamark, T., Brech, A., Bruun, J. A., Outzen, H., Øvervatn, A., Bjørkøy, G., and Johansen, T. (2007). p62/SQSTM1 binds directly to Atg8/LC3 to facilitate degradation of ubiquitinated protein aggregates by autophagy. *J. Biol. Chem.* **282,** 24131–24145.

Pickford, F., Masliah, E., Britschgi, M., Lucin, K., Narasimhan, R., Jaeger, P. A., Small, S., Spencer, B., Rockenstein, E., Levine, B., and Wyss-Coray, T. (2008). The autophagy-related protein beclin 1 shows reduced expression in early Alzheimer disease and regulates amyloid β accumulation in mice. *J. Clin. Invest.* **118,** 2190–2199.

Stumptner, C., Fuchsbichler, A., Heid, H., Zatloukal, K., and Denk, H. (2002). Mallory body-a disease-associated type of sequestosome. *Hepatology* **35,** 1053–1062.

Tokuyasu, K. T. (1986). Application of cryoultramicrotomy to immunocytochemistry. *J. Microsc.* **143,** 139–149.

Ueno, T., Watanabe, S., Hirose, M., Namihisa, T., and Kominami, E. (1990). Phalloidin-induced accumulation of myosin in rat hepatocytes is caused by suppression of autolysosome formation. *Eur. J. Biochem.* **190,** 63–69.

Wang, Q. J., Ding, Y., Kohtz, D. S., Mizushima, N., Cristea, I. M., Rout, M. P., Chait, B. T., Zhong, Y., Heintz, N., and Yue, Z. (2006). Induction of autophagy in axonal dystrophy and degeneration. *J. Neurosci.* **26,** 8057–8068.

Wooten, M. W., Hu, X., Babu, J. R., Seibenhener, M. L., Geetha, T., Paine, M. G., and Wooten, M. C. (2006). Signaling, polyubiquitination, trafficking, and inclusions: Sequestosome 1/p62's role in neurodegenerative disease. *J. Biomed. Biotechnol.* **2006,** 62079.

Zatloukal, K., Stumptner, C., Fuchsbichler, A., Heid, H., Schnoelzer, M., Kenner, L., Kleinert, R., Prinz, M., Aguzzi, A., and Denk, H. (2002). p62 Is a common component of cytoplasmic inclusions in protein aggregation diseases. *Am. J. Pathol.* **160,** 255–263.

CHAPTER TEN

Analyzing Autophagy in Clinical Tissues of Lung and Vascular Diseases

Hong Pyo Kim,[*,†] Zhi-Hua Chen,[†] Augustine M. K. Choi,[†] *and* Stefan W. Ryter[*,†]

Contents

Abstract

Autophagy, a process by which organelles and cellular proteins are encapsulated in double-membrane vesicles and subsequently degraded by lysosomes, plays a central role in cellular and tissue homeostasis. In various model systems, autophagy may be triggered by nutrient deprivation, oxidative stress, and other insults such as endoplasmic reticulum stress, hypoxia, and pathogen infection. The role of autophagy in lung physiology and homeostasis, however, has not been well studied. Even less is known of the role of autophagy

* Division of Pulmonary, Allergy and Critical Care Medicine, Department of Medicine, University of Pittsburgh, Pittsburgh, Pennsylvania, USA

† Division of Pulmonary and Critical Care Medicine Brigham and Women's Hospital, Harvard Medical School, Boston, Massachusetts, USA

Methods in Enzymology, Volume 453
ISSN 0076-6879, DOI: 10.1016/S0076-6879(08)04010-X

in the pathogenesis of chronic lung disease. Autophagy may act essentially as a protective mechanism in lung cells, by removing dysfunctional organelles, and recycling essential nutrients. On the other hand, excessive autophagy may also contribute to cell death pathways, resulting in the depletion of critical cell populations, and thus may also contribute to the disease pathogenesis. An understanding of the cell-type specific regulation and function of autophagy in the lung may facilitate the development of therapeutic strategies for the treatment of lung pathologies. This chapter provides protocols for the isolation of distinct lung cell types, such as epithelial, endothelial, macrophages, and fibroblasts; as well as protocols for the analysis of autophagy in lung cells and tissues.

1. Introduction

Autophagy is a regulated pathway for internal organelle or protein degradation (Kelekar, 2006). This dynamic process is characterized by the formation of distinct cytoplasmic double membrane-bound vesicles, which sequester organelles for delivery to the lysosome or vacuole where proteins and other cell constituents are degraded and recycled (Kondo *et al.*, 2005; Mizushima *et al.*, 2008). Autophagy plays a central role in cell homeostasis and adaptation to environmental stress conditions such as oxidative stress, serum starvation, endoplasmic reticulum stress, hypoxia, and pathogen infection. The morphological and biochemical features of autophagy and apoptosis are distinct. Cells undergoing autophagy display an increase in autophagic vesicles (autophagosomes and autolysosomes) (Kondo *et al.*, 2005; Mizushima *et al.*, 2008; Kiffin *et al.*, 2006). Although partial chromatin condensation appears in autophagic cells, DNA fragmentation does not occur (Kondo *et al.*, 2005). The Bcl-2 interacting protein Beclin 1 and the microtubule-associated protein 1 light chain 3 (LC3), represent major regulators of autophagy and autophagosome formation in mammalian cells. Beclin 1 and LC3 are the mammalian homologues of the yeast autophagy-associated (Atg) genes *ATG6/VPS30* (*Liang et al.*, 1999) and *ATG8* (He *et al.*, 2003), respectively. In addition, a number of other autophagy-associated proteins have been characterized (e.g., Atg5, Atg12, Atg7) (Mizushima *et al.*, 2008).

In general, autophagy may be considered a major homeostatic function of most organ tissues, though currently little is known of the specific role of this process in the lung. Autophagy may promote survival under stress conditions through the removal of damaged organelles and by recycling essential metabolic building blocks. On the other hand, excessive autophagy has been also associated with promotion of various forms of cell death, including autophagic cell death (Type II programmed cell death) and/or apoptosis (Type I programmed cell death). The increased expression of autophagic vesicles in dying cells may indicate that autophagy is involved with cell death; however, the complex relationships between autophagy and cell death mechanisms remain incompletely understood (Cuervo 2004; Shimizu *et al.*, 2004). Furthermore, the specific roles of autophagy in the pathogenesis of lung injury and disease remain unknown, and are currently under intensive investigation in our laboratory.

The lung is a complex organ consisting of over forty specialized cell types. Among these are the epithelial cells of the airways, bronchii, and alveoli, interstitial fibroblasts of the lung parenchyma, endothelial cells of the pulmonary vasculature, and smooth muscle cells that provide the contractility of the airway and pulmonary vasculature. Furthermore, the lung contains resident inflammatory cells such as alveolar macrophages, and other specialized cells involved in immune function or secretion (Menzel and Amdur, 1986). To study adaptive or pathological responses of the lung in response to environmental stress and disease, it is of critical importance to understand the cell-type specificity of their origin, regulation, and function. To this end, the morphological and biochemical features of autophagy can be analyzed in individual lung cell types exposed to various environmental stimuli, as well as in whole lung tissue from animal models of disease or clinical specimens. This chapter provides protocols for the isolation of distinct pulmonary cell types, as well as for the analysis of autophagy and autophagic markers in lung cells and tissues.

2. Methods for Preparation of Lung and Vascular Cells

The following protocols describe the isolation of major pulmonary cell types, to facilitate the analysis of pulmonary cell autophagy.

2.1. Endothelial cells

1. For isolation of mouse lung endothelial cells (MLEC) (Tang *et al.*, 2002; Wang *et al.*, 2005), mice are sacrificed by standard methods according to individual institutional approval.
2. Prepare a collagenase digest of isolated lung tissue.
   a. Dip the mouse in 70% ethanol.
   b. Open the chest to expose the heart and the lung.
   c. Cut off the left atrium, and perfuse the lung with 4 ml of Hank's Balanced Salt Solution (HBSS) containing heparin (10 U/ml) through the left ventricle.
   d. Repeat the perfusion step through the right ventricle with 4 ml of HBSS containing heparin.

e. Carefully remove and place the lung in a culture plate (60 mm) containing 2 mL of HBSS with heparin.
f. Rinse the lung 3 times with HBSS/heparin and trim the unwanted connective tissues.
g. Digest the lung with 5 ml of collagenase solution (collagenase Type 1A, Sigma Chemical Co., St. Louis MO, #C9722), 2 mg/ml in HBSS, fortified with 2 m*M* $CaCl_2$ or commercially available HBSS containing Ca^{2+} and Mg^{2+} supplement) for 20–25 min in a 37 °C incubator.
h. Stop the enzymatic reaction by adding 10 ml Opti-MEM containing 10% fetal bovine serum (FBS) (Gibco Invitrogen, Carlsbad, CA, #31985-062).
i. Disrupt the lung by pushing through a 30-ml syringe with an 18-G needle. Run cell slurry through the syringe 5 times.
j. Successively filter the slurry through 100-μm and then 70-μm cell strainers (BD Falcon, Franklin Lakes, NJ, #352350) into a fresh 50-ml conical tube.
k. Centrifuge the cell suspension at 1000×*g* for 5 min and save the pellet fraction, discarding the supernatant fraction.
l. Resuspend the cell pellet in 3 ml of red blood cell lysis buffer (Sigma, #R7757) for 1 min and dilute the solution with 30 ml of Opti-MEM containing 10% FBS.
m. Centrifuge the cell suspension at 1000×*g* for 5 min and save the pellet fraction.

3 Purify the endothelial cell fraction by immunoisolation techniques.
a. [Note: Perform steps a–c the day before harvesting of tissue]. For immunoisolation, prepare a 10X binding buffer (1.81 g $NaH_2PO_4 \cdot 2H_2O$, 7.93 g Na_2HPO_4, 81 g of NaCl, 4 ml of 0.5 *M* EDTA stock solution, H_2O to 1 L, and adjust to pH 7.4).
b. Prepare 1X binding buffer (200 ml per isolation) by diluting 10X binding buffer, and adjust the pH to 7.4. Add bovine serum albumin (BSA) (final concentration, 0.5%; Sigma, #A9418) just before use.
c. For isolation of cells from 5 mice, prepare a mixture of 80 μl of rabbit antimouse PECAM (M-185) antibody (Santa Cruz Biotechnology, Santa Cruz, CA, #sc-28188) with 130 μl of Dynabeads (M-280 sheep antirabbit IgG, Dynal, Invitrogen Corporation, Carlsbad, CA, #112-03D) in 2 ml of 1X binding buffer with BSA. Incubate for 18 h on a rotator at 4 °C. Coat tissue culture plates with Gelatin (Sigma, #48723). Prepare a 0.2% gelatin solution in double distilled H_2O and filter sterilize. Add 1 ml of gelatin solution per 100-mm plate and gently coat the surface in a tissue culture hood. Remove excess gelatin and let the plates dry overnight at 37 °C.
d. Wash cells once with 1X binding buffer.
e. Wash beads 3 times in 1X binding buffer. Resuspend the cells in of the prepared bead solution (2 ml of 1X binding buffer containing Dynabeads

conjugated with antimouse PECAM antibody). Incubate for 3 h on rotator at 4 °C.

f. Transfer suspension in a 15-mL conical tube and place on a magnetic tube rack (e.g., Dynamag-15, Dynal, Invitrogen, Carlsbad, CA, #123-01D). Wash the beads 5 times with Opti-MEM containing 10% FBS to remove unbound cells.

g. Remove the tubes from the magnetic rack and collect the beads bound to cells with 2 ml of complete endothelial cell medium. Endothelial cell medium consists of high glucose DMEM (Gibco Invitrogen, #10313-039), 3.16 g of HEPES/500 ml, containing 15% FBS, endothelial cell growth supplement (30 μg/ml) (Sigma, catalog number E2759), L-glutamine (2 m*M*) (Gibco Invitrogen, #25030-081), and penicillin/streptomycin solution (1%) (Gibco Invitrogen, #15140-122).

h. Plate cell suspension on gelatin-coated culture plates.

i. For further purification, a flow cytometric cell sorter may be used to select cells conjugated with diacetyl-low density lipoprotein-fluorophore (optional) (Tang *et al.*, 2002).

2.2. Smooth muscle cells

The sources of lung smooth muscle cells are upper airway, pulmonary artery and aorta (Sreejayan *et al.*, 2007, Mahabeleshwar *et al.*, 2006). Here, we describe the isolation methods from aorta since it is considered to represent the general case.

1. Isolate the aorta.
   a. Remove the skin from the thorax and open the thorax to expose the heart and lung.
   b. Cut out the thoracic aorta from its origin just above the heart to the iliac bifurcation.
   c. Place the aorta in a culture plate containing prewarmed HBSS solution with penicillin/streptomycin.
   d. Gently trim the connective tissue and adventitia from the aorta under a dissection microscope. The adventitia will peel off the aorta as a single unit.
   e. Wash aorta with HBSS by pipetting out with a wide-bore glass pipette.

2. Enzymatic digestion.
   a. Digest the chopped aorta into small pieces with digestion enzymes (15 U/ml elastase (Sigma, #E0127), 200 U/ml collagenase Type IA (Sigma, #C9722), 0.4 mg/ml trypsin inhibitor (Sigma, #T0256) and 0.2% BSA in Mg^{2+}, Ca^{2+} -fortified HBSS) for 60 min at 37 °C with gentle shaking.

b. Stop the enzymatic reaction by centrifugation (1000×*g*, 5 min).
c. Wash the pellet fraction 3 times by swirling gently with a 5-ml glass pipette.
d. Dissociate to single cells by pipetting the pellet up and down with a glass pipette.
e. Collect the cells in the supernatant by spinning at 100×*g* for 30 s.
f. Repeat the dissociation and collecting step 3 times, and discard the supernatant.
g. Plate the cells at a density of $2.5 \sim 3\times 10^4$ cells/cm^2, on 60-mm cell culture plates.

3. Isolation of smooth muscle cells from aortic explants
   a. Collect and trim the aorta as described in step 1.
   b. Open the aorta longitudinally and rub the intimal surface vigorously using a sterile cotton swab (endothelial cells are the major contaminant in smooth muscle culture; therefore, they should be removed before implanting the aorta into Matrigel).
   c. Remove the intimal layer with sterile forceps and mince into small pieces (1 mm × 1 mm).
   d. Place the minced pieces evenly on culture plates in a humidified incubator at 37 °C, to allow adhesion of the tissues to the plates.
   e. Once firmly attached to the plates, add prewarmed media very gently (cells usually sprout from the explants with 4–7 days).

2.3. Type I/type II epithelial cells

The methods for isolating pulmonary epithelial cell cultures are adapted from previously published methods (Chen *et al.*, 2004; Corti *et al.*, 1996).

1. Perfuse the lung.
   a. Anaesthetize mice and secure to a dissecting board. Open the peritoneum, clip the left renal artery and remove the ventral rib cage.
   b. Perfuse the lung with 0.9% saline, using a 10-ml syringe fitted with a 21-G needle through the right ventricle of the heart until free of blood.

2. Prepare a dispase digestion of the lung.
   a. Insert a 20-G intravenous catheter into the trachea and secure tightly with a suture.
   b. Fill the lung with approximately 1–2 ml of sterile dispase solution (Gibco, Invitrogen #17105), 2 U/mL in Ca^{2+}/Mg^{2+}-free phosphate buffered saline (PBS), through the tracheal catheter and allow the lung to collapse.
   c. Slowly infuse low-melting agarose (Sigma, #A194), (1%, 0.45 ml prewarmed at 45 °C) through catheter. Immediately cover the lung

with crushed ice and let stand for 2 min. (Due to the similar physical properties [size, density, and adherent behavior to substrate] of Clara cells with alveolar type II cells, an agarose solution is utilized to prevent the release of the cells into the crude cell suspension by filling the tracheobronchial tree, preventing the release of unwanted cells or to cause clumping of the cells so that the contaminants are removed during subsequent filtration steps).

d. Remove and transfer the lung to 2 ml of dispase in a polypropylene tube, further incubate for an additional 45 min at room temperature. Stop the reaction by standing the tube on ice.

e. Transfer the lung to 7 ml of DMEM with 0.01% DNase (Sigma #D5025) in a 60-mm culture plate.

f. Carefully tease the digested tissue with the curved edge of a forceps and gently swirl for approximately 5–10 min by rocking.

g. Successively filter through 100- and 40-μm cell strainers and save the pellet after centrifugation (130×*g* for 8 min). Resuspend the pellet in complete medium (DMEM supplemented with 25 m*M* HEPES buffer, pH 7.4, 10% FBS and 1% penicillin and streptomycin, Gibco Invitrogen).

2. Negative immunoselection of alveolar type II cells.

a. Incubate the cell suspension with biotinylated anti-CD32 (0.65 $\mu g/10^6$ cells) (R&D systems, Minneapolis, MN, #BAF1460) and biotinylated CD-45 (1.5 $\mu g/10^6$ cells) for 30 min at 37 °C (R&D systems, #BAM1217). (Anti-CD32 antibody targets to macrophages and anti-CD45 reacts with leukocytes in the crude cell suspension.)

b. Wash streptavidin-coated magnetic beads (i.e., Dynabeads M-270 Streptavidin, Dynal Invitrogen, #653.05) twice with PBS (10 min each wash) in a polypropylene tube using a magnetic separator.

c. Harvest the cells and remove the unreacted antibodies by centrifugation (130×g, 8 min at 4 °C; discard the supernatant fraction). Resuspend the cells in 7 ml of DMEM with washed magnetic beads for 30 min at room temperature (with gentle rocking).

d. Collect the unbound cells by standing the tube onto the magnetic separator for 15 min. The cell suspension is aspirated from the bottom of the tube, centrifuged, and resuspended in a complete medium.

e. Inoculate the cell suspension into noncoated culture plates for 4–6 h and transfer the nonadherent cells to fibronectin-coated plates (i.e., BD biosciences, Bedford, MA, #354411). This step ensures the elimination of adherent endothelial cells or fibroblasts.

Note: In some species such as rats, alveolar type II cells are isolated by enzymatic dispersion with elastase followed by panning on IgG-coated bacteriological plates (negative selection). The enriched alveolar epithelial cell preparation is maintained in defined serum-free medium consisting of

DMEM and Ham's F-12 nutrient medium (1:1, v/v ratio) (Gibco Invitrogen, #11330-099) supplemented with bovine serum albumin, HEPES, non-essential amino acids, glutamine, and penicillin/streptomycin. To obtain freshly isolated preparations partially enriched for alveolar type I cells, lungs from adult male rats are digested with elastase and collagenase type 1, chopped, and filtered through 100-μm strainers. Approximately 20% of the cells obtained in this fashion are type I cells, which are further selected by positive immunoisolation with anti-T1α (podoplanin) antibody (R&D systems, #AF3244) (Chen *et al.*, 2004; Corti *et al.*, 1996).

2.4. Lung fibroblasts

The general method for isolating murine lung fibroblasts has been previously described (Wang *et al.*, 2003).

1. Place the lung on the culture plates (lung perfusion/removal of blood cell would be desirable) and remove blood clots or trachea and upper respiratory tract.
2. Mince the lung into small pieces (1 mm $\times$ 1 mm) and transfer the tissue to a conical tube containing 5–10 ml of HBSS with penicillin/streptomycin).
3. Wash those small pieces with a wide-bore pipette and remove the supernatant fraction by centrifugation (500$\times g$, 5 min, at 4 °C) (repeat 3 times).
4. Place the tissue evenly onto 100-mm tissue culture plates (5–10 mm apart).
5. Add 1 drop of FBS (50–100 μl/lung piece) gently onto the single piece (to trigger the adherence of tissue explants to the plastic and prevent tissue dryness).
6. Incubate overnight at 37 °C in a humidified incubator and confirm the firm attachment of explants.
7. Add prewarmed medium (10 ml DMEM supplemented with 10% FBS and 1% penicillin/streptomycin) very gently.
8. The cells sprout out from the attached tissue within approximately 3–4 days.

2.5. Alveolar macrophages

1. Insert a 20-G intravenous catheter into the trachea and secure tightly with suture as described in the epithelial cell culture protocol.
2. Lavage the lung with 5 ml of 1-ml aliquots of saline (0.9% NaCl). 3–4 ml of lavaged solution is recovered from one mouse (In the unexposed mice, over 99% of the lavaged cells are shown to be macrophages.

If treated or operated, the cell composition might be altered) (Dolinay *et al.*, 2008).

3. Harvest the cells from centrifugation (500×*g*, 5 min at 4 °C) and culture with DMEM (10% FBS and 1% penicillin/streptomycin) 10 ml, in 100-mm culture plates, in a humidified incubator at 37 °C.

Note: Lung-derived immune cells (T cells, dendritic cells) can be isolated by Ficoll-Hypaque purification and further sorted using a flow cytometric cell sorter (Arora *et al.*, 2006).

3. Analysis of Autophagy

For investigating the incidence of autophagy in lung injury and disease, human lung tissues (from lung transplantation, organ donor programs, and clinical specimen banks), mouse or rat lung tissues from *in vivo* lung disease modes, and cells derived from human or mouse lungs can be subjected to the following analyses.

3.1. Transmission electron microscopy

Electron microscopy is the gold standard for the analysis of autophagy (Klionsky *et al.*, 2008). Two kinds of autophagic vacuoles (AV, autophagosomes/autolysosomes) can be observed during autophagy. Double-membrane autophagosomes indicate immature autophagic vacuoles, whereas single-membrane autolysosomes indicates degradative, late autophagic vacuoles. EM analysis is semiquantitative. It might be necessary to quantify a number of representative images and show the increases of the number of AV per cell or per identified area (see also the chapter by Ylä-Anttila *et al.*, in volume 452).

Note: In lung type II cells, there are onion-shaped secretory vacuoles with similar structures as are seen during autophagy. The vacuoles contain surfactant, which is a protein-lipid complex that is synthesized and modified in the rough endoplasmic reticulum, transported through the Golgi complex and packaged into lamellar bodies before secretion. After exocytosis of the lamellar bodies, surfactant forms a lipid-rich layer at the air-liquid interface of the alveoli. Most of the extracellular surfactant is taken up by alveolar type II cells, and transported to lamellar bodies for recycling (Whitsett *et al.*, 2002). Therefore, there are many vacuoles undergoing exocytosis or uptake in lung type II cells. The important feature to distinguish these vacuoles from true autophagic vacuoles is that the secretory vacuoles have the lamellar structure without any inclusion of other organelles that are destined for degradation by autophagy, whereas

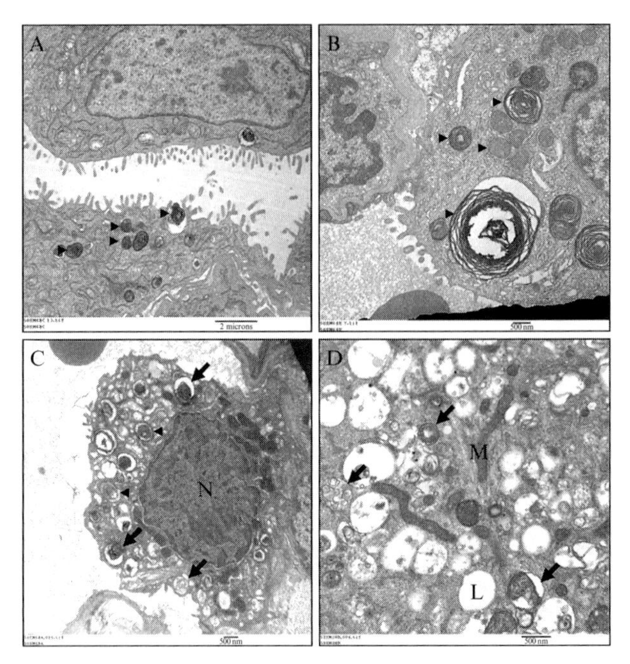

Figure 10.1 Secretory vesicles and autophagosomes in lung alveolar type II cells. (A) In normal human lung alveolar type II cells, there are onion-shaped secretory vacuoles (arrowhead) with similar structures as autophagosomes, whose surfactant is a protein-lipid complex synthesized in the rough endoplasmic reticulum and Golgi complex and stored in the lamellar bodies. These structures are continuously secreted via exocytosis (plasma membrane regions). As shown in (B), different sizes of secretory vesicles are formed in these cells. (C) In lung obtained from chronic obstructive pulmonary disease patients, autophagic vacuoles (AV) have been observed under electron microscopy. The two types of vesicles are sometimes confused for one another due to the similar size (500 nm to 1 μm) and morphology. Although it is not clear whether autophagy is involved in secretory vesicle degradation, careful attention should be paid to discriminate the secretory vesicle (arrowheads) and autophagosome (arrows). The important feature to distinguish these secretory vesicles from AV is that the AV (autophagosomes) should have a clear double-membrane structure. In starved fibroblasts (D), we can observe well-developed autophagosomes (arrows) in the cytoplasm. More careful analysis is needed for cells and tissues that are enriched in secretory vesicles, such as in endocrine cells or tissue. L; lysosome, M; mitochondria, N; nucleus.

autophagosomes have double-membrane layers with encapsulated organelles. See Fig. 10.1 for examples.

1. Human lung tissues:
   a. Cut tissues into approximately 1-mm cubes for fixing.
   b. Fix the lung tissue sections in formalin (37% solution) at room temperature for up to 8 h.
   c. Wash with 0.1 *M* sodium cacodylate buffer for 1–2 h.

d. Postfix in 1% osmium tetroxide in 0.2 *M* sodium cacodylate buffer for 1 h.
e. Rinse in 0.2 *M* sodium cacodylate buffer for 10 min, 3 times.
f. Dehydrate in 70% ethanol, 20 min, 2 times.
g. Dehydrate in 90% ethanol, 10 min, 2 times.
h. Dehydrate in 100% ethanol, 20 min, 2 times.
i. Incubate with propylene oxide (epoxy propane) (Sigma, #240397) for 10 min, 2 times.
j. Incubate with propylene oxide/epoxy resin mixture (50:50) for 1 hour. The epoxy resin mixture is: 24 g of Agar 100 resin, 13 g of dodecenylsuccinic anhydride, 13 g of methylnadic anhydride, and 1 ml of N-benzyldimethylamine (Sigma, #45346, #45347, #185582). Mix thoroughly in a disposable beaker using a wooden spatula.
k. Incubate with epoxy resin overnight in uncapped vials (this allows any remaining propylene oxide to evaporate).
l. Embed in labeled capsules with freshly prepared resin.
m. Polymerize at 60 °C for 48 h.
n. Photograph tissue sections using a JEOL JEM 1210 transmission electron microscope (JEOL, Peabody, MA) at 80 or 60 kV onto electron microscope film (ESTAR thick base; Kodak, Rochester, NY).
o. Print the images onto photographic paper.

2. Mouse lung tissues.
   a. Immediately following necropsy, inflate the right lung by gravity with 4% paraformaldehyde in PBS (USB corporation, Cleveland, OH, #19943) and hold at a pressure of 30 cm H_2O for 15 min.
   b. Gently dissect the right lung from the thorax and place in 4% paraformaldehyde in PBS for up to 8 h.
   c. Follow the steps after the primary fixation described previously for human lung tissues.

3. Pulmonary cells.
   a. Following experimental manipulations, rinse the cells 3 times with ice-cold phosphate buffered saline (PBS).
   b. Fix cells in 2.5% glutaraldehyde in PBS for up to 8 h.
   c. Follow the steps after primary fixation described above for human lung tissues.

3.2. Western immunoblot analysis

Western immunoblot analysis can be used for the analysis of autophagic marker proteins in lung cells and tissue. The ratio of LC3-II/LC3-I is widely accepted as the best biomarker of autophagy. Nevertheless, the amount of LC3-II can be also compared between samples because that LC3-II tends to be much more sensitive to detection by immunoblotting than LC3-I (Klionsky *et al.*, 2008). The autophagic proteins are differentially regulated during autophagy induction, depending on the cell types and the stimuli. During autophagy, the ratio of LC3-II/LC3-I, or the amount of LC3-II, is generally increased, whether in human lung tissues, mouse lung tissues, or pulmonary cells. However, other autophagy related proteins, such as Beclin 1, Atg5–Atg12, Atg4, and Atg7 may not necessarily be increased. For example, in most cases, Beclin 1 levels are not changed in lung epithelial cells, in fibroblasts, or in lung tissues during autophagy, whereas this protein is dramatically induced by hypoxia in lung endothelial cells (unpublished observations).

1. Human or animal tissues:
   a. Cut tissue sections into small pieces on dry ice and transfer into a 1.5-ml microcentrifuge tube.
   b. Rinse the samples with ice-cold PBS (to remove the blood which interferes with the Western blot resolution due to the similar migratory behavior of light IgG chain as that of LC3). Ideally the removal of blood cells must be done before sample processing for Western, however, to minimize red blood cell contamination, wash the small pieces of tissue in 1 ml erythrocyte lysis buffer (0.16 *M* NH_4Cl, 0.17 *M* Tris-HCl, pH 7.65). Stop the reaction with 9 ml of DMEM containing 5% FCS and centrifuge (500×*g*, 5 min, at 4 °C). Discard the supernatant fractions.
   c. Add 1 ml of RIPA buffer (25 m*M* Tris-HCl pH 7.6, 150 m*M* NaCl, 1% NP-40, 1% Sodium deoxycholate, and 0.1% SDS, Pierce, Rockford, IL #89900) to samples and homogenize the samples using a tissue homogenizer for 15 s, 3 times (samples in small amount can also be sonicated).
   d. Centrifuge the samples at 13,000 rpm, 4 °C for 15 min. Discard the pellet fraction and remove the supernatant fraction to a new tube. Complete removal of undigested and tissue extracellular matrix by repeated centrifugation will lead to better blot resolution.
   e. Quantify the protein concentration of the samples and adjust the samples to the same concentration using RIPA buffer.
   f. Add an equal amount of 2x SDS-polyacrylamide gel electrophoresis (PAGE) sample loading buffer and heat the samples at 99 °C for 5 min.
   g. Run protein samples on an SDS–PAGE gel.
   h. Transfer the blots to polyvinylidene difluoride membranes at 300 mA for 3 h.

Block the membrane for 1 h at room temperature in fresh blocking buffer (TBST, 1X Tris buffered saline (TBS; 25 m*M* Tris, 150 m*M* NaCl, 2 m*M* KCl, pH 7.4) and 0.1% Tween 20, containing 5% nonfat dry milk).

Blocking is also important for better blot resolution. For tissue samples, membranes can be also blocked in either 1% BSA, or 5% goat serum, or both for better Western blot results. Moreover, before adding sample loading buffer, samples can be pre-cleared with protein A/G agarose beads to reduce the appearance of IgG reactivity.

i. Incubate membrane with primary antibodies (in TBS with 3% nonfat dry milk) overnight at 4 °C.
j. Wash blots with TBST buffer 3 times for 10 min.
k. Incubate membrane with horseradish peroxidase-conjugated secondary antibodies in TBS with 3% nonfat dry milk for 1 h at room temperature.
l. Wash membrane in TBST buffer for 10 min, 3 times.
m. Incubate membrane with ECL solution (GE Healthcare, Piscataway, NJ, #RPN2124) for 1 min.
n. Visualize blots with radiographic film.

2. Cell samples:
   a. Following experimental manipulations, rinse the cells twice with ice-cold PBS. Add RIPA buffer to the culture plates and harvest the samples into microcentrifuge tubes.

Boiled sample buffer might be used to minimize the inevitable protein degradation during routine sample preparation, especially for labile proteins, such as LC3.

   c. Sonicate samples for 15 s, 3 times.
   d. Follow the steps after sample homogenization (step 1.d) as described for tissues.

3.3. Immunofluorescence staining

1. Tissue paraffin sections. Paraffin-embedding hardens tissue to allow for very thin sections (usually 4–5 μm). After fixation, tissues that are to be paraffin embedded are dehydrated by first using graded ethanol solutions, graded xylene solutions, then finally liquid paraffin. The graded solutions gradually expose the sample to changes in hydrophobicity, minimizing damage to cells. After a short time in the liquid paraffin, the tissue is placed into a mold with more paraffin. The wax is allowed to solidify, forming a block that can be held in a microtome.
   a. Deparaffinize sections with xylene for 5 min, 3 times
   b. Wash with 100% ethanol for 10 min, 2 times.
   c. Wash with 95% ethanol for 10 min, 2 times.
   d. Rinse with dH_2O for 5 min, 2 times.
   e. Place sections in 10 m*M* sodium citrate buffer, pH 6.0.
   f. Boil sections using water bath or microwave, and maintain at 95–99 °C for 10 min.

g. Cool down slides for 30 min at room temperature.
h. Rinse with dH_2O for 5 min, 3 times.
i. Rinse with PBS for 5 min (hereafter all washing solutions are ice cold).
j. Block sections with blocking serum (5% goat serum in PBS) for 45 min.
k. Rinse with BSA solution (0.5% BSA and 0.15% glycine in PBS) for 5 times.
l. Incubate sections with primary antibody (1:100 dilution) overnight at 4 °C.
m. Rinse again with BSA solution 5 times.
n. Incubate sections with fluorescence-conjugated secondary antibody (1:100–500 dilution) for 1 h.
o. Rinse with BSA solution 5 times.
p. Rinse with PBS solution 5 times.
q. Hoechst 33342 stain (50 μM in PBS) for 30 s (nuclear staining).
r. Wash with PBS twice.
s. Place on a cover slip with a small amount of polyvinyl alcohol solution. (Polyvinyl alcohol solution is prepared as follows: Add 21 g of polyvinyl alcohol (Sigma, #P8136) to 42 mL of glycerol. Add 52 mL of dH_2O. Add a few crystals of sodium azide. Add 106 ml of 0.2 *M* Tris buffer, pH 8.5. Stir on low heat for a few hours or until the reagents are dissolved. Clarify the mixture by centrifugation at 5000*g* for 15 min. Aliquot and store at 4 °C).
t. Dry slides or coverslips overnight at 4 °C in the dark.
u. View samples with an Olympus Fluoview 300 confocal laser-scanning head with an Olympus IX70 inverted microscope.

2. Cryostat sections.
   a. Submerge the slides in paraformaldehyde solution (4% paraformaldehyde in PBS, make fresh solution) for 20 min at room temperature.
   b. Wash the slides in PBS for 5 min, 3 times.
   c. Using a PAP pen (Invitrogen, #00-8888), encircle the tissue section on each slide. This will help prevent excessive spreading of the reagents allowing one to use less. Do not allow the sections to dry, by incubating the slide in a humidified chamber or by covering the slide with an adequate-sized piece of Parafilm (Pechiney, Chicago, IL).
   d. Permeabilize and block tissue sections with ice-cold solution (0.1% Triton X-100, 2% paraformaldehyde in PBS) for 60 min at room temperature using a micropipettor; cover each tissue section (volume will depend on tissue size) with the solution.
   e. Perform serum blocking and subsequent steps (step 1.j).

3. Cell samples
   a. Fix and permeabilize cells with ice-cold fix solution (0.1% Triton X-100, 2% paraformaldehyde in PBS) for 15 min.
   b. Rinse the cells with PBS 5 times at room temperature to remove the fixative.
   c. Perform serum blocking and subsequent steps (step 1.j).

3.4. GFP-LC3 for cells

GFP-LC3 is another widely used method for measuring autophagy in cells (Kabeya *et al.*, 2000). Under basal conditions, GFP-LC3 should be distributed uniformly in the cytosol, whereas under certain types of stimulation, the formation of GFP-LC3-labeled puncta represents the formation of autophagosomes. GFP-LC3 mice have recently been developed, and these mice might be used for monitoring the formation of autophagy *in vivo* (see the chapter by N. Mizushima in volume 452). Similar to EM, the GFP-LC3 method is also semiquantitative. Therefore, it may be necessary to quantify and show the increased number of puncta per cell.

Note: Transfection itself sometimes induces cellular stress, which is accompanied by autophagosome formation (false positive).

a. Plate cells in 6- or 12-well plates with micro cover slides.
b. Transfect 1–2 μg of GFP-LC3B into 2×10^5 cells using lipofectamine 2000 according to the supplier's protocol (Invitrogen, CA). 48 h after transfection, the cells can be used for experimental manipulations (for example, we have treated human bronchial epithelial cells with cigarette smoke extract, CSE, 0%–20% in complete medium, for 6 h).
c. After the experimental procedures, remove the culture medium and wash with ice-cold PBS 3 times.
d. Fix and permeabilize the cells with ice-cold fix solution (0.1% Triton X-100, 2% paraformaldehyde in PBS) for 15 min.
e. Rehydrate with PBS 5 times.
f. Wash cells with BSA solution 5 times.
g. Place on a coverslip with a small amount of polyvinyl alcohol solution.
h. Dry slides or cover slips overnight at 4 °C in the dark.
i. View samples with an Olympus Fluoview 300 confocal laser-scanning head with an Olympus IX70 inverted microscope.

4. Chromatin Immunoprecipitation

To date, little is known about the transcriptional regulation of autophagic genes. It has been only recently reported that transcription factor E2F1 regulates four autophagy related genes, LC3, Atg1, Atg5, and DRAM (Polager *et al.*, 2008).

Chromatin immunoprecipitation (ChIP) is a powerful tool for identifying proteins, including histone proteins and nonhistone proteins, associated with specific regions of the genome by using specific antibodies that recognize a specific protein or a specific modification of a protein. The initial step of ChIP is the cross-linking of protein-protein and protein-DNA in live cells with formaldehyde. After cross-linking, the cells are lysed and crude extracts are sonicated to shear the DNA to appropriate-sized fragments. Proteins together with cross-linked DNA are subsequently immunoprecipitated with transcription factor–specific antibodies. Protein-DNA cross-links in the immunoprecipitated material are then reversed and the DNA fragments are purified and PCR amplified.

The ChIP procedures are principally based on the kit protocol (Active Motif, Cat# 53008) using protein G-coated magnetic beads, with slight modifications (Active motif, 2008).

1. Chromatin isolation.
   a. Grow cells to 70%–80% confluency on 3 10-cm plates for each sample.
   b. The cells can now be subjected to experimental treatment protocols (for example, we have treated human bronchial epithelial cells with cigarette smoke extract [CSE, 0%–20% in complete medium], for 1 h).
   c. When the cells are ready to harvest, cross-link the protein to the DNA by adding formaldehyde directly to the culture medium to a final concentration of 1% (270 μl of a 37% formaldehyde solution into 10 ml of growth medium). Incubate on a shaking platform at room temperature for 10 min.
   d. Stop the cross-linking reaction by adding glycine to a 125 mM final concentration (500 μl of 2.5 M stock in 10 ml of medium), rocking at room temperature for 5 min.
   e. Aspirate away medium and rinse twice with ice-cold PBS.
   f. Add 3 ml of PBS (with protease inhibitors) to each plate and scrape cells into a 15-ml conical tube (harvest 3 plates together for each sample). Pellet cells by centrifuging at 2500 rpm for 10 min at 4 °C, carefully remove the supernatant fraction, and discard (the pellet can be frozen at –80° C at this time and used later if desired).
   g. Resuspend the cell pellets in 350 μl of lysis buffer with protease inhibitor cocktail (Active Motif, Cat# 53008) and transfer to a 1.5-ml microcentrifuge tube. Incubate the samples on ice for 30 min.
   h. Centrifuge at 5000 rpm for 10 min at 4 °C to pellet the nuclei.
   i. Carefully remove and discard the supernatant fraction, and resuspend the nuclei pellets in 350 μl of shearing buffer (Active Motif, Cat# 53008) with protease inhibitors.

j. Shear the DNA using optimized conditions (generally, 20 s × 10 pulses sonication at 25% power using a sonicator with a 3-mm stepped microtip).
k. Remove cell debris by centrifuging 20 min at 13,000 rpm at 4 °C. Transfer the supernatant fraction into a fresh tube. (Samples can be frozen at −80 °C at this time if desired.)

2. ChIP reaction.
   a. Transfer 10 μl of the supernatant fraction to a microcentrifuge tube, which will be input DNA used as controls in the PCR analysis. Store the chromatin on ice.
   b. Set up the ChIP reactions by adding 25 μl of protein G magnetic beads, 10 μl of ChIP buffer 1, 60 μl of sonicated chromatin, 1 μl of protease inhibitor cocktail, and 1–3 μg of antibody (an IgG control is necessary using the same amount of chromatin; i.e., ChIP-IT Control Kit –Human or Mouse, Active Motif #53010, 53011).
   c. Mix thoroughly and incubate on a rocking shaker for 4 h at 4 °C.
   d. Centrifuge the tube briefly (13,000 rpm, 30 s) and put the tube on a magnetic stand to pellet the beads.
   e. Carefully remove and discard the supernatant fraction.
   f. Wash the beads once with 800 μl of ChIP buffer 1 (using a 1-ml pipette, pipetting up and down several times).
   g. Wash the beads 2 times with 800 μl of ChIP buffer 2. After the final wash, completely remove the supernatant fraction, taking away as much as possible without disturbing the beads.

3. Elution of chromatin
   a. Resuspend the washed beads with 50 μl of elution buffer AM2, and incubate for 15 min at room temperature on a rotator.
   b. Briefly centrifuge the tubes (13,000 rpm, 30 s) and add the reverse cross-link buffer to the eluted chromatin and immediately place the tubes in the magnetic stand to pellet the beads.
   c. Quickly transfer the chromatin (the supernatant fraction) into a fresh tube.
   d. Remove the input DNA from ice, and add 88 μl of ChIP buffer 2 and 2 μl of 5 *M* NaCl.
   e. Incubate all samples at 94 °C for 15 min.
   f. Return the tubes to room temperature and add 2 μl of proteinase K (0.5 μg/μl).
   g. Mix well and incubate at 37 °C for 1 h.
   h. Return the tubes to room temperature and add 2 μl of the proteinase K stop solution. The DNA can be used immediately for PCR or stored at −20 °C.

4. PCR analysis
   1. PCR conditions:
      10 μl of Premix G (failsafe PCR 2x premix G, Epicentre)
      0.5 μl of Taq DNA polymerase
      0.4 μl of Primer 1
      0.4 μl of Primer 2
      2 μl of DNA
      RNase-free water to total 20 μl
   2. Cycling conditions:
      95 °C, 4 min
      95 °C, 30 s
      56 °C, 30 s
      72 °C, 30 s
      30–32 cycles (depending on the intensity)
      72 °C, 5 min
      4 °C holding
   3. Primers are custom designed according to the experimental design. We provide the following PCR primer sequences for LC3A, LC3B and Beclin 1 designed to evaluate E2F and Egr-1 factor binding.

LC3A-F	335bp	5′ TGC CTT CTC ACC TCA TCC TC 3′
LC3A- R		5′ CAG CCC GAA CTC AAG GTA AG 3′
LC3A-F2	217bp	5′ GCC TCC CCT TTA AGG AAT GT 3′
LC3A-R2		5′ CTT GAA AGG CCG GTC TGA G 3′
LC3B-F2	387bp	5′ CCA CAA CCG TCA CCT CAG 3′
LC3B-R2		5′ CGC TGC TTG AAG GTC TTC TC 3′
LC3B-F	427bp	5′ GCT CGG GAC AAA AGC AGT T 3′
LC3B-R		5′ CCC TGA GGT GAC GGT TGT 3′
Beclin1-F	352bp	5′ GCC TTG GCT CCT ACA CTT CC 3′
Beclin1-R		5′ TAA GAG CCG TGA GGG TTC C 3′

   4. Agarose gel analysis
      a. Use approximately 5–10 μl of sample to run on a 1.5% agarose gel, using an appropriate DNA ladder (e.g., Sigma, # P1473-1VL).
      b. Stain and analyze the gel, looking for bands at the expected fragment size.

5. Conclusions

Methods for the analysis of autophagy in tissues continue to evolve. The general techniques to date include EM, GFP-LC3 immunofluorescence, and Western blotting. A recent comprehensive review has been written by leaders in the field to describe the current consensus for interpretation and pitfalls of the available techniques (Klionsky *et al.*, 2008). The analysis of autophagy in the lung in general, and in the context of pulmonary disease, however, is currently underrepresented in the biomedical literature. This chapter has placed methods for the preparation of lung cell and tissue samples in the context of these basic autophagy detection methods to provide a starting point to further the investigation of this critical cellular function in the pulmonary system.

REFERENCES

Active, Motif (2008). CHIP-IT Express Magnetic Chromatin Immunoprecipitation Kits (Active Motif, Cat# 53008, online protocolhttp://www.activemotif.com/documents/1574.pdf).

Arora, M., Chen, L., Paglia, M., Gallagher, I., Allen, J. E., Vyas, Y. M., Ray, A., and Ray, P. (2006). Simvastatin promotes Th2-type responses through the induction of the chitinase family member Ym1 in dendritic cells. *Proc. Natl. Acad. Sci.* **103,** 7777–7782.

Chen, J., Chen, Z., Narasaraju, T., Jin, N., and Liu, L. (2004). Isolation of highly pure alveolar epithelial type I and type II cells from rat lungs. *Lab. Invest* **84,** 727–735.

Corti, M., Brody, A.R, and Harrison, J.H (1996). Isolation of primary culture of murine alveolar type II cells. *Am. J. Respir. Cell Mol. Biol.* **14,** 309–315.

Cuervo, A. M. (2004). Autophagy: In sickness and in health. *Trends. Cell Biol.* **14,** 70–77.

Dolinay, T., Wu, W., Kaminski, N., Ifedigbo, E., Kaynar, A. M., Szilasi, M., Watkins, S. C., Ryter, S. W., Hoetzel, A, and Choi, A. M. (2008). Mitogen-activated protein kinases regulate susceptibility to ventilator-induced lung injury. *PLos One* **3,** e1601.

He, H., Dang, Y., Dai, F., Guo, Z., Wu, J., She, X., Pei, Y., Chen, Y., Ling, W., Wu, C., Zhao, S., Liu, J. O., *et al.* (2003). Post-translational modifications of three members of the human MAP1LC3 family and detection of a novel type of modification for MAP1LC3B. *J. Biol. Chem.* **278,** 29278–29287.

Kabeya, Y., Mizushima, N., Ueno, T., Yamamoto, A., Kirisako, T., Noda, T., Kominami, E., Ohsumi, Y., and Yoshimori, T. (2000). LC3, a mammalian homologue of yeast Apg8p, is localized in autophagosome membranes after processing. *EMBO J.* **19,** 5720–5728.

Kelekar, A. (2006). Autophagy. *Ann. N. Y. Acad. Sci.* **1066,** 259–271.

Kiffin, R., Bandyopadhyay, U., and Cuervo, A. M. (2006). Oxidative stress and autophagy. *Antioxid. Redox Signal.* **8,** 152–162.

Klionsky, O. J., Abeliovich, H., Agostinis, P., Agrawal, O. K., Aliev, G., Askew, D. S., Baba, M., Baehrecke, E. H., Bahr, B. A., Ballabio, A., Bamber, B. A., Bassham, O. C., *et al.* Guidelines for the use and interpretation of assays for monitoring autophagy in higher eukaryotes *Autophagy* **4,** 151–175.

Kondo, Y., Kanzawa, T., Sawaya, R., and Kondo, S. (2005). The role of autophagy in cancer development and response to therapy. *Nat. Rev. Cancer* **5,** 726–734.

Liang, X. H., Jackson, S., Seaman, M., Brown, K., Kempkes, B., Hibshoosh, H., and Levine, B. (1999). Induction of autophagy and inhibition of tumorigenesis by *beclin 1*. *Nature* **402,** 672–676.

Mahabeleshwar, G. H., Somanath, P. R., and Byzova, T. V. (2006). Methods for isolation of endothelial and smooth muscle cells and *in vitro* proliferation assays. *Methods Mol. Med.* **129,** 197–208.

Menzel, D. B., and Amdur, M. O. (1986). Toxic response of the respiratory system. *In* "Casarett and Doull's Toxicology, the basic science of poisons" (K. Klaassen,

M. O. Amdur, and J. Doull, eds.) 3rd ed., pp. 330–358. NY, MacMillan Publishing Company, New York.

Mizushima, N., Levine, B., Cuervo, A. M., and Klionsky, D. J. (2008). Autophagy fights disease through cellular self-digestion. *Nature* **451,** 1069–1075.

Polager, S., Ofir, M., and Ginsberg, D. (2008). E2F1 regulates autophagy and the transcription of autophagy genes. *Oncogene* Apr 14. [E-pub ahead of print].

Shimizu, S., Kanaseki, T., Mizushima, N., Mizuta, T., Arakawa-Kobayashi, S., Thompson, C. B., and Tsujimoto, Y (2004). Role of Bcl-2 family proteins in a non-apoptotic programmed cell death dependent on autophagy genes. *Nat. Cell Biol.* **6,** 1221–1228.

Sreejayan, N., and Yang, X. (2007). Isolation and functional studies of rat aortic smooth muscle cells. *Methods Mol. Med.* **139,** 283–292.

Tang, Z. L., Wasserloos, K. J., Liu, X., Stitt, M. S., Reynolds, I. J., Pitt, B. R., and St Croix, C.M (2002). Nitric oxide decreases the sensitivity of pulmonary endothelial cells to LPS-induced apoptosis in a zinc-dependent fashion. *Mol. Cell. Biochem.* 234–235, 211–217.

Wang, X., Ryter, S. W., Dai, C., Tang, Z. L., Watkins, S. C., Yin, X.M, Song, R., and Choi, A.M (2003). Necrotic cell death in response to oxidant stress involves the activation of the apoptogenic caspase-8/bid pathway. *J. Biol. Chem.* **278,** 29184–29191.

Wang, X., Wang, Y., Zhang, J., Kim, H. P., Ryter, S. W., and Choi, A. M. (2005). FLIP protects against hypoxia/reoxygenation-induced endothelial cell apoptosis by inhibiting Bax activation. *Mol. Cell. Biol.* **25,** 4742–4751.

Whitsett, J. A., and Weaver, E. T. (2002). Hydrophobic surfactant proteins in lung function and disease. *N. Engl. J. Med.* **347,** 2141–2148.

CHAPTER ELEVEN

Autophagy in Neurite Injury and Neurodegeneration: *In Vitro* and *In Vivo* Models

Charleen T. Chu,* Edward D. Plowey,* Ruben K. Dagda,* Robert W. Hickey,† Salvatore J. Cherra III,* *and* Robert S. B. Clark†

Contents

Abstract

Recent advances indicate that maintaining a balanced level of autophagy is critically important for neuronal health and function. Pathologic dysregulation of macroautophagy has been implicated in synaptic dysfunction, cellular stress,

* Department of Pathology, Division of Neuropathology, University of Pittsburgh School of Medicine and Center for Neuroscience (CNUP), Pittsburgh, Pennsylvania, USA
† Departments of Critical Care Medicine and Pediatrics, Safar Center for Resuscitation Research University of Pittsburgh School of Medicine, Pittsburgh, Pennsylvania, USA

Methods in Enzymology, Volume 453
ISSN 0076-6879, DOI: 10.1016/S0076-6879(08)04011-1

and neuronal cell death. Autophagosomes and autolysosomes are induced in acute and chronic neurological disorders including stroke, brain trauma, neurotoxin injury, Parkinson's, Alzheimer's, Huntington's, motor neuron, prion, lysosomal storage, and other neurodegenerative diseases. Compared to other cell types, neuronal autophagy research presents particular challenges that may be addressed through still evolving techniques. Neuronal function depends upon maintenance of axons and dendrites (collectively known as neurites) that extend for great distances from the cell body. Both autophagy and mitochondrial content have been implicated in regulation of neurite length and function in physiological (plasticity) and pathological remodeling. Here, we highlight several molecular cell biological and imaging methods to study autophagy and mitophagy in neuritic and somatic compartments of differentiated neuronal cell lines and primary neuron cultures, using protocols developed in toxic and genetic models of parkinsonian neurodegeneration. In addition, mature neurons can be studied using *in vivo* protocols for modeling ischemic and traumatic injuries. Future challenges include application of automated computer-assisted image analysis to the axodendritic tree of individual neurons and improving methods for measuring neuronal autophagic flux.

1. Introduction

The neuron is the most highly polarized postmitotic cell, consisting of a soma and specialized axonal and dendritic projections (collectively called *neurites*) that form networks of arborizing intercellular synaptic connections. Maintenance of these long neuritic structures is required for the propagation of electrochemical signals across vast cellular distances. Basal physiological levels of autophagy play a critical role in maintaining neuronal health, presumably by removing effete, oxidized, or aggregated proteins and organelles [Reviewed in (Boland and Nixon, 2006; Cherra and Chu, 2008; Ventruti and Cuervo, 2007)], whereas dysregulated autophagy contributes to neurite degeneration and neuronal cell death, as shown in several *in vitro* and *in vivo* systems (Koike *et al.*, 2008; Plowey *et al.*, 2008; Yang *et al.*, 2007; Yue *et al.*, 2008; Zhu *et al.*, 2007).

Synaptic/neuritic pathology is a prevalent theme in neurodegenerative diseases [Reviewed in (Wishart *et al.*, 2006)] and neurite degeneration and remodeling are elicited in acute brain injuries (Ito *et al.*, 2006). These neuritic pathologies include protein and organelle accumulation, beading/fragmentation of neurites, and disruption of microtubule networks involved in axonal transport (Fiala *et al.*, 2007). Neuritic pathologies highlighted by silver stains or immunohistochemistry of disease-related proteins include Lewy neurites in Parkinson/Lewy body diseases (Spillantini *et al.*, 1997), dystrophic neurites in Alzheimer disease (Nixon, 2007), huntingtin protein in cortical neurites (DiFiglia *et al.*, 1997), amyloid precursor protein in axonopathy of head trauma (Gentleman *et al.*, 1993), and spongiform change in prion diseases (Ironside, 1998). Dystrophic neurites, which likely exhibit abnormal retrograde and anterograde transport, are associated with dysfunctional synapses. Alzheimer disease brains exhibit decreased synaptic density (Masliah *et al.*, 2001), even in the earliest stages of clinical disease (Scheff *et al.*, 2006), and inhibition of hippocampal long-term potentiation has been demonstrated in animal models (Chapman *et al.* 1999; LaFerla & Oddo 2005; Walsh *et al.* 2002). Alterations in synaptic morphology are also observed in the basal ganglia of Parkinson's disease patients (Lach *et al.*, 1992; Machado-Salas *et al.*, 1990), with synaptic dysfunction as a prominent feature in models based on mutations in α-synuclein [Reviewed in (Cookson and van der Brug, 2008)] or PINK1 (Kitada *et al.*, 2007). In models of Huntington disease, the medium spiny striatal neurons demonstrate abnormal dendritic spine morphology and electrophysiology (Di Filippo *et al.*, 2007). Synaptic pathology has also been implicated in models of prion diseases (Chiesa *et al.*, 2005; Clinton *et al.*, 1993; Fournier, 2008; Jeffrey *et al.*, 2000; Kitamoto *et al.*, 1992). As synaptic contacts form the morphological substrate for neuronal function, neuritic indices of injury should be considered as carefully as historically emphasized cell death endpoints in studies of neurological diseases.

Unlike other eukaryotic cells, neurons are almost devoid of autophagic vacuoles (AVs) and lysosomes under basal conditions (Boland and Nixon, 2006). Nevertheless, autophagy has been implicated in normal physiological neuritic/synaptic function. In the nematode, autophagy regulates levels of $GABA_A$ receptors at inhibitory synapses (Rowland *et al.*, 2006). Autophagy-deficient knockout mice show severe axonal dystrophy with relative sparing of dendrites (Hara *et al.*, 2006; Komatsu *et al.*, 2007; Yue *et al.*, 2008). Neuritic differentiation is impaired with either too much or too little mTOR activity (Zeng and Zhou, 2008), suggesting a possible role for autophagy in synaptogenesis. There is also evidence that mTOR modulates activity-dependent synaptic plasticity in the dendrites of hippocampal neurons (Gong *et al.*, 2006), although the specific role of autophagy was not investigated.

Pathological increases in AVs and lysosomes are observed in Parkinson's, Alzheimer's, and Huntington's diseases, transmissible spongiform encephalopathies, and in toxicity due to MPP+, methamphetamine, dopamine and mutations in the familial parkinsonian gene *LRRK2* which encodes leucine rich repeat kinase 2 (Chu *et al.*, 2007; Gomez-Santos *et al.*, 2003; Larsen *et al.*, 2002; Liberski *et al.*, 2002; Nixon *et al.*, 2005; Orr, 2002; Plowey *et al.*, 2008; Rudnicki *et al.*, 2008; Zhu *et al.*, 2007). Brain tissue from Parkinson and Lewy body disease patients display increased mitochondrial autophagy (Zhu *et al.*, 2003). The accumulation of autophagosomes, often in neurites, is observed following stroke and brain trauma (Adhami *et al.*, 2006; Lai *et al.*, 2008; Liu *et al.*, 2008). Increased autophagosomes observed in disease states could also reflect impaired autophagosome clearance, as suggested for Alzheimer and lysosomal storage diseases (Koike *et al.*, 2005; Nixon *et al.*, 2005). These findings implicate disruption of the balance between degradative and biosynthetic mechanisms as a key regulator of neurite remodeling and dysfunction elicited by acute and chronic insults (Cherra and Chu, 2008). In the following sections, we will discuss protocols developed to study neuritic and somatic autophagy in neuronal cell lines, primary neuron cultures and *in vivo* models of brain injury.

2. Studying Neuronal Autophagy *In Vitro*

2.1. Differentiation of SH-SY5Y neuroblastoma cells

SH-SY5Y is a third-generation cell line cloned from a metastatic neuroblastoma removed from a 4-year-old girl (Ross *et al.*, 1983). Dividing SH-SY5Y cells have an undifferentiated appearance and tend to aggregate in culture. We induce neuron-like differentiation by seeding and incubating the cells in 10 μM retinoic acid (RA; Sigma, St. Louis, MO, USA). RA induces a fusiform to triangular soma accompanied by elongation of slender neuritic processes (Pahlman *et al.*, 1984). Advantages of RA-differentiation include markedly reduced cell division, and spreading out of differentiated cells uniformly across the culture surface. We store 1000X aliquots of RA in dimethyl sulfoxide (DMSO; Sigma) at 4 °C. Under these conditions, RA is stable for about 3 months; with longer storage times, it loses efficacy resulting in decreased neurite lengths in treated cells. We have also employed a differentiation protocol involving sequential exposure of cells to RA followed by brain-derived neurotrophic factor (BDNF) (Encinas *et al.*, 2000). BDNF-differentiated SH-SY5Y cells demonstrate further elongation of neurites accompanied by increased neuritic branching compared to RA alone. However, these cultures are more cumbersome to study due to increased neurite overlap between cells, extension of neurites in and out of the focal plane, and greater variability in neurite lengths in control differentiated cultures (sometimes requiring piecing together multiple high-power fields). Consequently, we conduct the majority of our experiments in RA-differentiated SH-SY5Y cells.

2.1.1. Plating and Differentiation Protocol (Plowey *et al.*, 2008)

1. SH-SY5Y cells are maintained in a humidified incubator with 5% CO_2 at 37 °C in antibiotic-free Dulbecco's modified Eagle's medium (DMEM; Lonza, Walkersville, MD, USA) supplemented with 10% heat-inactivated fetal calf serum, 10–15 mM HEPES and 2 mM

glutamine (Biowhitaker, Walkersville, MD), referred to henceforth as DMEM/FBS.

2. Culture chambers are seeded on Day 1 at a relatively low density of $5.0 \times 10^4/\text{cm}^2$ in DMEM/FBS supplemented with 10 μM RA. Uncoated 10-cm^2 plastic culture chambers are used for biochemical analyses and uncoated Lab-Tek II chambered cover glasses (#1.5 German borosilicate; Nalge Nunc International, Naperville, IL, USA) for imaging experiments.
3. After 3 days differentiation (Day 4), the media is refreshed and experiments are initiated. For BDNF-differentiation, the media is changed to serum-free DMEM with 50 ng/ml BDNF and experiments are initiated following an additional two days of incubation.

Because of reduced proliferation, RA cultures can be maintained for about 10 days and BDNF cultures for about 12 days at this seeding density. Maximal neurite lengths are reached by 4–5 days in RA and 2–3 days in BDNF. Because neurite length is maintained in the weeks following differentiation, these cultures can be used for chronic studies of 2–4 weeks if seeded at a lower density, although eventually an undifferentiated flat population may overgrow the plate.

2.2. NIH ImageJ assisted analysis of neurite length and arborization

The length and complexity of the axodendritic arbor in neurons can be markedly altered by pathological stresses and physiologic activity-dependent remodeling. To analyze the impact of autophagy on neurites, rapid, reproducible, and unbiased methods for measuring neurite lengths and branching complexity are desirable. A variety of approaches have been used in cell culture studies of outgrowth/differentiation (Charych *et al.*, 2006; Dominguez *et al.*, 2004; Hynds *et al.*, 2002; Price *et al.*, 2006), degeneration/injury (Bilsland *et al.*, 2002), or dystrophy (Pigino *et al.*, 2001). Imaging methods developed to study neurite outgrowth from neuronal explants involve counting the number of neurites that extend beyond the circumference of a circle centered at the explant (Lagreze *et al.*, 2005). However, this method is not readily translated to studying injury-induced neurite retraction in dispersed neuronal cultures, as neuritic processes associated with an individual neuron are not separately delineable.

Commercial software such as Metamorph (Molecular Devices, Sunny-Vale, CA) and Image Pro Plus/Express (Media Cybernetics, Silver Springs, MD) have been employed to measure neurites and branching complexity in primary neurons, which involve manually drawing polygons around the neurite(s) and soma of neurons to be measured (Gerecke *et al.*, 2004). In addition, many researchers use NIH ImageJ software for data quantification.

Advantages of ImageJ include the open access nature of its source code, availability of an online manual in macro programming and maintenance of a large repository of macros available in the public domain (http://rsb.info.nih.gov/ij/macros/).

In the subsections here, we discuss two protocols that we have developed using the NeuroJ plug-in for the public access software ImageJ (NIH, Bethesda, MD, USA), along with their advantages and limitations. Both protocols involve transient transfection of a random subset of neuronal cells in culture to delineate the neuritic arbor of individual cells. The first involves intensity-assisted tracing of neurites of individual neurons showing little overlap (Plowey *et al.*, 2008), and the second reflects our early efforts towards automated measurements that can be used to analyze more densely transfected cells in a higher throughput manner.

2.2.1. GFP transfection to delineate neuronal cells for quantitative analysis

Overlap of neuritic processes from multiple cells hampers confident documentation of the neuritic arbors of individual neurons and differentiated cell lines (Fig. 11.1A). While one approach may involve plating cells at very low densities, we find that neuronal cells are not well maintained in the absence of sufficient cellular contacts. In our studies, quantitative analysis of neuritic extensions of individual cells is made possible by transfection with plasmids encoding soluble, monomeric GFP to highlight the cell bodies and neuritic arbors. This approach capitalizes on relatively low transfection efficiencies that neuronal cells exhibit with cationic lipid-based methods (Lipofectamine 2000; Invitrogen, Carlsbad, CA, USA), allowing us to achieve adequate spatial resolution between transfected cells for confident image analysis at plating densities supportive of neuronal health in culture. We routinely use the pEGFP-C1 (Invitrogen) vector (Fig. 11.1B–C). Although pMAX GFP (Amaxa Biosystems) exhibits more intense fluorescence amenable to low-power analysis, transient expression result in cytosolic GFP granules in the soma of differentiated cells. These may reflect either aggregation or accumulation in lysosomes, both of which appear similar by fluorescence microscopy (Katayama *et al.*, 2008). Thus, we prefer using the pEGFP-C1 plasmid, which remains diffusely distributed (Katayama *et al.*, 2008; Plowey *et al.*, 2008).

2.2.2. *Lipofectamine Transfection Protocol* (Plowey *et al.*, 2008; Dagda *et al.*, 2008)

The following protocol is effective for transient transfection of differentiated or undifferentiated SH-SY5Y cells with a variety of DNA plasmids (pEGFP-C1, pMAX GFP, GFP-LC3, GFP-ERK2). See section 2.5.1 for copy ratio considerations when cotransfection of more than one plasmid is desired, and Section 2.6 for combining with siRNA protocols.

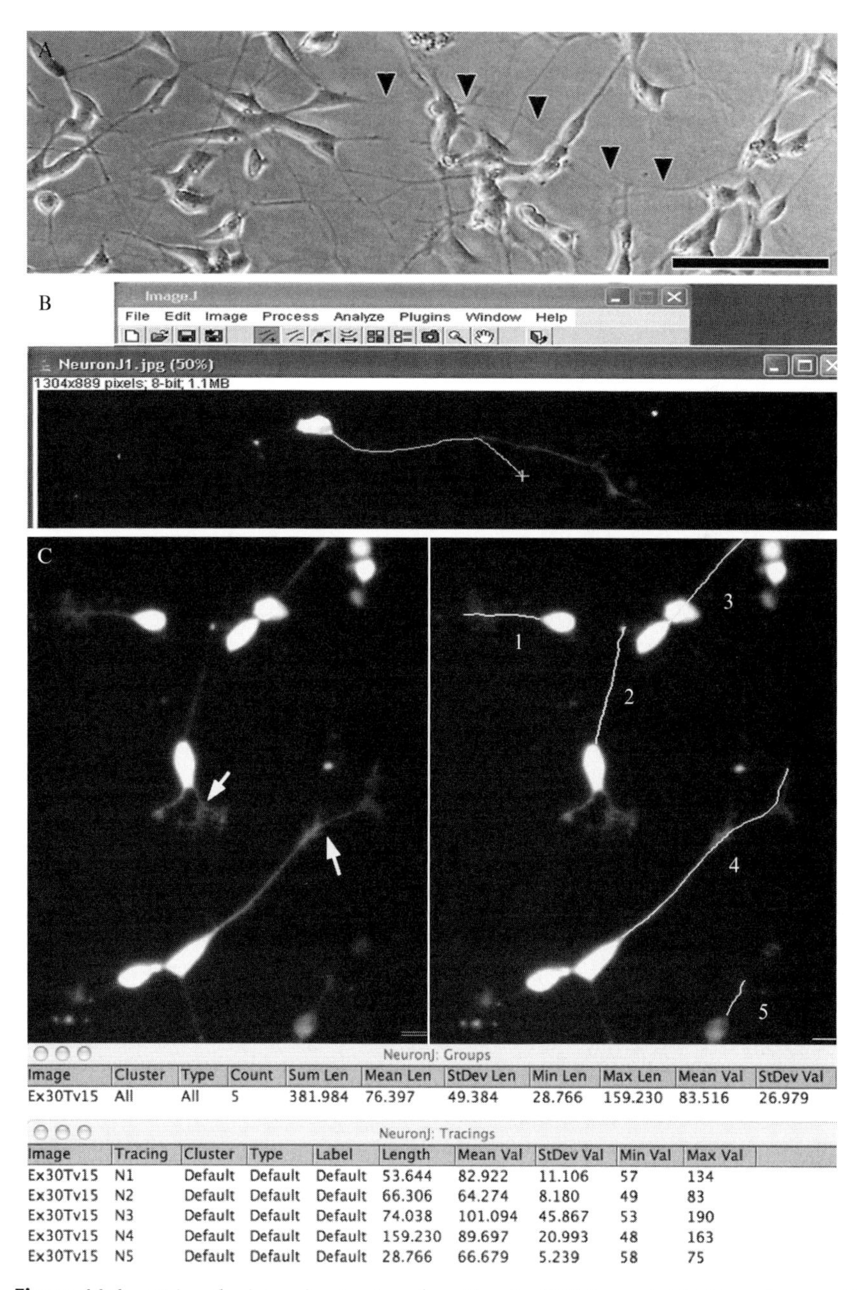

Image	Cluster	Type	Count	Sum Len	Mean Len	StDev Len	Min Len	Max Len	Mean Val	StDev Val
Ex30Tv15	All	All	5	381.984	76.397	49.384	28.766	159.230	83.516	26.979

Image	Tracing	Cluster	Type	Label	Length	Mean Val	StDev Val	Min Val	Max Val
Ex30Tv15	N1	Default	Default	Default	53.644	82.922	11.106	57	134
Ex30Tv15	N2	Default	Default	Default	66.306	64.274	8.180	49	83
Ex30Tv15	N3	Default	Default	Default	74.038	101.094	45.867	53	190
Ex30Tv15	N4	Default	Default	Default	159.230	89.697	20.993	48	163
Ex30Tv15	N5	Default	Default	Default	28.766	66.679	5.239	58	75

Figure 11.1 Using the intensity-tracing algorithm in NIH ImageJ. (A) Phase image of differentiated SH-SY5Y cells showing extensive overlap of processes making it difficult to determine the end of a neuritic process (arrowheads) that merges with processes of other cells. Scale: 100 μm. Transfection of RA/BDNF (B) or RA-differentiated (C) cultures with EGFP-C1 allows visualization of the neuritic arbors associated with

1. Cells should be seeded on 2-well Lab-Tek II chambered cover glasses as described previously. They can be differentiated prior to or after transfection, and we often design experiments so that the final days of differentiation occur during the 2–3 day posttransfection period necessary to allow time for protein expression.
2. We have found that delivery of 1–1.6 μg of DNA/well in a final lipofectamine concentration of 0.1% is effective at transfecting both SH-SY5Y and primary neurons without causing measurable injury to the cells. For 4-well chambered slides, 0.5–0.8 μg is used per well.

For example, 1 μg of high quality DNA (260 nm:280 nm absorbance ratio > 1.5) in Tris-EDTA is mixed in 100 μl of OptiMEM (Invitrogen), and 1% Lipofectamine is prepared separately in the same volume of OptiMEM.

3. The Lipofectamine in OptiMEM is incubated for 5 min prior to the addition of the DNA/OptiMEM. The transfection mixture is then allowed to sit for 20 min at room temperature prior to being added to the cells.
4. Because the cells are grown in 1-ml volumes, 200 μl of DMEM/FBS is removed, and replaced with the entire contents of the transfection mixture, distributed in a dropwise fashion around the well in a spiral pattern. If concurrent RA differentiation is desired, we reduce the OptiMEM volumes in step 2 to 50 μl each (doubling the initial DNA and Lipofectamine concentrations) such that 100 μl of total DNA/Lipofectamine/OptiMEM is added to 900 μl of DMEM/FBS/RA in order to minimize dilution of the RA.
5. The cells are incubated in the DNA/Lipofectamine/OptiMEM-containing medium for at least 48 h prior to imaging to allow time for protein expression.

individual neurons. (B) Using the NeuronJ plug-in for NIH ImageJ, the user clicks at the base of the neurite and as the mouse (cyan X) is moved toward the distal end, the software automatically follows the subtle curves of the structure allowing the user to simply move the mouse to the end of the structure without having to manually trace it. (C) The arrows in the left panel illustrate examples of branch points, and as seen on the right, the user has input in deciding whether to end the measurement at the branch point for primary segment determinations, or select the longer branch to determine the maximal extent of neurite elongation from the soma. As shown in the data table, individual NeuronJ tracings can be grouped to obtain summated and average lengths of all traced neurites in a group. Multiple neurites extending from a single multipolar neuron may be defined as a group to obtain summated neurite lengths per cell (not shown). The software is compatible with both Macintosh and PC operating systems (Panel B: Windows XP running ImageJ 1.36b and NeuronJ 1.1.0; Panel C: MacBook Pro OSX 10.4.11 running ImageJ 1.38X with Java 1.5.0_13 and NeuronJ 1.2.0). (See Color Insert.)

This protocol yields transfection efficiencies of 25%–35% in undifferentiated SH-SY5Y cells, 10% in RA-differentiated cells, and 2%–5% in primary mouse neurons. Forty to fifty random low-power fields are captured blind to the experimental conditions using an inverted epifluorescence microscope and images are quantified as described subsequently.

2.2.3. NeuroJ tracing of neurites for quantification of length and branchpoints

The NeuroJ/ImageJ system provides the ability to measure neurite lengths and branching with accuracy by manually defining neurites for a small sample size of neurons. The JAVA source code of NeuroJ allows the user to interactively analyze running averages and statistics of branch point numbers, neurite lengths and fluorescence intensities during the analysis. We define maximal neurite length as the length from the hillock to the tip of the most distant terminal branch, beginning the measurement at the hillock of each primary neurite and selecting the branch at each branchpoint that would yield the most distant terminal end. For RA-differentiated SH-SY5Y cells, the majority are unipolar or bipolar. In contrast, BDNF-differentiated cells are occasionally multipolar and show more elaborate neurite branching.

We employ the NeuroJ extension (Meijering *et al.*, 2004) of NIH ImageJ to rapidly quantify neurite lengths in our images. This shareware plug-in utilizes an intensity-tracing algorithm that facilitates measurements of neuritic processes in gray-scale images (Meijering *et al.*, 2004). This is an important feature that significantly reduces variability between users and fatigue-related error, as the user needs only to identify the beginning and end of the neurite rather than attempting to trace the whole structure (Fig. 11.1B). Occasionally, a few extra mouse clicks may be required to assist the program in spanning regions that are out of focus or exhibit ambiguous signal intensity.

Neurite branching is another parameter of neurite morphology (Dominguez *et al.*, 2004). We calculate neurite branchpoint number by summing up the number of times a secondary or tertiary neurite branch diverges from the longest primary path. Treatment induced modulation of neurite branchpoints is observed as an index of altered complexity in BDNF-differentiated SH-SY5Y cells. However, in RA-differentiated cells, neurite branchpoints are relatively few even under control conditions (Fig. 11.1C). For analysis of more complex primary neurons, our pEGFP-C1 transfection protocol would also yield images suitable for Sholl analysis of individual neurons (Gutierrez and Davies, 2007) following creation of a binary mask for each transfected neuron.

2.2.4. Semiautomated Measure Neurites macro for high-throughput analysis of neurites

Whereas NeuroJ-assisted tracing is an excellent method to assess neurite morphology, this procedure still requires manually defining neurites of individual neurons. We have been developing alternative methods for measuring the effects of autophagy on neurite abundance (incorporating both length and complexity), using the automated custom Record Macro function of ImageJ (Fig. 11.2). We created the Measure Neurites macro, a custom ImageJ macro for measuring neurite lengths in a high-throughput manner, which is available for download at the ImageJ website (http://imagejdocu.tudor.lu/author/rkd8/). The macro uses the Outlines Algorithm component of the Analyze Particles function of ImageJ, an algorithm used to outline interconnected pixels around the object to be analyzed.

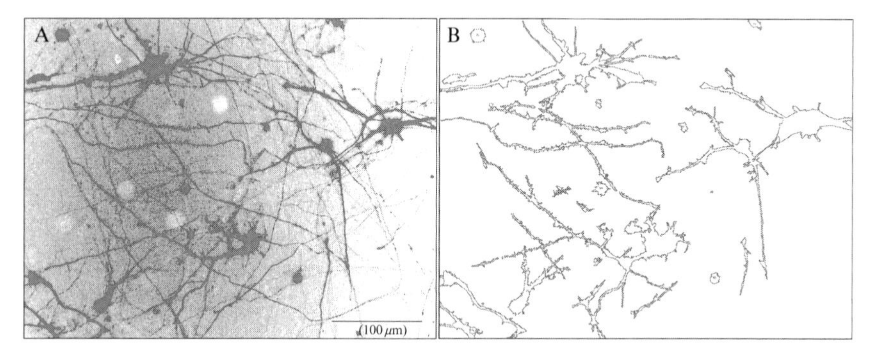

Figure 11.2 Using the Measure Neurites macro. The Measure Neurites macro was used to analyze an epifluorescent micrograph captured at 20X magnification of mouse cortical neurons (7 days *in vitro;* DIV7). Mouse primary neurons were seeded at a relatively high density of $1.5 \times 10^5/\text{cm}^2$ on DIV0, and transiently transfected with GFP on DIV5. RGB images of GFP-expressing neurons were processed by the Measure Neurites macro at 48 h following transfection. The green channel was extracted from the RGB image and inverted, so the highly fluorescent pixels appear black. The image is manually thresholded such that pixels that exceed the threshold intensity are highlighted in red, while those falling below the threshold remain in greyscale (A). The macro traces the thresholded neurites using the Outlines algorithm (B) and computes the number of neurites measured, average neuritic area, and average neuritic perimeter per image, and the longest neuritic perimeter. Notice that background fluorescence (left side of panel (A) or regions of neurites that are incompletely thresholded due to low fluorescence or uneven illumination (right side of panel A) are not traced by the macro and are thus excluded from the analyses (B). The macro computes average neuritic length by substracting the average soma perimeter from the average total perimeter (soma plus neurites) divided by the number of objects being counted to produce summated neuritic perimeters, which can be divided by 2 to approximate neurite lengths. Alternatively, one can normalize neuritic perimeters or lengths by the number of transfected neurons included in the thresholded image (e.g., GFP neurons containing DAPI stain). (See Color Insert.)

Once it is downloaded from the ImageJ website, the text file should be moved to the macro folder of the ImageJ program (c:/programs/ImageJ/macro).

The Measure Neurites macro is optimized to measure neurites of GFP transfected, differentiated SH-SY5Y cells, and primary cortical neurons. However, the macro can be customized depending on the neuronal population being analyzed by changing the value of the average soma perimeter (line 34) that is subtracted from the total average cellular perimeter lengths to obtain average neurite perimeters. In brief, the macro opens up RGB images, splits them into individual channels, and performs photographic inversion of the green channel. The macro allows the user time to manually threshold the pixels (set a minimum pixel value for the structures to be analyzed in a given epifluorescent field) to trace the neurites (Fig. 11.2A–B). The macro then computes the number of neurites, summation of the areas and perimeters of soma and neurites, and the average neuritic area and perimeter per epifluorescent field. The results can be pasted onto an Excel spreadsheet. The summated neurite perimeters for each image is divided by 2, assuming negligible contributions of neurite thicknesses, to yield the summed total of neurite lengths, then normalized to the number of cells analyzed in the image (determined by counting the number of green neurons with DAPI stained nuclei), prior to statistical analysis.

Prior to running the macro, it is important to set the correct scale of the image being analyzed using the Set Scale command under the Analyze pull-down menu. The scale is to be assigned as pixels/μm. It is also important to determine if the average soma perimeter of the cells being studied is accurately reflected in line 34 of the macro as discussed previously. The quality and uniformity of the background in the image to be analyzed will determine whether the majority of neurites will be traced by thresholding. As the macro obtains summated lengths of primary, secondary, and tertiary neurites, it estimates a higher neurite number compared to manual identification; although average lengths in RA differentiated, predominantly unipolar SH-SY5Y cells show only ~1 micron differences when the same set of images is used to directly compare the two methods. In terms of time efficiency, the Measure Neurites macro is best used for images with higher neuron densities or complex neuritic arbors because more technical replicates are required to develop consistent user thresholding. For images with fewer neurons with minimal overlap, the manual identification method is still faster and more precise.

Although this is still a work in progress, a strength of the macro is that it can be used as a template for further optimization by modifying the code to fit new conditions. Updates and detailed instructions on how to use our most current version of the macro will be available on the website (http://imagejdocu.tudor.lu/author/rkd8/).

2.3. Studying neuritic and somatic autophagy in culture models of Parkinsonian injury

The conversion of cytosolic LC3-I to LC3-II, which is covalently bound to membranes of AVs, is considered to be the most specific marker of autophagy to date (Klionsky *et al.*, 2008). As illustrated elsewhere in this volume, the repertoire of available tools used to measure autophagy in cells includes counting LC3 puncta, measuring conversion of LC3-I to LC3-II by western blot, electron microscopy, and lysosomal markers for later stages of maturation.

2.3.1. Fluorescent markers of autophagic vacuoles

Visualization of GFP-LC3 puncta has become popular due to its technical simplicity, convenience, and reliability, though potential caveats relating to GFP aggregation and/or autophagy induced by transfection-related injury need to be considered (Klionsky *et al.*, 2008). The average number of GFP-LC3 puncta per cell is frequently used to quantify autophagic activity in the soma. In neurites, determining the number of axonal GFP-LC3 puncta per μm length requires high-resolution, high-magnification epifluorescence imaging, as autophagic puncta in neurites are generally smaller than those observed in the soma and fluoresce to a lesser degree (Fig. 11.3).

Using the protocol described in section 2.2.1, approximately 10% of RA differentiated SH-SY5Y cells are transfected with GFP-LC3 using 0.1% lipofectamine in DMEM media containing 10% fetal bovine serum. If transfection is performed prior to RA differentiation, 25%–35% of differentiated cells will express GFP-LC3. These transfection efficiencies result in little neuritic overlap between cells, ideal for image-based counting of GFP-LC3 puncta, yet is sufficient for detection of a GFP-LC3-II shift as a biochemical measure of autophagy in transfected cells (Plowey *et al.*, 2008) (see sections 2.3.2–2.3.3). In primary cultures, only 2%–5% of neurons are transfected with GFP-LC3 using Lipofectamine 2000. While this is sufficient for image analysis of GFP-LC3 puncta and the axodendritic arbor of individual neurons, transgenic mice or viral transduction methods may be necessary for biochemical studies of autophagy in genetically altered primary neurons.

We have also constructed stable cell lines of GFP-LC3-expressing SH-SY5Y cells (Dagda *et al.*, 2008), which can be RA differentiated and maintain the same pattern of autophagic responses as the parent line. While these cells are useful for somatic studies or biochemical analysis, they are not as amenable to quantitative image analysis of neuritic autophagy due to extensive overlap of fluorescent structures. If stable cell lines are used, multiple clones should be studied, as the stable clones vary with regard to the degree of neuronal versus fibroblastic versus epithelioid differentiation.

In terms of studying early versus late AVs, combined studies using GFP-LC3 and LC3 N-terminally fused to monomeric red fluorescent protein

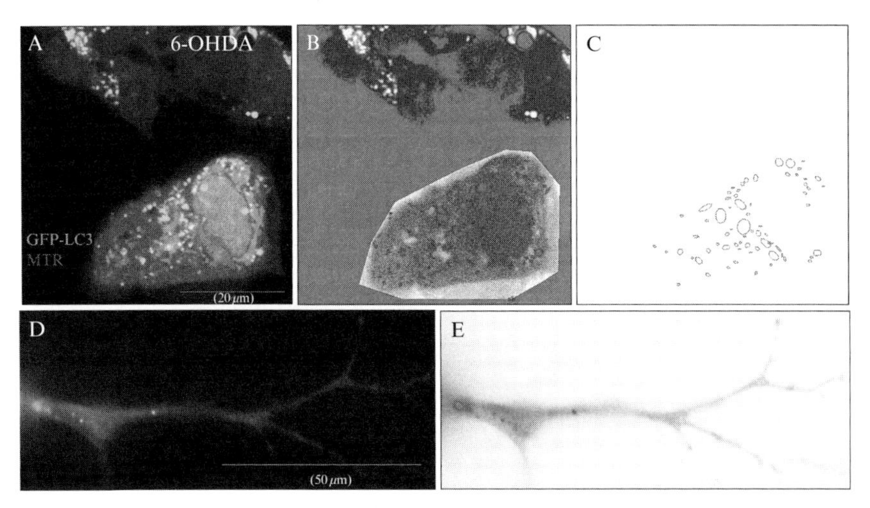

Figure 11.3 Features of the GFP-LC3 macro. A representative image of four SH-SY5Y cells transfected with GFP-LC3 and loaded with MitoTracker dye to label mitochondria. Cells were treated with 6-hydroxydopamine for four hours to induce autophagy (A). The cell of interest is traced using the polygonal tracing tool of ImageJ for analysis. The green channel is then extracted from the RGB image and manually thresholded to trace the majority of the somatic GFP-LC3 puncta (colored red within the grey region of interest (B). The macro then traces the thresholded GFP-LC3 puncta by employing the Outlines algorithm (C) and computes the total number of GFP-LC3 puncta analyzed in the field, the average GFP-LC3 puncta size (μm), and percent area occupied by GFP-LC3 puncta. Notice that cytosolic and nuclear GFP-LC3 fluorescence, and background pixelation are not thresholded and are therefore excluded from analysis. (D) Representative neurite of a RA-differentiated SH-SY5Y cell transfected with GFP-LC3. Note that GFP-LC3 puncta are smaller and less numerous than somatic puncta. (E) The GFP-LC3 macro identifies neuritic GFP-LC3 puncta (colored red) for further analysis as described earlier. (See Color Insert.)

(RFP-LC3) may be needed. Neurons show higher basal levels of RFP-LC3 puncta per cell than seen with GFP-LC3, probably because the RFP moiety is more stable to lysosomal conditions (Kimura *et al.*, 2007) Other markers to analyze later stages of autophagy include monodansylcadaverine (MDC) (50 μM final concentration in media, SIGMA, St. Louis, MO) or Lyso-Tracker Red DND-99 (100 nM, Molecular Probes, Eugene, CA) (also see the chapter by Vázquez and Colombo in volume 452). Briefly, we use the following protocol:

1. 10 mM working stocks of MDC in DMSO is added directly onto cells in medium (50 μM final MDC concentration) and incubated for 30 min at 37 °C, 5% CO_2.
2. Cells are washed with PBS containing glucose prior to imaging of MDC-labeled vacuoles.

3. To visualize lysosomes, a 1 m*M* stock of LysoTracker Red DND-99 in DMSO is prepared in medium (100 n*M*, or a 10,000X dilution from 1 m*M* stocks). The medium of cells is replaced with medium containing LysoTracker Red DND-99 and cells are incubated for 30 min at 37 °C in a 5% CO_2 tissue culture incubator.
4. The medium is refreshed with DMEM/FBS prior to imaging.

MDC-labeled structures colocalize with lysosomes but not GFP-LC3 (Munafo and Colombo, 2001), consistent with observations that GFP-LC3 fluorescence is diminished in lysosomal environments (Kimura *et al.*, 2007). MDC labels a subset of mCherry-LC3 puncta *in vivo* and does not label any structures not labeling with mCherry-LC3, indicating specificity for autolysosomes (Iwai-Kanai *et al.*, 2008). A limitation is that the fluorescence of lysosomal markers tends to quench rapidly. The combined use of early and late AV markers such as GFP-LC3 and MDC form a useful complement in studying autophagic maturation (Chu, 2006).

2.3.2. Semiautomated GFP-LC3 macro for measuring autophagy in fluorescent images (Dagda *et al.*, 2008)

Taking a similar approach that we used previously to create a MetaMorph journal to analyze the size distribution of MDC-labeled autolysosomes (Zhu *et al.*, 2007), we have made an NIH ImageJ macro called GFP-LC3 available for download (http://imagejdocu.tudor.lu/author/rkd8/), which performs automated quantification of the number and size of GFP-LC3 puncta (Dagda *et al.*, 2008) (Fig. 11.3). It is important to analyze GFP-LC3 puncta only from high-resolution, high-quality (low autofluorescence) TIFF images captured at 60X–90X magnification when using this macro. Images captured at 40X magnification are inadequate, as the decreased resolution of GFP-LC3 puncta in these images will result in identification of overlapping or clustered puncta as a single large particle by the macro. This is particularly evident in somatic measures of puncta, where clustering is more frequent.

Prior to using this macro, it is critical to assign the correct scale (pixels/μm) of the epifluorescence micrograph images being analyzed, using the Set Scale command found in the Analyze pull-down menu. The macro has been optimized to measure specific GFP-LC3 puncta larger than 0.2 but less than 10 μm, a size range that is consistent with phagophores and autophagosomes while excluding background pixels and potential nuclear GFP fluorescence from the analysis. The macro performs the following commands:

1. First, the macro allows the user time to select the region of interest or the cell to be analyzed using the polygon selection tool. The macro then splits the RGB image into individual channels (blue, green, and red), closes the blue and red channels, and extracts the green channel to

grayscale followed by photographic inversion (GFP fluorescence converted to black pixels over a white background).

2. The macro will allow the user time to assign the regions of interest to be measured by manually thresholding for minimal and maximal pixel values using the Threshold function under the Image pull-down menu, Adjust submenu (Fig. 11.3A–C). Consistency in the application of thresholding across cells is important. A threshold value of two standard deviations from the background fluorescence is enough to trace the majority of GFP-LC3 puncta.

Please note that GFP-LC3 puncta in axons and dendrites must be thresholded and measured independently of somatic GFP-LC3 puncta for each neuron to be analyzed. This is due to the fact that GFP-LC3 puncta in neurites are smaller and less bright, and cannot be detected using thresholds set according to somatic puncta.

3. Once pixels are thresholded, the macro employs the Measure Particles algorithm to record GFP-LC3 puncta number, area, size (expressed as radii length in μm, calculated using the formula r = square root (area/π)), and fraction of cellular area occupied by GFP-LC3 puncta. Results are displayed in the Results window and can be transferred to an Excel spreadsheet using the copy and paste functions of ImageJ. The macro is conveniently looped to restart as many times as necessary to allow the user to analyze multiple neurons in the same field without closing the image being analyzed. This macro can also be adjusted to quantify number and area occupied by other organelles such as mitochondria (Dagda *et al.*, 2008).

2.3.3. LC3 shift Western blot

As the analysis of LC3 gel shift has been extensively reviewed (Mizushima and Yoshimori, 2007) (see also the chapter by Kimura *et al.*, in volume 452), this section will focus on highlighting considerations for transfected cells or neurons. The lipidated form of LC3 (LC3-II) exhibits faster electrophoretic mobility by SDS-PAGE compared to LC3-I, and the LC3-II/LC3-I ratio is a measure of steady state autophagosome content in cells. Some suggest that the LC3-II/β-actin ratio may be more reliable, as LC3-I expression can be transcriptionally regulated and there is variability in how well LC3-I is recognized by different antibodies (reviewed in (Klionsky *et al.*, 2008)).

Because neurons typically show reduced transfection efficiency compared to other cell types, analysis of genetic models of neurodegeneration by western blot may be limited to virally transduced or stable cell lines. In transiently transfected differentiated cell lines, however, a disease associated mutant protein can be cotransfected with GFP-LC3 (see section 2.5). The GFP tagged LC3-I and GFP-LC3-II bands can be detected by immunoblotting with a polyclonal rabbit anti-GFP IgG (1:5000; Invitrogen,

Carlsbad, CA) (Plowey *et al.*, 2008). This GFP-LC3-II shift is a biochemical measure of autophagy in the transfected subpopulation, and is inhibited by RNAi knockdown of Atg7 (Plowey *et al.*, 2008).

2.3.4. Measurements of autophagic flux

To measure autophagic flux, as compared to steady-state levels of AVs, LC3-II turnover or GFP-LC3 puncta are measured at different time points in the presence and absence of bafilomycin A_1 (10 n*M* for SH-SY5Y cells), which inhibits lysosomal acidification and fusion of autophagosomes with the lysosome, or with cell permeable lysosomal protease inhibitors E64-D (40 μM for SH-SY5Y cells) and pepstatin (25 μM for SH-SY5Y cells). For *in vivo* studies of autophagic flux, chloroquine has been employed to block degradation (Iwai-Kanai *et al.*, 2008). The fraction of AVs that is elevated in response to the presence of fusion/degradation inhibitors represents an estimate of flux, but the length of the bafilomycin treatment must be titrated to ensure that the assay is not approaching saturation under basal control conditions. Autophagic flux in axons and neurites is more difficult but may be possible using the image-based methods of sections 2.3.1–2.3.2 for quantifying GFP-LC3 puncta in the presence or absence of degradation inhibitors. The additional factor of AV trafficking into and out of the region of quantification would also need to be considered. Clearly, additional methods for studying autophagic flux in neuronal cells are needed both *in vitro* and *in vivo*.

2.4. Methods for measuring mitochondrial autophagy

Mitochondria undergo selective autophagic degradation in a process called mitophagy (Kissova *et al.*, 2004; Rodriguez-Enriquez *et al.*, 2004). In neurons, autophagy is critical for the turnover of aged or subtly impaired mitochondria (Kiselyov *et al.*, 2007), whereas pathological mechanisms may cause excess degradation of neuronal mitochondria (Xue *et al.*, 2001; Zhu *et al.*, 2007). To measure the effects of environmental or genetic factors on neuronal mitophagy, one can employ both image-based and biochemical techniques.

2.4.1. Image-based methods for analyzing mitophagy (Dagda *et al.*, 2008)

1. To analyze for mitophagy using GFP-LC3, RA differentiated SH-SY5Y cells are prepared on uncoated chambered Lab-tek II cover glasses. Cover glasses coated with 100 μg/ml per well of poly-D-lysine (SIGMA, St. Louis, MO) are used to culture primary neurons.
2. The cultures are transiently transfected with GFP-LC3 (1 μg of DNA in 0.10% Lipofectamine) in neurobasal media as described in sections 2.2.1 and 2.3.

3. To visualize mitochondria, GFP-LC3 expressing cultures are labeled with 100 n*M* MitoTracker Red dye 580 (diluted from 1 m*M* stocks in DMSO directly into medium overlying cells) (Molecular Probes, Eugene, CA) or 200 n*M* tetramethylrhodamine methyl ester (TMRM) (mixed into media to create a working stock that is used to replace the growth media) (Sigma, St. Louis, MO) for at least 30 min in 37 °C, 5% CO_2.
4. The cultures are washed with warm medium prior to imaging mitochondria using an epifluorescence or laser confocal microscope.

Alternatively, cells can be cotransfected with a mitochondrially targeted monomeric RFP (mito-RFP) to visualize mitochondria. A word of caution is that transient overexpression of inner mitochondrial membrane targeted RFP may result in cell injury/increased basal autophagy/mitophagy in neurons (Dagda & Chu, unpublished observations). Statistical significance for measuring effects of an environmental or genetic factor on mitophagy can be achieved by analyzing high quality images captured at a high magnification (60X), taking images from at least 25 cells per condition (Dagda & Chu, unpublished observations). Mitophagy in axons is expressed as the percent of GFP-LC3 or RFP-LC3 puncta that colocalize with mitochondria within a given unit length.

Late stages of mitophagy (mitochondria undergoing lysosomal degradation) are followed by delivery of mitochondria to lysosomes. To this end, neurons can be colabeled with 250 n*M* MitoTracker Green FM dye (Molecular Probes, Eugene, CA) or transiently transfected with mitochondrially targeted GFP (mito-GFP) to visualize mitochondria and loaded with LysoTracker Red DND-99 (Molecular Probes, Eugene, CA) to label lysosomes followed by a wash with warm media. To determine whether lysosomal degradation of mitochondria depends on autophagy, mitophagy is analyzed in neurons treated with bafilomycin A (cannot be used with LysoTracker Red) or RNAi targeting the autophagic machinery (Zhu *et al.*, 2007) (See section 2.6, below). The use of 3-methyladenine (3-MA) to inhibit autophagy is not optimal for this kind of analysis as not all forms of mitophagy are dependent on Beclin 1/class 3 phosphoinositide 3-kinase signaling (Chu *et al.*, 2007).

2.4.2. Western blot of mitochondrial proteins

Mitophagy can be analyzed by quantifying cellular levels of mitochondrial proteins by immunoblotting in the presence or absence of autophagy inhibitors or autophagy-related gene (Atg) RNAi (Dagda *et al.*, 2008). Depending on the time frame and context of the experiment, mitochondrial biogenesis may or may not play a confounding role to be considered. For example, a study of the Parkinsonian toxin MPP+ used immunofluorescence and Western blot analysis of proteins residing in different compartments of the mitochondria, and the effects of MAP kinase inhibitors were confirmed using electron microscopy (Zhu *et al.*, 2007). The 60-kDa mitochondrial complex IV protein and p110 mitochondrial membrane antigen antibodies can also be used to stain mitochondria in paraffin embedded human brain tissues (Zhu *et al.*, 2003). Intermembrane space proteins such as cytochrome c or apoptosis inducing factor are less reliable as they are lost from mitochondria that have undergone permeability transition during apoptosis (Chu *et al.*, 2005).

2.5. Autophagy in genetic models of Parkinson's disease

In addition to toxin models of parkinsonian injury (Dagda *et al.*, 2008; Zhu *et al.*, 2007), we found that autophagy actively contributes to the neurodegenerative phenotype in a genetic model of Parkinson's disease based on mutation in the *LRRK2* gene (Plowey *et al.*, 2008). Mutations in *LRRK2* have been detected in familial and apparently sporadic cases of parkinsonism (Kay *et al.*, 2006). The protocols described herein would be generally applicable to other genetic models of neurological diseases that use transient transfection to achieve expression of a disease-associated mutant protein.

2.5.1. Optimizing cotransfection conditions (copy ratio)

To measure autophagy in mutant-*LRRK2* transfected cells, we cotransfect the pEGFP-LC3 plasmid with either the wild type full-length HA-tagged *LRRK2* cDNA subcloned into the pcDNA 3.1 vector backbone or one of several mutant *LRRK2* constructs. Following media refreshment, we modify the basic transfection protocol in Section 2.2.1 to cotransfect the pEGFP-LC3 and pcDNA3-LRRK2 plasmids in a 1:15 mass ratio (MacLeod *et al.*, 2006) (equivalent to a 1:4 molar ratio) in OptiMEM medium (Invitrogen) (5% of culture volume) mixed 1:1 with 2% Lipofectamine prepared separately in the same volume (final lipofectamine concentration of 0.1%). These conditions yield an overall transfection rate of 10%–15% and a 75%–90% coexpression rate as determined by immunolabeling for HA and GFP.

2.5.1.1. Visualization of GFP-LC3 and LRRK2 co-transfected cells (Plowey et al., 2008)

1. Following 2 days of expression after transfection, the cells are briefly and gently rinsed with warmed Dulbecco's Phosphate Buffered Saline (DPBS) (Invitrogen).
2. After aspiration of washing medium, 3.9% paraformaldehyde in phosphate buffered saline [137 m*M* NaCl, 2.7 m*M* KCl, 4.3 m*M*

$Na_2HPO_4 \cdot 7H_2O$, and 1.4 m*M* KH_2PO_4] (PFA) is added to the wells, and the cells are incubated for 20 min at room temperature.

3. The fixed cells are permeabilized by washing in PBST (PBS containing 0.1% Triton X-100) for 15 min on a rocker at room temperature.
4. *LRRK2*-transfected cells are incubated overnight at 4 °C on a rocker table with a mouse monoclonal HA-tag primary IgG (1:1000 dilution; Covance, Emeryville, CA, USA). The cells are then brought to room temperature for 30 min, washed 4 times with PBST for 5 min each, followed by incubation for 1 h at room temperature on a rocker with a polyclonal Cy3-labeled donkey antimouse IgG (1:1000 dilution; Jackson Immunolabs).
5. The cells are then washed twice in PBST and twice in PBS. Nuclei are counterstained in a 1:500 dilution of DAPI in PBS (2.5 mg/ml stock solution in DMSO stored at −20 °C) for 5 min. The cells are then washed 3 times for 5 min each with PBS prior to imaging. Fifty random 60X objective oil-immersion images of *LRRK2*-positive cells, or in the case of vector transfected cells, 50 random images of GFP-LC3 transfected cells are digitally captured for data analysis.

2.5.2. Quantitative analysis of somatic and neuritic autophagy

The basic methods of identification and quantification of GFP-LC3 puncta in mutant-*LRRK2* cotransfected cells are similar to the threshold-based ImageJ techniques described for toxin models in section 2.3.2. Because of the diffuse nature of GFP-LC3-I distribution in cells, neurites are well highlighted in these cells and the brighter puncta are well visualized against this background (Plowey *et al.*, 2008). Cotransfection of mutant LRRK2 plasmid with pEGFP-C1 is used as a control to ensure that the mutant plasmid does not induce puncta through aggregation of GFP, and no protein aggregates are observed by electron microscopy. We find no significant differences in mean neurite lengths as measured in pEGFP-C1 versus GFP-LC3 transfected cells.

With mutant *LRRK2* expression, we see significant increases in the numbers of AVs per neurite, the number of AVs per unit length of neurite, the number of AVs per cell body and the percent cytoplasmic area of the cell body occupied by GFP-LC3-labeled AVs (Plowey *et al.*, 2008). The latter parameter reflects the observation that somatic AVs are increased in number as well as size, whereas neuritic AVs are small and increased in number alone. Because only 10% of cells express *LRRK2* constructs, the standard LC3 mobility shift assay is not sensitive enough to detect increased autophagy in the transfected cell population among a background of untransfected cells. However, we use GFP immunoblotting to detect a GFP-LC3 mobility shift indicative of increased autophagy in *LRRK2*-co-expressing cells as described in Section 2.3.3.

2.5.3. Special considerations for protein aggregation disease

A recent study demonstrated that cytosolic protein aggregates can be labeled with GFP-LC3 even in the absence of progression to autophagy (Kuma *et al.*, 2007). This phenomenon has the potential to be misinterpreted. Although the authors note differences between transient and stable transfectants, the phenomenon is most likely related to levels of overexpression or using high levels of lipofectamine rather than representing an intrinsic advantage of stable lines, which can be confounded by clonal selection bias for cells equipped to survive chronic expression of a toxic gene product. As lipofectamine-related GFP-LC3 puncta are typically transient (Klionsky *et al.*, 2008), including appropriate time-matched transfection controls or verification of findings using other methods are recommended. Use of lipidation-deficient mutants of GFP-LC3 (Tanida *et al.*, 2008), or demonstration that the observed increases in GFP-LC3 puncta are inhibited by siRNA knockdown of core autophagy machinery proteins, or confirmatory transmission electron microscopy (TEM) studies may be used to verify that the observed increases in GFP-LC3 puncta correspond to increases in AVs (Plowey *et al.*, 2008).

2.6. Autophagy modulation using RNA interference against Atg proteins

Because short double-stranded siRNAs are effectively transfected into >90% of differentiated neuronal cells (Plowey *et al.*, 2008), standard Western blot analysis can be used to assess efficacy of RNAi knockdown at the time that toxin is administered, or the disease-related genes are expressed, and for the duration of the injury (Zhu *et al.*, 2007). Note that increased plating density, higher lipofectamine concentrations and/or use of antibiotics in the maintenance of cells may all serve to reduce siRNA transfection efficiency, potentially due to enhanced cytotoxicity. We typically maintain lines under antibiotic-free conditions, as antibiotics can impair mitochondrial metabolism or mask suboptimal sterile technique, increasing the potential for mycoplasma contamination. Even when the population of interest reflects a smaller percentage of randomly distributed cells transfected to initiate injury a few days after RNAi, as in our study of *LRRK2* (G2019S)-induced neurite remodeling, standard western blot analysis can still be used to assess efficacy of the siRNA. Control experiments using fluorescently labeled dsRNA confirm that siRNA remains uniformly distributed in >90% of the smaller population of cells expressing HA-tagged *LRRK2* (G2019S) (Plowey *et al.*, 2008). Thus, there is no evidence of preferential transfection of cells depending upon RNAi status, nor that subsequent DNA plasmid transfection might affect the stability of the siRNA. Alternatively, the effects of Atg7 siRNA on autophagy can be monitored specifically in transfected cells by co-transfecting mutant *LRRK2* with GFP-LC3 and following the inhibition of the GFP-LC3-II shift as described above (Plowey *et al.*, 2008).

Interpretation of 3-MA in the context of neurite remodeling is complex as 3-MA (5 m*M*) effectively inhibits Akt phosphorylation, an important neuronal prosurvival factor, as well as several other death regulatory protein kinases (Xue *et al.*, 1999; Zhu & Chu, unpublished observation). The DMF or DMSO concentrations necessary to achieve 5–10 m*M* doses of this drug that are typically used in autophagy studies must also be considered. Moreover, phosphoinositide 3-kinase signaling plays an important role in regulating neurite outgrowth (Kimura *et al.*, 1994), potentially by preventing the inhibition of anterograde axonal transport by glycogen synthase kinase 3 (Pigino *et al.*, 2003). These effects would confound the analysis of experiments seeking to inhibit autophagy using pharmacological methods.

We highlight herein a general timeline for using siRNA to study the role of autophagy during neurite remodeling induced by transient expression of a disease-causing mutant protein. During the entire timeline, cells are maintained in 10% fetal bovine serum DMEM containing retinoic acid (10 μM). Western blot is used to study efficacy of protein knockdown, and the techniques described previously to confirm reduction of injury-induced neuritic GFP-LC3 puncta, prior to analysis of neurite length and complexity.

Cells are plated on day 1 in DMEM/FBS/RA, and transfected with siRNA targeting either Atg7 or LC3 using 0.1% lipofectamine on day 2. On day 3, the DMEM/FBS/RA media is refreshed. On day 4, the insult being studied is applied in DMEM/FBS/RA (e.g., neurotoxin or transfection with the mutant gene of interest; typically cotransfection with a fluorescent marker is used to delineate neuritic processes of the transfected subpopulation). On day 6, the cells are fixed with paraformaldehyde for GFP-LC3 imaging and neurites are measured as described in sections 2.5.1–2.5.2. The effects of Atg7 knockdown on autophagy are confirmed by monitoring GFP-LC3 puncta or GFP-LC3 gel shift, while MDC is used to confirm effects of reduced LC3 expression on autophagy.

2.7. Use of GFP-LC3 transgenic mice for primary neuron culture studies of autophagy

To develop sensitive and specific tools for the detection and quantification of autophagy in primary cortical neurons, our initial studies employed the classic stimulus of nutrient deprivation. Primary neurons from rats or mice are cultured using standard methods (Du *et al.*, 2004). Nutrient deprivation is induced by replacing standard culture media with custom media lacking D-glucose, sodium pyruvate, L-glutamine, L-glutamate, and L-aspartate, and without the fatty acid–containing supplement B27 (Gibco). Neurons can be cultured in 96-well plates for rapid-throughput analysis of the effect of pharmacological or molecular interventions using the 3-[4,5-dimethyl thiazol]-2,5-diphenyltetrazolium bromide (MTT) assay, on larger plates or dishes for analysis of proteins by Western blot (Du *et al.*, 2004), or on glass chamber slide coverslips for immunofluorescence analysis (Zhu *et al.*, 2007). We have used primary cortical neurons from GFP-LC3$^{+/-}$ transgenic reporter mice, identified at birth using a headset equipped for epifluorescent illumination, to track dynamic formation of autophagosomes under conditions of nutrient deprivation and following *in vitro* stretch-induced traumatic injury (Fig. 11.4). The fluorescent signal can be enhanced using immunolabeling with anti-GFP antibody, or visualized directly for live cell assays.

3. Studying Brain Autophagy *In Vivo*

3.1. Models of traumatic brain injury

Although the role for autophagy has yet to be established, it is clear that autophagy is increased in brains after acute traumatic injury. Upregulation of *beclin 1* was reported within injured hippocampal and cortical neurons after closed head injury in mice (Diskin *et al.*, 2005). We recently reported increased LC3-II and formation of autophagosomes and secondary lysosomes after focal traumatic brain injury (TBI) in mice (Lai *et al.*, 2008). In addition, increased formation of Atg12-5 conjugates is observed after fluid percussion injury in rats (Liu *et al.*, 2008).

For our studies, adult male C57BL/6J mice are subjected to moderate-severe TBI to the left parietal cortex using a controlled cortical impact device (Satchell *et al.*, 2003). For ultrastructural studies, mice are perfused with 2% paraformaldehyde/0.01% glutaraldehyde in phosphate buffered saline (PBS) and prepared for electron microscopy using standard methods (Lai *et al.*, 2008). In control mice, ultrastructural evidence for autophagy is rarely detected; however, observation of normal brain tissue raises an important caveat relevant specifically to brain tissue. Researchers should be aware that neurites cut in cross section within brain tissues often appear as double membrane, organelle and/or vesicle containing, structures (examples in Figs. 11.5A–B). Autophagosomes can therefore be difficult to discriminate from cell processes in coronal section with dissolution of intralumenal contents. Identification of true autophagosomes requires criterion based on size (generally 400–1000 nm in diameter) and lack of synaptic clefts or myelin sheaths. Location within a larger, clearly defined profile representing a dendrite or myelinated axon also helps (examples in Figs. 11.5C–D). Autophagosomes with double membranes containing cellular debris and multilamellar bodies are readily detectible in the injured cortex, hippocampus, and thalamus 2 h and 24 h after TBI. Interestingly, after TBI, AVs are most prominent in cell processes and axons, raising the

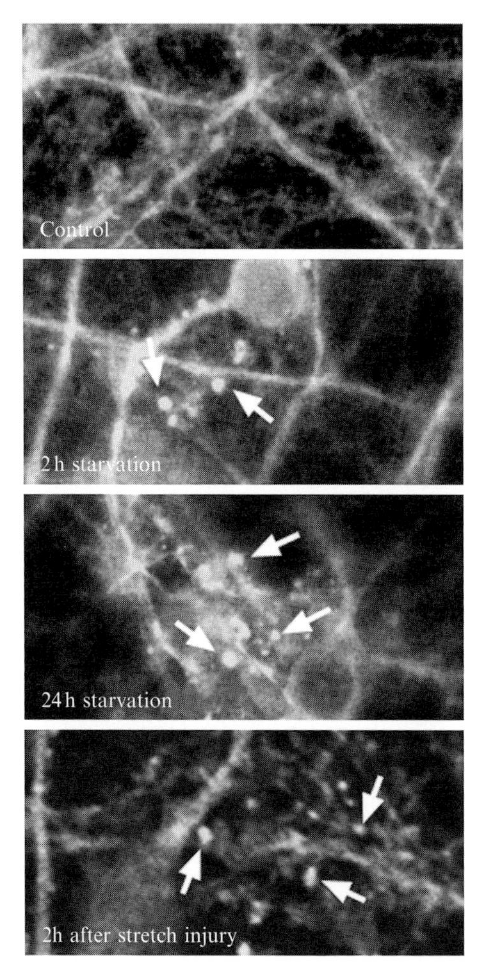

Figure 11.4 Detecting autophagy in neurons during nutrient deprivation and after dynamic stretch-induced injury *in vitro*. Primary cortical neurons from GFP-LC3$^{+/-}$ transgenic mice were subjected to nutrient deprivation (media without glucose, pyruvate, glutamate, glycine, aspartate, or fatty acids) or dynamic stretch (4.7 psi × 100 ms, ~50% strain). Autophagosomes are identified by discrete, high fluorescent intensity, punctuate labeling (arrows).

possibility that autophagy may be involved in axonal damage and/or dendritic pruning (or play roles in glial processes) after TBI, consistent with autophagy-related neurite degeneration described in other injuries (Plowey *et al.*, 2008; Yang *et al.*, 2007).

Autophagosomes in brain can be evaluated using confocal immunohistochemistry and antibodies against LC3. Because current antibodies identify both LC3-I and -II, discrimination relies on identification of discrete, punctate labeling. Again, the caveat of misidentifying LC3-I enriched neurites cut in cross-section applies. Nonetheless, using stringent criteria perhaps with the addition of other labels, such as Fluorojade C, which labels degenerating neurons, one can track autophagy by immunofluorescence in brain with a reasonable degree of confidence. The example shown in Fig. 11.5E is an ipsilateral CA3 hippocampal neuron 24 h after TBI in a postnatal day 17 rat immunostained for LC3 (red) and colabeled with Fluorojade C (green) and DAPI (blue). Punctate LC3 labeling within degenerating neurons in vulnerable regions after TBI is suggestive of autophagosome formation, and supports a role for autophagic stress in acute neurodegenerative diseases such as TBI. Due to fluorescence overlap, biomarkers other than Fluorojade C are required for colabeling of autophagy in dying or stressed cells when examining tissues from GFP-LC3 transgenic mice, unless one has access to a microscope equipped with a spectral analyzer.

One can also use GFP-LC3$^{+/-}$ mice to track autophagy in brain; however, we have found this to be more complicated than use *in vitro*. First, the endogenous GFP signal appears less robust in brain tissue sections (10 micron thick), and typically needs to be enhanced using anti-GFP immunofluorescence. The second, and perhaps more important reason, is that after TBI the milieu within injured brain tissue appears to quench the endogenous GFP signal. Shown in Fig. 11.5F–G are coronal brain sections from normal and injured GFP-LC3$^{+/-}$ mice, respectively. The sections were colabeled using an antibody against GFP (red) to demonstrate that the lack of green fluorescence was not due to a lack of GFP. The quenching of GFP in injured tissue may be related either to increased oxidative stress and/or reduced pH, both of which are known to occur in damaged tissue after TBI.

An LC3 Western gel shift can be used to detect and semiquantify autophagy after TBI. For Western blot examination of LC3-II after TBI, mice are perfused with ice-cold saline. The left dorsal hippocampus and overlying cortex are dissected and processed for subcellular fractionation as described previously (Lai *et al.*, 2008). We find in cultured neurons and after *in vivo* TBI that LC3-II is concentrated in the mitochondria-enriched, P2 fraction (Fig. 11.5H). TBI results in increased LC3-II at 2, 24, and 48 h, peaking at 24 h, versus control mice (Lai *et al.*, 2008). Similar subfractionation techniques are also used by other investigators to detect changes in LC3-II and Atg12-5 conjugates in rats after TBI (Liu *et al.*, 2008).

3.1.1. Western blot analysis for LC3-II *in vivo* (Lai *et al.*, 2008)

1. Animals are anesthetized then transcardially perfused with ice-cold saline.
2. Brains are removed and regions of interest are dissected, placed in microcentrifuge tubes and homogenized in 10 volumes of lysis buffer

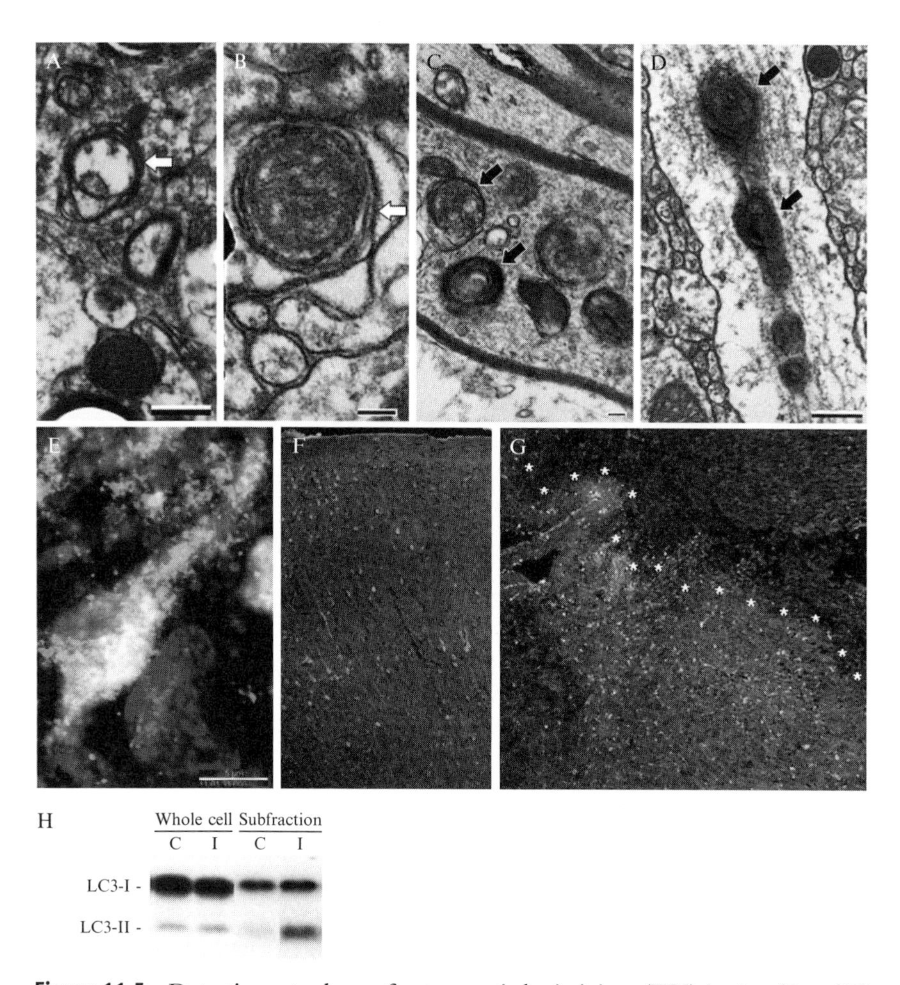

Figure 11.5 Detecting autophagy after traumatic brain injury (TBI) *in vivo*. (A and B) Transmission EM from naive mice showing double-membrane structures likely representing neurites in cross section (white arrows). (C and D) Transmission EM from mice 24 h after TBI showing secondary lysosomes and autophagosomes within axons and dendrites (black arrows). (E) Confocal immunohistochemical labeling for LC3 (red), Fluorojade C (green) and DAPI (blue) in a male PND 17 rat 24 h after TBI. Punctate LC3 labeling is shown in a Flurojade-C-positive neuron in the CA3 hippocampus. (F and G) Immunohistochemical labeling for GFP in GFP-LC3$^{+/-}$ mice. Note the loss of green fluorescence but retention of immunolabeling for GFP (red) in the injured region (demarcated by asterisks) from a mouse 24 h after TBI (G) compared to the control (F). (H) Western blot for LC3 in brain tissue. LC3-II is enriched in the P2 subfraction vs. the whole cell lysate (C = control, I = injured). (See Color Insert.)

(20 m*M* HEPES-KOH, pH 7.4, 10 m*M* NaCl, 1.5 m*M* MgCl2, 1 m*M* EDTA, 1 m*M* EGTA, 250 m*M* sucrose, 1 μM DTT, 1 m*M* PMSF, 2 μg/ml aprotinin).

3. Lysates are centrifuged at 1025×*g* for 15 min at 4 °C to pellet nuclei (P1). The S1 supernatant fractions are centrifuged at 735×*g* at 4 °C for 10 min; the resulting supernatant fractions are further centrifuged at 10,000×*g* at 4 °C for 15 min to pellet autophagosomes, mitochondria, and small organelles (P2). Samples are stored at −80 °C in 10% glycerol.
4. Proteins are loaded into 15% acrylamide gels (30 μg/well) and separated electrophoretically, then transferred to a polyvinyl difluoride membrane overnight. The transferred membranes are incubated in a 1:1000 dilution of a monoclonal antibody that recognizes both LC3-I and -II (Clone 51-11; MBL International, Woburn, MA) at room temperature for 1 h. Membranes are washed in PBS containing 0.1% Tween 20, then incubated in the appropriate secondary antibody (1:3000) for 1 h, incubated in chemiluminescence reagents, and exposed to X-ray film.
5. The relative optical densities for LC3-I and LC3-II can be semiquantified using standard image analysis. Membranes are stripped and probed for cytochrome oxidase as a loading control for the P2 subfraction.

3.1.2. Is autophagy increased after TBI in humans?

To begin to address the question, we performed LC3 and beclin 1 Western blot analysis on brain tissue samples obtained from 5 adult patients resected for management of refractory intracranial hypertension after TBI (Clark *et al.*, 2008). We also examined control brain tissue from 5 adult patients dying of causes unrelated to CNS trauma. An LC3 shift could be detected in 4/5 TBI and 3/5 control patients. Signal intensity for LC3-II was less robust than that observed in cultured neurons or experimental TBI, perhaps related to the fact that whole-cell lysates were used, since subfractionated samples were not available. Punctate LC3 labeling consistent with autophagy was observed immunohistochemically. Autophagy has also been demonstrated in brain tissue from patients with Parkinson, diffuse Lewy body, Alzheimer, and Creutzfeldt-Jakob diseases (Anglade *et al.*, 1997; Nixon *et al.*, 2005; Sikorska *et al.*, 2004; Zhu *et al.*, 2003). The data suggest that autophagy occurs in adult human brain during a variety of diseases, including TBI, warranting further study.

3.2. Models of ischemic brain injury

Ischemic brain injury is either focal or global. Focal ischemic injury results from occlusion of one or more cerebral vessels as occurs in stroke. Focal ischemia is further divided into permanent or temporary. Temporary ischemia occurs when an occluded vessel regains patency (from thrombolytic therapy or percutaneous manipulation). Global injury is typically the result of complete cessation of total cerebral circulation as occurs in cardiac arrest, caused by asphyxia, electrically induced ventricular fibrillation, compression of the great vessels leaving the heart, or chemical cardiac paralysis using potassium (animal models of cerebral ischemia are reviewed in Durukan and Tatlisumak, 2007; Ginsberg and Busto, 1989; Kim, 1997; Traystman, 2003). Total cerebral ischemia secondary to cardiac arrest is an insult shared by both the brain and other major organs. Thus, cardiac arrest models more closely resemble clinical events, but important caveats include the need to consider effects of multisystem organ dysfunction and extracerebral toxins on progression of brain injury.

3.2.1. Autophagy in cerebral hypoxic-ischemic injury

In 1995, Nitatori *et al.*, found evidence of autophagy using electron microscopy in gerbils treated with temporary bilateral occlusion of the common carotid arteries. The next contribution to this field did not occur until 10 years later when increased LC3-II was reported following transient middle cerebral artery occlusion in mice (Degterev *et al.*, 2005) and in a mouse model of neonatal cerebral hypoxic ischemia (Zhu *et al.*, 2005, 2006). GFP-LC3 transgenic mice subjected to unilateral common carotid artery occlusion plus hypoxia (Levine/Vanucci procedure) (Adhami *et al.*, 2006) exhibit increased immunofluorescence with punctuate redistribution consistent with autophagy, autophagy-like vacuoles by EM and decreased cytoplasmic LC3-I by Western blot. Rats exposed to middle cerebral artery occlusion for 90 min (Rami *et al.*, 2008) exhibit increases in beclin 1 expression at 6–48 h following ischemia by Western blot and immunohistochemistry, with partial co-localization of caspase-3 and Beclin 1 and LC3 redistribution and colocalization with beclin 1 in ischemic neurons. Our unpublished data confirm that detection of LC3-II in models of global cerebral ischemia is enhanced by examination of the P2 fraction as described in section 3.1.1.

3.2.2. Causality, the next frontier of *in vivo* autophagy studies

The preceding studies clearly demonstrate that autophagy is invoked following cerebral ischemia. However, whether autophagy is playing a protective or harmful role cannot be inferred from these studies. Studies demonstrating that chronic inhibition of constitutive autophagy (e.g., in Atg knockout mice) creates a neurodegenerative phenotype suggest that inhibition of autophagy might exacerbate hypoxic ischemic injury. In contrast, other studies show that inhibition of autophagy may be protective following hypoxic ischemic injury, implicating excessive autophagy as a mechanism promoting cell death. Neonatal hypoxia-ischemia in mice elicits an increase in LC3-II (Western blot and immunohistochemistry) and formation of autophagosomes (electron microscopy) (Koike *et al.*, 2008). As expected, the same insult in Atg7 knockout mice does not show features of autophagy. More important, the Atg7 knockout mice show dramatically less neuronal death. Atg7-deficient mice also show decreased caspase-dependent neuronal death further supporting a link between autophagy and regulation of apoptosis. Thus, at least in immature rodents, autophagy appears to exacerbate hypoxic-ischemic neuronal cell death.

4. Future Perspectives and Challenges

Evidence supporting a role for autophagy in traumatic, hypoxic-ischemic, and neurodegenerative brain disorders is just emerging and necessarily incomplete. Additional tools, such as specific inhibitors of autophagy and better methods to measure autophagic flux are needed to advance this field. The mechanism of autophagy-dependent regulation of apoptosis and the relationship of autophagic processes with neurite remodeling, synaptic (dys)function and regenerative biosynthetic processes will also be important to characterize (Cherra and Chu, 2008). This is highlighted by the prominence of autophagosomes and dysfunction in neurites after TBI—an acute neurodegenerative disease—as well as in Alzheimer's disease and in neurotoxin and mutant *LRRK2*-mediated models of Parkinson's disease. Finally, improved nonbiased, automated algorithms for image-based quantification of neuritic health and complexity are needed. The development of new brainbow mice expressing different fluorophores in adjacent neurons (Livet *et al.*, 2007) coupled with advances in two photon imaging (St. Croix *et al.*, 2006) may pave the way toward visualizing the axodendritic arbor of individual neurons for studies of autophagy and neurodegeneration *in vivo*.

Acknowledgments

We thank Jason Callio for technical assistance and maintenance of transgenic mouse colonies. The authors are supported by funding from the National Institutes of Health (R01 AG026389, R21 NS053777, K18 DC009120 to CTC; F32 AG030821 to RKD; K08 HD040848 to RWH; R01 NS038620 to RSBC).

References

Adhami, F., Liao, G., Morozov, Y. M., Schloemer, A., Schmithorst, V. J., Lorenz, J. N., Dunn, R. S., Vorhees, C. V., Wills-Karp, M., Degen, J. L., Davis, R. J., Mizushima, N., *et al.* (2006). Cerebral ischemia-hypoxia induces intravascular coagulation and autophagy. *Am. J. Pathol.* **169,** 566–583.

Anglade, P., Vyas, S., Javoy-Agid, F., Herrero, M. T., Michel, P. P., Marquez, J., Mouatt-Prigent, A., Ruberg, M., Hirsch, E. C., and Agid, Y. (1997). Apoptosis and autophagy in nigral neurons of patients with Parkinson's disease. *Histol. Histopathol.* **12,** 25–31.

Bilsland, J., Roy, S., Xanthoudakis, S., Nicholson, D. W., Han, Y., Grimm, E., Hefti, F., and Harper, S. J. (2002). Caspase inhibitors attenuate 1-methyl-4-phenylpyridinium toxicity in primary cultures of mesencephalic dopaminergic neurons. *J. Neurosci.* **22,** 2637–2649.

Boland, B., and Nixon, R. A. (2006). Neuronal macroautophagy: From development to degeneration. *Mol. Aspects. Med.* **27,** 503–519.

Charych, E. I., Akum, B. F., Goldberg, J. S., Jornsten, R. J., Rongo, C., Zheng, J. Q., and Firestein, B. L. (2006). Activity-independent regulation of dendrite patterning by postsynaptic density protein PSD-95. *J. Neurosci.* **26,** 10164–10176.
Cherra, S. J., III, and Chu, C. T. (2008). Autophagy in neuroprotection and neurodegeneration: A question of balance. *Future Neurol.* **3,** 309–323.
Chiesa, R., Piccardo, P., Dossena, S., Nowoslawski, L., Roth, K. A., Ghetti, B., and Harris, D. A. (2005). Bax deletion prevents neuronal loss but not neurological symptoms in a transgenic model of inherited prion disease. *Proc. Natl. Acad. Sci. USA* **102,** 238–243.
Chu, C. T. (2006). Autophagic stress in neuronal injury and disease. *J. Neuropathol. Exp. Neurol.* **65,** 423–432.
Chu, C. T., Zhu, J., and Dagda, R. (2007). Beclin 1-independent pathway of damage-induced mitophagy and autophagic stress: Implications for neurodegeneration and cell death. *Autophagy* **3,** 663–666.
Chu, C. T., Zhu, J. H., Cao, G., Signore, A., Wang, S., and Chen, J. (2005). Apoptosis inducing factor mediates caspase-independent 1-methyl-4-phenylpyridinium toxicity in dopaminergic cells. *J. Neurochem.* **94,** 1685–1695.
Clark, R. S., Bayir, H., Chu, C. T., Alber, S. M., Kochanek, P. M., and Watkins, S. C. (2008). Autophagy is increased in mice after traumatic brain injury and is detectable in human brain after trauma and critical illness. *Autophagy* **4,** 88–90.
Clinton, J., Forsyth, C., Royston, M. C., and Roberts, G. W. (1993). Synaptic degeneration is the primary neuropathological feature in prion disease: A preliminary study. *Neuroreport.* **4,** 65–68.
Cookson, M. R., and van der Brug, M. (2008). Cell systems and the toxic mechanism(s) of α-synuclein. *Exp. Neurol.* **209,** 5–11.
Dagda, R. K., Zhu, J., Kulich, S. M., and Chu, C. T. (2008). Mitochondrially localized ERK2 regulates mitophagy and autophagic cell stress. *Autophagy* **4,** 770–782.
Degterev, A., Huang, Z., Boyce, M., Li, Y., Jagtap, P., Mizushima, N., Cuny, G. D., Mitchison, T. J., Moskowitz, M. A., and Yuan, J. (2005). Chemical inhibitor of nonapoptotic cell death with therapeutic potential for ischemic brain injury. *Nat. Chem. Biol.* **1,** 112–119.
Di Filippo, M., Tozzi, A., Picconi, B., Ghiglieri, V., and Calabresi, P. (2007). Plastic abnormalities in experimental Huntington's disease. *Curr. Opin. Pharmacol.* **7,** 106–111.
DiFiglia, M., Sapp, E., Chase, K. O., Davies, S. W., Bates, G. P., Vonsattel, J. P., and Aronin, N. (1997). Aggregation of huntingtin in neuronal intranuclear inclusions and dystrophic neurites in brain. *Science* **277,** 1990–1993.
Diskin, T., Tal-Or, P., Erlich, S., Mizrachy, L., Alexandrovich, A., Shohami, E., and Pinkas-Kramarski, R. (2005). Closed head injury induces upregulation of Beclin 1 at the cortical site of injury. *J. Neurotrauma.* **22,** 750–762.
Dominguez, R., Jalali, C., and de Lacalle, S. (2004). Morphological effects of estrogen on cholinergic neurons *in vitro* involves activation of extracellular signal-regulated kinases. *J. Neurosci.* **24,** 982–990.
Du, L., Bayir, H., Lai, Y., Zhang, X., Kochanek, P. M., Watkins, S. C., Graham, S. H., and Clark, R. S. (2004). Innate gender-based proclivity in response to cytotoxicity and programmed cell death pathway. *J. Biol. Chem.* **279,** 38563–38570.
Durukan, A., and Tatlisumak, T. (2007). Acute ischemic stroke: Overview of major experimental rodent models, pathophysiology, and therapy of focal cerebral ischemia. *Pharmacol. Biochem. Behav.* **87,** 179–197.
Encinas, M., Iglesias, M., Liu, Y., Wang, H., Muhaisen, A., Cena, V., Gallego, C., and Comella, J. X. (2000). Sequential treatment of SH-SY5Y cells with retinoic acid and brain-derived neurotrophic factor gives rise to fully differentiated, neurotrophic factor-dependent, human neuron-like cells. *J. Neurochem.* **75,** 991–1003.

Fiala, J. C., Feinberg, M., Peters, A., and Barbas, H. (2007). Mitochondrial degeneration in dystrophic neurites of senile plaques may lead to extracellular deposition of fine filaments. *Brain. Struct. Funct.* **212,** 195–207.

Fournier, J. G. (2008). Cellular prion protein electron microscopy: Attempts/limits and clues to a synaptic trait. Implications in neurodegeneration process. *Cell Tissue Res.* **332,** 1–11.

Gentleman, S. M., Nash, M. J., Sweeting, C. J., Graham, D. I., and Roberts, G. W. (1993). β-amyloid precursor protein (βAPP) as a marker for axonal injury after head injury. *Neurosci. Lett.* **160,** 139–144.

Gerecke, K. M., Wyss, J. M., and Carroll, S. L. (2004). Neuregulin-1β induces neurite extension and arborization in cultured hippocampal neurons. *Mol. Cell Neurosci.* **27,** 379–393.

Ginsberg, M. D., and Busto, R. (1989). Rodent models of cerebral ischemia. *Stroke* **20,** 1627–1642.

Gomez-Santos, C., Ferrer, I., Santidrian, A. F., Barrachina, M., Gil, J., and Ambrosio, S. (2003). Dopamine induces autophagic cell death and α-synuclein increase in human neuroblastoma SH-SY5Y cells. *J. Neurosci. Res.* **73,** 341–350.

Gong, R., Park, C. S., Abbassi, N. R., and Tang, S. J. (2006). Roles of glutamate receptors and the mammalian target of rapamycin (mTOR) signaling pathway in activity-dependent dendritic protein synthesis in hippocampal neurons. *J. Biol. Chem.* **281,** 18802–18815.

Gutierrez, H., and Davies, A. M. (2007). A fast and accurate procedure for deriving the Sholl profile in quantitative studies of neuronal morphology. *J. Neurosci. Methods* **163,** 24–30.

Hara, T., Nakamura, K., Matsui, M., Yamamoto, A., Nakahara, Y., Suzuki-Migishima, R., Yokoyama, M., Mishima, K., Saito, I., Okano, H., and Mizushima, N. (2006). Suppression of basal autophagy in neural cells causes neurodegenerative disease in mice. *Nature* **441,** 885–889.

Hynds, D. L., Takehana, A., Inokuchi, J., and Snow, D. M. (2002). L- and D-threo-1-phenyl-2-decanoylamino-3-morpholino-1-propanol (PDMP) inhibit neurite outgrowth from SH-SY5Y cells. *Neuroscience* **114,** 731–744.

Ironside, J. W. (1998). Prion diseases in man. *J. Pathol.* **186,** 227–234.

Ito, U., Kuroiwa, T., Nagasao, J., Kawakami, E., and Oyanagi, K. (2006). Temporal profiles of axon terminals, synapses and spines in the ischemic penumbra of the cerebral cortex: Ultrastructure of neuronal remodeling. *Stroke* **37,** 2134–2139.

Iwai-Kanai, E., Yuan, H., Huang, C., Sayen, M. R., Perry-Garza, C. N., Kim, L., and Gottlieb, R. A. (2008). A method to measure cardiac autophagic flux *in vivo*. *Autophagy* **4,** 322–329.

Jeffrey, M., Halliday, W. G., Bell, J., Johnston, A. R., MacLeod, N. K., Ingham, C., Sayers, A. R., Brown, D. A., and Fraser, J. R. (2000). Synapse loss associated with abnormal PrP precedes neuronal degeneration in the scrapie-infected murine hippocampus. *Neuropathol. Appl. Neurobiol.* **26,** 41–54.

Katayama, H., Yamamoto, A., Mizushima, N., Yoshimori, T., and Miyawaki, A. (2008). GFP-like proteins stably accumulate in lysosomes. *Cell Struct. Funct.* **33,** 1–12.

Kay, D. M., Zabetian, C. P., Factor, S. A., Nutt, J. G., Samii, A., Griffith, A., Bird, T. D., Kramer, P., Higgins, D. S., and Payami, H. (2006). Parkinson's disease and LRRK2: Frequency of a common mutation in U.S. movement disorder clinics. *Mov. Disord.* **21,** 519–523.

Kim, H. K. (1997). Experimental models of cerebral ischemia. *Acta Anaesthesiol. Scand. Suppl.* **111,** 91–92.

Kimura, K., Hattori, S., Kabuyama, Y., Shizawa, Y., Takayanagi, J., Nakamura, S., Toki, S., Matsuda, Y., Onodera, K., and Fukui, Y. (1994). Neurite outgrowth of PC12 cells is suppressed by wortmannin, a specific inhibitor of phosphatidylinositol 3-kinase. *J. Biol. Chem.* **269,** 18961–18967.

Kimura, S., Noda, T., and Yoshimori, T. (2007). Dissection of the autophagosome maturation process by a novel reporter protein, tandem fluorescent-tagged LC3. *Autophagy* **3,** 452–460.

Kiselyov, K., Jennigs, J. J., Jr., Rbaibi, Y., and Chu, C. T. (2007). Autophagy, mitochondria and cell death in lysosomal storage diseases. *Autophagy* **3,** 259–262.

Kissova, I., Deffieu, M., Manon, S., and Camougrand, N. (2004). Uth1p is involved in the autophagic degradation of mitochondria. *J. Biol. Chem.* **279,** 39068–39074.

Kitada, T., Pisani, A., Porter, D. R., Yamaguchi, H., Tscherter, A., Martella, G., Bonsi, P., Zhang, C., Pothos, E. N., and Shen, J. (2007). Impaired dopamine release and synaptic plasticity in the striatum of PINK1-deficient mice. *Proc. Natl. Acad. Sci. USA* **104,** 11441–11446.

Kitamoto, T., Shin, R. W., Doh-ura, K., Tomokane, N., Miyazono, M., Muramoto, T., and Tateishi, J. (1992). Abnormal isoform of prion proteins accumulates in the synaptic structures of the central nervous system in patients with Creutzfeldt-Jakob disease. *Am. J. Pathol.* **140,** 1285–1294.

Klionsky, D. J., Abeliovich, H., Agostinis, P., Agrawal, D. K., Aliev, G., Askew, D. S., Baba, M., Baehrecke, E. H., Bahr, B. A., Ballabio, A., Bamber, B. A., Bassham, D. C., *et al.* (2008). Guidelines for the use and interpretation of assays for monitoring autophagy in higher eukaryotes. *Autophagy* **4,** 151–175.

Koike, M., Shibata, M., Tadakoshi, M., Gotoh, K., Komatsu, M., Waguri, S., Kawahara, N., Kuida, K., Nagata, S., Kominami, E., Tanaka, K., and Uchiyama, Y. (2008). Inhibition of autophagy prevents hippocampal pyramidal neuron death after hypoxic-ischemic injury. *Am. J. Pathol.* **172,** 454–469.

Koike, M., Shibata, M., Waguri, S., Yoshimura, K., Tanida, I., Kominami, E., Gotow, T., Peters, C., Figura, K. V., Mizushima, N., Saftig, P., and Uchiyama, Y. (2005). Participation of autophagy in storage of lysosomes in neurons from mouse models of neuronal ceroid-lipofuscinoses (Batten disease). *Am. J. Pathol.* **167,** 1713–1728.

Komatsu, M., Wang, Q. J., Holstein, G. R., Friedrich, V. L., Jr., Iwata, J., Kominami, E., Chait, B. T., Tanaka, K., and Yue, Z. (2007). Essential role for autophagy protein Atg7 in the maintenance of axonal homeostasis and the prevention of axonal degeneration. *Proc. Natl. Acad. Sci. USA* **104,** 14489–14494.

Kuma, A., Matsui, M., and Mizushima, N. (2007). LC3, an autophagosome marker, can be incorporated into protein aggregates independent of autophagy: Caution in the interpretation of LC3 localization. *Autophagy* **3.**

Lach, B., Grimes, D., Benoit, B., and Minkiewicz-Janda, A. (1992). Caudate nucleus pathology in Parkinson's disease: Ultrastructural and biochemical findings in biopsy material. *Acta Neuropathol.* **83,** 352–360.

Lagreze, W. A., Pielen, A., Steingart, R., Schlunck, G., Hofmann, H. D., Gozes, I., and Kirsch, M. (2005). The peptides ADNF-9 and NAP increase survival and neurite outgrowth of rat retinal ganglion cells *in vitro*. *Invest. Ophthalmol. Vis. Sci.* **46,** 933–938.

Lai, Y., Hickey, R. W., Chen, Y., Bayir, H., Sullivan, M. L., Chu, C. T., Kochanek, P. M., Dixon, C. E., Jenkins, L. W., Graham, S. H., Watkins, S. C., and Clark, R. S. (2008). Autophagy is increased after traumatic brain injury in mice and is partially inhibited by the antioxidant gamma-glutamylcysteinyl ethyl ester. *J. Cereb. Blood Flow Metab.* **28,** 540–550.

Larsen, K. E., Fon, E. A., Hastings, T. G., Edwards, R. H., and Sulzer, D. (2002). Methamphetamine-induced degeneration of dopaminergic neurons involves autophagy and upregulation of dopamine synthesis. *J. Neurosci.* **22,** 8951–8960.

Liberski, P. P., Gajdusek, D. C., and Brown, P. (2002). How do neurons degenerate in prion diseases or transmissible spongiform encephalopathies (TSEs): Neuronal autophagy revisited. *Acta Neurobiol. Exp. (Wars).* **62,** 141–147.

Liu, C. L., Chen, S., Dietrich, D., and Hu, B. R. (2008). Changes in autophagy after traumatic brain injury. *J. Cereb. Blood Flow Metab.* **28,** 674–683.
Machado-Salas, J., Ibarra, O., Martinez Fong, D., Cornejo, A., Aceves, J., and Kuri, J. (1990). Degenerative ultrastructural changes observed in the neuropil of caudate nuclei from Parkinson's disease patients. *Stereotact. Funct. Neurosurg* **54-55,** 297–305.
MacLeod, D., Dowman, J., Hammond, R., Leete, T., Inoue, K., and Abeliovich, A. (2006). The familial Parkinsonism gene LRRK2 regulates neurite process morphology. *Neuron.* **52,** 587–593.
Masliah, E., Mallory, M., Alford, M., DeTeresa, R., Hansen, L. A., McKeel, D. W., Jr., and Morris, J. C. (2001). Altered expression of synaptic proteins occurs early during progression of Alzheimer's disease. *Neurology.* **56,** 127–129.
Meijering, E., Jacob, M., Sarria, J. C., Steiner, P., Hirling, H., and Unser, M. (2004). Design and validation of a tool for neurite tracing and analysis in fluorescence microscopy images. *Cytometry A.* **58,** 167–176.
Mizushima, N., and Yoshimori, T. (2007). How to Interpret LC3 Immunoblotting. *Autophagy.* **3,** 542–545.
Munafo, D. B., and Colombo, M. I. (2001). A novel assay to study autophagy: Regulation of autophagosome vacuole size by amino acid deprivation. *J. Cell Sci.* **114,** 3619–3629.
Nitatori, T., Sato, N., Waguri, S., Karasawa, Y., Araki, H., Shibanai, K., Kominami, E., and Uchiyama, Y. (1995). Delayed neuronal death in the CA1 pyramidal cell layer of the gerbil hippocampus following transient ischemia is apoptosis. *J. Neurosci.* **15,** 1001–1011.
Nixon, R. A. (2007). Autophagy, amyloidogenesis and Alzheimer disease. *J. Cell Sci.* **120,** 4081–4091.
Nixon, R. A., Wegiel, J., Kumar, A., Yu, W. H., Peterhoff, C., Cataldo, A., and Cuervo, A. M. (2005). Extensive involvement of autophagy in Alzheimer disease: An immuno-electron microscopy study. *J. Neuropathol. Exp. Neurol.* **64,** 113–122.
Orr, H. T. (2002). Lurcher, nPIST, and autophagy. *Neuron.* **35,** 813–814.
Pahlman, S., Ruusala, A. I., Abrahamsson, L., Mattsson, M. E., and Esscher, T. (1984). Retinoic acid-induced differentiation of cultured human neuroblastoma cells: A comparison with phorbolester-induced differentiation. *Cell Differ.* **14,** 135–144.
Pigino, G., Morfini, G., Pelsman, A., Mattson, M. P., Brady, S. T., and Busciglio, J. (2003). Alzheimer's presenilin 1 mutations impair kinesin-based axonal transport. *J. Neurosci.* **23,** 4499–4508.
Pigino, G., Pelsman, A., Mori, H., and Busciglio, J. (2001). Presenilin-1 mutations reduce cytoskeletal association, deregulate neurite growth, and potentiate neuronal dystrophy and tau phosphorylation. *J. Neurosci.* **21,** 834–842.
Plowey, E. D., Cherra, S. J., 3rd, Liu, Y. J., and Chu, C. T. (2008). Role of autophagy in G2019S-LRRK2-associated neurite shortening in differentiated SH-SY5Y cells. *J. Neurochem.* **105,** 1048–1056.
Price, R. D., Oe, T., Yamaji, T., and Matsuoka, N. (2006). A simple, flexible, nonfluorescent system for the automated screening of neurite outgrowth. *J. Biomol. Screen.* **11,** 155–164.
Rami, A., Langhagen, A., and Steiger, S. (2008). Focal cerebral ischemia induces upregulation of Beclin 1 and autophagy-like cell death. *Neurobiol. Dis.* **29,** 132–141.
Rodriguez-Enriquez, S., He, L., and Lemasters, J. J. (2004). Role of mitochondrial permeability transition pores in mitochondrial autophagy. *Int. J. Biochem. Cell Biol.* **36,** 2463–2472.
Ross, R. A., Spengler, B. A., and Biedler, J. L. (1983). Coordinate morphological and biochemical interconversion of human neuroblastoma cells. *J. Natl. Cancer Inst.* **71,** 741–747.
Rowland, A. M., Richmond, J. E., Olsen, J. G., Hall, D. H., and Bamber, B. A. (2006). Presynaptic terminals independently regulate synaptic clustering and autophagy of GABAA receptors in Caenorhabditis elegans. *J. Neurosci.* **26,** 1711–1720.

Rudnicki, D. D., Pletnikova, O., Vonsattel, J. P., Ross, C. A., and Margolis, R. L. (2008). A comparison of huntington disease and huntington disease-like 2 neuropathology. *J. Neuropathol. Exp. Neurol.* **67,** 366–374.

Satchell, M. A., Zhang, X., Kochanek, P. M., Dixon, C. E., Jenkins, L. W., Melick, J., Szabo, C., and Clark, R. S. (2003). A dual role for poly-ADP-ribosylation in spatial memory acquisition after traumatic brain injury in mice involving NAD+ depletion and ribosylation of 14-3-3gamma. *J. Neurochem.* **85,** 697–708.

Scheff, S. W., Price, D. A., Schmitt, F. A., and Mufson, E. J. (2006). Hippocampal synaptic loss in early Alzheimer's disease and mild cognitive impairment. *Neurobiol. Aging.* **27,** 1372–1384.

Sikorska, B., Liberski, P. P., Giraud, P., Kopp, N., and Brown, P. (2004). Autophagy is a part of ultrastructural synaptic pathology in Creutzfeldt-Jakob disease: A brain biopsy study. *Int. J. Biochem. Cell Biol.* **36,** 2563–2573.

Spillantini, M. G., Schmidt, M. L., Lee, V. M., Trojanowski, J. Q., Jakes, R., and Goedert, M. (1997). Alpha-synuclein in Lewy bodies. *Nature.* **388,** 839–840.

St Croix, C. M., Leelavanichkul, K., and Watkins, S. C. (2006). Intravital fluorescence microscopy in pulmonary research. *Adv. Drug. Deliv. Rev.* **58,** 834–840.

Tanida, I., Yamaji, T., Ueno, T., Ishiura, S., Kominami, E., and Hanada, K. (2008). Consideration about negative controls for LC3 and expression vectors for four colored fluorescent protein-LC3 negative controls. *Autophagy* **4,** 131–134.

Traystman, R. J. (2003). Animal models of focal and global cerebral ischemia. *Ilar J.* **44,** 85–95.

Ventruti, A., and Cuervo, A. M. (2007). Autophagy and neurodegeneration. *Curr. Neurol. Neurosci. Rep.* **7,** 443–451.

Wishart, T. M., Parson, S. H., and Gillingwater, T. H. (2006). Synaptic vulnerability in neurodegenerative disease. *J. Neuropathol. Exp. Neurol.* **65,** 733–739.

Xue, L., Fletcher, G. C., and Tolkovsky, A. M. (1999). Autophagy is activated by apoptotic signalling in sympathetic neurons: An alternative mechanism of death execution. *Mol. Cell Neurosci.* **14,** 180–198.

Xue, L., Fletcher, G. C., and Tolkovsky, A. M. (2001). Mitochondria are selectively eliminated from eukaryotic cells after blockade of caspases during apoptosis. *Curr. Biol.* **11,** 361–365.

Yang, Y., Fukui, K., Koike, T., and Zheng, X. (2007). Induction of autophagy in neurite degeneration of mouse superior cervical ganglion neurons. *Eur. J. Neurosci.* **26,** 2979–2988.

Yue, Z., Wang, Q. J., and Komatsu, M. (2008). Neuronal autophagy: Going the distance to the axon. *Autophagy* **4,** 94–96.

Zeng, M., and Zhou, J. N. (2008). Roles of autophagy and mTOR signaling in neuronal differentiation of mouse neuroblastoma cells. *Cell Signal.* **20,** 659–665.

Zhu, C., Wang, X., Xu, F., Bahr, B. A., Shibata, M., Uchiyama, Y., Hagberg, H., and Blomgren, K. (2005). The influence of age on apoptotic and other mechanisms of cell death after cerebral hypoxia-ischemia. *Cell Death Differ.* **12,** 162–176.

Zhu, C., Xu, F., Wang, X., Shibata, M., Uchiyama, Y., Blomgren, K., and Hagberg, H. (2006). Different apoptotic mechanisms are activated in male and female brains after neonatal hypoxia-ischaemia. *J. Neurochem.* **96,** 1016–1027.

Zhu, J.-H., Guo, F., Shelburne, J., Watkins, S., and Chu, C. T. (2003). Localization of phosphorylated ERK/MAP kinases to mitochondria and autophagosomes in Lewy body diseases. *Brain Pathol.* **13,** 473–481.

Zhu, J. H., Horbinski, C., Guo, F., Watkins, S., Uchiyama, Y., and Chu, C. T. (2007). Regulation of autophagy by extracellular signal-regulated protein kinases during 1-methyl-4-phenylpyridinium-induced cell death. *Am. J. Pathol.* **170,** 75–86.

CHAPTER TWELVE

Monitoring the Autophagy Pathway in Cancer

Frank C. Dorsey,* Meredith A. Steeves,* Stephanie M. Prater,* Thomas Schröter,† *and* John L. Cleveland*

Contents

Abstract

Autophagy is an ancient cell survival pathway that is induced by metabolic stress and that helps prevent bioenergetic failure. This pathway has emerged as a promising new target in cancer treatment, where agents that inhibit autophagic degradation have efficacy in preventing cancer and in treating resistant disease when combined with conventional chemotherapeutics, which generally activate the pathway. However, agents that specifically target the autophagy pathway are currently lacking, and monitoring the effects of therapeutics on the autophagy pathway raises several challenges. Here we review the potential roles of the autophagy pathway in tumor progression and in maintenance of the malignant state, and introduce novel methods that we have developed that allow one to monitor autophagic activity *ex vivo* and *in vivo*.

* Department of Cancer Biology, The Scripps Research Institute, Scripps-Florida, Jupiter, Florida, USA
† Translational Research Institute, The Scripps Research Institute, Scripps-Florida, Jupiter, Florida, USA

Methods in Enzymology, Volume 453
ISSN 0076-6879, DOI: 10.1016/S0076-6879(08)04012-3

1. Introduction

Though originally cast as part of the cell death machinery, recent evidence suggests the autophagy pathway supports tumor cell survival during metabolic stress. This is significant because autophagy is induced in response to a wide array of chemotherapeutic agents. In fact, activation of this pathway is a hallmark of nearly all therapeutic interventions. The clinical relevance of autophagy to cancer prevention and therapeutics is quickly coming into focus. For example, lymphomas arising in Eμ-*Myc* transgenic mice, a model of human B cell lymphoma (Adams *et al.*, 1985), are sensitized to therapy-induced cell death following knockdown of the essential autophagy regulator *Atg5* or treatment with the drug chloroquine (CQ) (Amaravadi *et al.*, 2007), which impairs lysosomal-mediated degradation of cargo delivered by autophagosomes (Carew *et al.*, 2007; Maclean *et al.*, 2008). Furthermore, CQ treatment alone markedly delays the onset of B cell lymphoma in Eμ-*Myc* transgenics, and the development of T cell lymphoma in *Atm* null mice (Maclean *et al.*, 2008). In addition, when combined with FDA-approved therapeutics such as the histone deacetylase inhibitor suberoylanilide hydroxamic acid (SAHA), CQ demonstrates efficacy even in refractory, Gleevec-resistant chronic myelogenous leukemia (CML) (Carew *et al.*, 2007). Finally, recent clinical trials have demonstrated that CQ in combination with radiation and chemotherapy doubles the mean survival of patients suffering from glioblastoma multiforme (Briceno *et al.*, 2003; Sotelo *et al.*, 2006).

Despite mounting evidence that inhibition of the pathway enhances the efficacy of anti-cancer regimens, autophagy appears to have, paradoxically, tumor suppressor properties. For example, deletion of one allele of *beclin 1* (*Atg6*), a component of the class III PI-3-kinase complex required for autophagy, is common in breast and ovarian cancers (Liang *et al.*, 1999), and mice haploinsufficient for *beclin 1* are tumor-prone (Qu *et al.*, 2003; Yue *et al.*, 2003). Further, *beclin 1* heterozygosity also accelerates lymphomagenesis in Eμ-*Myc* mice (Dorsey and Cleveland, unpublished data), a characteristic of classic tumor suppressors such as Arf or p53 (Eischen *et al.*, 1999; Alt *et al.*, 2003). Finally, mice lacking *Atg4c*, one of four Atg4 protease family members required for the cleavage of the proform of LC3 (Atg8) that mediates autophagosome formation (Fig. 12.1A), are more sensitive to chemically induced fibrosarcoma (Marino *et al.*, 2007). Therefore, in some scenarios autophagy functions to suppress tumorigenesis, underscoring the need for important management and monitoring of the pathway before, during, and following therapy.

One potential explanation for the apparent functional dichotomy of this pathway is that, similar to apoptosis, autophagy may curb oncogene-induced

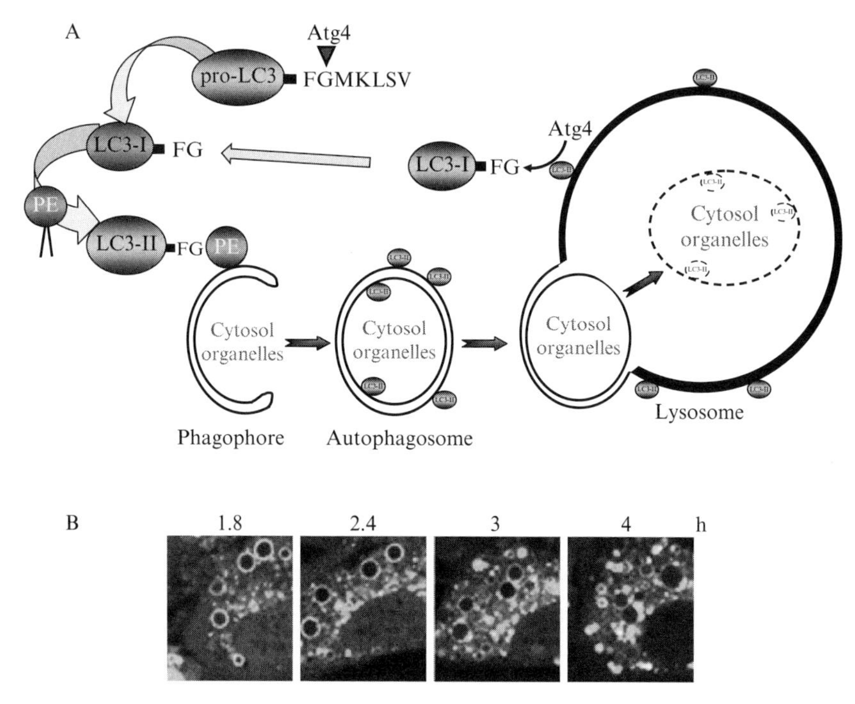

Figure 12.1 Monitoring autophagic vesicle formation by real-time microscopy using GFP-LC3. (A) Schematic depicting the steps of autophagic vesicle formation and maturation. The proform of LC3, a ubiquitin-like molecule required for fusion of isolation membranes of the phagophore to form the autophagosome, is first cleaved by the cysteine-dependent protease Atg4. This cleavage exposes a *C*-terminal glycine residue generating LC3-I. LC3-I is then activated by an E1 (Atg7), and is then transferred to an E2 (Atg3) resulting in its conjugation to phosphatidylethanolamine (PE) generating LC3-II. Then, LC3-II associates with phagophore membranes initiating the formation and maturation of autophagic vesicles. LC3-II remains associated with both the inner and outer membranes of autophagosomes and as a consequence is delivered along with the inner vesicle to the lysosome for degradation. LC3-II on the outer membrane of the autophagosome diffuses into the lysosomal membrane upon autophagic vesicle fusion with the lysosome. Here, Atg4 can then cleave PE from LC3-II, regenerating LC3-I that can then be used to form new autophagosomes. (B) Individual images of GFP-LC3–expressing mouse embryonic fibroblasts (MEFs) taken from a real-time microscopy analysis of autophagosome formation and maturation in response to 50 μM chloroquine (CQ). Note that GFP-LC3 autophagosomes form devoid of LysoTracker Red staining. After formation, these autophagosomes lose their GFP staining as they gradually accumulate LysoTracker Red suggesting that they are either fusing with lysosomes, or that they become acidic. (See Color Insert.)

transformation by compromising cell survival and/or impairing cell proliferation, in a fashion akin to Arf or p53 or the retinoblastoma (Rb) tumor suppressor (Lowe and Sherr, 2003; Sherr, 2006). Alternatively, the regulation of autophagy may be an intrinsic component of the tumor suppressor response. For example, though activation of the Arf/p53 tumor suppressor pathway by oncogenes generally leads to apoptosis, both Arf and p53 activation can induce autophagy (Feng *et al.*, 2005; Reef *et al.*, 2006; Abida and Gu, 2008). Thus, autophagy may also function to restrict oncogenesis. Importantly, autophagy may also limit accumulation of DNA damage, as *beclin* $1^{+/-}$ cells have increased DNA damage, centrosome abnormalities, and can become aneuploid (Mathew *et al.*, 2007).

Another unique attribute of the autophagy pathway is that it can play a significant cytoprotective role in established tumors undergoing nutrient limitation and metabolic stress, where it provides essential building blocks (*e.g.*, amino acids and fatty acids) and energy (ATP) sources (Degenhardt *et al.*, 2006). Thus, while dampening the tumor-suppressive properties of the autophagy pathway is advantageous to tumor cells, some level of the pathway must be sustained to respond to acute or chronic metabolic stress. Accordingly, the loss of both alleles of *beclin 1* has never been observed in tumors from either mice or men. Furthermore, autophagy also protects cancer cells from metabolic stress-induced necrosis (Mathew *et al.*, 2007), which often accompanies rapid tumor growth and provokes an inflammatory response that appears to be required for tumor progression (Luo *et al.*, 2004; Karin and Greten, 2005; Mathew *et al.*, 2007).

Collectively, these observations indicate that autophagy plays important roles in both tumor initiation and progression, and in creating and sustaining the proper tumor microenvironment. Thus, this pathway is a suitable target for anticancer therapeutics on two fronts. However, given its cancer cell-intrinsic versus non–tumor cell autonomous effects, a critical consideration is proper management and monitoring of this pathway. Herein, we summarize new reagents and methods that our laboratory has developed to understand the role of autophagy in the different phases of tumor development, maintenance, and treatment. These methods will support efforts focused on defining the complicated interactions between autophagy and cancer, and those that seek to define the activities of specific small molecules as future preventative or therapeutic agents that target this pathway.

2. LC3: A Phenotypic and Functional Marker of Autophagy

The formation of autophagic vesicles is dependent upon a unique ubiquitin-like modification system that results in the conjugation of phosphatidylethanolamine (PE) to the *C*-terminus of Atg4-cleaved LC3 (Ichimura *et al.*, 2000; Kabeya *et al.*, 2000; Kirisako *et al.*, 2000) (Fig. 12.1A). Consequently, GFP fused to the *N*-terminus of LC3

(GFP-LC3) can be used as a phenotypic reporter of autophagic vesicle formation (Kabeya *et al.*, 2000; Mizushima *et al.*, 2003) (see also the chapter by Kimura *et al.*, in this volume). Under normal growth conditions, GFP-LC3 is diffuse throughout the cytoplasm and nucleus. Upon the induction of autophagy, GFP-LC3 becomes covalently linked to PE and associates with the initiating phagophore membranes, redistributing GFP-LC3 to newly formed punctate, double-membrane autophagosomes (Fig. 12.1A). GFP-LC3 is an effective phenotypic marker where one can monitor autophagosome formation by fluorescence microscopy, and quantifying GFP-LC3 puncta is an important measure of the induction of autophagy in a wide variety of systems (Mizushima *et al.*, 2004).

Microscopy analysis alone, however, fails to distinguish between the induction of autophagy and the inhibition of turnover of autophagic vesicles. Specifically, following their formation, GFP-LC3-labeled autophagosomes fuse with lysosomes, an event that delivers their cargo, including GFP-LC3 localized to the inner membrane of the autophagosome, for degradation (Fig. 12.1B). We have exploited these properties and have developed a flow cytometric-based GFP-LC3 functional assay (see also the chapter by Shvets and Elazar in volume 452). In addition, we have developed several luciferase-LC3 (Luc2p-LC3) constructs that allow measurements of autophagic activity in a high-throughput format. Below we provide the methods where one can: (1) Use GFP-LC3 as a phenotypic marker; and (2) Use GFP-LC3 and Luc2p-LC3 fusion proteins as functional markers for monitoring the autophagy pathway. Importantly, these methods can be applied to both *in vitro* and *in vivo* models, allowing one to monitor autophagic activity in a wide array of biological systems.

2.1. Real-time imaging of GFP-LC3

We routinely use GFP-LC3 to monitor the formation of autophagosomes by real-time fluorescence microscopy. This allows one to temporally dissect and define the involvement of signaling pathways, autophagic effectors and inhibitors, and/or organelles in autophagic vesicle formation and maturation. Importantly, the overexpression of GFP-LC3 can result in the aggregation of GFP-LC3 independent of the autophagy pathway (Ciechomska and Tolkovsky, 2007; Kuma *et al.*, 2007). To avoid nonspecific aggregation, we routinely transduce cells with MSCV-based retroviruses (Hawley *et al.*, 1992) that allow one to express transgenes such as GFP-LC3 at more physiological levels and that also allow one to express a selectable marker such as the gene encoding *puromycin resistance* (Puro) through an internal ribosome entry site (IRES) that resides downstream of the transgene. The resulting MSCV-GFP-LC3-IRES-Puro retrovirus then allows one to generate stable cell lines expressing GFP-LC3 by growth in puromycin-containing medium. To further control for nonspecific aggregation of

GFP, we also employ a MSCV retroviral construct that directs the independent expression of GFP and LC3, using the IRES to express GFP (MSCV-LC3-IRES-GFP). Here one selects for transduced cells simply by sorting for GFP+ cells using a fluorescence-activated cell sorter (FACS).

Using real-time microscopy we have demonstrated that, although CQ inhibits autophagy by disrupting lysosomal-mediated degradation of autophagosomes (Maclean *et al.*, 2008), it actually induces the rapid (within four hours) formation of large, morphologically distinct autophagosomes (Fig. 12.1B). Sometime after their formation, these autophagosomes accumulate the pH sensitive dye LysoTracker Red indicating that they ultimately become acidic and then summarily lose GFP fluorescence (Fig. 12.1B). Although both bafilomycin A_1 (Baf-A1) and CQ both inhibit lysosomal-mediated degradation of autophagosomes, Baf-A1 treatment is characterized by a much slower accumulation of small GFP-LC3 puncta and an absence of any LysoTracker fluorescence, indicating that it is much more effective than CQ at raising the pH of the lysosome (data not shown). These observations underscore the importance of real-time microscopy in the study of autophagy, as steady state analysis of autophagic activity in response to these agents, for example, using long-lived protein degradation assays (see the chapter by Bauvy *et al.*, in volume 452) or by monitoring LC3-II (i.e., PE-conjugated LC3, Fig. 12.1A) formation by Western blot analyses would not distinguish these potentially important phenotypic differences.

For these analyses, we use an Olympus DSU spinning disc confocal microscope (http://www.olympusamerica.com/seg_section/product.asp?product=1009) equipped with a stage enclosed in a Weather Station incubator. Unlike traditional scanning confocal microscopes, the DSU passes laser light through a spinning disc, which contains small slits that act as a virtual pinhole that simultaneously illuminates the entire field. Two advantages to this method are that (1) images can be taken rapidly, up to 15 frames per second; (2) samples are exposed to less light, reducing phototoxic effects seen with traditional confocal microscopy. Glass-bottomed culture dishes (MatTek, Fisher, #P35G.5-14-C) are used for these analyses.

1. The day before imaging, GFP-LC3 expressing cells are plated at an appropriate density (so they are 60%–80% confluent the following day) in DMEM medium supplemented with 25 m*M* HEPES buffer, pH 7.4, 10% fetal calf serum (FCS), 2 m*M* L-glutamine, 100 units penicillin, 100 μg/ml streptomycin, 1 m*M* pyruvate, and 0.1 m*M* nonessential amino acids (for mouse embryonic fibroblasts [MEFs] also add 55 μM β-mercaptoethanol) and are incubated in a 37 °C incubator with 5% CO_2. 300,000 cells are plated in a 35-mm dish using Hep2 or HeLa cells, or MEFs. Multiple devices have been developed to maintain proper CO_2 concentration for pH stabilization, including an attached aerator, which bubbles 5% CO_2 mixed with normal atmosphere through prewarmed

water, thus delivering prewarmed, humidified CO_2 to the cell culture dish. The addition of HEPES buffer helps maintain the correct pH. Additionally, when using small volumes of cell culture media, one can overlay the media with a thin layer of mineral oil, to prevent evaporation.

2. Prior to imaging, cells are stained with LysoTracker Red (Invitrogen, #L-7528) to visualize the role of acidic lysosomal compartments in relation to autophagic vesicle formation and turnover. To stain the cells, LysoTracker is added to the medium at a concentration of 1 μM and incubated for 30–45 min in a 37 °C 5% CO_2 incubator.
3. Cells are then washed twice to remove excess LysoTracker from the medium. For adherent cells, the media is simply exchanged twice with fresh pre-warmed media. If cells are nonadherent, the cells are centrifuged and are suspended in prewarmed media, centrifuged again, and resuspended in fresh prewarmed media and are replated.
4. After LysoTracker staining, the plates are moved to the incubator on the microscope for 15–20 min to allow the plates to reach thermal equilibrium with the stage; this prevents thermal fluctuations, which can result in small movements in the plate and may cause cells to drift out of focus.
5. After equilibration, initial images are collected to document the state of GFP-LC3 vesiculation before, for example, amino acid starvation or drug treatment. The most commonly used agents that affect the autophagy pathway used in our laboratory are chloroquine (20–50 μM; Sigma, #C6628), bafilomycin A_1 (100 nM; Sigma, #11711), rapamycin (5 μg/ml; Sigma, #R0395), and 3-methyladenine (3-MA, 5 mM; Sigma, #08592). As 3-methyladenine is difficult to solubilize, it should be dissolved in cell culture medium at 37 °C just before treatment. For amino acid starvation, cells are rinsed with prewarmed Earl's balanced salt solution (EBSS; Invitrogen, #14155048) supplemented with 25 mM HEPES buffer, pH 7.4, twice before incubating the cells in EBSS for the duration of the experiment. When using 3-MA to inhibit autophagy, cells are washed twice with prewarmed medium containing 5 mM 3-MA just prior to culture in EBSS, or in normal medium containing reagents that affect the autophagy pathway.
6. To monitor vesiculation in response to different pharmacological agents, images are collected every 5 min for 4–6 h. To monitor vesicle formation and vesicle dynamics, images are typically acquired in the 15–30 ms range. Importantly, at the end of any acquisition, one should always move the objective to a field that has not been previously excited by the laser, to ensure that GFP-LC3 dynamics are not the result of excessive light exposure. Microscopes for real-time microscopy are now widespread and each manufacturer has their own individual software for image acquisition and analysis. We are currently using software developed by Intelligent Imaging Innovations (3i), which is standard for Olympus microscopes.

2.2. High content analysis of GFP-LC3 vesiculation

An important new arena for the autophagy field is the discovery and development of specific small molecule inhibitors and activators of the autophagy pathway. A key component of such campaigns is the ability to analyze large numbers of compounds using GFP-LC3 phenotypic screens. We have developed a medium-throughput method to measure GFP-LC3 vesiculation in response to different pharmacological agents using high-content screening. Excitingly, these methods are also applicable for siRNA and cDNA screens to identify new players in the autophagy pathway.

High content screening (HCS) is a method of screening fixed or living cells based on phenotype. It combines an automated microscope platform with image analysis software for the rapid acquisition and analysis of microscopy images in a microtiter plate format. Using this system, it is possible to accurately quantify GFP-LC3 vesiculation in a rapid and nonbiased manner. It is well accepted that amino acid starvation activates the autophagy pathway, yet using HCS we have demonstrated that amino acid starvation actually results in a slight reduction in GFP-LC3 vesiculation when compared to untreated cells (Figs. 12.2A–B). In contrast, CQ treatment resulted in the accumulation of a large number of autophagosomes at the same 6-h time point. Importantly, the accumulation of autophagic vesicles in response to CQ is autophagy gene-dependent (data not shown). CQ inhibits autophagic vesicle turnover by the lysosome, yet CQ also induces autophagic vesicle formation at early time points (Fig. 12.1B) (Maclean *et al.*, 2008). Regardless of the exact mechanism, CQ-induced vesiculation can be used to identify new agents that inhibit autophagic vesiculation and new genes that induce or suppress the autophagy pathway.

To measure autophagic vesiculation in response to CQ treatment by HCS, we use the following protocol:

1. We first generate stable cell lines using MSCV-GFP-LC3B-I-Puro or MSCV-LC3B-I-GFP retroviruses (see above) and then plate cells in DMEM medium (as previously) at 25,000 cells per well in a Packard View 96-well plate (PerkinElmer viewplate, #6005182).
2. The following day, the medium is replaced with pre-warmed complete medium, EBSS (amino acid [AA] starvation media), or medium containing 50 μM CQ and incubated for 6 h at 37 °C with 5% CO_2.
3. Cells are then washed twice with PBS and fixed with 4% paraformaldehyde for 10 min.
4. The cells are then washed twice with PBS and nuclei are stained with Hoechst 33442 (5 μg/ml; Invitrogen, #H3570) for 20 min and again washed twice with PBS.
5. Images are acquired on the InCell 1000 instrument (GE Healthcare) using a Q505LP dichroic mirror and corresponding UV/FITC filter sets with a 20x objective. Typically, we collect 8 independent images per well, which, for a 96-well plate, is a total of 768 images.

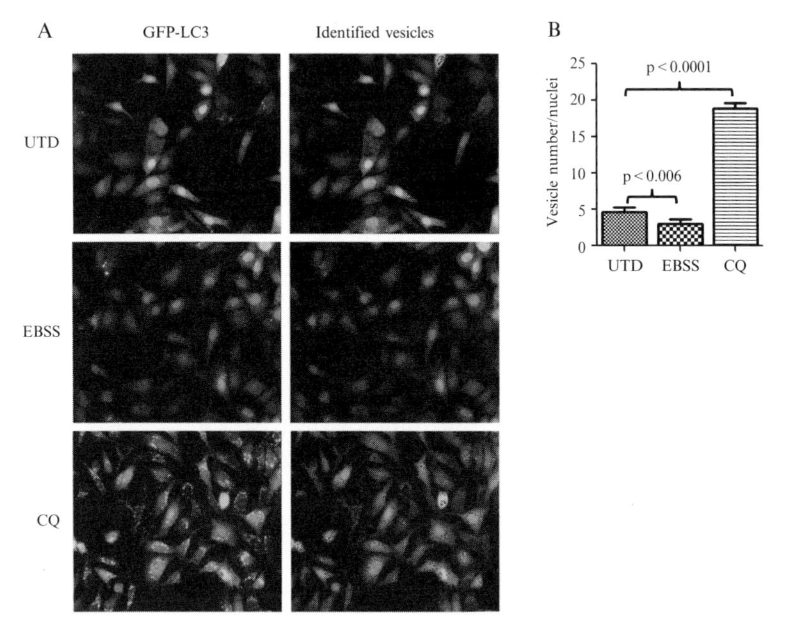

Figure 12.2 High content analysis of GFP-LC3 vesiculation. (A) GFP-LC3-expressing Hep2 cells were either left untreated (UTD), were starved of amino acids by incubation in EBSS, or were treated with 50 μM CQ for 6 h. The panels at left are representative images of GFP-LC3 vesiculation that occurred under each condition. The images in the panels at right correspond to those shown at left that have been analyzed using the InCell-based algorithm and shows the autophagosomes identified in blue. (B) Quantification of GFP-LC3 vesiculation in untreated, EBSS-treated, or CQ-treated Hep2 cells. To accurately quantify vesiculation, 48 individual images for each treatment were collected, containing approximately 30 cells/image, and GFP-LC3-associated vesicles were quantified using the InCell developer toolbox algorithm. (See Color Insert.)

Autophagosomes are quantified using the InCell Developer Toolbox software. First, numbers of nuclei are quantified by object segmentation using a Kernel size of 31 and a sensitivity of 1. Post-processing of images is performed using an erosion of 11 and sieve greater than 50-μm^2. GFP-LC3-positive autophagosomes are then quantified by object segmentation using a Kernel size of 5 and a sensitivity of 1. This algorithm accurately counts GFP positive vesicular structures in a nonbiased manner (Fig. 12.2A–B). Data shown in Fig. 12.2B are the number of autophagosomes per cell (represented by nuclei count), yet this software is also capable of providing a quantitative measure of both vesicle size, and fluorescent intensity represented by the fluorescence multiplied by vesicle area. These methods make it possible to quantitatively assess morphological differences in autophagic vesicles in response to a wide array of agents (data not shown).

One must be careful in interpreting quantification of autophagic vesiculation as a measure of autophagy. For example, though reductions in

GFP-LC3 vesiculation following amino acid starvation could suggest that this inhibits autophagy, a wealth of data has demonstrated that this is not the case. Thus one should assess autophagic activity by multiple methods to determine the control of autophagy in response to specific stimuli (Klionsky *et al.*, 2008). Indeed, the most likely explanation of these results is that GFP-LC3-positive autophagic vesicles are turned over at a much more rapid rate following amino acid starvation, which results in a steady state decrease in GFP-LC3-associated vesicles. Thus, while monitoring autophagy with this type of phenotypic screen can provide valuable of data, it is necessary to assess autophagy-specific degradation when screening for activators and inhibitors of this pathway.

2.3. Monitoring GFP-LC3 by flow cytometry

In addition to its use for microscopy analysis, GFP-LC3 can also be used to measure the turnover of autophagic vesicles by flow cytometry. Using primary nontransformed MEFs or Hep2 cells, we have demonstrated that activation of autophagy, for example, following amino acid deprivation, results in the rapid degradation of GFP-LC3, as documented by reductions in mean fluorescence measured by flow cytometry (Fig. 12.3A). Importantly, reduction in the mean fluorescence following the induction of autophagy is *Atg7*-dependent (Fig. 12.3B). In addition, using MSCV-LC3-IRES-GFP virus infected cells as a control, where GFP degradation is not linked to that of LC3, this method allows for the measure of autophagic function even when it is not possible to specifically silence an autophagy gene (Fig. 12.3A). Therefore, in addition to being a phenotypic marker for autophagy, GFP-LC3 can also be used as a functional marker for autophagic degradation. Finally, these constructs can also be used to monitor autophagic activity *in vivo*, for example in precancerous versus malignant tumor cells engineered to express GFP-LC3, or in GFP-LC3-expressing tumor cells treated with therapeutic agents.

For these experiments, we use the following protocol:

1. Hep-2 cells or primary MEFs are transduced with either MSCV-GFPLC3-IRES-PURO or MSCV-LC3-IRES-GFP retroviruses, and are then plated in a 24-well plate at 50,000 cells/well and cultured overnight in DMEM media (see previously).
2. The following day, cells are either left untreated, or are treated with various agents that effect autophagy, such as EBSS, 5 m*M* 3-MA, 3-MA in EBSS, 50 μM CQ, or 100 n*M* Baf-A1, in 0.5 ml of media. Treatment is anywhere from 2–24 h, and earlier time points are generally better to avoid complications arising from cell death.
3. Cells are then harvested at selected intervals after first collecting the medium from each well into 12×75 mm polystyrene tubes (BD Falcon,

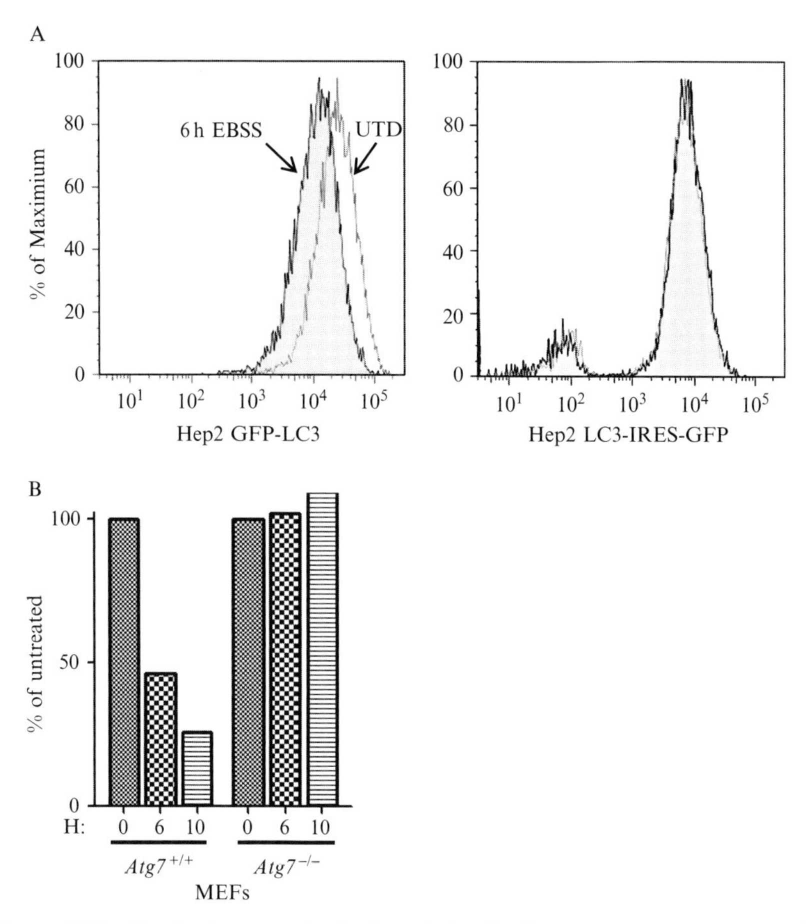

Figure 12.3 Monitoring autophagic degradation by flow cytometry. (A) Hep2 cells were transduced with either MSCV-GFP-LC3-IRES-Puro or MSCV-LC3-IRES-GFP retroviruses. These cells were either left untreated (blue) or were deprived of amino acids by incubating the cells in EBSS for 6 h (grey). GFP fluorescence was then quantified by flow cytometry. GFP-LC3 levels were reduced following amino acid deprivation, while GFP alone was unaffected. (B) To confirm that reductions in GFP-LC3 fluorescence were the result of the activation of autophagy, $Atg7^{+/+}$ and $Atg7^{-/-}$ MEFs were transduced with MSCV-GFP-LC3-IRES-Puro retrovirus and then cultured in EBSS for 6 or 10 h. The reduction of GFP-LC3 mean fluorescence only occurred in $Atg7^{+/+}$ MEFs; therefore, the reduction in GFP-LC3 fluorescence provoked by amino acid deprivation is autophagy dependent. (See Color Insert.)

#352058) on ice. Wells are subsequently washed with 0.5 ml of PBS/well; the wash is added to medium that was previously collected to avoid biasing the analysis to tightly adherent cells.

4. Fully adherent cells are trypsinized with 0.5 ml of 0.05% trypsin (Invitrogen, #25300-054) for 5 min at 37 °C and are examined under

the microscope to ensure complete trypsinization. The trypsinized cells are then pipetted several times to achieve a single-cell suspension, which is then added to the previously harvested supernatant fractions and washed; trypsin is neutralized by the addition of 3 ml of cold wash buffer (PBS containing 1% FCS) and cells are centrifuged at $300 \times g$ for 5 min.

5. The supernatant fraction is removed, and the cells are resuspended in approximately 100 μl of wash buffer. Cells are immediately analyzed on a FACSAria for forward versus side scatter and for GFP fluorescence with the collection of 10,000 live-gated events. Median fluorescence intensity (MFI) is then determined for GFP-expressing cells.

2.4. Luciferase LC3: A high-throughput method to monitor autophagic activity

Monitoring GFP-LC3 mean fluorescence in response to amino acid starvation confirms findings that LC3 is delivered along with autophagosomes to the lysosome for degradation (Tanida *et al.*, 2004). Therefore, we reasoned that a luciferase-LC3 fusion (Luc2p-LC3) would allow one to monitor and quantify autophagy-dependent degradation in a high-throughput manner. We chose to use the highly unstable *Luc2p luciferase* gene that has been engineered to contain the PEST domain from the *ornithine decarboxylase* gene (*ODC*), which has one of the shortest half-lives known for any protein (Fan and Wood, 2007). This PEST domain directs the rapid destruction of Luc2p in a proteasome-dependent manner, and the fusion of Luc2p to LC3 stabilizes this protein, effectively shuttling it into the autophagy pathway for degradation.

LC3 conjugation to PE and subsequent incorporation into autophagosomes requires a *C*-terminal glycine residue (Fig. 12.1A). Luc2p alone is rapidly degraded in response to amino acid starvation and this is not blocked by 3-MA, which inhibits autophagy (Fig. 12.4A). By contrast, Luc2p-LC3 degradation is largely autophagy-dependent, as its turnover is largely inhibited by 3-MA (Fig. 12.4A). As a control for autophagy-specific degradation, we also generated a Luc2p-LC3G120A mutant that cannot be covalently modified by PE, thus uncoupling its degradation from the autophagy pathway. Accordingly, the rate of turnover of Luc2p-LC3G120A is protracted relative to that of Luc2p-LC3 following the removal of amino acids (Fig. 12.4A).

To verify that turnover of Luc2p-LC3 is indeed mediated via the autophagy pathway, we also assessed its turnover in wild-type and *Atg7*$^{-/-}$ MEFs transduced with MSCV-Luc2p-LC3-IRES-Puro retrovirus. Notably, turnover of Luc2p-LC3 induced by withdrawal of amino acids was totally abolished in *Atg7*$^{-/-}$ MEFs (Fig. 12.4B).

This collection of LC3 fusion proteins provides powerful tools that allow one to monitor and quantify autophagy. Importantly, these assays

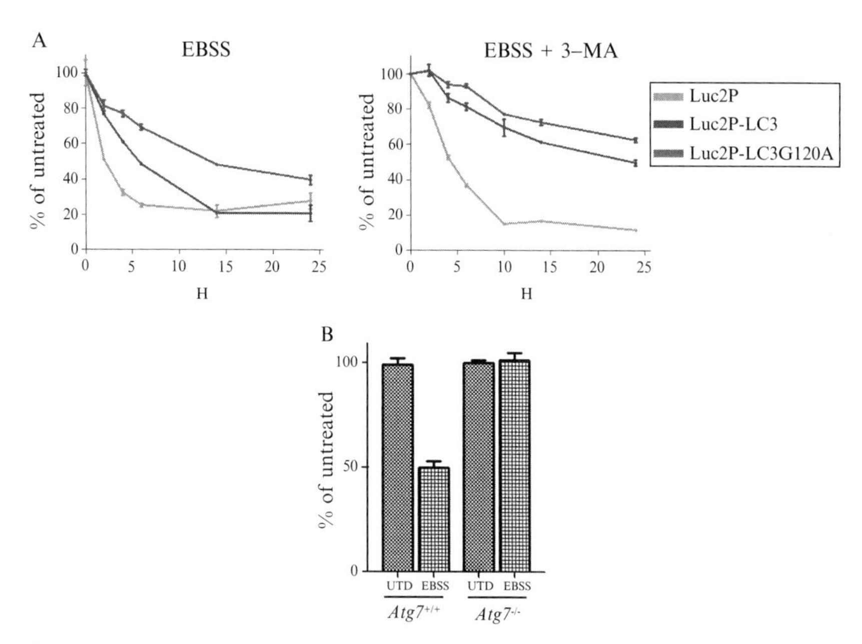

Figure 12.4 Measuring autophagic activity using luciferase-LC3 fusion proteins. (A) Luciferase activity was measured in Hep2 cells transduced with MSCV-Luc2p-IRES-GFP, MSCV-Luc2p-LC3-IRES-GFP, or MSCV-Luc2p-LC3G120A-IRES-GFP retroviruses. Transduced cells were starved of amino acids by incubation in EBSS alone or in combination with the autophagy inhibitor 3-MA. Interestingly, Luc2p alone is degraded in response to amino acid deprivation, yet its degradation was unaffected by the 3-MA. Luc2p-LC3 is degraded in response to amino acid deprivation, while the degradation of the Luc2p-LC3G120A mutant was markedly delayed. In contrast to Luc2p, the degradation of both Luc2p-LC3 and Luc2p-LC3G120A was inhibited by 3-MA; thus, these two fusion proteins are degraded by a mechanism different than that of Luc2p. (B) To confirm that Luc2p-LC3 is an accurate reporter of autophagy, luciferase activity was measured in *Atg7* wild-type and null MEFs that were transduced with MSCV-Luc2p-LC3-IRES-GFP virus. Transduced cells were cultured in EBSS for 6 h. The reduction in luciferase activity in response to EBSS treatment is autophagy dependent, as Luc2p-LC3-driven luciferase activity was not reduced in *Atg7*$^{-/-}$ MEFs when cultured in EBSS. (See Color Insert.)

also demonstrate suitable dynamic range and z-scores, which make them amenable to medium- or high-throughput screening for antagonists and agonists of the autophagy pathway.

To perform luciferase-based screening using the Luc2p-LC3 and Luc2p-LC3G120A constructs, it is necessary to generate stable cell lines, as these constructs failed to behave as accurate markers of autophagic activity in transient transfection experiments. Importantly, virtually all transfection reagents contain chloroquine, which affects the autophagy pathway in rather profound ways, and it is quite likely that the failure of transient transfection approaches is attributable to the presence of CQ.

1. Once stable cell lines are generated, cells are plated in a Costar™ white 96-well plate at 50,000 cells/well in 100 μl of DMEM medium (see previously). All treatments are done in triplicate, and the vehicles for all compounds under investigation, are included as controls.

As a negative control, 6 wells are left untreated, and as a positive control, 6 wells are incubated with EBSS after prewashing twice with prewarmed EBSS. Additionally, we use 5 m*M* 3-MA to inhibit autophagic degradation in response to amino acid starvation. These wells are first washed twice with EBSS containing 5 m*M* 3-MA and then incubated in EBSS + 3-MA for the indicated time. Usually, we follow luciferase activity with respect to time using one plate or set of plates per time point. To date, we have found that 6- or 10-h intervals give the best dynamic range between the Luc2p-LC3 and Luc2p-LC3G120A constructs (Fig. 12.4A).

Firefly luciferase produces light through the oxidation of the substrate D-luciferin in the presence of ATP and Mg^{2+}. This reaction has the highest quantum yield of all bioluminescent reactions (Fan and Wood, 2007).

2. To measure luciferase activity, we use PerkinElmer's Britelite plus reagent (catalog #6016767). For each 96-well plate, we thaw 10 ml of Britelite reagent and bring it to room temperature. An equal volume of reagent is then added to each well (100 μl).
3. The plate is incubated at room temperature for 2–4 min.
4. Plates are read using an Analyst GT plate reader (Molecular Devices). Data are analyzed by subtracting the RLU values for vehicle alone from experimental values and are then normalized to the untreated controls. As a result, data are expressed as percent of control luciferase activity.

3. Assessing the Role of Autophagy in Eμ-*Myc*-Driven Lymphoma

The Eμ-*Myc* transgenic mouse express increased levels of c-Myc in all B lymphoid cells by virtue of the Eμ immunoglobulin heavy chain enhancer (Adams *et al.*, 1985), a scenario that recapitulates *MYC* activation by the t (8;14) translocation of Burkitt lymphoma. Eμ-*Myc* mice develop rapid, clonal, pre B/immature B-cell lymphoma, with a mean survival of approximately 120 days (Eischen *et al.*, 1999). Similarly, *beclin 1*$^{+/-}$ mice are also tumor prone and can develop various types of mature B-cell lymphoma, but with a mean latency >13 months (Qu *et al.*, 2003; Yue *et al.*, 2003). Finally, *beclin 1* heterozygosity accelerates the rate of lymphoma development in Eμ-*Myc* mice (Dorsey and Cleveland, unpublished data), yet the loss of the remaining *beclin 1* allele in either mice or men has not been reported. Collectively, these data suggest that beclin 1 functions as a haploinsufficient tumor suppressor, but that the complete loss of the autophagy pathway may be selected against during tumorigenesis.

Studies using immortal, *Atg5*-null cells transformed with the E1A and Ras oncogenes have suggested that autophagy delays tumorigenesis by restricting necrosis, which induces an inflammatory response that promotes tumor progression (Degenhardt *et al.*, 2006). However, these experiments fail to address the role of autophagy in regulating the initial phases of tumor development en route to transformation. Importantly, it is not currently known whether autophagy is necessary, at some level, for transformation and tumor development. These studies are complicated by the fact that *beclin 1* knockout mice are embryonic lethals (E7.5–8.5), whereas *Atg5* and *Atg7* knockouts die soon after birth. Furthermore, the difference in the survival between *beclin* $1^{-/-}$ versus *Atg5*- and *Atg7*-null mice suggests that beclin 1 may have other functions in development unrelated to autophagy.

To address whether autophagy plays an *in vivo* role in tumor initiation and development, and to circumvent issues regarding lethality, we are using a transplant model, where one reconstitutes the hematopoietic system of lethally irradiated mice with hematopoietic cells derived from the fetal livers of Eμ-*Myc* transgenic mice that are either wild type, heterozygous, or null for *Atg7* (Fig. 12.5A). Generation of hematopoietic chimeras is a well-established and highly valuable tool in cancer research. Here we outline the method that we are currently employing to address these questions.

3.1. Hematopoietic cell isolation and transplantation

During vertebrate embryogenesis the fetal liver is the major site of hematopoiesis, from midgestation to the time of birth: This allows mouse hematopoietic stem cells to be easily isolated from embryonic day E11.5–E18.5 and used for transplant studies into lethally irradiated recipients (Schmitt *et al.*, 2002). To isolate fetal livers, timed matings are established to generate embryos with known gestation dates. To eliminate the need to look for transient vaginal plugs that indicate mating has occurred, mating pairs are left together for 48 h over 2 successive nights, after which males are separated from females. Therefore, embryos isolated on the fifteenth or sixteenth day after the initial day of mating will either be E14.5 or E15.5.

1. Pregnant females are humanely euthanized using CO_2, followed by cervical dislocation.
2. The abdomen is soaked in 70% ethanol, and surgically opened to expose the reproductive organs.
3. The uterine horns containing the embryos are collected, placed in a 50-ml tube containing 30 ml of cold 2% FCS/PBS, and kept on ice.

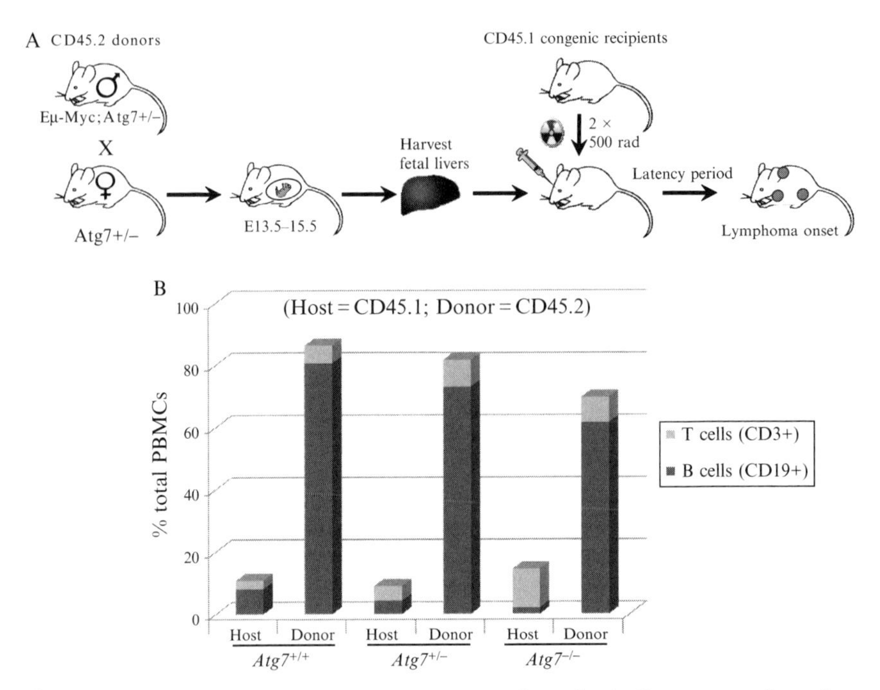

Figure 12.5 Hematopoietic reconstitution using fetal liver transplantation. (A) Hematopoietic reconstitution using embryonic fetal livers. Timed-matings of CD45.2 Eμ-*Myc*/*Atg7*$^{+/-}$ and *Atg7*$^{+/-}$ animals were performed and embryonic day 13.5–15.5 fetal livers were collected and single cell suspensions were isolated. CD45.1 recipients were exposed to 500 Rads of full body irradiation twice 6–8 h apart. After the second round of irradiation, 5 million fetal liver-derived hematopoietic cells were delivered to the congenic recipients via retro-orbital injection and observed for engraftment and subsequent lymphoma development. (B) Assessment of engraftment of donor *Atg7*$^{+/+}$, *Atg7*$^{+/-}$, and *Atg7*$^{-/-}$ lymphoid cells 6 weeks after transplant. Total peripheral blood mononuclear cells (PBMCs) were isolated, stained with antibodies for either CD45.1 (host) or CD45.2 (donor) along with CD3 (T-cells) and CD19 (B-cells), and analyzed by flow cytometry. Engraftment with all 3 donor genotypes occurred, but levels of engraftment were significantly reduced in mice receiving *Atg7*$^{-/-}$ hematopoietic cells. (See Color Insert.)

4. Isolated uterine horns are then taken to a sterile tissue culture hood, removed from the 50-ml tube, and placed in a 10-cm dish on ice containing cold 20 ml of 2% FCS/PBS.
5. Embryos are separated from one another with sterile scissors by cutting directly through the uterine horn between each embryo.
6. Individual embryos are then transferred to their own 10-cm dish where they are subsequently dissected away from both the uterine horn and amniotic sac, which are discarded.
7. The fetal liver is then removed with two curved forceps. Briefly, one forcep is used to puncture the embryo just in front of the spine, and the

other is used to pull back the skin on the abdomen. After the skin is pulled away, the fetal liver is removed from the abdomen by placing the curved forcep behind the fetal liver and gently moving it away from the rest of the embryo.

8. 10 ml of cold 2% FCS/PBS are added to the 10-cm dish containing the isolated fetal liver and pipetted up and down several times to achieve a single-cell suspension.
9. Unwanted connective tissue is removed by passing the cell suspension through a 70- to 100-μm filter.
10. Cells are diluted 1:4 and counted using a hemocytometer.
11. 20 μl of the cell suspension is added to 20 μl of a 2% acetic acid solution. Immediately after gentle mixing, 20 μl of PBS is added to neutralize the solution and 20 μl of 0.5% trypan blue (Invitrogen, #15250-061) is added as a viability dye.

   We typically recover 15–25 million hematopoietic cells per fetal liver with a viability of 70%–95%. In general, 5 million fetal liver-derived hematopoietic cells are transplanted per recipient, allowing for at least 2 transplants per embryo. Recipient mice are congenic for CD45.1, whereas donor hematopoietic cells express the CD45.2 alloantigen. This allotypic difference allows for later distinction between hematopoietic cells that are derived from the donor (CD45.2, mAB clone 104, BDBiosciences) and those remaining in the irradiated host (CD45.1, mAb clone A20, BDBiosciences) by specific antibody staining and FACS analyses.
12. CD45.1 recipients are irradiated with a total of 1000 Rads (10 Gy) split into two equal doses separated by 3–5 h. Split-dose irradiation prevents much of the gut toxicity normally associated with lethal irradiation (Alpdogan *et al.*, 2003).
13. Once mice are irradiated, isolated fetal liver hematopoietic cells are centrifuged at $300\times g$, at 4 °C for 5 min.
14. The supernatant fraction is removed and replaced with cold PBS containing 20 units of heparin per ml. Cells are resuspended in a volume to yield 5 million cells/100 μl.
15. Irradiated recipients are then anesthetized using an IMPAC[6] anesthesia machine from VetQuip (Pleasanton, CA) with settings at 2%–3% isoflurane and 2 liters/h of oxygen).
16. Five million cells in 100 μl are then injected retro-orbitally, using a ½ inch, 27-gauge needle. Cells may also be administered via tail vein injection, but retro-orbital injections are simple to master and give excellent success rates with high levels of engraftment.
17. Recipient mice are allowed to recover under observation until mobile. In some cases, a heat lamp may be used to assist in recovery and prevent hypothermia, but such use must be carefully monitored.

18. Mice are kept on 35 μg/ml Baytril™ (enrofloxacin, Bayer Health Care, #08713254-186599)-treated water for 4 weeks posttransplant to prevent infection that can occur during hematopoietic reconstitution.
19. At 6–8 weeks, mice are bled retro-orbitally and peripheral blood is analyzed to determine basal levels of engraftment (section 3.2).

With the above procedure, death due to failed engraftment is rare.

3.2. Assessing hematopoietic chimerism

1. At 6–8 weeks posttransplantation, blood is collected from the retro-orbital sinus of lethally irradiated CD45.1 animals that were reconstituted with hematopoietic stem cells from CD45.2 donor fetal livers as described previously. Mice are anaesthetized until they are unconscious with isoflurane/oxygen and approximately 100 μl of peripheral blood is collected using heparinized capillary glass tubes (Fisher, #22-362-566) and placed in EDTA-coated collection vials (BD Biosciences, #367835). Similarly harvested peripheral blood from a C57Bl/6 (CD45.2) mouse is used as a control.
2. Then, 75 μl of whole blood is transferred to a 5-ml 12×75-mm polystyrene tube. Red blood cells are lysed with 4 ml of Tris-ammonium chloride lysing buffer (144 m*M* NH_4Cl, 17 m*M* Tris-HCl, pH 7.2), and cells are immediately centrifuged at 300×*g* for 5 minutes, washed once with 1%FCS/PBS wash buffer, repelleted, then resuspended in 200 μl of 1%FCS/PBS and kept on ice.
3. Cell suspensions are split into 2 polystyrene tubes approximately 100 μl per tube, and blocked with 0.25 μg of BD Fc Block (BD Biosciences, #553141) for 5 min.
4. Cells are then immediately stained with a mixture of CD19-PE (0.1 μg; BD Biosciences, #553786), CD3e-PECy5.5 (0.1 μg; eBioscience, #35-0031) and either CD45.1-biotin (0.5 μg; BD Bioscience, #553774) or CD45.2-biotin (0.5 μg; BD Bioscience, #553771) for 30 minutes on ice. Unstained and single-color C57Bl/6 control tubes are generated using equal amounts of single antibodies for the purpose of setting the compensation of the flow cytometer.
5. Cells are then washed twice with 1%FCS/PBS and stained with 0.2 μg of streptavidin-APC-Alexa750 (Invitrogen, #SA1027) for 30 min on ice.
6. Cells are then washed twice with 1%FCS/PBS, and resuspended in 100 μl of 1%FCS/PBS.
7. Samples are then analyzed on a FACSCanto flow cytometer (BD Biosciences). Through gating, 20,000–50,000 live lymphocyte events are collected for subsequent analysis. Interestingly, we typically see a slight decline in engraftment when using *Atg7*$^{+/-}$ fetal liver cells, and a further reduction in the complete absence of *Atg7* (Fig. 12.5B).

Using this system recipient mice receiving Eμ-Myc transgenic stem cells will develop B cell lymphoma, usually within 3 months (Schmitt *et al.*, 2002), and here it is thus possible to evaluate the effects of heterozygosity or total loss of autophagy components on Myc-driven lymphomagenesis. Further, by evaluating the hematopoietic phenotypes of control non-transgenic hematopoietic recipients one can also determine the effects of heterozygosity or loss of *Atg7*, or *Atg5*, on hematopoietic development and homeostasis.

4. Concluding Remarks and Future Perspectives

Autophagy likely contributes to all aspects of cancer biology, including tumor initiation and progression, and in the maintenance of the malignant state. Cancer cells are, by their very nature, metabolically stressed. As they grow and expand, this problem becomes more acute with increasing hypoxia and nutrient deprivation, and autophagy is essential for surviving these stresses. Furthermore, autophagy also plays key roles in controlling the tumor microenvironment, the interaction of tumor cells with the immune system, and in the therapeutic response. Indeed, virtually all chemotherapeutic agents and radiotherapies induce such metabolic stress, and concomitant inhibition of autophagy potentiates the action of several known therapeutics. However, there are currently no small molecule inhibitors specific for the autophagy pathway. In addition to their use for understanding how autophagy contributes to the development and treatment of cancer, the methods presented herein provide key tools that allow for the development of screens for such small molecule autophagy antagonists and agonists. Finally, these methods can also be easily used to monitor and quantify the activity of the autophagy pathway in many other arenas of biology.

ACKNOWLEDGMENTS

We thank the Lehman Brothers Foundation–Dorothy Rodbell Cohen Cancer Research Fund for providing funding for the development of assays to monitor the autophagy pathway. We also thank Scot Ouellette for helpful discussion and critique regarding these methods. In addition, we thank Juliana Conkright for technical support and the Cell-Based Screening facility at the Scripps Research Institute-Florida and Mark A. Hall for generation of an MSCV-gateway-I-GFP vector, which facilitated cloning of the MSCV-Luc2pLC3-I-GFP construct. F.C.D. was also supported by a Ruth L. Kirschstein Fellowship from the National Cancer Institute, NIH and J.L.C. by NIH/NCI grant CA076379.

REFERENCES

Abida, W. M., and Gu, W. (2008). p53-Dependent and p53-independent activation of autophagy by ARF. *Cancer Res.* **68**(2), 352–357.

Adams, J. M., Harris, A. W., *et al.* (1985). The c-myc oncogene driven by immunoglobulin enhancers induces lymphoid malignancy in transgenic mice. *Nature* **318**(6046), 533–538.
Alpdogan, O., Muriglan, S. J., *et al.* (2003). IL-7 enhances peripheral T cell reconstitution after allogeneic hematopoietic stem cell transplantation. *J. Clin. Invest.* **112**(7), 1095–1107.
Alt, J. R., Greiner, T. C., *et al.* (2003). *Mdm2* haploinsufficiency profoundly inhibits Myc-induced lymphomagenesis. *EMBO J.* **22**(6), 1442–1450.
Amaravadi, R. K., Yu, D., *et al.* (2007). Autophagy inhibition enhances therapy-induced apoptosis in a Myc-induced model of lymphoma. *J. Clin. Invest.* **117**(2), 326–336.
Briceno, E., Reyes, S., *et al.* (2003). Therapy of glioblastoma multiforme improved by the anti-mutagenic chloroquine. *Neurosurg. Focus* **14**(2), e3.
Carew, J. S., Nawrocki, S. T., *et al.* (2007). Targeting autophagy augments the anticancer activity of the histone deacetylase inhibitor SAHA to overcome Bcr-Abl-mediated drug resistance. *Blood* **110**(1), 313–322.
Ciechomska, I. A., and Tolkovsky, A. M. (2007). Non-autophagic GFP-LC3 puncta induced by saponin and other detergents. *Autophagy* **3**(6), 586–590.
Degenhardt, K., Mathew, R., *et al.* (2006). Autophagy promotes tumor cell survival and restricts necrosis, inflammation, and tumorigenesis. *Cancer Cell* **10**(1), 51–64.
Eischen, C. M., Weber, J. D., *et al.* (1999). Disruption of the ARF-Mdm2-p53 tumor suppressor pathway in Myc-induced lymphomagenesis. *Genes Dev.* **13**(20), 2658–2669.
Fan, F., and Wood, K. V. (2007). Bioluminescent assays for high-throughput screening. *Assay Drug Dev. Technol.* **5**(1), 127–136.
Feng, Z., Zhang, H., *et al.* (2005). The coordinate regulation of the p53 and mTOR pathways in cells. *Proc. Natl. Acad. Sci. USA* **102**(23), 8204–8209.
Hawley, R. G., Fong, A. Z., *et al.* (1992). Transplantable myeloproliferative disease induced in mice by an interleukin 6 retrovirus. *J. Exp. Med.* **176**(4), 1149–1163.
Ichimura, Y., Kirisako, T., *et al.* (2000). A ubiquitin-like system mediates protein lipidation. *Nature* **408**(6811), 488–492.
Kabeya, Y., Mizushima, N., *et al.* (2000). LC3, a mammalian homologue of yeast Apg8p, is localized in autophagosome membranes after processing. *EMBO J.* **19**(21), 5720–5728.
Karin, M., and Greten, F. R. (2005). NF-κB: linking inflammation and immunity to cancer development and progression. *Nat. Rev. Immunol.* **5**(10), 749–759.
Kirisako, T., Ichimura, Y., *et al.* (2000). The reversible modification regulates the membrane-binding state of Apg8/Aut7 essential for autophagy and the cytoplasm to vacuole targeting pathway. *J. Cell. Biol.* **151**(2), 263–276.
Klionsky, D. J., Abeliovich, H., *et al.* (2008). Guidelines for the use and interpretation of assays for monitoring autophagy in higher eukaryotes. *Autophagy* **4**(2), 151–175.
Kuma, A., Matsui, M., *et al.* (2007). LC3, an autophagosome marker, can be incorporated into protein aggregates independent of autophagy: caution in the interpretation of LC3 localization. *Autophagy* **3**(4), 323–328.
Liang, X. H., Jackson, S., *et al.* (1999). Induction of autophagy and inhibition of tumorigenesis by beclin 1. *Nature* **402**(6762), 672–676.
Lowe, S. W., and Sherr, C. J. (2003). Tumor suppression by Ink4a-Arf: progress and puzzles. *Curr. Opin. Genet. Dev.* **13**(1), 77–83.
Luo, J. L., Maeda, S., *et al.* (2004). Inhibition of NF-κB in cancer cells converts inflammation-induced tumor growth mediated by TNFα to TRAIL-mediated tumor regression. *Cancer Cell* **6**(3), 297–305.
Maclean, K. H., Dorsey, F. C., *et al.* (2008). Targeting lysosomal degradation induces p53-dependent cell death and prevents cancer in mouse models of lymphomagenesis. *J. Clin. Invest.* **118**(1), 79–88.

Marino, G., Salvador-Montoliu, N., *et al.* (2007). Tissue-specific autophagy alterations and increased tumorigenesis in mice deficient in Atg4C/autophagin-3. *J. Biol. Chem.* **282** (25), 18573–18583.

Mathew, R., Karantza-Wadsworth, V., *et al.* (2007). Role of autophagy in cancer. *Nat. Rev. Cancer* **7**(12), 961–967.

Mathew, R., Kongara, S., *et al.* (2007). Autophagy suppresses tumor progression by limiting chromosomal instability. *Genes Dev.* **21**(11), 1367–1381.

Mizushima, N., Kuma, A., *et al.* (2003). Mouse Apg16L, a novel WD-repeat protein, targets to the autophagic isolation membrane with the Apg12-Apg5 conjugate. *J. Cell Sci.* **116** (Pt 9), 1679–1688.

Mizushima, N., Yamamoto, A., *et al.* (2004). *In vivo* analysis of autophagy in response to nutrient starvation using transgenic mice expressing a fluorescent autophagosome marker. *Mol. Biol. Cell* **15**(3), 1101–1111.

Qu, X., Yu, J., *et al.* (2003). Promotion of tumorigenesis by heterozygous disruption of the *beclin 1* autophagy gene. *J. Clin. Invest.* **112**(12), 1809–1820.

Reef, S., Zalckvar, E., *et al.* (2006). A short mitochondrial form of p19ARF induces autophagy and caspase-independent cell death. *Mol. Cell* **22**(4), 463–475.

Schmitt, C. A., Fridman, J. S., *et al.* (2002). Dissecting p53 tumor suppressor functions *in vivo*. *Cancer Cell* **1**(3), 289–298.

Sherr, C.J (2006). Divorcing ARF and p53: an unsettled case. *Nat. Rev. Cancer* **6**(9), 663–673.

Sotelo, J., Briceno, E., *et al.* (2006). Adding chloroquine to conventional treatment for glioblastoma multiforme: A randomized, double-blind, placebo-controlled trial. *Ann. Intern. Med.* **144**(5), 337–343.

Tanida, I., Ueno, T., *et al.* (2004). LC3 conjugation system in mammalian autophagy. *Int. J. Biochem. Cell Biol.* **36**(12), 2503–2518.

Yue, Z., Jin, S., *et al.* (2003). Beclin 1, an autophagy gene essential for early embryonic development, is a haploinsufficient tumor suppressor. *Proc. Natl. Acad. Sci. USA* **100**(25), 15077–15082.

CHAPTER THIRTEEN

Autophagy Pathways in Glioblastoma

Hong Jiang,* Erin J. White,* Charles Conrad,*
Candelaria Gomez-Manzano,* *and* Juan Fueyo*

Contents

Abstract

Glioma cells are more likely to respond to therapy through autophagy than through apoptosis. The most efficacious cytotoxic drugs employed in glioma therapy, such as temozolomide and rapamycin, induce autophagy. Oncolytic adenoviruses, which will soon be tested in patients with gliomas at the University of Texas M. D. Anderson Cancer Center, also induce autophagy. Autophagy in gliomas thus represents a promising mechanism that may lead to new glioma therapies. In this chapter, we present the methods for studying autophagy in glioma cells, including assessment of *in vitro* cellular markers acidic vesicle organelles, and green fluorescent protein (GFP)-LC3 punctation; biochemical

* Department of Neuro-Oncology, University of Texas M. D. Anderson Cancer Center, Houston, Texas, USA

Methods in Enzymology, Volume 453
ISSN 0076-6879, DOI: 10.1016/S0076-6879(08)04013-5

markers LC3-I/II conversion, p62 degradation, Atg12–Atg5 accumulation, and p70S6K dephosphorylation; and ultrastucture of the autophagosomes. In addition, we will address how LC3B and Atg5 up-regulation during autophagy can be examined through immunostaining in treated tumors and the potential of these proteins for use as surrogate markers to monitor therapeutic effects in clinical trials. Finally, we will discuss the challenges of studying autophagy in gliomas and the future directions of such use.

1. Introduction: Autophagy and Gliomas

Malignant astrocytic gliomas such as glioblastoma multiforme (GBM) are the most common and lethal intracranial tumors (Furnari *et al.*, 2007). Despite implementation of intensive therapeutic strategies and supportive care, the median survival duration of patients with GBM has remained at 12 months for the past decade (Furnari *et al.*, 2007). Gliomas are resistant to therapies that induce apoptosis (type I programmed cell death) (Furnari *et al.*, 2007; Lefranc and Kiss, 2006) (Ziegler *et al.*, 2008)). However, several lines of evidence indicate that GBM cells seem to be less resistant to therapies that induce autophagy (type II programmed cell death) (Lefranc *et al.*, 2007; Lefranc and Kiss, 2006). For example, rapamycin's disruption of the pathway controlled by mTOR induces marked autophagic processes in GBM cells (Iwamaru *et al.*, 2007). Temozolomide, currently the most efficacious cytotoxic drug employed to combat glioblastoma, exerts its cytotoxicity by inducing autophagic cell death (Kanzawa *et al.*, 2004). Recent reports show that oncolytic adenoviruses, a promising alternative therapy for gliomas, also cause autophagic cell death in glioma cells and even in brain tumor stem cells (Ito *et al.*, 2006; Jiang *et al.*, 2007). However, the role of autophagy in causing cell death, rather than occurring along with cell death, is still not clear. For example, controversial results have been reported on the effect of inhibiting autophagy on cytotoxicity induced by various treatments (Baird *et al.*, 2008; Ito *et al.*, 2006; Kanzawa *et al.*, 2004).

Autophagy is a dynamic subcellular process that degrades damaged or obsolete organelles and proteins (Rubinsztein *et al.*, 2007; Xie and Klionsky, 2007) and occurs as a cellular response to stresses such as starvation or pathogen infection (Kondo *et al.*, 2005; Levine, 2005). Autophagy has been implicated both in development and immunity (Levine and Deretic, 2007; Levine and Klionsky, 2004), and an abnormality in this autophagic process is related to diseases such as cancer (Kondo *et al.*, 2005; Levine and Kroemer, 2008; Mizushima *et al.*, 2008). Thus, for gliomas, understanding autophagy mechanisms at the molecular level will not only help identify efficacious points of intervention but also help explain the genesis of gliomas.

2. Prioritization of Methods to Characterize Autophagy in Gliomas

Autophagy is accompanied by the progressive development of vesicle structures from autophagosomes (not acidic) to amphisome and autolysosomes (acidic) (Klionsky *et al.*, 2008; Paglin *et al.*, 2001). With acidotropic dye acridine orange staining, the acidic compartments in the cell fluoresce bright red, whereas the nucleus and cytoplasm fluoresce bright green and dim red, respectively (Arvan *et al.*, 1984; Delic *et al.*, 1991; Mains and May, 1988). The intensity of the red fluorescence is proportional to the degree of acidity or the volume of the cellular acidic compartments, or both (Paglin *et al.*, 2001). Thus, it is possible to monitor the development of acidic vesicular organelles during autophagy with acridine orange staining in glioma cells (Ito *et al.*, 2006; Iwamaru *et al.*, 2007; Jiang *et al.*, 2007; Kanzawa *et al.*, 2004). However, it is important to keep in mind that this acidotropic dye can also be retained in other intracellular acidic compartments, such as lysosomes, endocytic vesicles, portions of the *trans*-Golgi apparatus, and certain secretory vesicles (Anderson and Orci, 1988). Therefore, the intensity of the red fluorescence is not exclusively an indication of autophagic vacuoles. The correlation between the intensity of the red fluorescence and autophagy development will vary by cell type and conditions. Thus, this methodology should be used with other autophagic markers.

The most commonly used biochemical marker to monitor autophagic flux is LC3, which can be monitored by examining GFP-LC3 dots and LC3 lipidation on a Western blot (Klionsky *et al.*, 2008) (also see the chapter by Kimura *et al.*, in volume 452). The LC3 protein is a ubiquitin-like protein that can be conjugated to phosphatidylethanolamine (PE) (Klionsky *et al.*, 2008). LC3 is initially synthesized in an unprocessed form, proLC3, which is cleaved from the C terminus, resulting in LC3-I, and is finally modified into the PE-conjugated form, LC3-II. LC3-II is the only protein marker that is reliably associated with completed autophagosomes (Klionsky *et al.*, 2008). LC3 with a green fluorescent protein (GFP) tag at the N terminus, GFP-LC3, is used to monitor autophagy through direct fluorescence microscopy, which is measured as an increase in punctate GFP-LC3 in glioma cells (Aoki *et al.*, 2007a,b; Aoki *et al.*, 2008; Iwamaru *et al.*, 2007; Kanzawa *et al.*, 2004; Klionsky *et al.*, 2008). Because LC3-II runs faster than LC3-I in sodium dodecyl-sulfate polyacrylamide gel electrophoresis (SDS-PAGE), autophagy can also be indicated by the increase in the conversion of LC3-I to LC3-II in glioma cells (Aoki *et al.*, 2007a,b; Aoki *et al.*, 2008; Ito *et al.*, 2006; Iwamaru *et al.*, 2007; Jiang *et al.*, 2007; Kanzawa *et al.*, 2004).

One of the hallmark biological aberrations in glioma cells is the alteration of the PI3K/AKT/mTor pathway (Furnari *et al.*, 2007). The activated pathway negatively regulates autophagy (Kondo *et al.*, 2005). Thus, the inhibition of mTOR activity, which is indicated by the dephosphorylation of its substrate p70S6K, is correlated with autophagy in glioma cells (Ito *et al.*, 2006; Iwamaru *et al.*, 2007). Other autophagic markers can also be used to detect autophagy in glioma cells. For example, when glioma cells are infected with oncolytic adenoviruses, the Atg12–Atg5 conjugate is dramatically up-regulated (Jiang *et al.*, 2007), whereas the long-lived protein p62 (Sequestosome 1 protein, SQSTM1), a ubiquitin- and LC3-binding protein (Komatsu *et al.*, 2007), was degraded.

One of the most sensitive and reliable methods used to detect autophagic compartments in mammalian cells is transmission electron microscopy (Eskelinen, 2008; Klionsky *et al.*, 2008) (see also the chapter by Ylä-Anttila in volume 452). This method can be used for both qualitative and quantitative analysis of changes in various autophagic structures (Klionsky *et al.*, 2008). The advantage of this method is that it does not depend on the availability of specific antibodies or probes (Eskelinen, 2008). The method's drawbacks include requiring an electron microscopy laboratory, taking longer than most other methods, and requiring special expertise that can only be gained by experience to correctly interpret the data (Eskelinen, 2008).

3. *In Vitro* Cellular Markers

3.1. Quantification of acidic vesicular organelles with acridine orange staining

i. Add acridine orange (Sigma-Aldrich, St. Louis, MO, USA, A8097) to glioma cells ($\approx 5 \times 10^5$ cells) cultured in DMEM/F12 supplemented with 10% fetal bovine serum, 100 μg/ml penicillin, and 100 μg/ml streptomycin (Invitrogen, Carlsbad, CA, USA, 10437-028, 10378-016) at 1.0 μg/ml. Note that nonadherent cells growing in an aggregated cluster, such as brain tumor stem cells, need to be digested with accutase solution (400-600 U/mL, Sigma-Aldrich, A6964) and dissociated by pipetting before adding acridine orange.

ii. Incubate for 15 min under normal culture conditions.

iii. Remove the medium and wash 1 time with phosphate-buffered saline (PBS).

iv. For adherent cells, detach the cells with 0.05% trypsin-ethylenediaminetetraacetic acid (EDTA) (Invitrogen, Carlsbad, CA, USA, 15400-054), and separate the cells with pipetting. Do not overdigest the cells,

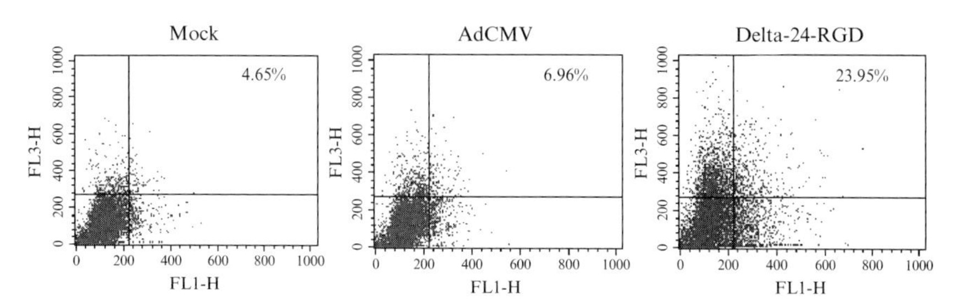

Figure 13.1 Induction of acidic vesicular organelles (AVOs) by Delta-24-RGD adenovirus in cell lines with properties of brain tumor stem cells (BTSCs). BTSC spheroids were dissociated and immediately infected with the indicated virus at 10 pfu/cell. Seventy-two hours later, cells were dissociated 2 h before acridine orange (1 μg/mL; Sigma-Aldrich) was added to the medium for 15 min. The percentage of cells presenting the formation of AVOs were quantified by flow cytometry. Empty adenoviral vector AdCMV was used as a control for the viral infection, and mock cells were treated without virus. The numbers indicate the percentage of cells with AVOs. Note that Delta-24-RGD induced a significant increase of the percentage of AVO-positive cells comparing to AdCMV.

as this will change the permeability of the cells and cause the cells to lose acridine orange staining.

v. Suspend the cells in 1.5 ml of PBS and pellet in a microcentrifuge at 14,000 rpm for 20 s.
vi. Resuspend the cells in 0.5 ml of PBS, and keep the samples on ice.
vii. Analyze the cells with a flow cytometer. Green (510–530 nm, FL1-H channel) and red (>650 nm, FL3-H channel) fluorescence emissions from 10^4 cells illuminated with blue (488 nm) excitation light are measured with a FACSCalibur from BD Biosciences using CellQuest software (San Jose, CA, USA). In the dot plot panel, set the bar for FL3-H in the control sample for your treatment so that the AVO-positive cells (the dots above the bar) are approximately 5% of the population (Fig. 13.1). Measure the test samples under the same condition. A significant increase in the percentage of AVO-positive cells may indicate the occurrence of autophagy (Fig. 13.1).

3.2. Assessing a punctate pattern of GFP-LC3 fluorescence (GFP-LC3 dots)

LC3-I is cytosolic; after LC3-I is processed into LC3-II, the latter is associated with the autophagosome membrane (Kabeya *et al.*, 2000). Therefore, the main population of GFP-LC3 is normally diffuse in the cell.

During autophagy, the tagged protein is recruited to the autophagosome membrane and presents a punctate pattern (Klionsky *et al.*, 2008). When the GFP-LC3 dots are countable, they can be quantified either by the number of dots per cell or the number of cells with GFP-LC3 dots exceeding the average number of dots in the control cells (Aoki *et al.*, 2008; Klionsky *et al.*, 2007). Both transient and stable expression of GFP-LC3 can be used for studies (Aoki *et al.*, 2008; Klionsky *et al.*, 2007). The protocol is as follows:

i. Seed 5×10^4 cells/well glioma cells in 4-well chamber slides. For stable expression, treat the cells directly with an autophagy-inducing agent (e.g., Delta-24-RGD adenovirus at 50 pfu/cell for 48 h).
ii. For transient expression, transfect the cells with a GFP-LC3 plasmid (Kabeya *et al.*, 2000) using Fugene 6 (Roche Molecular Biochemicals, Indianapolis, IN, USA, 11814443001) according to the manufacturer's instructions.
iii. Treat the cells with autophagy-inducing agents. If the treatment lasts longer than 48 h, treat the cells first with the agents, and then transfect the cells with the GFP-LC3 plasmid.
iv. 24–48 h after transfection, examine the cells directly or fix them with 4% paraformaldehyde in PBS for 30 min at 4 °C, and then examine the GFP-LC3 dots under a fluorescence microscope (Fig. 13.2). Determine the number of GFP-LC3 dots per cell in GFP-LC3-positive cells. Count a minimum of 50–100 cells per sample for triplicate samples per condition per experiment (Pattingre *et al.*, 2005).

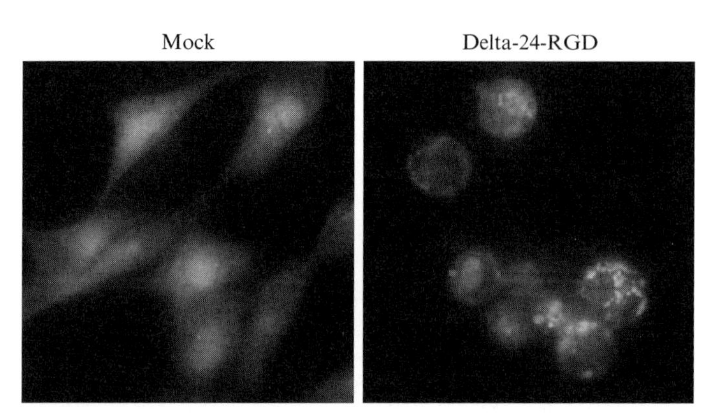

Figure 13.2 GFP-LC3 punctation induced by Delta-24-RGD adenovirus in U-87 MG-GFP-LC3 cells. The cells were infected with the virus at 50 pfu/cell. Forty-eight hours later, the green fluorescence was observed under a fluorescence microscope. Note the punctation of GFP-LC3 and rounding up of the cells showing the cytopathic effect caused by the virus.

4. *In Vitro* Biochemical Markers

4.1. Immunoblotting to determine the increase in LC3-II

LC3 expression in glioma cells is usually relatively abundant. An increase in LC3-II during autophagy has been confirmed in a panel of glioma cells and treatments (Aoki *et al.*, 2007a,b; Aoki *et al.*, 2008; Ito *et al.*, 2006; Iwamaru *et al.*, 2007; Jiang *et al.*, 2007; Kanzawa *et al.*, 2004). The protocol is as follows:

i. Wash and collect cells ($\approx 5 \times 10^6$ cells) in 4 ml of PBS.
ii. Pellet the cells with centrifugation at 500*g* for 5 min.
iii. Remove PBS, and resuspend the cells in approximately 4 cell-pellet volumes (typically approximately 80 μl) of PBS plus protease inhibitor cocktail (Sigma-Aldrich, P9599).
iv. Add an equal volume of 2 × SDS sample buffer (100 m*M* Tris-HCl, pH 6.8, 200 m*M* dithiothreitol, 4% SDS, 20% glycerol) to the cell suspension. Mix the sample with pipetting until the sample becomes viscous and clear.
v. Heat the samples at 95 °C for 10 min.
vi. Quantify the protein concentration of the samples with Bradford dye (Bio-Rad, Hercules, CA, USA, 500-0006) according to the manufacturer's instructions.
vii. Separate the proteins in a 12% SDS-PAGE gel at 50 V for stacking and 100 V for resolving in a Bio-Rad Mini-PROTEAN Electrophoresis System. Load approximately 40–60 μg of protein per lane.
viii. Transfer the proteins from the gel to a nitrocellulose or PVDF membrane.
ix. Block the membrane with 10% nonfat milk in Tris-Buffered Saline Tween 20 (TBST) (20 m*M* Tris-HCl, pH 7.6, 137 m*M* NaCl, 0.1% Tween 20) at room temperature for 1 h.
x. Incubate with rabbit polyclonal anti-LC3B antibody (1:3000 dilution, Novus Biologicals, Littleton, CO, USA, NB600-1384) diluted with 5% nonfat milk in TBST at 4 °C overnight.
xi. Remove the primary antibody. Rinse the membrane once with TBST, and wash the membrane with TBST for 5 min.
xii. Incubate the membrane in HRP-labeled secondary antibody goat antirabbit immunoglobulin G (IgG) (1:5000 dilution, Santa Cruz Biotechnology, Santa Cruz, CA, USA, sc-2004) diluted with 2% nonfat milk in TBST at room temperature for 1 h.
xiii. Remove the antibody, and rinse the membrane once with TBST. Wash the membrane with TBST three times for 15, 5, and 5 min at room temperature.

xiv. Visualize the protein-antibody complexes with the ECL Western blotting detection system (Amersham Pharmacia Biotech, Piscataway, NJ, USA). LC3-I and LC3-II are detected at a molecular mass of approximately 16 and 14 kDa respectively (Mizushima and Yoshimori, 2007) (Fig. 13.3).

4.2. Immunoblotting to examine changes in the Atg12–Atg5 conjugate, p62, and p70S6K dephosphorylation

During adenovirus-induced autophagy, we observe a dramatic increase of the Atg12–Atg5 conjugate (Jiang *et al.*, 2007) and a significant decrease in the p62 (SQSTM1/sequestosome 1) protein (Fig. 13.3). The procedure for the immunoblotting analysis of these proteins is similar to the procedure for the LC3 protein. Because the Atg12–Atg5 conjugate is approximately 55 kDa, p62 is approximately 62 kDa, and p70S6K is approximately 70 kDa, it is best to separate the proteins in a 10% SDS-PAGE gel. The primary antibody for Atg5 can be obtained from Cosmo Bio (Japan, CAC-TMD-PH-AT5), and the antibody for p62 is available from Santa Cruz Biotechnology (sc-25575). While autophagy occurs in glioma cells, the substrate of mTOR (p70S6K) is dephosphorylated as demonstrated by immunoblotting analysis (Ito *et al.*, 2006; Iwamaru *et al.*, 2007). Rabbit polyclonal anti-p70S6K and anti-phospho-p70S6K (Thr^{389}) antibodies are available from Cell Signaling Technology (Beverly, MA, USA, 9202, 9205).

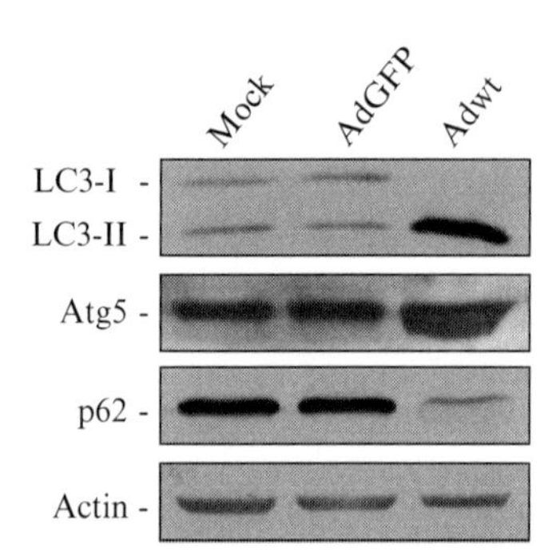

Figure 13.3 Expression of molecular markers of autophagy in malignant glioma U-87 MG cells after adenovirus infection. The cells were infected with replication-deficient adenovirus AdGFP or wild-type adenovirus (Adwt) at 10 pfu/cell. Seventy-two hours later, the cells were collected; the proteins were separated by sodium dodecyl-sulfate polyacrylamide gel electrophoresis (SDS-PAGE) and subjected to immunoblotting with rabbit polyclonal anti-LC3 (1:3000 dilution; Novus Biologicals), rabbit polyclonal anti-Atg5 (1:2000 dilution, Cosmo Bio), mouse monoclonal anti-p62 (1:400 dilution), and goat polyclonal antiactin (1:1000 dilution, Santa Cruz Biotechnology). AdGFP was used as the control for viral infection, and actin was used as a loading control.

5. Electron Microscopy to Monitor the Autophagic Vacuoles

i. Fix the cells ($\approx 1 \times 10^6$ cells) with a solution containing 3% glutaraldehyde plus 2% paraformaldehyde in 0.1 *M* cacodylate buffer, pH 7.3, for 1 h.
ii. Wash the cells with 0.1% Millipore-filtered cacodylate buffered tannic acid (Sigma-Aldrich, 403040).
iii. Postfix the cells with 1% phosphate buffered osmium tetroxide for 1 h.
iv. Stain the cells with 1% Millipore-filtered uranyl acetate.
v. Dehydrate the samples in increasing concentrations of ethanol (50%, 15 min; 70%, 15 min; 95%, 15 min; 100%, twice for 15 min).
vi. Infiltrate and embed the samples directly in Spurr's low-viscosity medium.
vii. Polymerize the samples in a 70 °C oven for 2 days.
viii. Cut the samples in ultrathin (50–100 nm) sections with a Leica Ultracut microtome (Leica, Deerfield, IL, USA).
ix. Stain the sections with 5% uranyl acetate solution for 15 min, rinse with distilled water, and then stain with Reynold's lead citrate Solution for 3–5 min, and rinse with distilled water in a Leica EM Stainer.
x. Examine the images in a JEM 1010 transmission electron microscope (JEOL, USA, Peabody, MA, USA) at an accelerating voltage of 80 kV. Digital images are obtained using AMT Imaging System (Advanced Microscopy Techniques, Danvers, MA, USA). Membrane-bound compartments accumulate in the cytoplasm (Fig. 13.4).

6. *In Vivo* Analysis of Biochemical Markers

There are very limited approaches to monitor autophagy *in vivo*. Atg5 and LC3B are successfully detected in the mouse brain in intracranially implanted gliomas treated with autophagy-inducing agents (Aoki *et al.*, 2008; Jiang *et al.*, 2007) (Fig. 13.5). The procedures to detect the Atg5 protein in the tumor xenografts through immunofluorescence are listed here.

i. Deparaffinize paraffin-embedded 5-μm sections of mouse brain tumor in xylene using 3 changes of 5 min each time.
ii. Hydrate the sections with 100% ethanol 2 times for 5 min each time.
iii. Hydrate the sections with 95% ethanol 2 times for 5 min each time.
iv. Rinse the sections in PBS 3 times for 5 min each time.
v. Block the samples with 3% normal goat serum (Santa Cruz Biotechnology, sc-2043) in PBS at room temperature for 30 min.

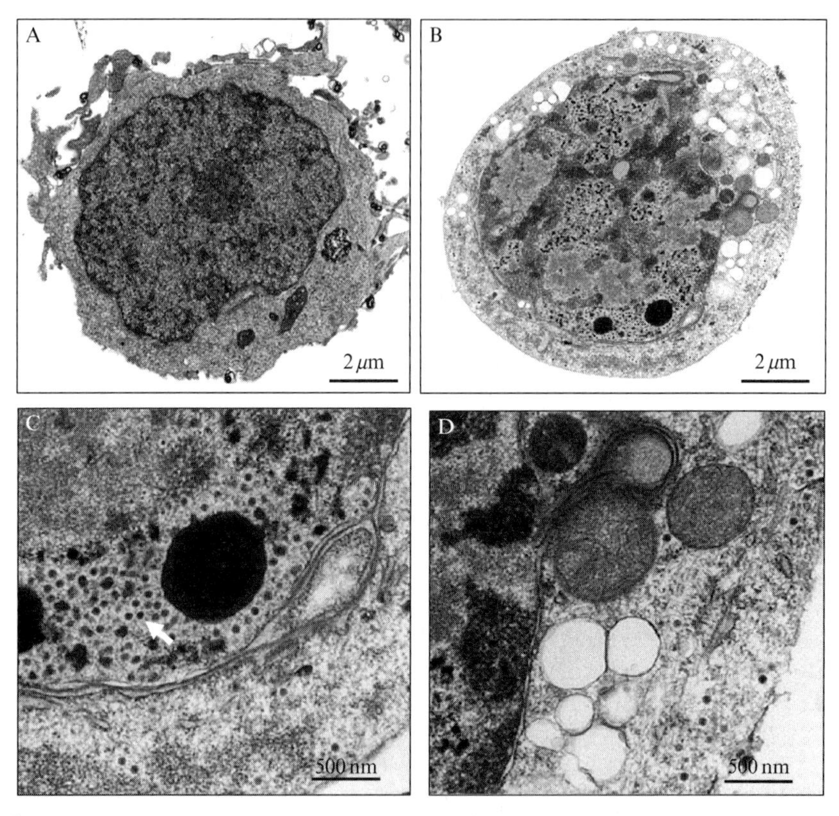

Figure 13.4 Representative electron micrographs showing the ultrastructure of the mock-infected (A) and Delta-24-RGD–infected (B) MDNSC11 cells. Note the vacuoles in the virus-infected cells but not in the untreated cells. Close-ups of Delta-24-RGD-infected cell illustrated in B show the cluster of the progenies of Delta-24-RGD (**white arrow**) in the nucleus (C) and complex autophagic multivacuolar bodies in the cytoplasm (D). This figure has been modified from a previous publication (Jiang *et al.*, 2007) and has been reproduced with permission from Oxford Journals.

vi. Incubate the samples overnight at 4 °C with anti-Atg5 antibody (1:200 dilution, Cosmo Bio, Japan, CAC-TMD-PH-AT5).
vii. Wash the samples 3 times with PBS plus 0.1% Tween 20 for 5 min each time.
viii. Incubate the samples with Alexa Fluor 488-conjugated secondary antibody (Molecular Probes, Eugene, OR, A-11008) for 50 min at room temperature.
ix. Wash the samples 3 times with PBS plus 0.1% Tween 20 for 5 min each time.

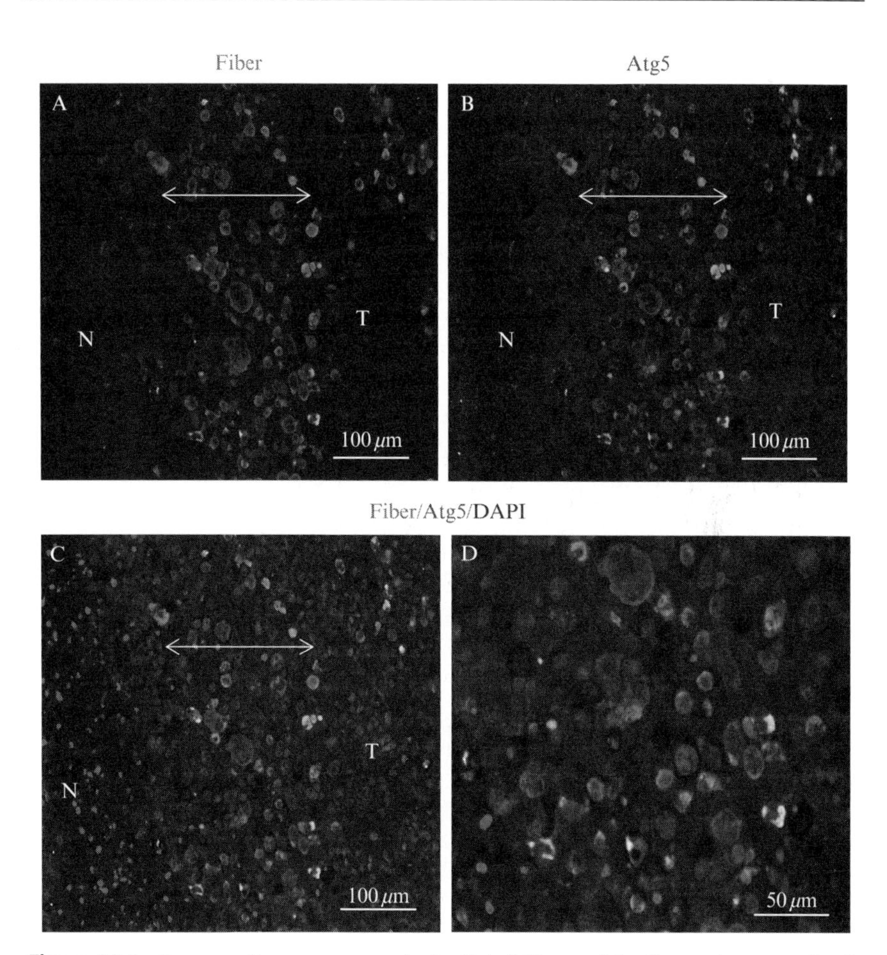

Figure 13.5 Immunofluorescence analysis of viral fiber and Atg5 protein expression in the brain of a mouse that was treated with Delta-24-RGD. The paraffin-embedded section of the mouse brain was double-immunostained with mouse monoclonal antibodies specific for adenoviral fiber protein (4D2, 1:500 dilution; Lab Vision, Fremont, CA, USA, MS-1027-P0) (A) or rabbit polyclonal Atg5 (1:200 dilution) (B) and then with Texas Red– (fiber) or Alexa Fluor 488– (ATG5) conjugated secondary antibodies (1:500 dilution; Molecular Probes, Eugene, OR, USA). Fluorescence for fiber and Atg5 were merged, and 4′,6-diamidino-2-phenylindole, dihydrochloride (DAPI) staining was used to visualize the cell nuclei (C). Expression of both proteins was positive double headed arrow) within the tumor (T) surrounding necrotic areas (N). Note that, in a close-up of C, the viral and cellular proteins are localized in the same cells around the cells that exhibit virally induced necrosis (D). This figure has been modified from a previous publication (Jiang *et al.*, 2007) and has been reproduced with permission from Oxford Journals. (See Color Insert.)

x. Mount the slides with ProLong Antifade Gold Reagent with DAPI (4′,6-diamidino-2-phenylindole, dihydrochloride, Molecular Probes, P-36931) to visualize the cell nuclei.

7. Autophagy Indicators as Surrogate Markers of Treatment Effect in Clinical Trials

Research indicates that autophagy is involved in the cell death induced by therapeutic agents for glioma, such as temozolomide, rapamycin, irradiation, and oncolytic adenoviruses (Ito *et al.*, 2006; Ito *et al.*, 2005; Iwamaru *et al.*, 2007; Jiang *et al.*, 2007; Kanzawa *et al.*, 2004). Thus, autophagy indicators can be used to monitor the cellular response to glioma treatment. As mentioned previously, Atg5 and LC3B can be detected in tumor tissues by immunostaining. After treating mice with temozolomide, an increased amount of LC3B is detected using immunofluorescence and immunohistochemistry in gliomas intracranially implanted in mouse brains (Aoki *et al.*, 2008). In the intracranial gliomas treated with oncolytic adenoviruses, Atg5 is readily detected with immunofluorescence, however, Atg5 is not detected in the untreated tumors (Jiang *et al.*, 2007) (Fig. 13.5). These markers show promise for evaluating how patients respond to glioma therapy and for helping predict whether the therapy is efficacious. This information will guide future clinical trials to optimize the therapeutic regimen to achieve the best results for patients with gliomas.

8. Future Directions

Intervention via autophagy is a promising approach for glioma therapy, as glioma cells are more sensitive to autophagy-inducing agents than to apoptosis-inducing agents (Lefranc *et al.*, 2007). Further studies dissecting autophagic pathways and molecules that are essential for autophagy in glioma cells, especially those involved in autophagic cell death, will help researchers identify potential targets for glioma therapies. To facilitate these studies, it is critical to develop reliable, specific, quantitative, and simple assays to examine the occurrence of autophagy in glioma cells. At the present time, multiple assays are needed because no single assay satisfies all the criteria for success (Klionsky *et al.*, 2008). Some assays are impossible or difficult to quantify for clinical samples. For example, LC3 punctation is impossible to be visualized by GFP tag in patient samples. In addition, quantification of immunostaining for autophagic markers, such as LC3B or Atg5, will be very difficult, if not impossible. Besides, we are not sure whether the markers can be applied in glioma treatments other than in the scenario reported in this chapter (Aoki *et al.*, 2008; Jiang *et al.*, 2007). In light of these challenges, we feel certain that our progressing knowledge of autophagy and new technologies in this area will help in the development of more optimal assays to be used in glioma research and treatment.

REFERENCES

Anderson, R. G., and Orci, L. (1988). A view of acidic intracellular compartments. *J. Cell Biol.* **106,** 539–543.

Aoki, H., Iwado, E., Eller, M. S., Kondo, Y., Fujiwara, K., Li, G. Z., Hess, K. R., Siwak, D. R., Sawaya, R., Mills, G. B., Gilchrest, B. A., and Kondo, S. (2007a). Telomere 3′ overhang-specific DNA oligonucleotides induce autophagy in malignant glioma cells. *FASEB J.* **21,** 2918–2930.

Aoki, H., Kondo, Y., Aldape, K., Yamamoto, A., Iwado, E., Yokoyama, T., Hollingsworth, E. F., Kobayashi, R., Hess, K., Shinojima, N., Shingu, T., Tamada, Y., *et al.* (2008). Monitoring autophagy in glioblastoma with antibody against isoform B of human microtubule-associated protein 1 light chain 3. *Autophagy* **4,** 467–475.

Aoki, H., Takada, Y., Kondo, S., Sawaya, R., Aggarwal, B. B., and Kondo, Y. (2007b). Evidence that curcumin suppresses the growth of malignant gliomas in vitro and in vivo through induction of autophagy: Role of Akt and extracellular signal-regulated kinase signaling pathways. *Mol. Pharmacol.* **72,** 29–39.

Arvan, P., Rudnick, G., and Castle, J. D. (1984). Osmotic properties and internal pH of isolated rat parotid secretory granules. *J. Biol. Chem.* **259,** 13567–13572.

Baird, S. K., Aerts, J. L., Eddaoudi, A., Lockley, M., Lemoine, N. R., and McNeish, I. A. (2008). Oncolytic adenoviral mutants induce a novel mode of programmed cell death in ovarian cancer. *Oncogene.* **27,** 3081–3090.

Delic, J., Coppey, J., Magdelenat, H., and Coppey-Moisan, M. (1991). Impossibility of acridine orange intercalation in nuclear DNA of the living cell. *Exp. Cell. Res.* **194,** 147–153.

Eskelinen, E.-L. (2008). To be or not to be? Examples of incorrect identification of autophagic compartments in conventional transmission electron microscopy of mammalian cells. *Autophagy* **4,** 257–260.

Furnari, F. B., Fenton, T., Bachoo, R. M., Mukasa, A., Stommel, J. M., Stegh, A., Hahn, W. C., Ligon, K. L., Louis, D. N., Brennan, C., Chin, L., DePinho, R. A., *et al.* (2007). Malignant astrocytic glioma: Genetics, biology, and paths to treatment. *Genes Dev.* **21,** 2683–2710.

Ito, H., Aoki, H., Kuhnel, F., Kondo, Y., Kubicka, S., Wirth, T., Iwado, E., Iwamaru, A., Fujiwara, K., Hess, K. R., Lang, F. F., Sawaya, R., and Kondo, S. (2006). Autophagic cell death of malignant glioma cells induced by a conditionally replicating adenovirus. *J. Natl. Cancer Inst.* **98,** 625–636.

Ito, H., Daido, S., Kanzawa, T., Kondo, S., and Kondo, Y. (2005). Radiation-induced autophagy is associated with LC3 and its inhibition sensitizes malignant glioma cells. *Int. J. Oncol.* **26,** 1401–1410.

Iwamaru, A., Kondo, Y., Iwado, E., Aoki, H., Fujiwara, K., Yokoyama, T., Mills, G. B., and Kondo, S. (2007). Silencing mammalian target of rapamycin signaling by small interfering RNA enhances rapamycin-induced autophagy in malignant glioma cells. *Oncogene* **26,** 1840–1851.

Jiang, H., Gomez-Manzano, C., Aoki, H., Alonso, M. M., Kondo, S., McCormick, F., Xu, J., Kondo, Y., Bekele, B. N., Colman, H., Lang, F. F., and Fueyo, J. (2007). Examination of the therapeutic potential of Delta-24-RGD in brain tumor stem cells: Role of autophagic cell death. *J. Natl. Cancer Inst.* **99,** 1410–1414.

Kabeya, Y., Mizushima, N., Ueno, T., Yamamoto, A., Kirisako, T., Noda, T., Kominami, E., Ohsumi, Y., and Yoshimori, T. (2000). LC3, a mammalian homologue of yeast Apg8p, is localized in autophagosome membranes after processing. *EMBO J.* **19,** 5720–5728.

Kanzawa, T., Germano, I. M., Komata, T., Ito, H., Kondo, Y., and Kondo, S. (2004). Role of autophagy in temozolomide-induced cytotoxicity for malignant glioma cells. *Cell Death Differ.* **11,** 448–457.

Klionsky, D. J., Abeliovich, H., Agostinis, P., Agrawal, D. K., Aliev, G., Askew, D. S., Baba, M., Baehrecke, E. H., Bahr, B. A., Ballabio, A., *et al.* (2008). Guidelines for the use and interpretation of assays for monitoring autophagy in higher eukaryotes. *Autophagy* **4,** 151–175.

Klionsky, D. J., Cuervo, A. M., and Seglen, P. O. (2007). Methods for monitoring autophagy from yeast to human. *Autophagy* **3,** 181–206.

Komatsu, M., Waguri, S., Koike, M., Sou, Y. S., Ueno, T., Hara, T., Mizushima, N., Iwata, J., Ezaki, J., Murata, S., Hamazaki, J., Nishito, Y., *et al.* (2007). Homeostatic levels of p62 control cytoplasmic inclusion body formation in autophagy-deficient mice. *Cell* **131,** 1149–1163.

Kondo, Y., Kanzawa, T., Sawaya, R., and Kondo, S. (2005). The role of autophagy in cancer development and response to therapy. *Nat. Rev. Cancer* **5,** 726–734.

Lefranc, F., Facchini, V., and Kiss, R. (2007). Proautophagic drugs: A novel means to combat apoptosis-resistant cancers, with a special emphasis on glioblastomas. *Oncologist* **12,** 1395–1403.

Lefranc, F., and Kiss, R. (2006). Autophagy, the Trojan horse to combat glioblastomas. *Neurosurg. Focus* **20,** E7.

Levine, B. (2005). Eating oneself and uninvited guests: Autophagy-related pathways in cellular defense. *Cell* **120,** 159–162.

Levine, B., and Deretic, V. (2007). Unveiling the roles of autophagy in innate and adaptive immunity. *Nat. Rev. Immunol.* **7,** 767–777.

Levine, B., and Klionsky, D. J. (2004). Development by self-digestion: Molecular mechanisms and biological functions of autophagy. *Dev. Cell* **6,** 463–477.

Levine, B., and Kroemer, G. (2008). Autophagy in the pathogenesis of disease. *Cell* **132,** 27–42.

Mains, R. E., and May, V. (1988). The role of a low pH intracellular compartment in the processing, storage, and secretion of ACTH and endorphin. *J. Biol. Chem.* **263,** 7887–7894.

Mizushima, N., Levine, B., Cuervo, A. M., and Klionsky, D. J. (2008). Autophagy fights disease through cellular self-digestion. *Nature* **451,** 1069–1075.

Mizushima, N., and Yoshimori, T. (2007). How to interpret LC3 immunoblotting. *Autophagy* **3,** 542–545.

Paglin, S., Hollister, T., Delohery, T., Hackett, N., McMahill, M., Sphicas, E., Domingo, D., and Yahalom, J. (2001). A novel response of cancer cells to radiation involves autophagy and formation of acidic vesicles. *Cancer Res.* **61,** 439–444.

Pattingre, S., Tassa, A., Qu, X., Garuti, R., Liang, X. H., Mizushima, N., Packer, M., Schneider, M. D., and Levine, B. (2005). Bcl-2 antiapoptotic proteins inhibit Beclin 1-dependent autophagy. *Cell* **122,** 927–939.

Rubinsztein, D. C., Gestwicki, J. E., Murphy, L. O., and Klionsky, D. J. (2007). Potential therapeutic applications of autophagy. *Nat. Rev. Drug Discov.* **6,** 304–312.

Xie, Z., and Klionsky, D. J. (2007). Autophagosome formation: Core machinery and adaptations. *Nat. Cell Biol.* **9,** 1102–1109.

Ziegler, D. S., Kung, A. L., and Kieran, M. W. (2008). Anti-apoptosis mechanisms in malignant gliomas. *J. Clin. Oncol.* **26,** 493–500.

CHAPTER FOURTEEN

Autophagy in Lung Cancer

Jerry J. Jaboin,* Misun Hwang,* *and* Bo Lu*

Contents

Abstract

Lung cancer is the leading cause of cancer-related deaths worldwide. The relatively poor cure rate in lung cancer patients has been associated with a resistance to chemotherapy and radiation that is at least in part related to defects in cellular apoptotic machinery. Exploitation of another form of cell death, autophagy, has the capacity to improve the therapeutic gain of current therapies. In an effort to develop novel treatment strategies to enhance the therapeutic ratio for lung cancer, we wish to better understand the role of autophagic cell death for the sensitization of lung cancer. This text reviews the most up to date protocols and techniques for the study of autophagic cell death in lung cancer models. Others may use these techniques as a framework for study within their experimental models.

* Department of Radiation Oncology, Vanderbilt Ingram Cancer Center, Vanderbilt University School of Medicine, Nashville, Tennessee, USA

Methods in Enzymology, Volume 453
ISSN 0076-6879, DOI: 10.1016/S0076-6879(08)04014-7

1. Introduction

Lung cancer is the most prevalent cancer worldwide. Despite improvements in multimodality therapy, it continues to be the leading cause of cancer-related death in the United States, with an estimated 161,840 deaths in 2008 (Jemal *et al.*, 2008). The majority of lung cancer patients present with advanced disease, and standard therapeutic regimens include radiation, platinum-based chemotherapy, and, rarely, surgery. Though there have been advances in radiation delivery, enhanced platinum pharmacological profiles, introduction of targeted biological agents and optimization of treatment schedules, there is still significant room for improvement in both the prognosis and side-effect profiles for this patient population (Rigas and Kelly, 2007).

A major barrier to curative therapy in lung cancer is the dysregulation of cell death signaling (Abend, 2003; Fesik, 2005; Melet *et al.*, 2008; Moretti *et al.*, 2007). Of the cellular death processes, apoptosis has been the best studied. Multiple investigators have demonstrated that deficits in apoptotic machinery can lead not only to abnormal proliferation but also to insensitivity to cytotoxic therapy. In fact, there have been efforts over the past decade to enhance apoptosis utilizing various activating antibodies, peptides and small molecules to restore the apoptotic machinery of tumor cells for the purpose of triggering cell death or rendering tumor cells more sensitive to chemotherapy and radiation (Fesik, 2005).

Autophagic (type II or macroautophagic) cell death was described as early as the 1970s, as a cell death mechanism that can occur independent of apoptosis (Clarke, 1990). It is an evolutionarily conserved process, which morphologically involves the formation of double-membrane-bound autophagic vacuoles, called autophagosomes (Baehrecke, 2002; Reggiori *et al.*, 2004a,b). These autophagosomes degrade and recycle proteins and cellular organelles by fusing with lysosomes to form autolysosomes (Levine and Klionsky, 2004). The role of autophagy in mammalian cell death is suggested by a study that implicates the autophagic genes, *ATG7* and *beclin 1* (Shimizu *et al.*, 2004; Yu, 2004). These studies not only demonstrate an antagonistic interplay between apoptosis and autophagic cell death but also suggest that they are nonexclusive events that can at times be observed in the same senescent cell (Gonzalez-Polo *et al.*, 2005; Shimizu *et al.*, 2004; Yu, 2004). The role of autophagy in cell death is still controversial, however, and a recent study finds that *ATG7* and *beclin 1* act in a protective manner in the same system (Wu *et al.*, 2008).

Autophagy has multiple roles in the promotion of carcinogenesis. It ensures survival in nutrient-poor conditions through lysosomal recycling of intracellular nutrients, which is proposed to allow time for the development of adaptive changes in gene expression and metabolic activity (Ogier-Denis and Codogno, 2003). Additionally, it may promote evasion of chemotherapy and radiation-induced apoptosis through removal of damaged organelles (Boya *et al.*, 2005; Lum, 2005; Ogier-Denis and Codogno, 2003; Paglin *et al.*, 2001). Although the autophagic response to starvation is less pronounced in cancer cells, it is still up-regulated in many tumor types suggesting it is an important survival mechanism (Gozuacik and Kimchi, 2004; Ogier-Denis and Codogno, 2003).

Suppression of carcinogenesis has also been proposed to be an important feature of autophagy. Cancer cells that are unable to undergo apoptosis secondary to genetic mutations are still susceptible to autophagic cell death (Paglin *et al.*, 2001). Additionally, some cancer cells are dependent on a blockade of autophagy for maintenance of their malignant phenotype (Liang *et al.*, 1999). This hypothesis was tested in several experimental models of hepatic carcinogenesis, where preneoplastic nodules and frank hepatocellular carcinomas demonstrate decreased autophagic capacity as compared to normal liver cells (Canuto *et al.*, 1993; Schwartz *et al.*, 1993; Schwarze and Seglen, 1985). It is proposed that this decreased autophagic activity associated with malignant cells may be related to the prevention of excessive protein loss upon starvation of tumor cells (Canuto *et al.*, 1993; Schwarze and Seglen, 1985).

Three are a number of cancer therapies that currently induce autophagy (Table 14.1). One of the most notable of these is temozolomide, which is a DNA alkylator that induces autophagy in malignant glioma cells (Kanzawa *et al.*, 2004). This is the first chemotherapeutic agent that has demonstrated efficacy in this difficult population. In addition to temozolomide, there are multiple therapies that are known to induce autophagy. These include the class of histone deacetylase inhibitors (Shao *et al.*, 2004), arsenic trioxide (Kanzawa *et al.*, 2003, 2005), tamoxifen (Bursch *et al.*, 1996, 2000), rapamycin (Takeuchi *et al.*, 2005), and irradiation (Ito *et al.*, 2005; Paglin *et al.*, 2001; Yao *et al.*, 2003). These drugs may actually cause autophagic cell death. For example, 3-methyladenine (3-MA) is an inhibitor of autophagosome formation and prevents tamoxifen-induced cell death in breast cancer cells (Paglin *et al.*, 2001). However induction of autophagy does not necessarily signify a causal therapeutic benefit. For example, 3-MA treatment results in radiosensitization with increased cell death observed in irradiated malignant glioma cells (Ito *et al.*, 2005). In the case of irradiation, induction of autophagy may represent a survival mechanism, or the balance of cellular decisions is tipped toward apoptotic death.

The rapid evolution in the study of autophagy has prompted a recent publication to establish guidelines to more clearly establish a basic foundation for the understanding of autophagic processes.(Klionsky *et al.*, 2008) The authors outline the current tools that at a minimum should be used by

Table 14.1 Characteristics of agents targeting autophagy in lung cancer

CLASS	AGENT	MECHANISM OF ACTION
Lysosomal Protease Inhibitors	E64d Leupeptin Pepstatin A	Lysosomal Inhibitor (proteases)
	Vinca alkaloids	Microtubule Inhibitors
	Bafilomycin A_1	Increases Lysosomal pH
Autophagy Sequestration Inhibitors	3-methyladenine (3-MA) LY294002 Wortmannin	Class I/III PI3K Inhibition
	Cycloheximide	Not established
Caspase Inhibitors	M687 (Kim *et al.*, 2008b); Merck	Not established
	Z-DEVD	Not established
Inducers of Autophagy	Arsenic Trioxide	Inhibition of mTOR (Kanzawa *et al.*, 2005)
	Butyrate & suberoylanilide hydroxamic acid	Histone Deacetylase Inhibitor (Shao *et al.*, 2004)
	Ceramide	Increased Beclin 1 (Daido *et al.*, 2004)
	Rapamycin, RAD001 (Everolimus) (Lefranc *et al.*, 2007)	Inhibition of mTOR (Albert *et al.*, 2006; Cao *et al.*, 2006; Kamada *et al.*, 2000; Kim *et al.*, 2006)
	Resveratrol (Opipari *et al.*, 2004)	Not established
	Tamoxifen (Bursch *et al.*, 1996, 2000)	Increased Beclin 1 expression (Scarlatti *et al.*, 2004)
	Temozolomide	DNA Alkylation (Kanzawa *et al.*, 2004)
	Endostatin (Chau *et al.*, 2003)	Not established

investigators to determine whether full autophagy (autophagic flux) has occurred, rather than accumulation of markers of autophagy, such as the autophagosome. Our text will outline these protocols with a special emphasis on the study of lung cancer models.

2. Methods

2.1. Cell culture

We utilize multiple cell lines in the study of cell death in lung cancer tumors. Primary mouse embryonic fibroblasts (MEFs) are a useful tool in the study of cell death mechanisms within various genetic backgrounds. We use MEFs with deficiencies in various key apoptotic proteins (e.g., caspases, Bcl-2 family members) to characterize cell death in response to irradiation with and without other cytotoxic agents (Kim *et al.*, 2008a,b). To complement these studies, we also evaluated well-characterized human lung cancer cell lines (e.g., H460) *in vitro* and *in vivo*. As neoangiogenesis also plays an important part in tumor control, we perform experiments with human umbilical vein endothelial cells (HUVECs) as a surrogate for tumor blood vessel response (Albert *et al.*, 2006; Cao *et al.*, 2005; Kim *et al.*, 2008a,b).

1. MEFs are derived from wild-type (WT), caspase-$3^{-/-}/7^{-/-}$ double knockout (DKO) mice and Bax-Bak DKO mice. They are then immortalized by transfection with a plasmid containing SV40-T-antigen.
   a. The MEFs are cultured in Dulbecco's Modified Eagle's Medium (DMEM) supplemented with 10% fetal bovine serum (FBS), 1% penicillin-streptomycin (Invitrogen, Cat. No. 10313-039) and 0.5 μmol/L 2-mercaptoethanol at 37 °C.
2. NCI-H460 (H460) lung cancer cells are obtained from ATCC (Cat. No. HTB-177).
   a. The H460 cells are cultured in RPMI 1640 (Invitrogen, Cat. No. 11875-119) supplemented with 10% fetal bovine serum and 1% penicillin-streptomycin at 37 °C and humidified 5% CO_2.
3. Human Umbilical Vein Endothelial Cells (HUVECs) are obtained from Clonetics (now Lonza, Cat. No. CC-2519).
   a. HUVECs are maintained in EBM-2 medium (Clonetics (now Lonza), Cat. No. CC-3156) supplemented with EGM-2 MV (Biowhittaker (now Lonza), Cat. No. CC-4147) single aliquots at 37 °C.

2.2. Drugs in autophagy analysis

We utilize various drugs in our studies and these agents serve various purposes (Table 14.1). The lysosomal protease inhibitors help verify the presence of autophagic flux. By blocking lysosomal degradation, we can evaluate the time-dependent accumulation of autophagosomes and LC3-II (Klionsky *et al.*, 2008). This allows for better differentiation of the presence of autophagosome accumulation versus changes in the rates of autophagic degradation (Klionsky *et al.*, 2008).

Inhibitors of autophagy (e.g., 3-methyladenine) are necessary in the validation of autophagy. They are most often used to determine a change in autophagosome production, as determined by GFP-LC3 puncta formation in response to therapy (Shintani, 2004). They can also be used to monitor and estimate the turnover rates of autophagic organelles (e.g., autophagosomes, autolysosomes) (Kirkegaard *et al.*, 2004; Shintani, 2004). There is also a complex interplay between apoptosis and autophagy (Jaboin *et al.*, 2007). We use inhibitors of apoptosis (e.g., caspase inhibitors) to better study this relationship.

The number of inducers of autophagy has been growing steadily over the past decade. These agents vary in their specificity and abilities to induce cell death, but some may prove extremely useful for the improvement of therapeutic ratio in a host of tumors. Listed in Table 14.1 are a few of the better characterized agents that either stimulate or inhibit autophagy (Lefranc *et al.*, 2007).

2.3. Clonogenic assay

The clonogenic assay is an *in vitro*cell survival assay that has been the standard for evaluating reproductive cell death following irradiation for many years (Franken *et al.*, 2006). It has the advantage of analysis of cell death without significant confounding factors of changes in cellular proliferation or metabolic changes. This tool is useful in the study of various cellular models, and multiple cytotoxic agents.

1. H460 cells and MEFs are treated with DMSO or drug (e.g. M867) at varying concentrations (1.4, 5, and 10 n*M* in H460 cells; 5 and 10 n*M* in MEFs) for a period 24 h at 37 °C in 60-mm tissue culture dishes.
2. Cells are subsequently treated with siRNAs against empty vector, *caspase-3*, *caspase-7*, *beclin 1*, and/or *ATG5*.
3. Cells are then irradiated with 0–6 Gy at a dose rate of 1.8 Gy/min using a ^{137}Cs irradiator (J.L. Shepherd and Associates, Glendale, CA).
4. After irradiation, cells are incubated at 37 °C to allow for at least 6 cell divisions, which typically is a period of 6–14 days.
5. The medium is removed, and cells are then fixed for 15 min with 3:1 methanol: acetic acid and stained for 15 min with 0.5% crystal violet (Sigma-Aldrich, Cat. No. C0775) in methanol.
6. Following staining, colonies are counted. A colony is defined as containing a minimum of 50 viable cells.
7. Surviving fraction is calculated as (mean colony counts)/(cells inoculated) × (plating efficiency (PE)), where PE is defined as (mean colony counts)/(cells inoculated for nonirradiated controls).

8. Dose enhancement ratio (DER) is calculated as the dose (Gy) for radiation alone divided by the dose (Gy) for radiation plus drug (normalized for drug toxicity) necessary for a surviving fraction of 0.25. Experiments are conducted in triplicate and mean, standard deviation, and P values are calculated.

2.4. Endothelial cell morphogenesis assay: tubule formation

Angiogenesis is an important facet of cancer therapy, as it is critical in maintaining nourishment after tumors reach a critical size (Folkman, 2007). As such, *in vitro*studies of novel drugs and combination therapies would not be complete without an analysis of tumor endothelial cells. Most of our studies have been performed utilizing human umbilical vein endothelial cells, as a surrogate for tumor endothelial cells (Albert *et al.*, 2007; Kim *et al.*, 2008a).

1. HUVECs are grown to ~70% confluency, and treated with DMSO, Z-DEVD (50 μM for 24 h), RAD001 (10 nM for 2 h) or combined Z-DEVD with RAD001 (at the same doses for a period of 24 h), and then cells are treated with 5 Gy.
2. The media is removed from the culture dish and washed once with PBS. The PBS is aspirated, and 1 ml of trypsin-EDTA is added to the plate. The plate is rocked to ensure that the entire surface is covered. Incubate the dish at room temperature for 1–3 min, and observe under a microscope. When the cells are completely round, gently dislodge the cells from the surface of the dish by rapping.
3. They are seeded at 48,000 cells per well on 24-well plates coated with 300 μl of Matrigel (BD Biosciences, Cat. No. 354234).
4. The cells undergo differentiation into capillary-like tube structures, and are periodically observed by microscopy.
5. Twenty-four h later, cells are stained with hematoxylin and eosin and photographs are taken via microscopy.
6. The average number of tubes is calculated from examination of three separate microscope fields (100X) and representative photographs are taken.

2.5. Analysis of autophagosomes (Immunofluorescence Microscopy)

Detection of fluorescence-tagged GFP is a useful tool for the microscopy detection of autophagosome production. LC3 is tagged with GFP at the amino terminus. The cells are subsequently transiently transfected to overexpress the GFP-LC3 proteins, and treated in various cellular conditions. At various time points, the cells are analyzed by confocal microscopy with images recorded for analysis of punctate GFP-LC3 as an indication of autophagosome production.

There are limitations to this strategy. A stable transfection may result in fewer artifacts, and decreased background over transient transfection (Klionsky *et al.*, 2008). Analysis would also be easier, because nearly every cell would express the GFP-LC3. Generally we prefer the transient transfection method, as the effect of LC3 overexpression is detected shortly after transfection, which theoretically reduces alterations of the cellular machinery over the time it would take to generate stable transfectants.

1. H460 cell and MEFs are seeded at a density of 2×10^5cells into 6-well tissue culture plates.
2. After 16 h, the cells are transfected with 2.5 μg of GFP-LC3 expression plasmid using the Lipofectamine 2000 reagent (Invitrogen Life Technologies, Cat. No. 12566-014) per the manufacturer's protocols.
3. Twenty-four h later, the cells are treated with 5 Gy of radiation, with or without drug.
4. After 24 and 48 h time points, the fluorescence of GFP-LC3 is observed using confocal microscopy.

2.6. Immunoblotting

Western immunoblotting has been useful for the evaluation of autophagic proteins in our models. In mammalian systems, determinations of the total levels of LC3 is not sufficient, as it does not account for the variations in conversion and degradation of LC3-I and LC3-II (Klionsky *et al.*, 2008). Thus, Western analysis of LC3-I and LC3-II, and determination of a LC3-II/LC3-I ratio has been considered a good marker for autophagy.

There are limitations to this approach. Changes in LC3-I and LC3-II levels vary with cell type and conditions, which can present significant challenges for heterogeneous populations (Mizushima *et al.*, 2004). Also there are reported variations in the sensitivity of various antibodies to LC3-I with less lability associated with LC3-II (Klionsky *et al.*, 2008). That combined with the fact that LC3-II is degraded by autophagy can confound assessments. Finally, LC3-II and the LC3-II/LC3-I ratio can represent autophagosome accumulation at a given point in time. It is better to also assess LC3-II in the presence and absence of lysosomal protease inhibitors to determine the amount of LC3-II delivered to lysosomes, which is a better determination of autophagic flux (Mizushima and Yoshimori, 2007).

1. H460 cells (0.5×10^6) are treated with varying doses of radiation and drugs.
2. They are subsequently collected at multiple time points, and washed with ice-cold PBS twice before the addition of lysis buffer (20 m*M* Tris-HCl, pH 7.4, 150 m*M* NaCl, 20 m*M* EDTA, 1% NP40, 50 m*M* NaF, 1 m*M* Na_3Vo_4, 1 m*M* $NaMO_4$and protease cocktail inhibitor (Sigma-Aldrich, Cat. No. P8340)).

3. Protein concentration is quantified using the Bio-Rad protein assay kit (Bio-Rad, Cat. No. 500-0001).
4. An equal amount of 2X SDS-polyacrylamide gel electrophoresis sample loading buffer is added to each sample, and the samples are heated at 100 °C for 5 min.
5. Equal amounts of protein are loaded into each well, and resolved on a 12.5% SDS-PAGE gel.
6. The blots are transferred to a polyvinylidene difluoride (PVDF) membrane at 300 mA for 3 h at 4 °C.
7. The PVDF membranes are blocked using 5% nonfat dry milk in TBS-T (Tris-buffered saline (TBS) and 0.1% Tween-20 with 5% nonfat dry milk) for 1 h at room temperature.
8. The blots are then incubated with various primary antibodies in TBS with 5% nonfat milk for 1 h at room temperature or overnight at 4 °C.
   a. Rabbit anti-LC3 polyclonal antibody: Medical & Biological Laboratories ITL, Cat. No. PD012
      i. LC3-I: 16 kDa
      ii. LC3-II: 18 kDa
   b. Rabbit anti-caspase-3 polyclonal antibody: Cell Signaling, Cat. No. 9662
      i. Full length caspase-3: 35 kDa
      ii. Cleaved caspase-3: 17 kDa
   c. Rabbit anti-caspase-7 polyclonal antibody: Cell Signaling, Cat. No. 9492
      i. Full length caspase-7: 35 kDa
      ii. Cleaved caspase-7: 20 kDa
   d. Rabbit anti-Akt polyclonal antibody: Cell Signaling, Cat. No. 4685
      i. Akt: 60 kDa
   e. Rabbit anti-phospho-Akt (Ser-473) monoclonal antibody: Cell Signaling, Cat. No. 4058
      i. Phospho-Akt: 60 kDa
   f. Rabbit anti-S6 ribosomal protein monoclonal antibody: Cell Signaling, Cat. No. 2217
      i. S6 ribosomal protein: 32 kDa
   g. Rabbit anti-phospho-S6 ribosomal protein (Ser-240/244) monoclonal antibody: Cell Signaling, Cat. No. 4838
      i. Phospho-S6 ribosomal protein: 32 kDa
   h. Rabbit anti-Actin polyclonal antibody: Santa Cruz Biotech, Cat. No. sc-10731
      i. Actin: 45 kDa

9. The membranes are washed with TBS-T 3 times for 10 min.
10. Then, the membranes are incubated with the appropriate secondary antibodies conjugated to horseradish peroxidase for 45 min at room temperature.
11. Immunoblots are developed by using the chemiluminescence detection system (PerkinElmer) according to the manufacturer's protocol and autoradiography.

2.7. Quantification of autophagic flux

To distinguish between autophagic induction and autophagic flux, we assay for the levels of LC3-II by Western analysis following treatment with either 3-MA or lysosomal protease inhibitors.

1. H460 cells are seeded in 6-well tissue culture plates at a density of 2×10^5cells.
2. Sixteen h later, the cells are treated with Z-DEVD (50 μM for 24 h).
3. Then 23 h later, if indicated, the cells are treated with RAD001 (10 nM for 1 h).
4. One h later, the cells are irradiated at 5 Gy, and then treated with either 200 μM 3-MA or a combination of pepstatin A (10 μg/ml) and E64d (10 μg/ml) for a period of 2 h.
5. Cells are then collected at 2 time points (24 and 48 h), and washed with ice-cold PBS twice before lysis and protein concentration determination (as described previously).
6. The lysate is prepared as described previously, and then probed with anti-LC3 antibody in milk for a period of 1 h.
7. The blots are washed 3 times with TBS-T for 10 min.
8. Then the blots are reprobed with antiactin antibody for standardization of LC3-II levels to actin.

In addition, the LC3-II/LC3-I ratio can be determined.

2.8. Analysis of apoptosis

There are multiple methods to detect and quantify apoptosis which include, but are not limited to, end labeling of DNA, detection of phosphatidylserine changes, DNA laddering, and Western analysis of apoptosis signaling proteins. In our studies, we find that utilizing flow cytometric analysis of phosphatidylserine changes provides reliable and reproducible quantitative analysis of early apoptosis change in response to our various cytotoxic conditions.

1. H460 cells (2.5×10^5) are plated into 10-mm dishes for each data point.
2. After 24 h of incubation at 37 °C, the cells are treated with drug and immediately irradiated with 3, 5, 10, or 20 Gy.

3. Twenty-four h later, the cells are treated with 1 ml of the cell detachment medium Accutase (Millipore, Cat. No. SCR005) for 4 min and then cell counts are recorded for each sample.
4. Cells are centrifuged at 2000 rpm at 4 °C for 15 min, and resuspended in 1X Annexin V Binding Buffer (BD Biosciences, Cat. No. 51-66121E) at a concentration of 1×10^6cells/ml.
5. One hundred microliters of each solution (1×10^5cells) is transferred into 12×75-mm ml FACS tubes (BD Biosciences, Cat. No. 340265) to which is added 1.2 μl of Annexin V-FITC (BD Biosciences, Cat. No. 556570) and 1.2 μl of propidium iodide.
6. After 30 min incubation at room temperature in the dark, 400 μl of 1x binding buffer is added to each tube.
7. The rate of apoptosis (as determined by external membrane translocation of phosphatidylserine) is measured using the Annexin V-Fluorescein Isothiocyanate Apoptosis Detection Kit II (BD Biosciences, Cat. No. 556570) with flow cytometry per the manufacturer's protocols.

2.9. Gene knockdown utilizing siRNA transfection

Genetic manipulation utilizing siRNA can be a relatively simple way to associate a given agent to an effect. The limitations of these experiments are similar to those discussed with transient versus stable transfections of other proteins. However in the study of autophagy, where causal relationships are difficult to establish, determining the mediators of a given effect is particularly important.

1. We use siRNAs against various proteins:
   a. siRNA mouse caspase-3; Santa Cruz Biotechnology, Cat. No. sc-29927
   b. siRNA mouse caspase-7; Santa Cruz Biotechnology, Cat. No. sc-29928
   c. siRNA mouse Beclin 1; Santa Cruz Biotechnology, Cat. No. sc-29798
   d. siRNA mouse ATG5
   e. 5′-AACUUGCUUUACUCUCUCAUCAUU-39 (Sense)
   f. 3′-UUUUGAACGAAAUGAGAGAUAGU-59 (Antisense)
   g. Control siRNA; Santa Cruz Biotechnology, Cat. No. sc-37007
2. Cells are transfected with 25 n*M* of siRNAs using Lipofectamine 2000 (Invitrogen Life Technologies, Cat. No. 12566-014).
3. The transfected cells are used for experiments 24 h later.

2.10. Human lung cancer xenograft

1. Human H460 lung cancer cells are used in a xenograft model in female athymic nude mice (nu/nu), 5–6 weeks old.
2. A suspension of 1×10^6cells in 50 μL volume is injected subcutaneously into the left posterior flank of mice using a 27½-gauge needle.

3. Tumors are grown for 6–8 days until the average tumor volume reaches $0.25\ cm^3$.
4. Treatment groups consist of vehicle control (DMSO), drug, vehicle plus radiation, and drug plus radiation. Each treatment group contains 5 mice.
5. DMSO or drug is given daily by intraperitoneal (i.p.) injection at doses of 2 mg/kg for 7 consecutive days.
6. In the case of combination treatment, drug or vehicle is given for 2 days prior to the first dose of irradiation.
7. Mice in radiation groups are irradiated 1 h after drug or vehicle treatment with daily 2 Gy fractions given over 5 consecutive days.
8. Tumors on the flanks of the mice are irradiated using an X-ray irradiator (Therapax, AGFA NDT).
9. The non-tumor-bearing parts of the mice are shielded by lead blocks.
10. Tumors are measured 2–3 times weekly in 3 perpendicular dimensions using a Vernier caliper and the volume is calculated using the modified ellipse volume formula (volume = (height × width × depth)/2).
11. Growth delay is calculated for treatment groups relative to control tumors.

2.11. Immunohistochemistry

1. Mice are implanted with H460 cells and treated as described previously in the tumor volume studies.
2. After 7 days of daily treatments, the mice are sacrificed and tumors are paraffin fixed.
3. Slides from each treatment group are then stained for von Willebrand factor (vWF) using anti-vWF polyclonal antibody (Millipore, Cat No. AB7356).
4. Blood vessels are quantified by randomly selecting 400X fields and counting the number of blood vessels per field.
5. This is done in triplicate and the average of the three counts is calculated. Ki67 and terminal deoxynucleotidyl transferase(TdT)-mediated dUTP nick end labeling (TUNEL) staining are performed in the Vanderbilt University pathology core laboratory according to the following protocols.

2.11.1. Protocol: TUNEL staining

2.11.1.1. Solutions

TdT Buffer Stock Solution:

Tris-HCl (MW: 157.6) .... 1.97 g
Sodium cacodylate, trihydrate (MW: 214.0) .... 21.4 g
Bovine serum albumin (BSA) .... 0.125 g

Distilled water .... 100 ml
Adjust pH to 6.6, and store aliquots at −20 °C.
Cobalt Chloride Stock Solution:
Cobalt chloride, hexahydrate (MW: 237.9) .... 0.6 g
Distilled water .... 100 ml
Mix to dissolve, and store aliquots at −20 °C.
TdT Reaction Buffer:
TdT buffer stock solution. . . . . 40 μl
Cobalt chloride stock solution.. 8 μl
Distilled water .... 160 μl
Mix well. Store at −20 °C.
TdT Storage Buffer:
Potassium phosphate (K_2HPO_4; MW: 174.18) .... 1.05 g
KCl (FW: 74.55) .... 1.12 g
Distilled water .... 50 ml

Stir to dissolve and adjust pH to 7.2 using concentrated HCl. Add 50 ml of glycerin (100% glycerol), 0.5 ml of Triton X-100, and 8 μl of 2-mercaptoethanol (99% solution. FW: 78.13). Store at −20 °C.

2.11.1.2. Reagents

PBS:
Sodium phosphate, dibasic (Na_2HPO_4) .... 1.44 g
Sodium chloride .... 8 g
Potassium phosphate (KH_2PO_4) .... 0.24 g
Potassium chloride .... 0.2 g
Distilled water .... 800 ml
Adjust pH to 7.4, and adjust to a final volume of 1 liter with additional distilled water.
PBS-T (PBS/Tween Solution):
0.1% volume of Tween 20 prepared in PBS
Triton X-100 (octylphenolpoly(ethyleneglycolether):
Roche Diagnostics, Cat. No. 11332481001.
Enzyme Reagent:
TdT (Roche Diag., Cat. No. 03333574001) .... 4 μl
TdT storage buffer.. 100 μl
Mix well and store at −20 °C.
Label Reagent:
Biotin-16-dUTP (Roche Diag., Cat. No. 11093070910) .... 4 μl
TdT reaction buffer .... 1 ml
Mix well and store at −20 °C.
TdT Reaction Mixture:
Enzyme reagent .... 100 μl
Label reagent .... 900 μl

Mix just before use. Use the remaining 100 μl of label solution as a negative control.

Stop Wash Buffer:

NaCl (MW: 58.44)........ 1.75 g
Sodium citrate, trihydrate (MW: 294.11)........ 0.88 g
Distilled water........ 100 ml

Mix to dissolve and store at room temperature.

2.11.1.3. Protocol

1. Deparaffinize sections in 2 changes of xylene for 5 min each in labeled 2.0 ml microcentrifuge tubes
2. Hydrate with 2 changes of 100% ethanol for 3 min each, followed by 95% ethanol for 1 min.
3. Rinse in distilled water.
4. Add 800 μl of lysis buffer, and add 9 μl of proteinase K (20 mg/ml; Promega, Cat. No. V302B).
5. Vortex samples for 15 s, and incubate at 55 °C until tissue is completely lysed (may need to be overnight). Vortex samples occasionally.
6. Add 180 μL of 5 *M* NaCl and vortex well. The solution will become frothy.
7. Spin tubes at 13,000 rpm for 5 min, and the salted out debris will pellet.
8. Transfer the supernatant fractions to cryotubes (screw cap).
9. Add 420 μL of ice-cold isopropanol (2-propanol) to the supernatant fractions.
10. Mix slowly by inversion 5–10 times. CAUTION: Do not vortex.
11. DNA fibers may be seen at this time.
12. Centrifuge the tube at 13,000 rpm for 10 min.
13. The DNA pellet should be visible
14. Pour out the supernatant to discard.
15. Add 400 μL of 70% ethanol to wash the DNA pellet.
16. Wash for 20 min on a cell rotator at room temp.
17. Centrifuge the tubes at 13,000 rpm for 5 min and pour out the ethanol carefully! Note that the pellet may be loose. If the pellet is loose, pipette the ethanol out, being careful to not disturb the pellet.
18. Dry the DNA pellet in a speed vac on high for 10 min.
19. Resuspend the pellet in distilled H_2O.
    a. If a small pellet add approximately 50 μL
    b. If a large pellet add approximately 100 μL
20. Let the tubes stand at room temp overnight.
21. Perform 2 washes of 2 min each with PBS-T.
22. Incubate the sections in TdT Reaction Buffer for 10 min.
23. Incubate the sections in TdT Reaction Mixture for 1–2 h at 37 °C in a humidified chamber.

24. Rinse the sections in stop wash buffer for 10 min.
25. Perform 3 washes of 2 min each with PBS-T.
26. Counterstain with propidium iodide or DAPI for 20 min.
27. Rinse in PBS for 5 min.
28. Mount the sections with antifading mounting medium.

The number of positive cells per field are scored and graphed by averaging 3 repeated assessments.

3. Conclusion

Techniques for the analysis of autophagy have been rapidly evolving over the past decade. As a result, there has been a concerted effort by the leaders in this field to maintain strict guidelines regarding the interpretation and appropriate methods for analysis of autophagy (Klionsky *et al.*, 2008). Though there have been relatively few studies into autophagy and its role in lung cancer, we hope that this chapter will serve as the basis for further investigations. As these techniques continue to evolve, it will be important to continue investigating this process to develop optimal therapeutic combinations for the improvement of therapeutic ratio in lung cancer.

References

Abend, M. (2003). Reasons to reconsider the significance of apoptosis for cancer therapy. *Int. J. Radiat. Biol.* **79,** 927–941.

Albert, J. M., Cao, C., Kim, K. W., Willey, C. D., Geng, L., Xiao, D., Wang, H., Sandler, A., Johnson, D. H., Colevas, A. D., Low, J., Rothenberg, M. L., *et al.* (2007). Inhibition of poly(ADP-ribose) polymerase enhances cell death and improves tumor growth delay in irradiated lung cancer models. *Clin. Cancer Res.* **13,** 3033–3042.

Albert, J. M., Kim, K. W., Cao, C., and Lu, B. (2006). Targeting the Akt/mammalian target of rapamycin pathway for radiosensitization of breast cancer. *Mol. Cancer Ther.* **5,** 1183–1189.

Baehrecke, E. H. (2002). How death shapes life during development. *Nat. Rev. Mol. Cell Biol.* **3,** 779–787.

Boya, P., Gonzalez-Polo, R. A., Casares, N., Perfettini, J. L., Dessen, P., Larochette, N., Metivier, D., Meley, D., Souquere, S., Yoshimori, T., Pierron, G., Codogno, P., *et al.* (2005). Inhibition of macroautophagy triggers apoptosis. *Mol. Cell Biol.* **25,** 1025–1040.

Bursch, W., Ellinger, A., Kienzl, H., Torok, L., Pandey, S., Sikorska, M., Walker, R., and Hermann, R. S. (1996). Active cell death induced by the anti-estrogens tamoxifen and ICI 164 384 in human mammary carcinoma cells (MCF-7) in culture: The role of autophagy. *Carcinogenesis* **17,** 1595–1607.

Bursch, W., Hochegger, K., Torok, L., Marian, B., Ellinger, A., and Hermann, R. S. (2000). Autophagic and apoptotic types of programmed cell death exhibit different fates of cytoskeletal filaments. *J. Cell Sci.* **113**(Pt7), 1189–1198.

Canuto, R. A., Tessitore, L., Muzio, G., Autelli, R., and Baccino, F. M. (1993). Tissue protein turnover during liver carcinogenesis. *Carcinogenesis* **14,** 2581–2587.

Cao, C., Shinohara, E. T., Niermann, K. J., Donnelly, E. F., Chen, X., Hallahan, D. E., and Lu, B. (2005). Murine double minute 2 as a therapeutic target for radiation sensitization of lung cancer. *Mol. Cancer Ther.* **4,** 1137–1145.

Cao, C., Subhawong, T., Albert, J. M., Kim, K. W., Geng, L., Sekhar, K. R., Gi, Y. J., and Lu, B. (2006). Inhibition of mammalian target of rapamycin or apoptotic pathway induces autophagy and radiosensitizes PTEN null prostate cancer cells. *Cancer Res.* **66,** 10040–10047.

Chau, Y. P., Lin, S. Y., Chen, J. H., and Tai, M. H. (2003). Endostatin induces autophagic cell death in EAhy926 human endothelial cells. *Histol. Histopathol.* **18,** 715–726.

Clarke, P. G. (1990). Developmental cell death: Morphological diversity and multiple mechanisms. Anat. *Embryol. (Berl.)* **181,** 195–213.

Daido, S., Kanzawa, T., Yamamoto, A., Takeuchi, H., Kondo, Y., and Kondo, S. (2004). Pivotal role of the cell death factor BNIP3 in ceramide-induced autophagic cell death in malignant glioma cells. *Cancer. Res.* **64,** 4286–4293.

Fesik, S. W. (2005). Promoting apoptosis as a strategy for cancer drug discovery. *Nat. Rev. Cancer* **5,** 876–885.

Folkman, J. (2007). Angiogenesis: An organizing principle for drug discovery? Nat. *Rev. Drug Discov.* **6,** 273–286.

Franken, N. A., Rodermond, H. M., Stap, J., Haveman, J., and van Bree, C. (2006). Clonogenic assay of cells in vitro. *Nat. Protoc.* **1,** 2315–2319.

Gonzalez-Polo, R. A., Boya, P., Pauleau, A. L., Jalil, A., Larochette, N., Souquere, S., Eskelinen, E. L., Pierron, G., Saftig, P., and Kroemer, G. (2005). The apoptosis/autophagy paradox: Autophagic vacuolization before apoptotic death. *J. Cell Sci.* **118,** 3091–3102.

Gozuacik, D., and Kimchi, A. (2004). Autophagy as a cell death and tumor suppressor mechanism. *Oncogene* **23,** 2891–2906.

Ito, H., Daido, S., Kanzawa, T., Kondo, S., and Kondo, Y. (2005). Radiation-induced autophagy is associated with LC3 and its inhibition sensitizes malignant glioma cells. *Int. J. Oncol.* **26,** 1401–1410.

Jaboin, J. J., Shinohara, E. T., Moretti, L., Yang, E. S., Kaminski, J. M., and Lu, B. (2007). The role of mTOR inhibition in augmenting radiation induced autophagy. *Technol. Cancer Res. Treat* **6,** 443–447.

Jemal, A., Siegel, R., Ward, E., Hao, Y., Xu, J., Murray, T., and Thun, M. J. (2008). Cancer statistics, 2008. *CA Cancer J. Clin.* **58,** 71–96.

Kamada, Y., Funakoshi, T., Shintani, T., Nagano, K., Ohsumi, M., and Ohsumi, Y. (2000). Tor-mediated induction of autophagy via an Apg1 protein kinase complex. *J. Cell Biol.* **150,** 1507–1513.

Kanzawa, T., Germano, I. M., Komata, T., Ito, H., Kondo, Y., and Kondo, S. (2004). Role of autophagy in temozolomide-induced cytotoxicity for malignant glioma cells. *Cell Death Differ.* **11,** 448–457.

Kanzawa, T., Kondo, Y., Ito, H., Kondo, S., and Germano, I. (2003). Induction of autophagic cell death in malignant glioma cells by arsenic trioxide. *Cancer Res.* **63,** 2103–2108.

Kanzawa, T., Zhang, L., Xiao, L., Germano, I. M., Kondo, Y., and Kondo, S. (2005). Arsenic trioxide induces autophagic cell death in malignant glioma cells by upregulation of mitochondrial cell death protein BNIP3. *Oncogene* **24,** 980–991.

Kim, K. W., Hwang, M., Moretti, L., Jaboin, J. J., Cha, Y. I., and Lu, B. (2008a). Autophagy upregulation by inhibitors of caspase-3 and mTOR enhances radiotherapy in a mouse model of lung cancer. *Autophagy* **4,** 659–668.

Kim, K. W., Moretti, L., and Lu, B. (2008b). M867, a novel selective inhibitor of caspase-3 enhances cell death and extends tumor growth delay in irradiated lung cancer models. *PLoS ONE* **3,** e2275.

Kim, K. W., Mutter, R. W., Cao, C., Albert, J. M., Freeman, M., Hallahan, D. E., and Lu, B. (2006). Autophagy for cancer therapy through inhibition of proapoptotic proteins and mTOR signaling. *J. Biol. Chem.* **3,** 142–144.

Kirkegaard, K., Taylor, M. P., and Jackson, W. T. (2004). Cellular autophagy: Surrender, avoidance and subversion by microorganisms. *Nat. Rev. Microbiol.* **2,** 301–314.

Klionsky, D. J., Abeliovich, H., Agostinis, P., Agrawal, D. K., Aliev, G., Askew, D. S., Baba, M., Baehrecke, E. H., Bahr, B. A., and Ballabio, B. A. (2008). Guidelines for the use and interpretation of assays for monitoring autophagy in higher eukaryotes. *Autophagy* **4,** 151–175.

Lefranc, F., Facchini, V., and Kiss, R. (2007). Proautophagic drugs: a novel means to combat apoptosis-resistant cancers, with a special emphasis on glioblastomas. *Oncologist* **12,** 1395–1403.

Levine, B., and Klionsky, D. J. (2004). Development by self-digestion: molecular mechanisms and biological functions of autophagy. *Dev. Cell* **6,** 463–477.

Liang, X. H., Jackson, S., Seaman, M., Brown, K., Kempkes, B., Hibshoosh, H., and Levine, B. (1999). Induction of autophagy and inhibition of tumorigenesis by beclin 1. *Nature* **402,** 672–676.

Lum, J., Deberardinis, R. J., and Thompson, C. B. (2005). Autophagy in metazoans: Cell survival in the land of plenty. *Nat. Rev. Mol. Cell Biol.* **6,** 439–448.

Melet, A., Song, K., Bucur, O., Jagani, Z., Grassian, A. R., and Khosravi-Far, R. (2008). Apoptotic pathways in tumor progression and therapy. *Adv. Exp. Med. Biol.* **615,** 47–79.

Mizushima, N., Yamamoto, A., Matsui, M., Yoshimori, T., and Ohsumi, Y. (2004). *In vivo* analysis of autophagy in response to nutrient starvation using transgenic mice expressing a fluorescent autophagosome marker. *Mol. Biol. Cell* **15,** 1101–1111.

Mizushima, N., and Yoshimori, T. (2007). How to interpret LC3 immunoblotting. *Autophagy* **3,** 542–545.

Moretti, L., Attia, A., Kim, K. W., and Lu, B. (2007). Crosstalk between Bak/Bax and mTOR signaling regulates radiation-induced autophagy. *Autophagy* **3,** 142–144.

Ogier-Denis, E., and Codogno, P. (2003). Autophagy: A barrier or an adaptive response to cancer. Biochim. *Biophys. Acta* **1603,** 113–128.

Opipari, A. W. Jr,, Tan, L., Boitano, A. E., Sorenson, D. R., Aurora, A., and Liu, J. R. (2004). Resveratrol-induced autophagocytosis in ovarian cancer cells. *Cancer Res.* **64,** 696–703.

Paglin, S., Hollister, T., Delohery, T., Hackett, N., McMahill, M., Sphicas, E., Domingo, D., and Yahalom, J. (2001). A novel response of cancer cells to radiation involves autophagy and formation of acidic vesicles. *Cancer Res.* **61,** 439–444.

Reggiori, F., Tucker, K. A., Stromhaug, P. E., and Klionsky, D. J. (2004a). The Atg1-Atg13 complex regulates Atg9 and Atg23 retrieval transport from the pre-autophagosomal structure. *Dev. Cell* **6,** 79–90.

Reggiori, F., Wang, C.-W, Nair, U., Shintani, T., Abeliovich, H., and Klionsky, D. J. (2004b). Early stages of the secretory pathway, but not endosomes, are required for Cvt vesicle and autophagosome assembly in Saccharomyces cerevisiae. *Mol. Biol. Cell* **15,** 2189–2204.

Rigas, J. R., and Kelly, K. (2007).). Current treatment paradigms for locally advanced non-small cell lung cancer. *J. Thorac. Oncol.*(suppl12)**2,** S77–S85.

Scarlatti, F., Bauvy, C., Ventruti, A., Sala, G., Cluzeaud, F., Vandewalle, A., Ghidoni, R., and Codogno, P. (2004). Ceramide-mediated macroautophagy involves inhibition of protein kinase B and up-regulation of beclin 1. *J. Biol. Chem.* **279,** 18384–18391.

Schwartz, L. M., Smith, S. W., Jones, M. E., and Osborne, B. A. (1993). Do all programmed cell deaths occur via apoptosis? Proc. *Natl. Acad. Sci. USA* **90,** 980–984.

Schwarze, P. E., and Seglen, P. O. (1985). Reduced autophagic activity, improved protein balance and enhanced *in vitro* survival of hepatocytes isolated from carcinogen-treated rats. *Exp. Cell Res.* **157,** 15–28.

Shao, Y., Gao, Z., Marks, P. A., and Jiang, X. (2004). Apoptotic and autophagic cell death induced by histone deacetylase inhibitors. *Proc. Natl. Acad. Sci. USA* **101,** 18030–18035.

Shimizu, S., Kanaseki, T., Mizushima, N., Mizuta, T., Arakawa-Kobayashi, S., Thompson, C. B., and Tsujimoto, Y. (2004). Role of Bcl-2 family proteins in a non-apoptotic programmed cell death dependent on autophagy genes. *Nat. Cell Biol.* **6,** 1221–1228.

Shintani, T., and Klionsky, DJ. (2004). Autophagy in health and disease: A double-edged sword. *Science* **306,** 990–995.

Takeuchi, H., Kondo, Y., Fujiwara, K., Kanzawa, T., Aoki, H., Mills, G. B., and Kondo, S. (2005). Synergistic augmentation of rapamycin-induced autophagy in malignant glioma cells by phosphatidylinositol 3-kinase/protein kinase B inhibitors. *Cancer Res.* **65,** 3336–3346.

Wu, Y. T., Tan, H. L., Huang, Q., Kim, Y. S., Pan, N., Ong, W. Y., Liu, Z. G., Ong, C. N., and Shen, H. M. (2008). Autophagy plays a protective role during zVAD-induced necrotic cell death. *Autophagy* **4,** 457–466.

Yao, K. C., Komata, T., Kondo, Y., Kanzawa, T., Kondo, S., and Germano, I. M. (2003). Molecular response of human glioblastoma multiforme cells to ionizing radiation: Cell cycle arrest, modulation of the expression of cyclin-dependent kinase inhibitors, and autophagy. *J. Neurosurg.* **98,** 378–384.

Yu, L., Ajjai Alva, Helen Su, Parmesh Dutt, Eric Freundt, Sarah Welsh, Eric H. Baehrecke, Michael J. Lenardo. (2004). Regulation of an *ATG7-beclin 1* program of autophagic cell death by caspase 8. *Science* **304,** 1500–1502.

CHAPTER FIFTEEN

Signal-Dependent Control of Autophagy-Related Gene Expression

Fulvio Chiacchiera* and Cristiano Simone*

Contents

Abstract

Several tumors arise from deregulated signaling pathways leading to increased proliferation and impairment of differentiation. To bypass endogenous control mechanisms and to survive the environmental stress associated with increased growth, tumor cells acquire a plethora of modifications that ultimately tend to

* Laboratory of Signal-Dependent Transcription, Department of Translational Pharmacology, Consorzio Mario Negri Sud, Santa Maria Imbaro (Chieti), Italy

Methods in Enzymology, Volume 453
ISSN 0076-6879, DOI: 10.1016/S0076-6879(08)04015-9

down-regulate the ability to undergo apoptosis and exacerbate prosurvival mechanisms. Autophagy is an evolutionarily conserved mechanism through which cells recycle essential molecular constituents or eliminate damaged organelles under stress conditions imposed by nutrients or growth factors deprivation. As such, autophagy acts as a prosurvival mechanism for cancer cells. However, when overactivated, autophagy could also represent a cell death mechanism acting through self-cannibalization. Therefore, understanding the various signaling pathways that regulate autophagy could be of extreme importance. Indeed, the identification of specific molecular targets amenable to pharmacological manipulation to induce cancer cell self-cannibalization could represent a promising approach to treat apoptosis-resistant tumors.

1. Introduction

Cancer essentially arises upon deregulation of molecular pathways that control genome stability, cell growth, and differentiation and cell death. The integrity of the signaling network that preserves genome integrity during the cell cycle constitutes a barrier to malignant transformation. Moreover, further protection is provided by additional, and often redundant, cell-death mechanisms, such as apoptosis and autophagy (Chiacchiera and Simone, 2008).

Autophagy is a ubiquitous evolutionarily conserved catabolic process present in eukaryotic cells that works to protect them against metabolic stresses and to eliminate defective cellular constituents. It represents a form of self-cannibalism in which cells break down their own components not only to eliminate damaged or supernumerary organelles, but also to adapt their metabolism to starvation imposed by deprivation of nutrients or growth factors. Once the cell receives the appropriate signals, autophagy-execution proteins trigger a cascade of reactions that result in membrane rearrangements to form vesicles called autophagosomes, which enwrap molecules and organelles and then fuse with lysosomes to digest their content. Cells can then recycle the resulting degradation products, using them to provide energy and molecules necessary for survival (Levine and Klionsky, 2004).

This process is strongly required for maintaining genome stability (Mathew *et al.*, 2007), but paradoxically it is also a way in which tumor cells are able to survive under stress conditions imposed by insufficient vascularization or chemotherapy (Degenhardt *et al.*, 2006). Indeed, impaired autophagy can be among the events that promote cancer formation, but once the transformed phenotype is acquired, autophagy can represent an essential survival pathway for cancer cells. Impairing both apoptosis and autophagy leaves the cells unable to sustain metabolic stresses leading to necrotic death, a process also known as metabolic catastrophe

(Jin *et al.*, 2007). This in turn can be an advantage for surrounding tumor cells that receive cytokines and growth factors typically released during necrotic processes.

Many signaling pathways converge to regulate autophagy. The kinase TOR (target of rapamycin) plays a crucial role in sensing growth factors levels and negatively regulates autophagy induction (Sabatini, 2006). The LKB1/AMPK pathway also regulates autophagy induction (Meley *et al.*, 2006) through cross talk with the mTOR kinase. In fact, it has been shown that the energy sensor AMP-activated protein kinase (AMPK) inhibits the TORC1 complex by activating Tsc2 in response to high AMP/ATP ratios, thus promoting autophagy induction (Brugarolas *et al.*, 2004). The activation of Raf-1/MEK/ERK1/2 by a Ras mutant also promotes autophagy induction (Pattingre *et al.*, 2003). Even though a major role for the PI3K/AKT (growth factor sensor) and the LKB1/AMPK (energy sensor) pathways in autophagy regulation has been well characterized, many other signaling pathways modulating this important process still remain uncovered.

Therefore, the fine dissection of the signaling cascades that regulate autophagy is of crucial importance to shed light on its role in cell homeostasis and cancer.

2. Overview: Signal Transduction and Chromatin-Associated Kinases

As the activation of specific signaling pathways determines a specific cellular response, mainly dictated by characteristic gene expression profiles, it is extremely important to understand how external stimuli are transduced to mediate chromatin modification and gene expression. Indeed, cell fate is not determined by the activation of a single signaling pathway but is the result of the activation of a specific combination of many different signaling pathways. These in turn converge to activate or repress the activity of a defined number of chromatin binding proteins that ultimately dictate the transcription response (Simone, 2006).

In recent years, several already known kinases have been shown to possess a chromatin-associated kinase activity. For example, p38α/β associates with chromatin and phosphorylates the SWI/SNF complex, inducing chromatin remodeling and transcription of muscle related genes (Simone *et al.*, 2004). Upon estradiol treatment IKKα rapidly associates with the cyclin D1 promoter and is able to phosphorylate ERα and its co-activator AIB1/SRC-3 on residues critical for their activity, thus promoting estrogen-regulated gene transcription (Park *et al.*, 2005). The ATM kinase is activated upon DNA damage and associates with sites of double-strand breaks and the Aurora kinase binds to mitotic chromatin (Chow and Davis, 2006). Hence, this class of proteins represents a promising pharmacological target as it directly regulates gene expression by phosphorylating multi-protein transcriptional complex subunits bound to DNA. As a consequence, the effect of chromatin-associated kinases inhibition is limited to the promoters on which they are recruited without spreading to other signalling pathways.

3. The p38 Pathway in Colorectal Cancer Cells

The p38 MAP kinases are a family of serine/threonine protein kinases involved in stress response, cell growth, differentiation, and apoptosis. The p38α MAPK was initially identified as a 38-kDa polypeptide undergoing tyrosine phosphorylation in response to endotoxin treatment and osmotic shock. In parallel, other groups independently identified p38α as a target for pyridinyl imidazole anti-inflammatory drugs and as a stress-activated kinase. A few years after p38α identification, three other isoforms were described, p38β, p38γ, and p38δ (Raman *et al.*, 2007). p38α can be considered the prototype of chromatin-associated kinases. It regulates gene expression acting at different levels and modulates both transcription and translation. It is able to associate with and phosphorylate transcription factors and subunits of the SWI/SNF ATP-dependent remodeling complexes directly on DNA, thereby modulating chromatin structure and transcription. It is involved in the elongation step of transcription by associating with the RNA polymerase II and its inhibition affects the proper assembly of the cdk9/cyclin T2 complex. Last, it acts at the posttranscriptional level, being required for KSRP-mediated transcript stabilization (Chiacchiera and Simone, 2008). The fundamental role of p38 in chromatin remodeling and transcription is also highlighted by its evolutionary conservation. In yeast, the p38 homolog Hog1 is recruited to DNA regulatory regions of target genes by its physical interaction with transcription factors and can convert repressor complexes into activators through the engagement of histone-modifying enzymes together with SWI/SNF.

We have recently shown that p38α is required for colorectal cancer cell (CRC) proliferation and survival. Inhibition (using SB202190) or knock-down (using specific siRNAs) of p38α induce cyclin A2 and cyclin E transcriptional repression and up-regulation of GABARAP, one of the human homologs of the yeast *ATG8* gene, which codes for a member of a novel ubiquitin-like protein family playing an essential role in the activation of the autophagic machinery. Deficiency of p38α activity induces a perturbation in the expression profile of a subset of genes that ultimately leads to cell cycle arrest, autophagy and cell death in a cell type–specific fashion. In these cells, a complex network of intracellular kinase cascades controls autophagy and survival because the effect of p38α blockade is differentially affected by the pharmacological inhibition of MEK1 and PI3K class I and III. Our results show that the autophagic response to p38α blockade initially represents a survival pathway, whereas prolonged inactivation of the kinase leads to cell death. Moreover, inhibition of the autophagic process triggered by p38α inactivation promotes an apoptotic response. It is conceivable that p38α is involved in the management of cellular stress and/or in the regulation of essential metabolic cascades in colorectal cancer cells. In accordance with this hypothesis, early reactivation of p38α induces a significant time-dependent reduction in the autophagic process with a slow reentry into the cell cycle (Comes *et al.*, 2007).

As several tumors are resistant to apoptosis induction but require autophagy to survive environmental stress, dissection of the various signalling pathways that regulate this process is of extreme importance in order to identify suitable pharmacological targets making it possible to turn a pro-survival mechanism into a self-cannibalization process.

Here, we describe several commonly used techniques useful to dissect kinase-dependent signaling pathways, from the cytoplasm down to the nucleus.

4. Methods to Test Kinase Activity

Several extracellular signals are transduced through the activation of a phosphorylation cascade. A series of kinases are activated upon phosphorylation on specific residues to phosphorylate and activate their own targets. Therefore, the easiest way to test the involvement of a certain signaling pathway in a specific biological process is to test the activation of the kinase involved or its activity.

4.1. Immunoblot using phospho-specific antibodies

This approach allows the monitoring of the activation of a specific kinase by the upstream signaling pathway. The availability of a good phospho-specific antibody is of course a crucial issue together with the parallel use of an antibody raised against the nonphosphorylated form of the kinase to compare the amount of the activated fraction versus the total amount of the protein.

4.1.1. Procedure

1. Collect growth medium, wash cells twice with PBS, and collect wash buffer.
2. Collect cells by scraping or trypsin digestion.

3. Lyse cells, a 75% confluent 60-mm dish is sufficient, in 200-μl pre-chilled RIPA (150 m*M* NaCl, 50 m*M* Tris-HCl, pH 8, 1 m*M* EDTA, 1% NP40, 0.5% sodium deoxycholate, 0.1% SDS) supplemented with protease (1 m*M* PMSF; 1.5 μM pepstatin A; 2 μM leupeptin; 10 μg/ml aprotinin) and phosphatase inhibitors (5 m*M* NaF; 1 m*M* Na_3VO_4) at 4 °C for 30 min.
4. Centrifuge at maximum speed in a refrigerated centrifuge, transfer the supernatant fractions to a new tube and measure the protein concentration.
5. Add the proper amount of 2x Laemmli Sample buffer (Sigma cat. no. S3401) and denature at 90–100 °C for 5 min.
6. Load an SDS-PAGE gel and perform a Western blot.
7. Incubate the blot membranes for 1 h in TBS-T-Milk (TBS [50 m*M* Tris-HCl, pH 7.5, 150 m*M* NaCl]; 0.1% Tween 20; 5% nonfat dry milk) (add NaF to a final concentration of 5 m*M*).
8. Wash twice 5 min each in TBS-T-BSA (TBS-T; 5% BSA) (add NaF to a final concentration of 5 m*M*).
9. Add the primary antibody overnight in TBS-T-BSA (add NaF to a final concentration of 5 m*M*), at 4 °C with agitation.
10. Perform 4 washes of 5 min each in TBS-T-milk with agitation.
11. Add the proper horseradish peroxidase–conjugated secondary antibody, and incubate for 30 min at room temperature in TBS-T-milk with agitation.
12. Perform 4 washes of 5 min each in TBS-T-milk with agitation.
13. Perform 3 rapid washes with TBS with agitation.
14. Develop with chemoluminescent or chromogenic substrates as preferred.

We successfully used the following subsequent phospho-specific antibodies: p-p38 (Thr180/Tyr182) [Cell signalling #9211]; p-MAPKAPK2 (Thr334) [Cell signalling #3007]; p-JNK (Thr183/Tyr185) [Santa Cruz Biotechnology # SC-6254]; p-AKT (ser 473) [Cell signalling #9271]; p-AMPKα (Thr172) [Cell signalling #2531]; p-p70S6K (Ser371) [Cell signalling #9208].

4.2. Kinase assay

This approach is required to study the activity of the kinase of interest and can be performed using both a radioactive and nonradioactive (using a phospho-specific antibody) method, although the radioactive approach is much more quantitative and can be performed independently from the availability of other reagents. Here, we describe the method to perform a radioactive kinase assay.

4.2.1. Procedure

1. Collect cells by scraping or trypsin digestion, and neutralize trypsin with a small amount of serum.
2. Centrifuge cells 5 min at 1000 rpm and wash twice with cold PBS.
3. Lyse cells, a 75% confluent 100-mm dish is sufficient, in a prechilled lysis buffer: 400 μl RIPA buffer supplemented with protease (1 m*M* PMSF, 1.5 μM pepstatin A, 2 μM leupeptin, 10 g/ml aprotinin) and phosphatase (5 m*M* NaF, 1 m*M* Na_3VO_4) inhibitors 30 min at 4 °C.
4. Centrifuge at maximum speed in a refrigerated centrifuge, transfer the supernatant fractions to a new tube, and quantify the protein concentration.
5. Split the lysate into two parts and add at least 1 μg of specific antibody or control IgG. Incubate 1–2 h at 4 °C with rocking.
6. Pull down the immunocomplex using protein A (or protein G)-sepharose beads by incubating for 45 min at 4 °C with rocking, and then pelleting the beads by centrifugation for 2 min, at 2000 rpm.
7. Wash the resin 3 times with lysis buffer and twice with lysis buffer containing 400 m*M* NaCl.
8. Equilibrate the complex in kinase assay buffer (minus ATP) (20 m*M* HEPES, pH 7.4, 10 m*M* Mg acetate).
9. Add 0.2 μg of purified target protein (in the case of GST fusion proteins include also a negative control with GST alone) together with 5 μCi/sample of γ-ATP in a final volume of 20 μl. Incubate 30 min at 30 °C.
10. Stop the reaction by adding Laemmli sample buffer.
11. Run an SDS-PAGE gel, dry, and expose the gel.
12. If necessary, excise the bands and measure radioactivity by scintillation counting.

5. Profiling Gene Expression Pattern

Recent works points out the transcriptional control of autophagy-related genes. Several genes are demonstrated to be up-regulated during autophagy in a cell type–specific and signal-dependent manner, and this phenomenon is evolutionarily conserved from yeasts to humans. To study the involvement of subsets of autophagy-related genes, alone or in combination with other functional clusters of genes, as for cell death genes in the autophagic cell death process or programmed type II cell death, we propose two different approaches, one of which is biased, while the other is not. The first method is a gene candidate approach. Researchers should search the literature for genes involved in or related to autophagy and validate their expression by quantitative real-time PCR. The second method relies on large-scale screenings employing microarray techniques. Several companies offer the opportunity to create custom microarrays; alternatively, precast 96-well plates for Real-Time PCR with different sets of validated primers are also commercially available and represent a good option.

6. Transcriptional Control of ATG Genes

The ultimate step in the activation of a signaling cascade resides in the modulation of gene expression. This is principally achieved by modulating the ability of defined transcription factors to transactivate gene promoters. Mechanistically, a change in the protein-protein interaction pattern is observed, which determines the ability to recruit transcriptional cofactors or chromatin remodelers; or to engage or disengage transcriptional repressors. It is of crucial importance to understand which transcription factors are the ultimate target of a defined signaling cascade on a defined set of gene promoters and what kind of chromatin modifications are involved in the binding to their specific responsive elements. These multiprotein complexes are indeed the best pharmacological targets to modulate gene expression patterns.

Induction of autophagy is sustained by an increase in the mRNA levels of certain autophagy-related genes. This phenomenon is evolutionarily conserved from yeasts to humans. In yeast, Atg8 is up-regulated by rapamycin-dependent Tor inhibition and nitrogen starvation (Abeliovich *et al.*, 2000). Moreover, large modifications in autophagy gene expression have been described to occur just prior to *Drosophila melanogaster* salivary gland cell death. The observed increase in autophagy correlates with Atg2, Atg4, Atg5, and Atg7 upregulation, whereas no significant changes in Atg8a or Atg8b are detected (Gorski *et al.*, 2003; Lee *et al.*, 2003). However, upregulation of *D. melanogaster* Atg8a and Atg8b is described in fat bodies following induction of autophagy at the end of larval development, (Juhasz *et al.*, 2007) and an increase in Atg8b is detected in cultured *D. melanogaster l(2)mbn* cells following starvation (Klionsky *et al.*, 2008). With regard to mammalian cells, MAP1LC3, one of the Atg8 homologs, and Atg12 are transcriptionally induced by different stimuli in human HeLa cells and murine P19 cells, respectively (Kouroku *et al.*, 2007; Nara *et al.*, 2002). Furthermore, several autophagy-related genes are induced during murine muscle cells atrophy, including the Atg8 family members MAP1LC3 and Gabarapl1, Atg4b, Atg12, Atg6/Beclin 1, BNIP3, and BNIP3L (Mammucari *et al.*, 2007). In human HT29 cells, the expression levels of the GABARAP gene, another Atg8 homologue, are significantly augmented by p38α blockade and correlate with autophagic vacuoles and autophagosomes formation, whereas rapamycin fails to induce both autophagy and GABARAP up-regulation in these cells (Comes *et al.*, 2007). Interestingly, 3MA fails to impair the SB-mediated induction of GABARAP expression (Comes *et al.*, 2007). This is noteworthy because 3MA prevents autophagy at an early stage by inhibiting PI3K class III, which promotes the nucleation step of the autophagic process. Our data suggest that the p38α control of GABARAP transcription is independent from this first step.

6.1. Promoter analysis

The FoxO3a transcription factor is able to bind specific DNA cognate sequences (5′TTGTTTAC3′) on autophagy-related genes (Mammucari *et al.*, 2007). Again, this mechanism seems to be evolutionarily conserved, since FoxO overexpression induces autophagy also in Drosophila (Juhasz *et al.*, 2007). These studies highlight the importance of the study of DNA regulatory regions (promoters and enhancers) to identify other transcription factors and co-factors able to modulate the expression of autophagy-related genes. In particular, when a list of modulated genes is obtained, several bioinformatic tools are freely available online that can be used to predict potential binding sites for transcription factors. We use a software program called CONFAC (Karanam and Moreno, 2004) available at http://morenolab.whitehead.emory.edu/cgi-bin/confac.pl?id=1. This software is able to easily and simultaneously analyze the promoters of different genes experimentally known to be up- or down-regulated, in the search for transcription factors and cofactors binding sequences (position weight matrices present in TRANSFAC 4.0 database) evolutionarily conserved between human and mouse. It has to be considered that, as stated from the authors, comparing human and mouse sequences reduces the background noise, but may sometimes also reduce sensitivity.

6.2. Luciferase reporter gene assay

A widely used system to study a specific promoter is the dual-reporter system. This reporter gene system is used to study the regulation of gene expression, by means of an experimental reporter gene (e.g., firefly luciferase), cloned downstream of a promoter of interest. The expression of a second reporter gene (e.g., renilla luciferase, or beta-galactosidase) driven by a constitutive promoter is used as internal control, to normalize the activity of the primary reporter and minimize the variability caused by differences in cell viability or transfection efficiency.

As a first step, the promoter region of interest has to be cloned in a proper vector upstream to a firefly luciferase coding sequence. This construct has to be transfected with the second reporter construct at least 24 h before the assay. Several commercial kits are available that allow efficient cell lysis and the reading of primary and secondary reporters using a luminometer at the same time. The luciferase assay is very sensitive, and therefore it is strongly recommended to plate cells as homogeneously as possible and take care when performing transfections. Another parameter to be considered is the integration time: to be sure that the assay is read during the logarithmic phase, far from the saturation limit, it is crucial to accurately plan when to perform it.

6.3. Chromatin immunoprecipitation (ChIP)

This technique allows a thorough analysis of promoter occupancy by transcription factors and cofactors and an accurate examination of the chromatin modifications induced by their binding.

The ChIP analysis is a powerful tool to finely dissect the multiprotein complexes that assemble on and regulate a defined promoter. Also, histone modifications can be finely mapped to describe the status of chromatin around the multiprotein complex assembled. Very briefly, chromatin-protein complexes are cross-linked using formaldehyde, and after fragmentation (by sonication or enzymatic digestion) the DNA-protein complexes are pulled down using specific antibodies. The reversal of the cross-linking and the amplification using specific primers of the DNA fragments pulled-down make it possible to map the binding of a specific protein on a defined promoter region (see Fig. 15.1). Use of antibodies against modified histones (acetylated, phosphorylated, or methylated histones) can also map the status of the chromatin around the same sequence, according to what is just called the histones code (Jenuwein and Allis, 2001).

6.3.1. Procedure

1. Add formaldehyde directly to cells in tissue culture medium to a final concentration of 1% (from 37% stock). We generally use 2×10^7 cells per antibody per time point.
2. Stop the cross-linking reaction by adding glycine to a final concentration of 0.125 *M* (2.5 *M* stock). Continue to rock at room temp for 5 min.
3. For adherent cells, pour off the medium, and rinse plates twice with cold PBS.
4. Scrape adherent cells from dishes using 1–4 ml of ice-cold PBS. If you have used trypsin to detach the cells from the dish, inactivate the trypsin by adding a small amount of serum. Centrifuge scraped adherent or suspension cells 5 min at 1000 rpm and wash the pellet fraction once with PBS plus protease inhibitors.
5. Resuspend the cell pellets in cell lysis buffer (10 m*M* Tris-HCl, pH 8.0, 10 m*M* NaCl, 0.2% NP40) plus 1 m*M* PMSF and 1 m*M* leupeptin. The final volume of cell lysis buffer should be sufficient to avoid clumps of cells (for a 150-mm dish of confluent cells use 250–500 μl). Incubate on ice for 10 min.

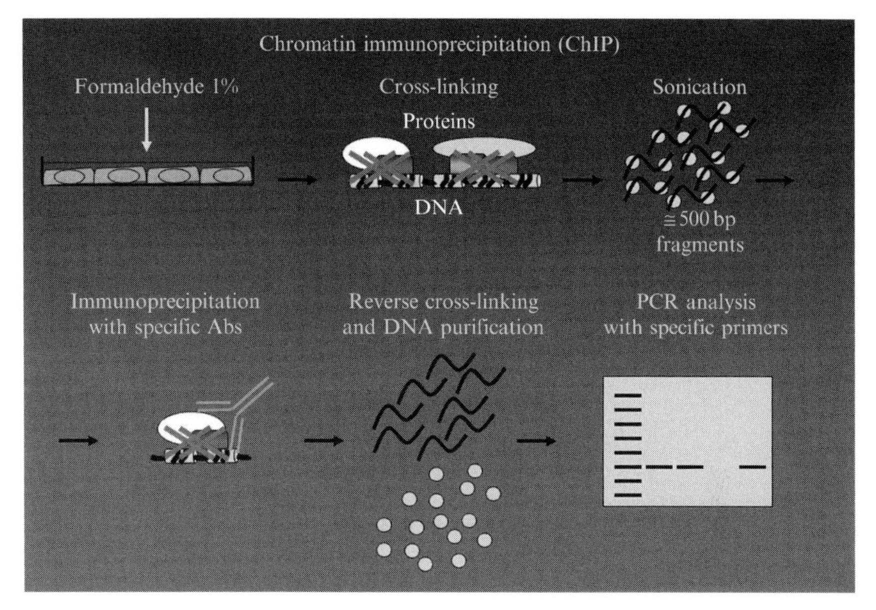

Figure 15.1 Scheme indicating the most important steps in the ChIP protocol. See text for details.

6. Centrifuge at 5000 rpm for 5 min at 4 °C to pellet the nuclei.
7. Resuspend nuclei in nuclei lysis buffer (50 m*M* Tris-HCl, pH 8.1, 10 m*M* EDTA, 1% SDS) plus protease inhibitor cocktail (500 μl). Try to resuspend the pellets of cells treated under the same conditions at the same time. Incubate on ice for 10 min.
8. Sonicate chromatin to an average length of approximately 500 bp while keeping the samples on ice (the time and number of pulses will vary depending on the sonicator, cell type and extent of cross-linking and must be determined empirically). Centrifuge at 14,000 rpm for 10 min at 4 °C.
9. Carefully remove the supernatant fraction and transfer to a new tube. At this point, the chromatin can be snap frozen in liquid nitrogen and stored at −70 °C for up to several months.
10. Preclear the chromatin by adding sepharose A or G beads. Use 40 μl of sepharose A or G beads.
11. Incubate by rocking at 4 °C for 3 h. Centrifuge at 14,000 rpm for 5 min.
12. Transfer the supernatant fractions to a clean tube and divide equally among your samples. Be sure to include a no-antibody sample as a control, and add the appropriate specific antibodies to the other sample. Set aside a volume equal to 1% of the input for later use as an input control. Adjust the final volume of each sample with IP dilution buffer

(0.01% SDS, 1.1% Trition X-100, 1.2 m*M* EDTA, 16.7 m*M* Tris-HCl, pH 8.1, 167 m*M* NaCl) plus protease inhibitors cocktail. The sample volumes should be 500 μl. To increase specificity add NaCl to 150 m*M* final concentration. Incubate with specific antibody at 4 °C overnight.

13. Add 40 μl of sepharose A or G precleared beads to each sample. Incubate by rocking at 4 °C for 1–2 h.
14. Centrifuge samples at no more than 2000 rpm for 5 min and discard the supernatant fractions.
15. Wash the pellet fractions sequentially with:
    a. 1 ml of wash buffer 1 (0.1% SDS, 1% Triton X-100, 2 m*M* EDTA, 20 m*M* Tris, ph 8.1, 150 m*M* NaCl),
    b. twice with 1 ml of wash buffer 2 (0.1% SDS, 1% Triton X-100, 2 m*M* EDTA, 20 m*M* Tris, pH 8.1, 500 m*M* NaCl),
    c. 4 times with wash buffer 1,
    d. once with 1 ml of wash buffer 3 (0.25 M LiCl, 1% NP-40, 1% deoxycholate, 1 m*M* EDTA, 10 m*M* Tris, pH 8.1),
    e. once with TE 1X (Sigma T9285).

For each wash, incubate samples on a rotating platform for 5 min at 4 °C and then centrifuge at 2000 rpm for 5 min at room temperature. Try to remove as much buffer as possible after each wash without aspirating the sepharose beads.

16. After the last wash, centrifuge and remove the last traces of buffer. Elute antibody/protein/DNA complexes by adding 100 μl of IP elution buffer 1 (1% SDS, 1 m*M* EDTA, 10 m*M* Tris-HCl, pH 8.1). Incubate at 65 °C for at least 15 min. Centrifuge at 14,000 rpm for 3 min. Transfer the supernatant fractions to clean tubes. Add 150 μl of elution buffer 2 (0.67% SDS, 1 m*M* EDTA, 10 m*M* Tris-HCl, pH 8.1) to the pellet fractions, incubate at 65 °C for at least 15 min. and finally combine both elutions in the same tube. For the input fractions from step 12 follow the procedure from step 10, add 20 μl of 10% SDS and enough TE to make a final volume of 250 μl.
17. Incubate the samples in a 65 °C water bath for 4–5 h (or overnight) to reverse the formaldehyde cross-links.
18. Centrifuge the samples at 14,000 rpm for 1 min.
19. Add to each sample 250 μl of TE and 10 μl of 10 mg/ml proteinase K. Incubate in a 42 °C waterbath for 1–2 h.
20. Add 55 μl of 4 M LiCl to each sample. Extract with 500 μl of phenol/chloroform.
21. Add 1 μl of glycogen to each sample. Mix well then add 900 μl of ethanol. Precipitate in a –20 °C freezer overnight.
22. Centrifuge the samples at 14,000 rpm for 30 min at 4 °C. Allow the pellets to air-dry. Resuspend the DNA in water or TE and analyze by PCR using primers based on the predicted DNA sequence. We generally resuspend the immunoprecipitates in 30 μl and then use 2–5 μl for each PCR reaction (25–30 cycles).

6.4. Chromatin remodelling: Endonuclease accessibility assay

This is a clear-cut assay to verify histone occupancy in a defined area of chromatin. Briefly, a unique site for endonuclease enzymes is identified (the site differ depending on the region analyzed), and used to verify, through subsequent PCR analysis, the presence (no cut) or the absence (cut) of nucleosomes (see Fig. 15.2).

6.4.1. Procedure

1. Start from one 100-mm dish of confluent cells (approximately 10 × 10^6 cells).
2. Scrape cells into 1 ml of cold PBS 1x plus 1 m*M* PMSF and 1 m*M* leupeptin.
3. Centrifuge at 2000 rpm for 2 min at 4 °C.
4. Wash once with 1 ml cold PBS 1x plus 1 m*M* PMSF and 1 m*M* leupeptin
5. Centrifuge at 2000 rpm for 2 min at 4 °C.
6. Resuspend the cell pellets in 1 ml cold RSB (10 m*M* Tris-HCl, pH7.4, 10 m*M* NaCl, 5 m*M* $MgCl_2$) 0.5 % NP40 plus 1 m*M* PMSF and 1 m*M* leupeptin and incubate on ice for 20 min with occasional gentle mixing.
7. Homogenize by passing through a syringe using a 27-G needle 5 times.
8. Centrifuge at 2000 rpm for 5 min at 4 °C to pellet the nuclei; discard the supernatant fraction.
9. Wash the nuclei with 1 ml of cold RSB plus 1 m*M* PMSF and 1 m*M* leupeptin.
10. Resuspend the nuclei in 200 μl of cold RSB plus 1 m*M* PMSF and 1 m*M* leupeptin.
11. Take 40 μl of nuclei and resuspend in 400 μl endonuclease buffer (New England Biolabs, the type of buffer depends on the endonuclease used)

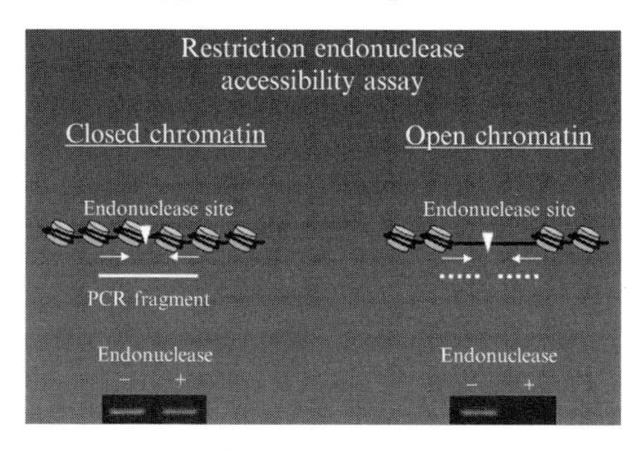

Figure 15.2 Scheme representing how the restriction endonuclease accessibility assay could reflect the chromatin status in amplified PCR DNA fragments. See text for details.

and add 5 μl (100 units) of endonuclease enzyme (New England Biolabs); also prepare a sample without enzyme added.
12. Incubate samples with or without the endonuclease for 1 h at 37 °C.
13. Add 10 μl of proteinase K (10 mg/ml stock; final concentration 0.2 μg/μl) and 50 μl of SDS 20% (final concentration 2%). The proteinase K should be made fresh to avoid self-degradation.
14. Incubate the samples for 2 h at 42 °C.
15. Extract twice with 500 μl of phenol/chloroform.
16. Centrifuge at maximum speed for 10 min.
17. Precipitate with 900 μl of ethanol, and 50 μl 3 *M* NaAc at −20 °C for 30 min to overnight.
18. Centrifuge at maximum speed for 15 min.
19. Wash with 500 μl of 70% ethanol.
20. Air-dry the pellets and resuspend in 100 μl of TE buffer then measure the DNA concentration.
21. Use 50 ng for PCR.

7. Analysis of Transcriptional Multiprotein Complexes

7.1. *In vitro* pull-down

Pull-down is an *in vitro* method used to assess physical interactions between two or more proteins. In a pull-down assay, a GST (or His)-tagged bait protein is immobilized on glutathione-sepharose beads (or Nickel resin for using a His tag) and subsequently incubated in batch with a solution containing putative prey proteins. Cell lysates, *in vitro* transcription/translation reactions, and purified proteins are commonly used as prey protein sources.

7.2. Procedure: GST pull-down from total cell extracts

1. Plate cells 2–3 days in advance to achieve 75% confluency and transfect them if required (include a negative control that does not express the prey protein).
2. Remove the growth medium and wash gently twice with cold PBS, harvest the cells by scraping in cold PBS plus protease inhibitors.
3. Lyse cells using Lysis buffer (50 m*M* Tris-HCl, pH 7.5, 0.5% NP40, 10% glycerol, 300 m*M* NaCl) supplemented with protease (1 m*M* PMSF, 1.5 μ*M* pepstatin A, 2 μ*M* leupeptin, 10 μg/ml aprotinin) and phosphatase inhibitors (5 m*M* NaF, 1 m*M* Na_3VO_4). Incubate for 20 min at 4 °C.
4. Centrifuge 10 min at 13,000 rpm at 4 °C.

5. Transfer the supernatant fractions to a new tube. Dilute 1:2 with pull-down buffer (50 m*M* Tris-HCl, pH 7.5, 0.5% NP40, 10% glycerol) supplemented with protease and phosphatase inhibitors as previously.
6. Take one-tenth of each lysate to be used as input (to check the total amount of prey protein in each reaction), add Laemmli sample buffer and incubate at 90–100 °C for 5 min. Store at −20 °C.
7. To preclear the lysates incubate each sample with 50 μl of GSH-sepharose beads (GE healthcare 17-5132-01) 1:2 in PBS, and leave for 20 min at 4 °C while rocking. Centrifuge 2–3 min at 5000 rpm at 4 °C. Transfer the supernatant fractions to a fresh tube.
8. Split each lysate into 2 tubes and add 2–5 μg of purified GST-bait protein to one and the same amount of GST-negative control to the other.
9. Incubate 2 h at 4 °C rocking on a wheel.
10. Centrifuge at 5000 rpm for 2 min at 4 °C. Carefully discard the supernatant fraction and resuspend the beads in 1 ml of chilled wash buffer (50 m*M* Tris-HCl, pH 7.5; 0.5% NP40; 10% glycerol; 150 m*M* NaCl) supplemented with protease and phosphatase inhibitors as above. Leave for 5–10 min rocking on a wheel at 4 °C. Repeat once.
11. Centrifuge at 5000 rpm for 2 min at 4 °C. Carefully discard the supernatant fractions. Resuspend the pellet fractions in 1 ml of wash buffer and mix by inverting 15–20 times.
12. Centrifuge at 5000 rpm for 2 min at 4 °C. Carefully discard the supernatant fractions, and thoroughly remove all buffer by using a vacuum pump and a needle. Resuspend the pellet fractions in Laemmli sample buffer and incubate at 90–100 °C for 5 min.
13. Load samples and inputs on a SDS-PAGE gel.
14. Proceed with Western blotting.

7.3. Procedure: GST pull-down from *in vitro* transcription/translation proteins

This protocol describes how to use radiolabeled proteins to perform protein-protein interaction experiments. Take care in handling radioactive material and work in accordance with the local laws and institute policy. Discard every solid or liquid waste in a proper radioactive waste.

1. Set up S^{35} radiolabeled *in vitro* transcription/translation (IVTT) reactions for the prey protein and for a negative control protein (we suggest using the Promega TNT Coupled Reticulocyte Lysate System with the desired RNA polymerase, following the protocol suggested by the manufacturer).

2. Prepare 150 m*M* NaCl buffer (50 m*M* Tris-HCl, pH 7.5, 0.5% NP40, 10% glycerol, 150 m*M* NaCl) supplemented with protease and phosphatase inhibitors as above (200 μl for each pull-down reaction) and keep on ice.
3. Thaw the IVTT reactions on ice.
4. Take one-tenth volume of each IVTT reaction to be used as input, add Laemmli sample buffer and incubate at 90–100 °C for 5 min. Store at −20 °C.
5. In a 1.5-ml tube dilute 10–15 μl of each IVTT reaction with 200 μl of chilled 150 m*M* NaCl buffer.
6. Add 2–5 μg of GST-bait protein or GST-negative control to each IVTT tube.
7. Incubate 2 h at 4 °C while rocking on a wheel.
8. Centrifuge at 2000 rpm for 5 min at 4 °C. Carefully discard the supernatant fraction and resuspend the beads in 1 ml of chilled 150 m*M* NaCl buffer. Leave for 5–10 min rocking on a wheel at 4 °C. Repeat once.
9. Centrifuge at 2000 rpm for 5 min at 4 °C. Carefully discard the supernatant fraction. Resuspend pellets in 1 ml of wash buffer and mix by inverting 15–20 times.
10. Centrifuge at 2000 rpm for 5 min at 4 °C. Carefully discard as much as possible of the supernatant fraction, add Laemmli sample buffer and incubate at 90–100 °C for 5 min.
11. Run an SDS-PAGE gel.
12. Dry the gel and proceed with autoradiography.

7.4. Mammalian two-hybrid

The mammalian two-hybrid is a relatively high throughput technique that allows protein-protein interaction analysis in cell culture by using the dual-luciferase system and a luminometer. Several kits are commercially available. Briefly, one of the two proteins is cloned in frame with the GAL4 DNA-binding domain to generate a GAL4 fusion protein, whereas the other one is fused in frame with the potent transactivation domain VP16 from herpes virus. These two constructs are transfected together with a third vector containing the firefly luciferase enzyme cloned under the control of a GAL4-responsive element. If the two proteins physically interact, the GAL4 fusion protein is able to tether on the DNA a protein with a potent transactivation domain and this induces luciferase transcription. Standardization is made using renilla luciferase located in the same vector of the GAL4-fusion protein. Positive and negative controls are strongly required.

7.5. Coimmunoprecipitation

Immunoprecipitation (IP) makes it possible to capture a protein with a specific antibody and purify the immune complex with protein G or protein A, linked to sepharose or agarose beads. This technique provides a rapid and simple means to separate a specific protein from whole cell lysates or culture supernatants. Additionally, one can use an IP to confirm the identity or study the biochemical characteristics, post-translational modifications and/or expression levels of a protein of interest.

Coimmunoprecipitation (co-IP) is a means to verify the presence of two proteins in the same complex in cells or tissues: a protein is immunoprecipitated and then the presence of another protein(s) in the immune complex is checked by Western blot analysis.

The success of IP depends also upon the affinity of the antibody for its antigen as well as for protein G or protein A. In general, polyclonal antibodies are the best option, but purified monoclonal antibodies (mAb), ascites fluids or hybridoma supernatants can also be used. In some cases, using a pool of antibodies increases the efficiency of the IP. Preferably, the antibody used for the IP should not be the same antibody used for the subsequent Western blot.

The protocol below offers a general guideline for IP. Optimization may be required for each specific antigen and antibody. The abundance of a given protein in a sample is variable and this is a critical factor for obtaining clear results. Fine adjustments of the cell amount, the Co-IP buffer composition and the volume used for cell lysis may be necessary.

7.5.1. Procedure

1. Chill a convenient amount (400 μl for each 100-mm plate that is 75% confluent) of lysis buffer (50 m*M* Tris-HCl, pH 7.5, 150 m*M* NaCl, 1 m*M* EDTA, 0.5% NP40) on ice; add protease and phosphatase inhibitors as described previously just before use.
2. For adherent cells, carefully remove the growth medium, and wash the cells, 75% confluent 100-mm dish, directly in the dish twice with PBS. Collect cells by scraping or trypsin digestion; neutralize the trypsin with a small amount of serum. Centrifuge for 5 min at 1000 rpm and wash twice with cold PBS. For cells growing in suspension: transfer the cells to a tube, centrifuge at 1000 rpm for 5 min, remove the medium and wash twice with cold PBS.
3. Lyse the cells in 400 μl of cold lysis buffer. Working in a cold room to preserve samples is recommended. Place the tube on ice for 10–20 min, with occasional mixing.
4. Centrifuge the cell lysate at 12,000 rpm for 10 min at 4 °C. Carefully collect the supernatant fractions, without disturbing the pellet fractions and transfer to a fresh tube. Discard the pellet fractions.

5. Take an aliquot of the lysate (usually one-tenth) to be loaded on the gel as "input". Add Laemmli sample buffer and incubate at 90–100 °C for 5 min. Store at −20 °C.
6. To preclear the cell lysate, equilibrate 50 μl of protein A (or G) beads slurry (for each IP) in 1 ml of cold lysis buffer.
7. Centrifuge at 2000 rpm for 2 min and remove the lysis buffer. Wash 2 or more times with 1 ml of cold lysis buffer. Never let the resin dry; leave a small amount of buffer.
8. Add the preceding prepared protein A (or G) slurry to the cell lysate in a tube and incubate for 30 min rocking at 4 °C.
9. Centrifuge at 12,000 rpm for 10 min at 4 °C and transfer the supernatant fractions to a fresh tube. Discard the beads.
10. Add specific antibody to the cold precleared lysate, start with 1 μg/IP working on ice.
11. Incubate at 4 °C for 1–4 h to overnight (this depends on the antibody and must be determined empirically) on a rocking platform or a rotator.
12. Add 15–20 μl of protein A or G slurry. Incubate for 40 min to 1 h at 4 °C on a rocking platform or a rotator.
13. Centrifuge the tubes at 2000 rpm for 2 min at 4 °C.
14. Carefully remove the supernatant fraction completely and discard.
15. Wash the bound beads by adding 1 ml of cold lysis buffer and incubate at 4 °C for 5–10 min on a rocking platform or a rotator.
16. Repeat steps 17–19 2 or more times.
17. After the last wash, remove the buffer completely, using a vacuum pump and a needle. Add Laemmli sample buffer and incubate at 90–100 °C for 5 min.
18. Analyze by SDS-PAGE and Western blot.

8. Concluding Remarks

It is of great relevance to understand how external stimuli are integrated and transduced into the nucleus to modulate autophagy-related gene expression. Indeed, if it is true that every step of a signaling cascade may represent a potential pharmacological target, it is also true that the closer the target is to chromatin, the less will the effect of the pharmacological treatment spread to other signaling pathways. As a consequence, the ultimate goal is to develop novel compounds able to influence the expression of only a subset of specific genes, such as autophagy-related genes. This result will be possible only after achieving a deep understanding of the various signaling pathways and their effects on autophagy and closely related process as metabolic regulation and apoptosis.

ACKNOWLEDGMENTS

We thank Dr. Francesco Paolo Jori for his helpful discussion during the preparation of the manuscript and editorial assistance. This work was partially supported by grants from the Italian Association for Cancer Research to CS.

REFERENCES

Abeliovich, H., Dunn, W. A., Jr., Kim, J., and Klionsky, D. J. (2000). Dissection of autophagosome biogenesis into distinct nucleation and expansion steps. *J. Cell Biol.* **151,** 1025–1034.

Brugarolas, J., Lei, K., Hurley, R. L., Manning, B. D., Reiling, J. H., Hafen, E., Witters, L. A., Ellisen, L. W., and Kaelin, W. G., Jr., (2004). Regulation of mTOR function in response to hypoxia by REDD1 and the TSC1/TSC2 tumor suppressor complex. *Genes Dev.* **18,** 2893–2904.

Chiacchiera, F., and Simone, C. (2008). Signal-dependent regulation of gene expression as a target for cancer treatment: Inhibiting p38alpha in colorectal tumors. *Cancer Lett.* **265,** 16–26.

Chow, C. W., and Davis, R. J. (2006). Proteins kinases: Chromatin-associated enzymes? *Cell* **127,** 887–890.

Comes, F., Matrone, A., Lastella, P., Nico, B., Susca, F. C., Bagnulo, R., Ingravallo, G., Modica, S., Lo Sasso, G., Moschetta, A., Guanti, G., and Simone, C. (2007). A novel cell type-specific role of p38alpha in the control of autophagy and cell death in colorectal cancer cells. *Cell Death Differ.* **14,** 693–702.

Degenhardt, K., Mathew, R., Beaudoin, B., Bray, K., Anderson, D., Chen, G., Mukherjee, C., Shi, Y., Gelinas, C., Fan, Y., Nelson, D. A., Jin, S., *et al.* (2006). Autophagy promotes tumor cell survival and restricts necrosis, inflammation, and tumorigenesis. *Cancer Cell* **10,** 51–64.

Gorski, S. M., Chittaranjan, S., Pleasance, E. D., Freeman, J. D., Anderson, C. L., Varhol, R. J., Coughlin, S. M., Zuyderduyn, S. D., Jones, S. J., and Marra, M. A. (2003). A SAGE approach to discovery of genes involved in autophagic cell death. *Curr. Biol.* **13,** 358–363.

Jenuwein, T., and Allis, C. D. (2001). Translating the histone code. *Science* **293,** 1074–1080.

Jin, S., DiPaola, R. S., Mathew, R., and White, E. (2007). Metabolic catastrophe as a means to cancer cell death. *J. Cell Sci.* **120,** 379–383.

Juhasz, G., Puskas, L. G., Komonyi, O., Erdi, B., Maroy, P., Neufeld, T. P., and Sass, M. (2007). Gene expression profiling identifies FKBP39 as an inhibitor of autophagy in larval Drosophila fat body. *Cell Death Differ.* **14,** 1181–1190.

Karanam, S., and Moreno, C. S. (2004). CONFAC: Automated application of comparative genomic promoter analysis to DNA microarray datasets. *Nucleic Acids Res.* **32,** W475–484.

Klionsky, D. J., Abeliovich, H., Agostinis, P., Agrawal, D. K., Aliev, G., Askew, D. S., Baba, M., Baehrecke, E. H., Bahr, B. A., Ballabio, A., Bamber, B. A., Bassham, D. C., *et al.* (2008). Guidelines for the use and interpretation of assays for monitoring autophagy in higher eukaryotes. *Autophagy* **4,** 151–175.

Kouroku, Y., Fujita, E., Tanida, I., Ueno, T., Isoai, A., Kumagai, H., Ogawa, S., Kaufman, R. J., Kominami, E., and Momoi, T. (2007). ER stress (PERK/eIF2alpha phosphorylation) mediates the polyglutamine-induced LC3 conversion, an essential step for autophagy formation. *Cell Death Differ.* **14,** 230–239.

Lee, C. Y., Clough, E. A., Yellon, P., Teslovich, T. M., Stephan, D. A., and Baehrecke, E. H. (2003). Genome-wide analyses of steroid- and radiation-triggered programmed cell death in Drosophila. *Curr. Biol.* **13,** 350–357.

Levine, B., and Klionsky, D. J. (2004). Development by self-digestion: Molecular mechanisms and biological functions of autophagy. *Dev. Cell* **6,** 463–477.

Mammucari, C., Milan, G., Romanello, V., Masiero, E., Rudolf, R., Del Piccolo, P., Burden, S. J., Di Lisi, R., Sandri, C., Zhao, J., Goldberg, A. L., Schiaffino, S., *et al.* (2007). FoxO3 controls autophagy in skeletal muscle *in vivo*. *Cell Metab.* **6,** 458–471.

Mathew, R., Kongara, S., Beaudoin, B., Karp, C. M., Bray, K., Degenhardt, K., Chen, G., Jin, S., and White, E. (2007). Autophagy suppresses tumor progression by limiting chromosomal instability. *Genes Dev.* **21,** 1367–1381.

Meley, D., Bauvy, C., Houben-Weerts, J. H., Dubbelhuis, P. F., Helmond, M. T., Codogno, P., and Meijer, A. J. (2006). AMP-activated protein kinase and the regulation of autophagic proteolysis. *J. Biol. Chem.* **281,** 34870–34879.

Nara, A., Mizushima, N., Yamamoto, A., Kabeya, Y., Ohsumi, Y., and Yoshimori, T. (2002). SKD1 AAA ATPase-dependent endosomal transport is involved in autolysosome formation. *Cell Struct. Funct.* **27,** 29–37.

Park, K. J., Krishnan, V., O'Malley, B. W., Yamamoto, Y., and Gaynor, R. B. (2005). Formation of an IKKalpha-dependent transcription complex is required for estrogen receptor-mediated gene activation. *Mol. Cell.* **18,** 71–82.

Pattingre, S., Bauvy, C., and Codogno, P. (2003). Amino acids interfere with the ERK1/2-dependent control of macroautophagy by controlling the activation of Raf-1 in human colon cancer HT-29 cells. *J. Biol. Chem.* **278,** 16667–16674.

Raman, M., Chen, W., and Cobb, M. H. (2007). Differential regulation and properties of MAPKs. *Oncogene.* **26,** 3100–3112.

Sabatini, D. M. (2006). mTOR and cancer: Insights into a complex relationship. *Nat. Rev. Cancer* **6,** 729–734.

Simone, C. (2006). SWI/SNF: The crossroads where extracellular signaling pathways meet chromatin. *J. Cell Physiol.* **207,** 309–314.

Simone, C., Forcales, S. V., Hill, D. A., Imbalzano, A. N., Latella, L., and Puri, P. L. (2004). p38 pathway targets SWI-SNF chromatin-remodeling complex to muscle-specific loci. *Nat. Genet.* **36,** 738–743.

CHAPTER SIXTEEN

Novel Methods for Measuring Cardiac Autophagy *In Vivo*

Cynthia N. Perry,* Shiori Kyoi,† Nirmala Hariharan,† Hiromitsu Takagi,† Junichi Sadoshima,† *and* Roberta A. Gottlieb*

Contents

Abstract

Autophagy, a highly conserved cellular mechanism wherein various cellular components are broken down and recycled through lysosomes, occurs constitutively in the heart and may serve as a cardioprotective mechanism in some situations. It has been implicated in the development of heart failure and is up-regulated following ischemia-reperfusion injury. Autophagic flux, a measure of autophagic vesicle formation and clearance, is an important measurement in evaluating the efficacy of the pathway, however, tools to measure flux *in vivo* have been limited. Here, we describe the use of monodansylcadaverine (MDC) and the lysosomotropic drug chloroquine to measure autophagic flux in *in vivo*

* San Diego State Research Foundation BioScience Center, San Diego State University, San Diego, California, USA
† Department of Cell Biology and Molecular Medicine, UMDNJ, New Jersey Medical School, Newark, New Jersey, USA

Methods in Enzymology, Volume 453
ISSN 0076-6879, DOI: 10.1016/S0076-6879(08)04016-0

model systems, specifically focusing on its use in the myocardium. This method allows determination of flux as a more precise measure of autophagic activity *in vivo* much in the same way that Bafilomycin A_1 is used to measure flux in cell culture. MDC injected 1 h before sacrifice, colocalizes with mCherry-LC3 puncta, validating its use as a marker of autophagosomes. This chapter provides a method to measure autophagic flux *in vivo* in both transgenic and nontransgenic animals, using MDC and chloroquine, and in addition describes the mCherry-LC3 mouse and the advantages of this animal model in the study of cardiac autophagy. Additionally, we review several methods for inducing autophagy in the myocardium under pathological conditions such as myocardial infarction, ischemia/ reperfusion, pressure overloading, and nutrient starvation.

1. Introduction

Autophagy occurs constitutively in the normal myocardium and is up-regulated after ischemia reperfusion (Decker *et al.*, 1980; Sybers *et al.*, 1976). Since the early electron microscopy observations of autophagy more than 40 years ago, relatively little work has been done in the heart, largely due to the lack of suitable molecular reagents to facilitate mechanistic studies. However, a recent proteomic study reveals the up-regulation of autophagic proteins in chronically ischemic myocardium (Yan *et al.*, 2005), and autophagy is up-regulated in the hearts of mice subjected to starvation (Pattingre *et al.*, 2005), thus raising interest in the topic among cardiovascular investigators. Despite growing interest in the field, studies have been hampered by the lack of available molecular tools to investigate autophagy especially *in vivo*.

LC3, the mammalian homolog of yeast Atg8, is regularly used as a marker for autophagy both in Western blotting and as a fluorescently conjugated marker of autophagic vesicle (AV) formation. A transgenic mouse expressing GFP-LC3, created in Japan by Mizushima's group, has been used to demonstrate the occurrence of autophagy in the heart and provided great insight into the role of autophagy *in vivo* after starvation (Mizushima *et al.*, 2005). Our publications and preliminary data with cardiac derived HL-1 cells subjected to simulated ischemia/reperfusion (sI/R) demonstrate that autophagy serves a protective role in ischemia/ reperfusion (Hamacher-Brady *et al.*, 2006). We believe this is a salvage response, because suppression of autophagy increases cell death. Whereas autophagy plays an important role in facilitating replacement of damaged organelles and promoting survival during nutrient deprivation, excessive autophagy can result in caspase-independent cell death. Therefore in the context of the heart, low-level autophagy may be beneficial, whereas excessive autophagy may be deleterious. Moreover, the up-regulation of autophagy in the context of compensated cardiac hypertrophy may contribute to the transition to failure (Hein *et al.*, 2003). To further investigate cardiac autophagy *in vivo*, we produced a line of transgenic mice expressing mCherry-tagged LC3 (mCherry-LC3) under the control of the cardiomyocyte-specific alpha-myosin heavy chain promoter. Here we describe the production of the mCherry-LC3 mouse line and its advantages in studying cardiac autophagy.

Despite the complex nature of cardiac autophagy, it is clear that data based on a snapshot of the cell at a given time are insufficient to accurately understand the behavior of autophagy. Static images of cells with numerous AVs could reflect increased autophagy but could also indicate reduced autophagic flux due to impaired fusion with lysosomes. A more thorough examination of the process requires researchers to measure autophagic flux directly. Autophagic flux is the measurement of the rate of autophagosome formation and clearance through the pathway (Fig. 16.1). Bafilomycin A_1, a potent inhibitor of vacuolar H^+-ATPase, is regularly used in the measurement of flux by preventing downstream clearance of autophagosomes (Yamamoto *et al.*, 1998). By inhibiting vacuolar acidification, Bafilomycin A_1 results in the accumulation of autophagosomes by possibly preventing lysosome-autophagosome fusion, or by blocking intralysosomal degradation, which is dependent on lowered intralysosomal pH. A comparison of steady-state autophagosome levels with accumulated autophagosomes following Bafilomycin A_1 treatment provides a good idea of the rate of production and clearance in the pathway. Differences between steady state and accumulated AV levels can be interpreted in several different ways as lower AV levels may indicate high turnover or decreased AV production. Alternatively, less significant differences in steady state and accumulated AV levels could denote that autophagy is unaffected, inhibited at both production and clearance stages or AV production is increased combined with a high rate of clearance.

Whereas Bafilomycin A_1 provides a tool for studying autophagic flux in cell culture, it is costly and unsuitable for studying flux in animals. We present here a method using chloroquine to measure flux *in vivo*. Chloroquine is an anti-inflammatory drug that has been used in the treatment of malaria for more than 60 years (O'Neil *et al.*, 1998). It is believed to work by raising lysosomal pH (Poole and Ohkuma, 1981; Kawai *et al.*, 2007) and thereby inhibiting lysosomal activity (Ohkuma and Poole, 1978; Sewell *et al.*, 1983). Because autophagosome-lysosome fusion is pH-dependent, the alkalinizing effects of chloroquine on lysosomes inhibit fusion and hydrolase activity, thus preventing AV clearance.

Additionally, we provide two methods for evaluating autophagy using monodansylcadaverine (MDC) and Alexa Fluor 488 Cadaverine (Invitrogen), which is known to label acidic endosomes, lysosomes and autophagosomes (Munafo and Colombo, 2001; Yan *et al.*, 2007) (also see the chapter by Vázquez and Colombo in volume 452). MDC labeling *in vivo* colocalizes

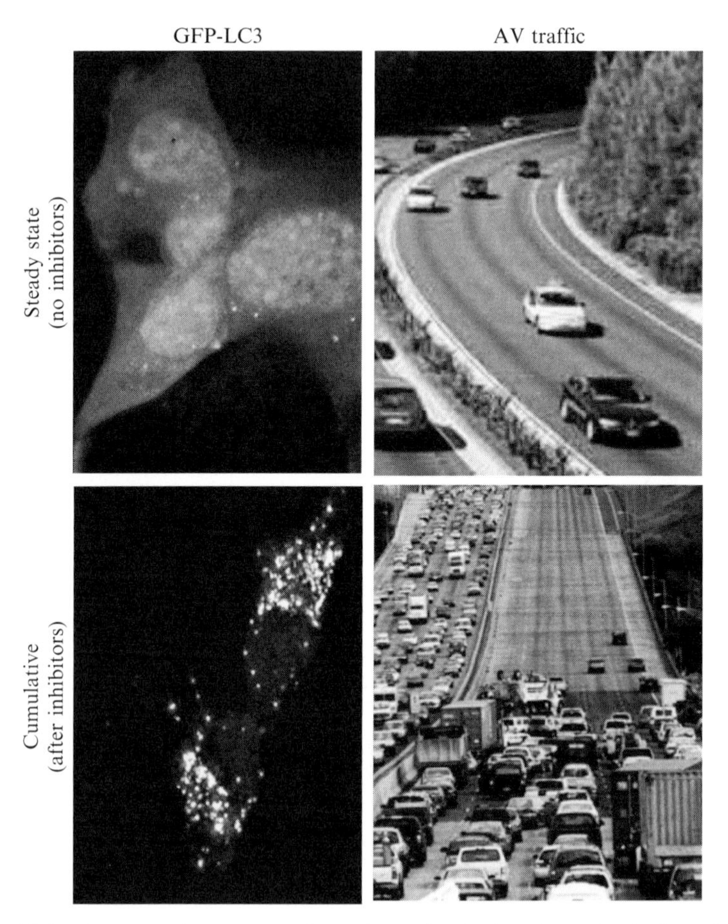

Figure 16.1 Conceptual representation of autophagic flux. Autophagic flux, the rate at which material is cleared by autophagy, is demonstrated here by the analogy of cars in a traffic jam. Lysosomal inhibitors such as Bafilomycin A_1 and chloroquine act as blockades, resulting in the backup of vesicles as they are processed through the autophagic pathway. Right column: GFP-LC3-expressing cardiac-derived HL-1 cells subjected to 4-h serum starvation incubated with and without Bafilomycin A_1.

with many mCherry-LC3 puncta and the number of labeled structures increases in parallel with induction of autophagy, validating its potential as a marker of autophagy *in vivo* (Iwai-Kanai *et al.*, 2008). Our plate-based assay provides a quantitative measurement of autophagy using isolated autophagosomes from both fresh and frozen tissue samples.

We begin with an overview of common *in vivo* models of autophagy and methods to induce autophagy in the myocardium including nutrient starvation, ischemia/reperfusion, chronic ischemia, myocardial infarction and pressure overload by transverse aortic constriction (TAC). Specific precautions to studying autophagy in the heart are noted within each method.

2. *In Vivo* Models of Autophagy in the Myocardium

2.1. Nutrient starvation

The heart is among the organs where rapid (within 30 min) and strong induction of autophagy is observed during the neonatal starvation period in mice (Kuma *et al.*, 2004). In adult mice, autophagy is induced in the heart by food starvation (Mizushima *et al.*, 2004; Pattingre *et al.*, 2005). Induction of autophagic vacuoles is greater in the heart than in the liver, in the rat model of calorie restriction (Wohlgemuth *et al.*, 2007). For induction of autophagy in the adult mouse heart, 48 h of food starvation has been commonly used (Mizushima *et al.*, 2004; Pattingre *et al.*, 2005). During starvation, mice have free access to water and their temperature and blood pressure should be checked periodically.

2.2. Ischemia/reperfusion

Both ischemia and ischemia followed by reperfusion (I/R) induce accumulation of autophagosomes in the heart. Decker et al report that 40-min ischemia causes up-regulation of autophagy and that subsequent reperfusion induces a drastic enhancement of autophagy in Langendorff perfused rabbit hearts (Decker and Wildenthal, 1980). In this model, increases in autophagy correlate with recovery of cardiac function and salvage of myocardium after I/R. Interestingly, 60-min ischemia causes lysosomal dysfunction during reperfusion, suggesting that extended ischemia may impair autophagy (Decker and Wildenthal, 1980). In another study, I/R was applied to the mouse heart by transiently occluding the coronary artery *in situ*. In this model, autophagy is induced by ischemia alone and it is further enhanced by reperfusion (Matsui *et al.*, 2007). I/R experiments can be performed in two forms. One uses the *ex vivo* isolated perfused heart (Langendorff) preparation and the other the *in situ* coronary artery ligation. The effect of global I/R on autophagy can be studied using the Langendorff preparation without need for an extensive surgical setup (Decker and Wildenthal 1980; Hamacher-Brady *et al.*, 2007). Changes in high energy phosphates and glycolysitic flux, as well as left ventricular (LV) cardiac function, can be monitored relatively easily during I/R (Luptak *et al.*, 2007). On the other hand, *in situ* coronary artery ligation in experimental animals induces focal ischemia, which mimics human pathological conditions. Experiments can be conducted even in conscious animals after initial instrumentation, which allows one to obtain continuous monitoring of LV cardiac function during I/R. The technique of *in situ* I/R can be applied to genetically altered mouse models to monitor the extent of autophagy or to evaluate the functional significance of autophagy during I/R. In the following, methods to apply I/R to the mouse heart *in situ* are discussed.

2.2.1. Ischemia/Reperfusion

1. The extent of autophagosome accumulation in the myocardium after I/R is evaluated most conveniently using transgenic mice harboring GFP-LC3 (GFP-LC3 mice) made by Dr. N. Mizushima (C57BL/6J background) (Mizushima *et al.*, 2004) or the mCherry-LC3 mouse as described later. In order to evaluate the effect of genetic interventions upon autophagy in the heart, genetically altered mice of interest can be crossed with the GFP-LC3 mice. It is important to use mice with a homogeneous genetic background. If cross breeding of mice with different genetic backgrounds is needed, sufficient generations of backcross should be conducted. Pathogen-free mice are housed in a temperature-controlled environment with a 12-h light/dark cycle where they receive food and water *ad libitum*.
2. Anesthetize mice by intraperitoneal injection of pentobarbital sodium (60 mg/kg). Use a rodent ventilator (model 683; Harvard Apparatus) with 65% oxygen during the surgical procedure.
3. Keep the mice warm using heat lamps and heating pads. Monitor the rectal temperature and maintain it between 36.8 °C and 37.2 °C, which is essential because the severity of ischemia depends in part on the rate of metabolism, which is temperature sensitive. The chest is opened by a horizontal incision through the muscle between the ribs at the third intercostal space.
4. Apply ischemia by ligating the anterior descending branch of the left coronary artery (LAD) using an 8–0 nylon suture, which is threaded through a silicon tubing (1-mm OD) propylene tube to form a snare on top of the LAD, 2 mm below the border between left atrium and LV. Confirm regional ischemia by ST-T changes in an electrocardiogram. The susceptibility of the mouse heart to ischemic injury is different from strain to strain. For example, the heart of FVB mice is more resistant to ischemia than that of C57BL/6 mice. Thus, longer ischemia is needed in FVB mice to achieve ischemic injury comparable to that in C57BL/6 mice. In order to observe induction of autophagy by ischemia alone, we use 30–45 min of ischemia for FVB mice and 20 min for C57BL/6. Longer ischemia may inhibit autophagosome formation since ATP-dependent steps are involved.
5. After transient occlusion of the coronary artery, remove silicon tubing to achieve reperfusion. An increase in autophagosome formation in the previously ischemic area is observed as early as 20 min after ischemia and it is further enhanced 2 h after reperfusion in C57BL/6 mice (Matsui *et al.*, 2007).

6. To observe LC3 dots representing autophagosomes, harvest the heart using the following method:
   a. After dissection of the heart, wash the heart several times with PBS.
   b. Cut the heart into several slices and fix them with 4% paraformaldehyde at 4 °C overnight.
   c. Replace 4% paraformaldehyde with 15% sucrose solution and keep the sample at 4 °C for at least 4–5 h (or overnight).
   d. Replace the 15% sucrose with a 30% sucrose solution and keep the sample at 4 °C overnight.
   e. Place the sample in the Tissue-Tek (Sakura Finetek USA, CA). Overlay the sample with Tissue-Tek O.C.T. Compound (Sakura Finetek USA, CA) and place it in ethanol-dry ice.
   f. Keep the sample at −80 °C.
7. To determine whether certain interventions increase or decrease the extent of myocardial damage, set the duration of ischemia to induce myocardial infarction with intermediate sizes, such as 20%–50% of the area at risk. It is difficult to see the protective effect of interventions when the infarction size is less than 10% without interventions. Conversely, it is also difficult to see the detrimental effect of interventions when the infraction size is greater than 60% without intervention. We have shown previously that heterozygous deletion of *beclin 1* (C57BL/6 background) reduces the size of myocardial infarction after I/R from 45 to 20% (Matsui Y *et al.*, 2007). In this case, we applied 20 min of ischemia and the size of myocardial infarction was determined 24 h after reperfusion.
8. After I/R, reanaesthetize and intubate the animals, and open the chest. Arrest the heart at the diastolic phase by injecting 0.5 mL of 100 m*M* KCl. Canulate the ascending aorta and perfuse the heart with saline to wash out blood. Occlude the LAD again with the same suture, which has been left at the site of the ligation. To demarcate the ischemic area at risk (AAR), perfuse 1% Evans blue dye (Sigma Aldrich, 206334) into the aorta and coronary arteries. Excise the heart and slice the LV into 1-mm thick cross sections. Incubate the heart sections with a 1% triphenyltetrazolium chloride (Sigma Aldrich, T8877) solution at 37 °C for 10 min. Measure the infarct area (pale), the AAR (not blue), and the total LV area from both sides of each section using Adobe Photoshop (Adobe Systems), and average the values obtained. Multiply the percentage of area of infarction and AAR of each section by the weight of the section and then obtain the total from all sections. Express AAR/LV and infarct area/AAR area as a percentage.

2.3. Chronic ischemia

Although the myocardium under chronic hypoxia exhibits reduced contractility, the condition termed myocardial hibernation, it can show a significant recovery when hypoxia is eased. Chronic hypoxia in the myocardium not only up-regulates a series of cell survival mechanisms but induces autophagy in the myocardium (Yan *et al.*, 2005; May *et al.*, 2008). A previous study shows an inverse correlation between the occurrence of autophagy and apoptosis in the hibernating myocardium. Thus, autophagy in the myocardium caused by chronic hypoxia may contribute to survival of cardiac myocytes. Thus far, several animal models of myocardial hibernation have been reported. One is a large animal model, in which 6 episodes of repetitive reduction in coronary flow are applied to instrumented conscious pigs (Yan *et al.*, 2005) The other is transgenic mice with conditional expression of a VEGF-sequestrating soluble receptor, which allows tetracycline-regulated VEGF blockade and fully reversible induction of the globally hypo-perfused heart with significant reduced contractility, mimicking the hibernating myocardium (May *et al.*, 2008). Autophagy is induced by chronic ischemia in both pig and mouse models of hibernation. The pig model is useful for translational research because of the similarity in the anatomy of coronary arteries between pigs and humans. The mouse model is useful for mechanistic studies because they can be crossed with other genetically altered mouse models of autophagy.

2.4. Myocardial infarction

In humans, occlusion of the coronary artery by plaque rupture causes myocardial infarction, which alone exhibits a very high mortality. Even though patients manage to survive the acute event, the area of the infarction is replaced with a scar and the heart undergoes structural and functional remodeling, which eventually leads to cardiac dilation and heart failure. Myocardial infarction triggers inflammation and a tissue remodeling process through up-regulation of cytokines, proteases, and lysosomal enzymes. Autophagy is observed in the surviving myocardium after myocardial infarction.

To create myocardial infarction in mice, the same procedures as those for I/R can be used except that the coronary artery is permanently ligated. The size of the myocardial infarction can be determined by TTC staining at early stages (up to 3–4 days), whereas the proportion of the MI area/total left ventricle can be speculated by measuring the proportion of the circumference occupied by a thin scar visualized by Masson Trichrome staining (Odashima *et al.*, 2007). Rupture of the LV wall due to vulnerability of the infarction area tends to occur at the acute phase, whereas LV dysfunction due to cardiac remodeling gradually develops after 2–4 weeks.

2.5. Pressure overload

Whether autophagy is stimulated in the heart under pathologically relevant stresses other than ischemia and, if so, whether autophagy is protective or detrimental in the failing heart are important issues. Transverse aortic constriction (TAC) is one of the most commonly used methods to induce pathological hypertrophy and heart failure by mimicking increased afterload caused by elevated blood pressure (Sadoshima *et al.*, 2002). In one report, accumulation of autophagosomes is observed as early as 24 h after TAC and remains elevated at least 2 weeks (Zhu *et al.*, 2007). In another report, autophagy is suppressed at 1 week after TAC, but up-regulated together with LV dysfunction at 4 weeks (Nakai *et al.*, 2007).

2.5.1. TAC

1. Anesthetize mice at 3–6 months of age by intraperitoneal injection of pentobarbital sodium (65 μg/kg). Intubate mice and ventilate them with a tidal volume of 0.2 ml and a respiratory rate of 110 breaths per minute using a rodent ventilator (model 683; Harvard Apparatus) with 65% oxygen during the surgical procedure.
2. Place mice in the supine position. Under a dissecting microscope, open the left side of the chest at the second intercostal space and expose the transverse aorta. Place a 7–0 prolene ligature around the aorta between the innominate and the left carotid artery. Place a 27-gauge needle on the aorta and remove after the ligature is tied. Use needles with a smaller diameter, such as a 28-gauge needle, to apply greater levels of pressure overload.
3. Close the chest in layers and maintain mechanical ventilation until the mice are able to breathe spontaneously. After extubation, keep the mice in an oxygenated warm chamber and monitor them until they recover from anesthesia.
4. Cardiac hypertrophy is developed within a week. If the purpose of the experiment is to induce heart failure, TAC is applied for more than 4 weeks in FVB mice, whereas 2 weeks are sufficient in C57BL/6 mice. To assess the severity of aortic constriction, measure the pressure gradient across the constriction, using two high-fidelity catheter tip transducers (1.4F; Millar Instruments, Houston). Insert one into the right carotid artery and the other into the right femoral artery. Advance them carefully to the ascending aorta and the abdominal aorta, respectively, where pressures are measured simultaneously. When the effect of the interventions upon cardiac responses, such as induction of autophagy, is compared, it is important to confirm that equal levels of pressure overload are applied to each animal. When some interventions facilitate the progression of heart failure, the pressure gradient measured after 2–4 weeks of TAC could be lower than that measured just after imposition of TAC due to the reduced contractility of the LV. In this case, the initial levels of pressure gradient could be assessed in a separate group of mice at earlier stages before the mice develop cardiac dysfunction.

2.6. Genetically altered mouse models

The functional significance of autophagy in the heart under various pathophysiological conditions can be studied in genetically modified mouse models where autophagy is impaired. These include *beclin 1*$^{+/-}$ mice (Qu *et al.*, 2003), cardiac specific *atg5*$^{-/-}$ mice (Nakai *et al.*, 2007), and *atg7*$^{-/-}$ mice (Komatsu *et al.*, 2005). *Beclin 1*$^{+/-}$ mice have been used to examine the role of autophagy in mediating survival and death of cardiac myocytes in response to I/R (Matsui *et al.*, 2007). Conditional deletion of *atg5* has been used to examine the role of autophagy during pressure overload (Nakai *et al.*, 2007).

2.7. Generation of mCherry-LC3 mouse line

2.7.1. Construction of plasmids

A 1.2-kbp DNA fragment containing rat LC3 cDNA fused to mCherry at the N-terminus (mCherry-LC3) was excised from mCherry-C1-LC3, originally cloned by replacement of EGFP with mCherry in pEGFP-C1-LC3 (gift from Tamotsu Yoshimori, Osaka, Japan) and inserted into the murine α-myosin heavy chain promoter expression vector C26-JM (Baines CP *et al.*, 2005) to generate mCherry-LC3-mHC (Fig. 16.2).

Cardiac-specific mCherry-LC3 transgenic mice were created in the FVB/NJ strain (Jackson Laboratories, Sacramento, CA) by pronuclear injection of murine alpha myosin heavy chain promoter-driven mCherry-LC3 transgene (mCherry-LC3-mHC) located proximal to the human growth hormone polyadenylation signal. Mice were screened for incorporation of the transgene by PCR using primers to the Human Growth Hormone poly A sequence (5′-GTCTGACTAGGTGTCCTTCT-3′ and 5′-CGTCCTCCTGCTGG-TATAG-3′). The PCR is programmed for: 96 °C for 25 s, 56 °C for 25 s, 72 °C for 60 s, and repeats for a total of 30 cycles. The Platinum PCR Super-MIX (Invitrogen, Carlsbad, CA) was used according to manufacturer's instructions and positive DNA samples produce a 410-bp product.

Positive mice were crossed with wild-type FVB/NJ mice and maintained as heterozygotes for the mCherry-LC3 transgene. Production of the mCherry-LC3 line resulted in generation of three founder lines: J8138, J8139, J8295. One line, J8139, was chosen to maintain for studies due to its low expression level and reduced fluorescent protein aggregation (available upon request to Dr. Roberta Gottlieb, San Diego State University). mCherry-LC3 mice accurately reflect induction of autophagy under conditions commonly known to up-regulate the process (Fig. 16.3).

2.7.2. Tissue processing and scoring in mCherry-LC3 mice

1. Excise the heart from pentobarbital anesthetized animals.
2. Rinse the heart in ice-cold 1x phosphate-buffered saline (PBS), pH 7.4. Rinse in fresh PBS until the heart is cleared of red blood cells.

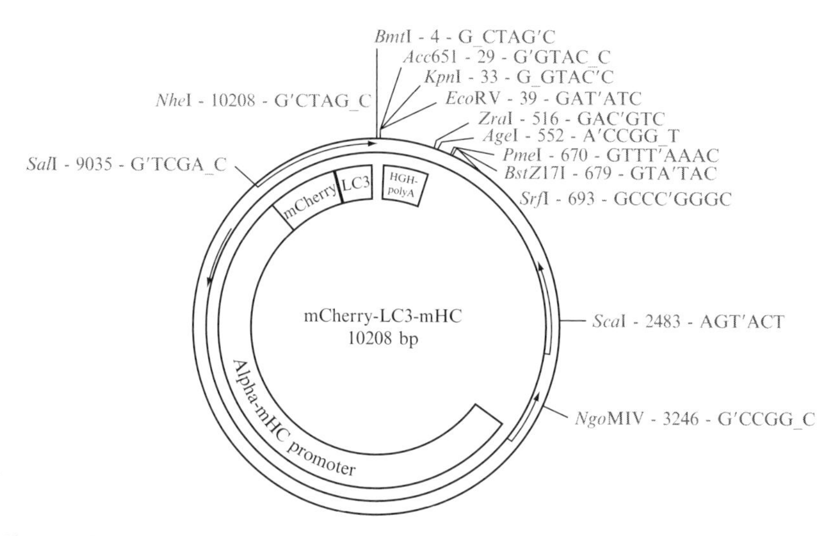

Figure 16.2 Map of mCherry-LC3 expression vector. mCherry-LC3 inserted into the cardiac-specific α-mHC promoter-driven expression vector proximal to the human growth hormone polyadenylation signal.

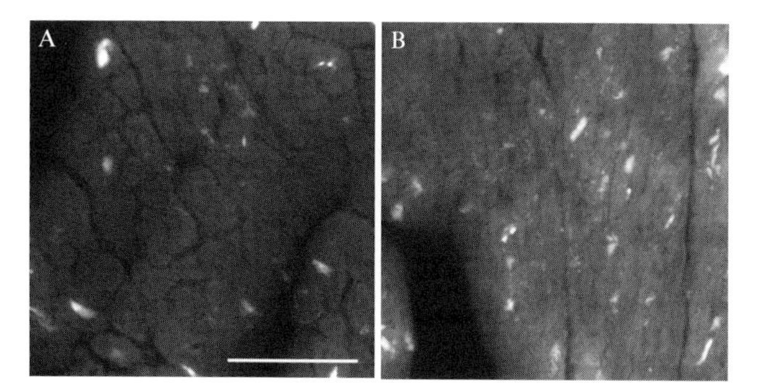

Figure 16.3 mCherry-LC3 transgenic mouse heart sections. Representative images of cardiac tissue cryosections prepared from mCherry-LC3 overexpressing mice following 30-min ischemia and 2-h reperfusion (B) or no ischemic period (A). Increased fluorescent mCherry-LC3 puncta reflect induction of autophagy following IR. (Bar 50 μm)

3. Embed the heart in Tissue Tek OCT compound (Fisher Sci, Pittsburg, PA) and freeze in liquid nitrogen.
4. Store at −80 °C until ready to section.
5. Prepare tissue section slides on a cryostat using Fisherbrand Superfrost Plus glass slides (Fisher Sci).
6. Rinse sections in 1x PBS 5 min to remove the OCT compound.
7. Fix sections with 4% Formaldehyde (Ted Pella)/1x PBS for 15 min, covered at room temperature.

8. Immerse slide in 1x PBS to wash.
9. Dilute Hoechst 33342 (Invitrogen, H3570) 1:1000 in 1x PBS.
10. Cover tissue section with this solution using 100 μL (or as needed) and incubate covered for 30 min.
11. Rinse slide in 1x PBS and mount coverslip (Fisherbrand Premium cover glass, 22×50mm) with Aquapolymount (Polysciences, Warrington, PA) or other mounting media.
12. Image sections using 4x air objective lens on a fluorescence microscope (Nikon TE300, Nikon, Melville, NY) equipped with a 4x lens and cooled CCD camera (Orca-ER, Hamamatsu, Bridgewater, NJ) and automated excitation emission filter wheels controlled by a LAMBDA 10-2 (Sutter instruments) operated by MetaMorph Version 6.2r (Molecular Devices). Select the appropriate dichroic filters for visualization: DAPI (D360/40x) for Hoechst 33342 and MDC staining, and Texas Red (D560/40x) for mCherry-LC3. Quickly focus on the sample by eye to prevent bleaching of the fluorescent signal.
13. Capture images of both mCherry-LC3 and Hoechst staining of the identical optical fields. Fluorescent light is collected via a polychromic beam splitter (61002bs) and an emission filter for DAPI (D460/50m) or Texas Red (D630/60m). All filters are from Chroma Technology group (Rockingham, VT).
14. Score number of AVs per cell by counting total Hoechst-positive nuclei and mCherry-LC3-positive dots. The ratio of AVs/cell gives a measurement of autophagic activity.

Note: Alternatively to quantify the autophagic flux *in vivo*, the percentage of surface area covered by mCherry-LC3 fluorescence can be measured using ImageJ software (http://rsb.info.nih.gov/ij) (see Fig. 16.4 for example). However, users should be cautious of this method in instances where there are numerous fluorescent aggregates that may not represent true mCherry-LC3-positive autophagosomes but rather protein aggregates. In these cases, it is preferential to manually count AV/cell ratio as described.

2.8. Chloroquine method

1. Weigh mice to determine dosage.
2. Prepare 100 μl chloroquine solution (EMD Bioscience, San Diego, CA) in sterile saline to administer at 10 μg/kg (★★2 ng/μl for average 20 g mouse).
3. Inject I.P. 100 μl solution using {1/2}-cc, 31-gauge BD Ultra fine II insulin syringe and needle (Becton Dickinson, Franklin Lakes, NJ). (★★This step can be done in conjunction with additional experimental treatments).
4. Wait a minimum of 2 h.
5. Sacrifice animals and harvest tissue as described previously.
6. Analyze tissue samples for autophagy (Fig. 16.4).

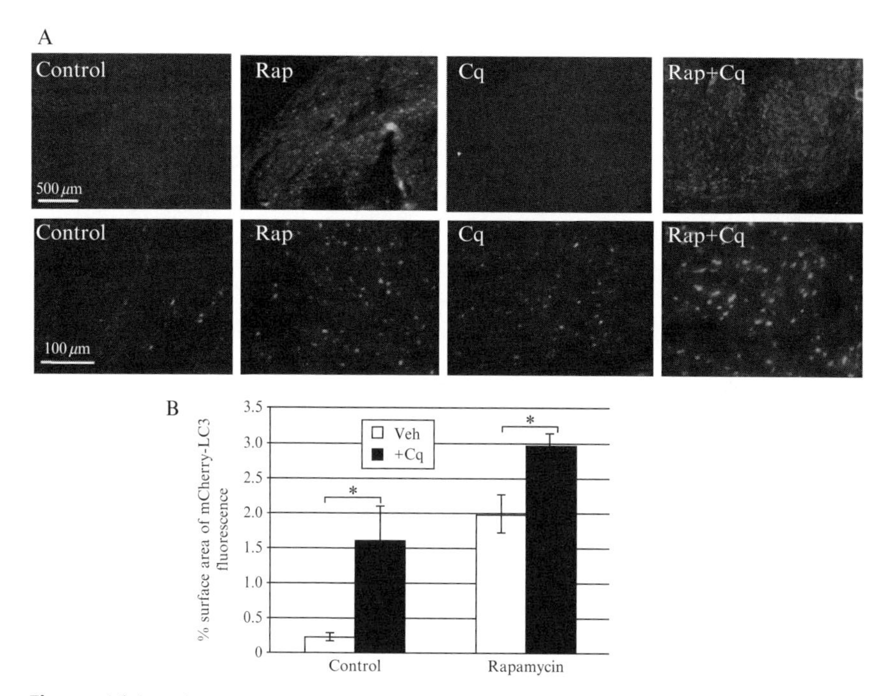

Figure 16.4 Flux in mCherry-LC3 mice treated with rapamycin. Cryosections of mCherry-LC3 mice injected with rapamycin +/− chloroquine (A). Quantification of punctate mCherry fluorescence in tissue sections (B). (Bars 500 μm and 100 μm, respectively) Reprinted from Iwai-Kanai *et al.*, 2008, with permission from Landes Bioscience.

2.9. MDC method

1. Weigh animals to determine correct dosage.
2. Prepare 100 μl of MDC solution (Sigma Aldrich, St. Loius, MO, 30532) in sterile saline to administer at 1.5 mg/kg (**0.3 μg/ml for average 20-g mouse).
3. Inject I.P. 100 μl solution using {1/2}-cc, 31-gauge BD Ultra fine II insulin syringe and needle (Becton Dickinson). (**This step can be done in conjunction with additional experimental treatments).
4. Wait a minimum of 1 h.
5. Sacrifice animals and harvest tissue as described previously.
6. Analyze tissue samples for autophagy (Fig. 16.5). MDC excitation/emission is 365/525 nm respectively.

Note: MDC is best visualized with a filter equipped for DAPI staining, however, its fluorescence diminishes quickly so great care must be taken to protect samples from light and to evaluate immediately after processing. Tissue fixation also reduces signal intensity so it is recommended not to process samples for fixation if possible.

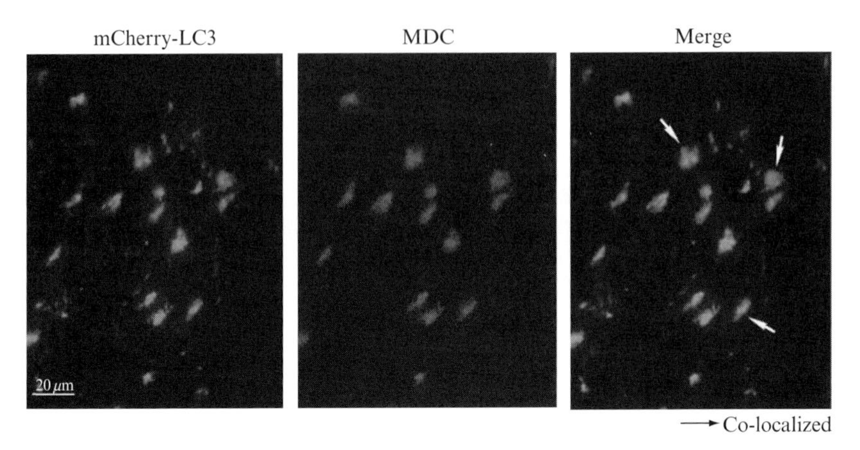

Figure 16.5 Monodansylcadaverine (MDC) fluorescence in mCherry-LC3 mice. mCherry-LC3 transgenic heart tissue (after rapamycin stimulation and injection with monodansylcadaverine) showing colocalization. (Bar 20 μm.) Reprinted from Iwai-Kanai *et al.*, 2008, with permission from Landes Bioscience. (See Color Insert.)

2.10. Quantitative cadaverine plate reader assay

1. Mince 1- to 5-mm^3 tissue sample in 1–2 mL of homogenization buffer in a 35-mm dish (adjust the volume depending on the starting tissue size).
2. Polytron at half speed, 5 s, on ice in a 15-mL round bottom polypropylene tube.
3. Spin out the nuclei and heavy membranes at 1000×*g*, 5 min, at 4 °C in a 15-mL Falcon tube.
4. Move the postnuclear supernatant fraction into a 1.5-mL microcentrifuge tube.
5. Add Alexa Fluor 488 Cadaverine (Invitrogen, A-30676) to a final concentration of 25 μ*M* from a 5 m*M* stock.
6. Incubate on ice 10 min protected from light.
7. Centrifuge the sample at 20,000×*g*, 20 min, at 4 °C.
8. Aspirate the supernatant fraction and rinse the pellet fraction with 1 mL of cold resuspension buffer two times.
9. Completely resuspend the pellet in 350 μL of resuspension buffer, pipetting well.
10. Add 100 μL per well in triplicate to a black 96-well plate (Corning, 3915).
11. Read on a fluorescence plate reader at excitation/emission 495/519 nm.
12. Subtract the readings from wells blanked with resuspension buffer alone.
13. Use the remaining sample to quantify the protein concentration with Coomassie Plus Better Bradford Reagent per the manufacturer's instructions (Thermo Fisher, 23238).

14. Calculate the results as relative fluorescent units (RFUs) per mg of protein.
15. Homogenization Buffer Recipe:
    1g of sucrose
    2mL of 100 m*M* Na_2EDTA
    0.477 g of Hepes free acid
    Bring volume up to 200 mL with distilled H_2O, pH 7.0
16. Resuspension Buffer Recipe:
    1.044 g of KCl
    0.203 g of $MgCl_2$
    0.208 g of MOPS, pH 7.4
    68 mg of KH_2PO_4
    38 mg of EGTA
    Bring final volume up to 100mL with distilled H_2O

3. Discussion

Microtubule-associated protein 1 light chain 3 (LC3), an 18-kDa mammalian homolog of autophagy-related protein 8 (Atg8) in yeast, is processed and conjugated to the nascent autophagosome membrane at the initiation of autophagy (Kabeya *et al.*, 2000). A major obstacle to the study of autophagy *in vivo* is the difficulty of quantifying autophagosomes in tissue. This is greatly aided by the introduction of the transgenic mouse expressing GFP-LC3 in all tissues (Mizushima *et al.*, 2004). However, due to the difficulty of obtaining this mouse from Japan, we created our own transgenic line, in which the red fluorescent protein, mCherry, is fused to LC3 and driven by the alpha-myosin heavy chain promoter for cardiac-restricted expression (Iwai-Kanai *et al.*, 2008). For studies of cardiac autophagy, this eliminates the potentially confusing contribution of endothelial cells and fibroblasts present in numbers equal to cardiomyocytes. The use of mCherry-LC3 offers several advantages over GFP-LC3: it retains fluorescence even in the acidic environment of the lysosome, and there is very little red autofluorescence background. Additionally it will be possible to cross these mice with others expressing GFP-tagged proteins. Our characterization of the αMHC-targeted mCherry-LC3 mice indicates no apparent effects on cardiac function and the mCherry-LC3 reports autophagy as expected. Our comparisons of wild-type and mCherry-LC3 hearts thus far indicate that the transgene does not affect autophagy and functions as a reliable reporter of autophagosome formation (and accumulation).

Furthermore, this mouse model provides a rich source of fluorescently labeled autophagosomes that may be useful in future biochemical analysis requiring concentrated, pure AV preparations. For instance, we are currently developing protocols for the isolation and characterization of autophagosomes sorted by flow cytometry (unpublished data). This technique enables us to characterize associated proteins under various conditions such as starvation and organelle-targeted insults (e.g., rotenone) in a pure preparation of isolated AVs and most importantly provides quantitative data on particle numbers.

As discussed, increased numbers of autophagosomes do not necessarily mean increased autophagic flux, as autophagosomes may accumulate if they are not cleared through lysosomal degradation. To address this, we have established a method to assess flux *in vivo* using chloroquine in place of Bafilomycin A_1 as presented here. Studies in cell culture indicated that Bafilomycin A_1 and chloroquine were equally effective and that adding inhibitors of lysosomal proteases was not necessary (data not shown). Accordingly, we injected chloroquine i.p. into mice to assess effects on autophagy *in vivo*. In transgenic mice expressing mCherry-LC3, rapamycin administration caused an increase in the abundance of mCherry-LC3-labeled AVs in the myocardium (Fig. 16.4). This was further enhanced by the co-administration of chloroquine, indicating that the increase in AVs after rapamycin administration was due to increased flux, not diminished clearance. These findings clearly indicate that chloroquine is useful in evaluation of autophagic flux *in vivo*. Moreover, these studies show that chloroquine can be used to suppress the late phase of autophagy, which may be of therapeutic value in certain conditions.

Whereas both the GFP-LC3 mice and our mCherry-LC3 mice provide a great advantage in the study of autophagy, we also wanted to develop a method to measure autophagy in nontransgenic mice that can be more widely used and does not require maintenance of an additional colony. Monodansylcadaverine has been used to detect autophagosomes, although it has been criticized for being nonspecific. We assessed colocalization of MDC with mCherry-LC3 in mouse hearts, and found excellent colocalization, and that MDC labeling increased in parallel with mCherry-LC3 puncta (Fig. 16.5). Under our conditions, MDC represents a valid marker of autophagosomes. Whereas GFP-LC3 fluorescence is lost in the acidic environment of the lysosome, mCherry fluorescence is stable at an acidic pH allowing detection of both early and late autophagosomes (Kimura *et al.*, 2007). Thus it is conceivable that the colocalization of MDC with many (but not all) LC3-mCherry puncta is an indication of the number of autolysosomes. If true, MDC would represent an accurate indicator of flux, as it would only label autophagosomes that have fused with lysosomes. However, MDC is also suggested to incorporate into membranes based on lipid characteristics independent of pH (Neimann *et al.*, 2000). If the incorporation is related to the double-membrane structure of the autophagosome, then MDC incorporation would be expected even when chloroquine is used to block autophagosome-lysosome fusion, as we have observed. In further validation studies, we have found that this method is suitable for nontransgenic animals and potentially other tissue types and therefore will advance the study of autophagy. In addition, we have investigated the efficacy of alternative fluorescently labeled cadaverine compounds such as Alexa Fluor 488 Cadaverine and Cadaverine Texas Red and provide here a completely novel quantitative assay for examining autophagy using isolated autophagosomes (manuscript in preparation). These innovative methods provide necessary tools to advance the study and understanding of cardiac autophagy in animal models.

ACKNOWLEDGMENTS

This work was funded by San Diego State Research Foundation and grants from NIH/NHLBI P01 HL08557701-02, NIH/NHLBI F31 HL091723-01, NIH/NHLBI RO1 HL69020 and NIH/NIA R01 AG33283. Mice were treated in accordance with the guidelines of the National Institutes of Health Guide for the Care and Use of Laboratory Animals. All animal protocols were approved by the Animal Care Committee of the San Diego State University in San Diego, CA.

REFERENCES

Baines, C. P., Kaiser, R. A., Purcell, N. H., *et al.* (2005). Loss of cyclophilin D reveals a critical role for mitochondrial permeability transition in cell death. *Nature* **434,** 658–662.

Decker, R. S., Poole, A. R., Crie, J. S., *et al.* (1980). Lysosomal alterations in hypoxic and reoxygenated hearts. II. Immunohistochemical and biochemical changes in cathepsin D. *Am. J. Pathol.* **98**(2), 445–456.

Hamacher-Brady, A., Brady, N. R., and Gottlieb, R. A. (2006). Enhancing macroautophagy protects against ischemia/reperfusion injury in cardiac myocytes. *J. Biol. Chem.* **281**(40), 29776–29787.

Hamacher-Brady, A., Brady, N. R., Logue, S. E., *et al.* (2007). Response to myocardial ischemia/reperfusion injury involves Bnip3 and autophagy. *Cell Death Differ.* **14,** 146–157.

Hein, S., Arnon, E., Kostin, S., *et al.* (2003). Progression from compensated hypertrophy to failure in the pressure-overloaded human heart: Structural deterioration and compensatory mechanisms. *Circulation* **107**(7), 984–991.

Iwai-Kanai, E., Yuan, H., Huang, C., *et al.* (2008). A Method to Measure Cardiac Autophagic Flux *in vivo*. *Autophagy* **4**(3), 322–329.

Kabeya, Y., Mizushima, N., Ueno, T., Yamamoto, A., Kirisako, T., Noda, T., Kominami, E., Ohsumi, Y., and Yoshimori, T. (2001). LC3, a mammalian homologue of yeast Apg8p, is localized on autophagosome membranes after processing. *EMBO J.* **19,** 5720–5728.

Kawai, A., Uchiyama, H., Takano, S., Nakamura, N., and Ohkuma, S. (2007). Autophagosome-lysosome fusion depends on the pH in acidic compartments in CHO cells. *Autophagy* **3,** 154–157.

Kimura, S., Noda, T., and Yoshimori, T. (2007). Dissection of the autophagosome maturation process by a novel reporter protein, tandem fluorescent-tagged LC3. *Autophagy* **3**(5), 452–460.

Komatsu, M., Waguri, S., Ueno, T., *et al.* (2005). Impairment of starvation-induced and constitutive autophagy in Atg7-deficient mice. *J. Cell Biol.* **169,** 425–434.

Kuma, A., Hatano, M., Matsui, M., *et al.* (2004). The role of autophagy during the early neonatal starvation period. *Nature* **432,** 1032–1036.

Luptak, I.., Yan, J., Cui, L., *et al.* (2007). Long-term effects of increased glucose entry on mouse hearts during normal aging and ischemic stress. *Circulation* **116,** 901–909.

Matsui, Y., Takagi, H., Qu, X., *et al.* (2007). Distinct roles of autophagy in the heart during ischemia and reperfusion. Roles of AMP-activated protein kinase and Beclin 1 in mediating autophagy. *Circ. Res.* **100,** 914–922.

May, D., Gilon, D., Djonov, V., *et al.* (2008). Transgenic system for conditional induction and rescue of chronic myocardial hibernation provides insights into genomic programs of hibernation. *Proc. Natl. Acad. Sci. USA* **105,** 282–287.

Mizushima, N., Yamamoto, A., Matsui, M., *et al.* (2004). *In vivo* analysis of autophagy in response to nutrient starvation using transgenic mice expressing a fluorescent autophagosome marker. *Mol. Biol. Cell* **15**(3), 1101–1111.

Munafo, DB., and Colombo, MI. (2001). A novel assay to study autophagy: Regulation of autophagosome vacuole size by amino acid deprivation. *J. Cell Sci.* **114,** 3619–3629.

Nakai, A., Yamaguchi, O., Takeda, T., *et al.* (2007). The role of autophagy in cardiomyocytes in the basal state and in response to hemodynamic stress. *Nat. Med.* **13**(5), 539–541.

Niemann, A., Takatsuki, A., and Elsasser, H. P. (2000). The lysosomotropic agent monodansylcadaverine also acts as a solvent polarity probe. *J. Histochem. Cytochem.* **48**(2), 251–258.

Odashima, M., Usui, S., Takagi, H., *et al.* (2007). Inhibition of endogenous Mst1 prevents apoptosis and cardiac dysfunction without affecting cardiac hypertrophy after myocardial infarction. *Circ. Res.* **100,** 1344–1352.

Ohkuma, S., and Poole, B. (1978). Fluorescence probe measurement of the intralysosomal pH in living cells and the perturbation of pH by various agents. *Proc. Natl. Acad. Sci. USA* **75,** 3327–3331.

O'neil, P. M., Bray, P. G., Hawley, S. R., Ward, S. A., and Park, B. K. (1998). 4-Amnoquinolines-past, present and future: A chemical perspective. *Pharmacol. Ther.* **77,** 29–58.

Pattingre, S., Tassa, A., Qu, X., *et al.* (2005). Bcl-2 anti-apoptotic proteins inhibit beclin 1-dependent autophagy. *Cell* **122**(6), 927–939.

Poole, B., and Ohkuma, S. (1981). Effect of weak bases on the intralysosomal pH in mouse peritoneal macrophages. *J. Cell Biol.* **90,** 665–669.

Qu, X., Yu, J., Bhagat, G., *et al.* (2003). Promotion of tumorigenesis by heterozygous disruption of the *beclin 1* autophagy gene. *J. Clin. Invest.* **112,** 1809–1820.

Sadoshima, J., Montagne, O., Wang, Q. M., *et al.* (2002). The MEKK1-JNK pathway plays a protective role in pressure overload, but does not mediate cardiac hypertrophy. *J. Clin. Invest.* **110,** 271–279.

Sewell, R. B., Barham, S. S., and LaRusso, N. F. (1983). Effect of chloroquine on the form and function of hepatocyte lysosomes: Morphologic modifications and physiologic alternations related to the biliary excretion of lipids and proteins. *Gastroenterology* **85,** 1146.

Sybers, HD., Ingwall, J., and DeLuca, M. (1976). Autophagy in cardiac myocytes. *Recent Adv. Stud. Cardiac Struct. Metab.* **12,** 453–463.

Wohlgemuth, S. E., Julian, D., Akin, D. E., *et al.* (2007). Autophagy in the heart and liver during normal aging and calorie restriction. *Rejuvenation Res.* **10,** 281–292.

Yamamoto, A., Tagawa, Y., Yoshimori, T., Moriyama, Y., *et al.* (1998). Bafilomycin A1 prevents maturation of autophagic vescicles by inhibiting fusion between autophagosomes and lysosomes in rat hepatoma cell line, H-4-II-E cells. *Cell Struct. Funct.* **23**(1), 33–42.

Yan, L., Vatner, D. E., Kim, S. J., *et al.* (2005). Autophagy in chronically ischemic myocardium. *Proc. Natl. Acad. Sci. USA* **102**(39), 13807–13812.

Yan, C. H., Yang, Y. P., Qin, Z. H., Gu, Z. L., Reid, P., and Liang, Z. Q. (2007). Autophagy is involved in cytotoxic effects of crotoxin in human breast cancer cell line MCF-7 cells. *Acta Pharmacol. Sin.* **28,** 540–548.

Zhu, H., Tannous, P., Johnstone, J. L., *et al.* (2007). Cardiac autophagy is a maladaptive response to hemodynamic stress. *J. Clin. Invest.* **117,** 1782–1793.

CHAPTER SEVENTEEN

Autophagy in Load-Induced Heart Disease

Hongxin Zhu,* Beverly A. Rothermel,* *and* Joseph A. Hill*,†

Contents

* Department of Internal Medicine (Cardiology), University of Texas Southwestern Medical Center, Dallas, Texas, USA
† Department of Molecular Biology, University of Texas Southwestern Medical Center, Dallas, Texas, USA

Methods in Enzymology, Volume 453
ISSN 0076-6879, DOI: 10.1016/S0076-6879(08)04017-2

Abstract

The heart is a highly plastic organ capable of remodeling in response to changes in physiological or pathological demand. When workload increases, the heart compensates through hypertrophic growth of individual cardiomyocytes to increase cardiac output. However, sustained stress, such as occurs with hypertension or following myocardial infarction, triggers changes in sarcomeric protein composition and energy metabolism, loss of cardiomyocytes, ventricular dilation, reduced pump function, and ultimately heart failure. It has been known for some time that autophagy is active in cardiomyocytes, occurring at increased levels in disease. Yet the potential contribution of cardiomyocyte autophagy to ventricular remodeling and disease pathogenesis has only recently been explored. This latter fact stems largely from the recent emergence of tools to probe molecular mechanisms governing cardiac plasticity and to define the role of autophagic flux in the context of heart disease. In this chapter, we briefly review prominent mouse models useful in the study of load-induced heart disease and standard techniques used to assess whether a molecular or cellular event is adaptive or maladaptive. We then outline methods available for monitoring autophagic activity in the heart, providing detailed protocols for several techniques unique to working with heart and other striated muscles.

1. Introduction

Heart disease is the leading cause of death in the industrialized world. In many instances, heart disease culminates in heart failure, a syndrome characterized by the heart's inability to meet the metabolic demands of the body. Currently, 5 million Americans suffer from chronic heart failure, the leading hospital discharge diagnosis in Medicare (Rosamond *et al.*, 2008). Mortality from heart failure is approximately 50% at 5 years (Rosamond *et al.*, 2008).

In response to increases in cardiac workload, a compensatory hypertrophic growth response ensues (Hill and Olson, 2008). However, under conditions of sustained load, such as occurs in patients with uncontrolled hypertension, the heart progresses to a state of decompensated heart failure (load-induced heart failure). Consistent with this, it has been shown clinically, epidemiologically, and experimentally that left ventricular hypertrophy is a predictor of increased risk of heart failure (Hein *et al.*, 2003; Levy *et al.*, 1996). As such, recent work has focused on the hypertrophic phenotype itself as a therapeutic target (Frey *et al.*, 2004). However, cellular mechanisms underlying pathological cardiac remodeling remain poorly understood.

Autophagy is a highly conserved mechanism of protein degradation whereby damaged proteins and organelles can be removed and recycled through lysosomes, a process that supports cell survival during times of stress. In the context of certain developmental and disease states, however, autophagy promotes cell death (autophagic programmed cell death, type II PCD) (Levine and Kroemer, 2008; Shintani and Klionsky, 2004). It has been known for 30 years that lysosomal pathways are activated in a variety of models of heart disease (Dammrich and Pfeifer, 1983; Decker and Wildenthal, 1980; Decker *et al.*, 1980; Pfeifer *et al.*, 1987; Sybers *et al.*, 1967). Indeed, a number of groups report increases in lysosomal activity in tissue samples from diseased and failing human hearts (Elsasser *et al.*, 2004; Hein *et al.*, 2003; Knaapen *et al.*, 2001; Kostin *et al.*, 2003; Lockshin and Zakeri, 2002; Shimomura *et al.*, 2001; Yamamoto *et al.*, 2000). More recent studies have revealed an important role for autophagy in the cardiomyocyte response to numerous types of stress, and compelling evidence has emerged that this autophagic response participates in the pathogenesis of disease. Recent advances in our understanding of the molecular mechanisms controlling autophagy (Klionsky, 2007; Kundu and Thompson, 2008; Mizushima, 2007) have led to the emergence of novel tools and techniques to probe this process during development and in the progression of disease. Our laboratory has adapted many of these techniques to monitor autophagy specifically in the heart (Tannous *et al.*, 2008a,b; Zhu *et al.*, 2007), allowing us to explore its contribution to the pathogenesis of load-induced heart failure.

The aim of this chapter is to review standard models of load-induced heart disease, to discuss strategies for assessing whether a molecular event is adaptive or maladaptive, and to outline methods available for monitoring autophagic activity in the heart. As part of this, we provide detailed protocols for several techniques unique to working with heart and other striated muscles.

2. Mouse Models of Load-Induced Heart Disease

2.1. Thoracic aortic banding (TAB)

The most commonly used surgical intervention to promote pressure overload-induced cardiac hypertrophy in mice is thoracic aortic banding (TAB; also termed *thoracic aortic constriction*, TAC). This procedure entails constricting the transverse aortic arch, between the innominate and left common carotid arteries, using a 27-gauge needle as a guide (Rockman *et al.*, 1991). At 3 weeks, the hypertrophic response has reached steady-state (Hill *et al.*, 2000), and the integrity of aortic banding can be confirmed by inspection of the surgical constriction and by visualization of marked differences in caliber of the right and left carotid arteries. In our hands, working with adult C57BL6 mice, this operation induces a stable compensated hypertrophic response with an approximately 40% increase in the heart weight to body weight ratio at 3 weeks (Hill *et al.*, 2000). The mice display no clinical signs of heart failure or malignant ventricular arrhythmia.

2.2. Severe thoracic aortic banding (sTAB)

Tighter constriction of the aorta to the diameter of a 28-gauge needle leads to pressure-overload failure (Rothermel *et al.*, 2005). Mice subjected to sTAB gradually develop signs of circulatory failure, including lethargy, impaired mobility, diminished appetite, and peripheral edema. Importantly, sTAB mice have elevated serum levels of tumor necrosis factor (TNFα), a marker of heart failure (Rothermel *et al.*, 2005). There is a doubling of the heart weight to body weight ratio over the course of three weeks and a substantial increase in mortality. It is important to be aware that, unlike cardiac remodeling in hypertensive patients, the onset of pressure overload is abrupt in the TAB and sTAB surgical models. Despite this important difference, these two models have proven to be invaluable techniques for dissecting pathways leading to load-induced hypertrophy and heart failure.

3. Analysis of Ventricular Remodeling

3.1. Assessing changes in cardiac function

Echocardiography is a standard approach for assessing changes in cardiac size and function, such as contractile performance, left ventricular end diastolic dimension (LVEDD), and left ventricular end systolic dimension (LVESD). Percentage fractional shortening, a measure of ventricular contraction, is significantly decreased in sTAB hearts, and LVEDD and LVESD are each increased. Technical details will depend on the equipment available; however, many machines designed for use in humans can be adapted for use on mice provided a small enough scan head of sufficient frequency is used. Because anesthesia can depress ventricular function, it is important to carry out these measurements on unanesthetized mice. This can be facilitated by acclimating the mice to handling and using a depilatory agent the day before study to remove chest hair. Invasive measurements of left ventricular pressures, along with estimates of ventricular volume using a conductance catheter, provide fundamental information regarding changes in ventricular compliance and inotropy (the force of muscle contraction). These data are typically analyzed as pressure-volume (PV) loops. Whereas this technique is powerful, it requires a high level of expertise and is not as readily available as echocardiography. An elevation in the lung weight to body weight ratio can be a useful indication of pulmonary venous congestion, but care must be taken to remove and weigh the lungs rapidly after sacrifice of the mouse to avoid postmortem edema of the lungs. Changes in circulating TNFα may be measured in serum or plasma using commercially available ELISA kits.

3.2. Assessing changes in ventricular morphology

Changes in cardiac mass are reported as total heart weight or LV weight normalized to body weight or tibia length. It is often important to report both, as differences in body weight due to fat content or edema often occur. Changes in morphology can be viewed in histological 4-chamber views, and the presence of fibrosis visualized using trichrome (or picrosirius red) staining for collagen. Hypertrophy of individual cardiomyocytes is measured as changes in cell cross-sectional area. This technique requires careful controls to assure measurement of cardiomyocytes sectioned perpendicularly to the plane of evaluation. Care must also be taken to minimize shrinkage of the tissue during processing. Alternatively, individual cardiomyocytes can be enzymatically dissociated from the heart and the two dimensional surface area measured. This approach, however, is limited in the number of cells that can be analyzed, as well as the possibility of selection bias (e.g., differential survival) during cell isolation. Assessment of changes in fine structural detail, such as sarcomere disarray or the accumulation of autophagic vesicles, requires electron microscopy. When assessing any of these morphological parameters it is essential that the hearts be in a similar state of relaxation. This can be achieved by perfusing hearts first with PBS then Krebs-Henseleit solution (120 m*M* NaCl, 4.5 m*M* KCl, 2.5 m*M* $CaCl_2$, 1.0 m*M* $MgCl_2$, 27 m*M* $NaHCO_3$, 1.0 m*M* KH_2PO_4, and 10 m*M* glucose, pH 7.4) prior to fixation.

3.3. Assessing changes in gene expression and intracellular signaling cascades

Cardiac hypertrophy and heart failure are associated with reactivation of a so-called fetal gene program in the ventricle. Many of these genes are expressed both during embryogenesis and in adult atria so it is essential to remove the atria before analysis of the LV. There is an increase in expression of the natriuretic (i.e., involved in the discharge of sodium through urine) peptides ANP (*Nppa*) and BNP (*Nppb*) as well as alpha skeletal actin (*Acta1*) and the fetal cardiac myosin heavy chain βMHC (*Myh7*). There is a corresponding decrease in expression of the sarcoplasmic reticulum calcium

ATPase SERCA2 (*Atp2a2*) and the adult cardiac myosin heavy chain αMHC (*Myh6*). Of course, many other changes in transcriptional profiles occur during cardiac remodeling, but changes in expression of these genes are widely used as markers of pathological remodeling. Sequences of several useful real-time PCR primers are provided in Table 17.1.

Activation of a variety of intracellular signaling cascades has been implicated in cardiac remodeling (Heineke and Molkentin, 2006); many of these cascades are also involved in the regulation of autophagy. Whereas these pathways are too numerous to detail here, several recent reviews are available which provide links to techniques for monitoring these pathways (Frey and Olson, 2003; Heineke and Molkentin, 2006) (see also the chapters by Chiacchiera and Simone and by Ikenoue *et al.*, in volume 452). Many of these signaling cascades involve changes in phosphorylation that are rapidly reversed once the heart is

Table 17.1 PCR primer sequences for mouse tissue

Protein (*Gene*)	Primer Sequence
LC3b (Map1lc3b)	5′-CGT CCT GGA CAA GAC CAA GT-3′ 5′-ATT GCT GTC CCG AAT GTC TC-3′
Atg12	5′-GGC CTC GGA ACA GTT GTT TA-3′ 5′-CAG CAC CGA AAT GTC TCT GA-3′
Atg4b	5′-ATT GCT GTG GGG TTT TTC TG-3′ 5′-AAC CCC AGG ATT TTC AGA GG-3′
Beclin (Becn1)	5′-GGC CAA TAA GAT GGG TCT GA-3′ 5′-CAC TGC CTC CAG TGT CTT CA-3′
Bnip3	5′-TTC CAC TAG CAC CTT CTG ATG A-3′ 5′-GAA CAC CGC ATT TAC AGA ACA A-3′
Cathepsin L (Ctsl)	5′-GTG GAC TGT TCT CAC GCT CAA G-3′ 5′-TCC GTC CTT CGC TTC ATA GG-3′
Gabarap	5′-CAT CGT GGA GAA GGC TCC TA-3′ 5′-ATA CAG CTG GCC CAT GGT AG-3′
β-MHC (Myh7)	5′-GCA TTC TCC TGC TGT TTC CTT-3′ 5′-TGG ATT CTC AAA CGT GTC TAG TGA-3′
ANF (Nppa)	5′-GTA CAG TGC GGT GTC CAA CA-3′ 5′-TCT CCT CCA GGT GGT CTA GCA-3′
GAPDH	5′-CAA GGT CAT CCA TGA CAA CTT TG-3′ 5′-TCT CCT CCA GGT GGT CTA GCA-3′
18S rRNA	5′-ACC GCA GCT AGG AAT AAT GGA-3′ 5′-TCT CCT CCA GGT GGT CTA GCA-3′
CyclophilinA (Ppia)	5′-CAG ACG CCA CTG TCG CTT T-3′ 5′-TGT CTT TGG AAC TTT GTC TGC AA-3′

removed and stops beating. Thus, when harvesting hearts to examine changes in protein phosphorylation, the heart should be freeze-clamped in liquid nitrogen immediately upon removal from the chest without taking time to weigh it or washing in either PBS or Krebs-Henseleit solution.

3.4. Distinguishing between adaptive and maladaptive remodeling of the heart

A striking feature of the heart is its capacity for structural remodeling in response to changes in environmental demand (Hill and Olson, 2008). Indeed, the heart is a remarkably plastic organ, capable of growing or shrinking in the setting of a variety of physiological or pathological stimuli. In some instances, cardiac growth is beneficial, facilitating the organism's response to exercise, postnatal development, or pregnancy. In other instances, such as the stress of hypertensive disease or infarction, cardiac remodeling is maladaptive, predisposing to arrhythmia and heart failure. In yet other instances, such as prolonged bed rest or weightlessness, the heart shrinks.

It is impossible to distinguish between pathological and physiological heart growth on the basis of heart weight to body weight ratios alone. Pathological hypertrophy can occur without any overt clinical signs or measurable changes in cardiac function. This is where analysis of fetal gene expression and the shift in myosin heavy chain isoform from αMHC to βMHC is extremely useful. For instance, mice subjected to a month of exercise training may have an increase in heart weight body weight ratio comparable to that elicited by TAB surgery. Neither heart manifests a decrease in cardiac function or elevated TNFα levels, such as occurs in mice subjected to sTAB. However, fetal gene expression and βMHC protein levels are increased in the TAB heart and not in the exercise-induced hypertrophic heart. Thus, it is critical to analyze a full range of parameters, from changes in transcription to functional clinical assessments, to distinguish between adaptive and maladaptive hypertrophy.

4. *In Vitro* Models of Load-Induced Hypertrophy

It is not possible to model load-induced hypertrophy in cultured cells. However, cultured ventricular myocytes isolated from neonatal rat pups (NRVM) are the standard model for studying the biology of cardiac myocytes in culture. We include protocols for examining autophagy in NRVM, because they can be used to test the direct effect of downstream mediators, such as TNFα, on autophagic processes in culture. Detailed procedures for their isolation and culture have been published (Maass and Buvoli, 2007).

5. Techniques to Analyze Cardiomyocyte Autophagy

The heart poses a number of unique challenges for analyzing autophagy. Although hypertrophy involves an increase in the size of cardiomyocytes, the heart is composed of many other cell types including fibroblasts, endothelial cells, smooth muscle cells and leukocytes. Indeed, it is estimated that these other cell types account for between 40% and 60% of the cell population in the heart (Brown *et al.*, 2005), although the bulk mass of protein is contributed by cardiomyocytes. Here, we describe cardiomyocyte-specific GFP-LC3 reporter mice we have developed that can be used to monitor autophagic activity specifically in cardiomyocytes. The presence of a densely packed sarcomere also poses unique challenges for immunohistochemistry and for isolation of soluble and insoluble proteins from heart. We outline some of the techniques that have proven useful.

5.1. The αMHC-GFP-LC3 cardiomyocyte-specific autophagy reporter mouse

GFP fused to microtubule-associated protein 1 light chain 3 (LC3), the mammalian homologue of yeast Atg8, provides a useful tool for monitoring autophagic activity *in vivo*. LC3 is an intermembrane component of the early autophagosome, and its redistribution from a diffuse cytosolic signal to punctate dots is a sensitive and specific indicator of autophagy (Kabeya *et al.*, 2000). Transgenic mice carrying a *GFP-LC3* transgene driven by a constitutive CAG-CMV promoter/enhancer have proven to be a powerful tool for *in vivo* analysis of autophagy (Mizushima *et al.*, 2004), where autophagosome abundance can be measured as an increase in GFP-LC3 puncta using fluorescence microscopy. Expression of the *CAGpCMV-GFP-LC3* transgene does not increase the level of autophagic activity. In these mice, however, the GFP-LC3 fusion protein is expressed in all cell types. Therefore, to study autophagy specifically in cardiomyocytes, we used the alpha myosin heavy chain (αMHC) promoter to drive cardiomyocyte-specific expression of a *GFP-LC3* transgene in the adult heart. (The *αMHC-GFP-LC3* transgene is not expressed during embryogenesis, so it cannot be used to study the role of autophagy during cardiac development.) Two independent lines of *αMHC-GFP-LC3* transgenic mice were obtained. Neither line shows any changes in heart development, cardiac growth, or cardiac function when compared to wild-type littermates. The *αMHC-GFP-LC3* transgene can be easily identified by PCR using the forward and reverse primers (5′-CATCGAGCTGAAGGGCATCG-3′ and 5′-CTATAATCACTGGGATCTTGGTG-3′). These reporter mice can be subjected to TAB and sTAB stress as well as mated to genetic backgrounds or transgenic mice of interest.

5.2. Detection of GFP-positive autophagic vesicles in the αMHC-GFP-LC3 reporter mouse

1. Retrograde perfuse heart with ice-cold PBS (10 mL for an adult mouse).
2. Perfuse with freshly prepared ice-cold 4% paraformaldehyde (PFA)/PBS (15–20 mL for an adult mouse).
3. Remove the heart and rinse with PBS.
4. Fix in 4% PFA/PBS for 4 h at 4 °C.
5. Section the heart in half for a 4-chamber view.
6. Equilibrate the heart in 30% sucrose/PBS overnight at 4 °C.
7. Mount the sucrose-equilibrated hearts in freezing matrix (TFM, Triangle Bioscience, Raleigh, NC, catalog number: TFM-C).
8. Section with a cryostat (8 μm, Leica CM3000) and mount on glass slides.
9. Air-dry the slides (the sections can be stored at −80 °C until use).
10. Rinse slides with PBS for 5 min.
11. Mount slides with Vectashield (Vector Laboratories, Burlingame, CA, catalog number: H1000) mounting media and a coverslip.
12. Cover-slipped slides can be store at −20 °C until photographing.
13. The GFP signal can be viewed using an Endow GFP Bandpass Emission filter set (Chroma 41017; Rockingham, VT).

Notes

1. Results can be expressed as the number of GFP-LC3 dots per field of high magnification or the number of GFP-LC3 dots per cell. Because the size of individual cardiomyocytes increases with hypertrophy, we suggest expressing the results as the number of GFP-LC3 dots per field of high magnification.
2. Multiple fields should be photographed and quantified. Autophagic activity is not homogeneous across the heart, neither is transgene expression driven by the αMHC promoter. For unclear reasons, autophagy activity is greater in the basal septum both at rest and in response to starvation or pressure overload (Zhu *et al.*, 2007). Therefore it is important to quantify different regions of the heart independently.
3. The heart has a high level of autofluorescence. For photographic images, a Texas Red filter set (red) can be used to exclude the autofluorescent signal.

6. Immunohistochemistry for LAMP-1 or Cathepsin D to Monitor Changes in Lysosomal Abundance

Quantification of autophagosome abundance is insufficient to study autophagy, a flux pathway of protein and organelle recycling. (Increases in autophagosome abundance could derive simply from defects in the pathway downstream, like a traffic tie-up occurring due to blockage of the road in the distance.) Thus, it is critical to evaluate autophagy using a variety of strategies and to test downstream mechanisms of autophagosome processing.

Autophagosomes fuse with lysosomes to form autolysosomes in which the cargo within in the autophagosome is degraded. To evaluate possible changes in the processing of autophagosomes, it is important to monitor changes in lysosome abundance. This section describes the procedure for immunohistochemical detection of the lysosome-associated membrane protein 1 (LAMP-1) and the lysosomal protease cathepsin D in the heart. Immunohistochemisty of striated muscle can be difficult because of the abundance of sarcomeric proteins. Because these proteins represent such a high percentage of the total protein present in a cardiomyocyte, a low affinity cross-reaction of an antibody to a sarcomeric protein could result in a non-specific banded sarcomeric pattern of staining. We have established reliable protocols for LAMP-1 and cathepsin D staining in the heart.

6.1. Immunohistochemistry for LAMP-1 or cathepsin D

1. Perfuse and fix hearts as for GFP-LC3 (steps 1–3 previously).
2. Fix in 4% PFA/PBS overnight at 4 °C.
3. Dehydrate and paraffin embed: Take tissue through a series of alcohol dehydration steps at room temperature (70% for 30 seconds, followed by 80%, 90%, 96%, and 100% each for an hour) followed by 3 complete changes of xylene (i.e., the xylene is removed and replaced every hour for a total of 3 times). The samples are then immersed in Paraplast Plus (McCormick Scientific catalog number 502004) for 1 h 10 min. Then, the Paraplast Plus is removed and replaced with fresh Paraplast Plus for 1 h 20 min, and the process is repeated once more for 1 h 30 min. The embedding in paraffin wax should be performed at 60 °C with agitation and under vacuum.
4. Section with microtome for 4-chamber view (5 μm, Leica RM2255 rotary microtome), slides may be stored at room temperature until needed.
5. Bake slides in 56 °C oven for 1 h.

6. Dewax sections in xylene for 2 changes, 5 min each.
7. Remove xylene from sections by washing in 100% ethanol, 3 changes, 2 min each.
8. Hydrate sections through 3 changes of xylene (5 min each), 3 changes of 100% ethanol (1.5 min each), into 95% ethanol (once for 1.5 min), and then into water.
9. Wash in 1x PBS for 5 min.
10. Fix in 4% PFA for 10 min.
11. Wash 3 times in PBS, 5 min for each wash.
12. Outline sections with PAPPEN (Invitrogen, Carlsbad, CA, catalog number 00-8877).
13. Permeabilize sections in 0.3% Triton X-100 in PBS for 10 min.
14. Wash three times in PBS, 5 min for each wash.
15. Incubate in blocking buffer (1xPBS, 3% goat serum (Vector Biolabs, S1000), 0.5% BSA (Fisher Scientific, BP6755)) for 1 h.
16. Incubate with rat anti-LAMP-1 (1:500 dilution in blocking buffer; Santa Cruz Biotechnology, SC-19992) or rabbit anti-cathepsin D (1:100 dilution; Santa Cruz Biotechnology, SC-10725) in blocking buffer overnight at 4 °C in a humidified environment.
17. Wash 3 times in PBS, 5 min for each wash.
18. Incubate with Cy3 conjugated donkey anti-rat IgG (1:100 dilution in blocking buffer; Jackson Immunology, 712-166-153) or donkey anti-rabbit IgG (1:100 dilution in blocking buffer; Jackson Immunology, 711-165-152) for 30 min.
19. Wash 3 times in PBS, 5 min each wash.
20. Mount with Vectashield antifade mounting medium (Vector Laboratories, Burlingame, CA, catalog number: H1000); slides may be stored at −20 °C until viewing.
21. View using a Cy3 filter set (Chroma 41007; Rockingham, VT).

6.2. Immunohistochemistry for ubiquitinated aggregates

Severe pressure overload triggers accumulation of aggregates of polyubiquitinated protein in the perinuclear zone of the cardiomyocyte (Rothermel and Hill, 2008; Tannous *et al.*, 2008b). Immunohistochemical imaging of the aggregates is similar to the procedure described previously but requires addition of a microwave-mediated antigen retrieval step and an additional blocking step, because of the use of a mouse monoclonal anti-ubiquitin antibody on mouse tissue sections. The microwave-mediated antigen retrieval step involves microwaving the sample in acidic citrate-based buffer, a process that breaks protein-protein aldehyde cross-links initially established by tissue fixation. With cleavage of these aldehyde bonds, antibodies are able to access more of their specific antigenic epitopes and anneal with affinity otherwise unachievable.

1. Perfuse, fix, embed, and section as previously (steps 1–4).
2. Air-dry slides.
3. Dewax sections in xylene for two changes, 5 min each.
4. Remove xylene from sections in 100% ethanol for 3 changes, 2 min each.
5. Hydrate sections through graded ethanol concentrations to H_2O as described earlier.
6. Perform microwave-induced epitope retrieval using 1xBiogenex Citra (Biogenex, San Ramon, CA, HK0869K) 10 min at 95 °C.
7. Cool slides with partial change-out of H_2O for 20 min, where partial change-out means pouring out {1/3} of the water and replacing the volume with room temperature water, thus gradually cooling down the water and the slides.
8. Wash in PBS 3 times, for 2 min each.
9. Permeabilize sections with 0.3% Triton X-100/PBS for 10 min.
10. Wash in PBS 3 times, for 2 min each.
11. Quench autofluorescence with 100 m*M* glycine (not pH adjusted) in PBS twice, for 8 min each.
12. Wash in PBS 3 times, for 2 min each.
13. Block endogenous mouse immunoglobulin activity by applying Mouse-on-Mouse (MOM) IgG Blocking Reagent diluted 1:27 in PBS and incubating for 1 h (Vector Laboratories, Burlingame, CA, catalog number BMK2202).
14. Wash in PBS 3 times, for 2 min each.
15. Equilibrate antigenic epitopes in MOM diluent by applying MOM Protein Concentrate (see previously) diluted in 1:12.5 in PBS and incubating for 5 min.
16. Decant excess MOM diluent, wipe, and apply anti-ubiquitin (UB-1) primary antibody (Santa Cruz, catalog number sc-6085) diluted 1:400 in MOM diluent/PBS and incubate overnight at 4 °C.
17. Wash in PBS 3 times, for 2 min each.
18. Apply MOM biotinylated anti-mouse IgG (see previously) diluted 1:250 in MOM diluent/PBS and incubate 10 min.
19. Wash in PBS 3 times, for 2 min each.
20. Apply fluorescein-avidin DCS (Vector Laboratories, Burlingame, CA, catalog number A2011) diluted 1:62 in PBS and incubate for 5 min.
21. Wash in PBS 3 times, for 2 min each.
22. Rinse in H_2O twice, for 5 min each.
23. Mask muscle specific autofluorescence with 0.1% Sudan black (Sigma-Aldrich; St. Louis, MO, catalog number S4261)/70% ethanol for 30 min (Baschong *et al.*, 2001).
24. Rinse in H_2O twice, for 5 min each.
25. Mount with a coverslip with Vectashield.

6.3. Immunohistochemistry of neonatal heart ventricular myocytes

We use the following protocol to image alpha-B crystallin (CryAB) and desmin localized to protein aggregates in cultured cardiomyocytes (Tannous *et al.*, 2008a). Neonatal heart ventricular myocytes (NRVM) should be grown on coverslips coated with laminin (Fisher Scientific 08774385). PBS used in this procedure should contain Ca^{2+} and Mg^{2+} ("PBS+") to maintain cell integrity. This can be obtained commercially or made using the recipe supplied here:

1. Rinse cells with PBS+ for 5 min, 3 times at room temperature (RT).
2. Fix cells with 4% PFA/PBS+ for 5 min on ice.
3. Rinse cells with PBS for 5 min 3 times at RT.
4. Permeabilize cells with 0.1% Triton/PBS for 2 min at RT.
5. Block nonspecific binding with 1.0% BSA/PBS for 15 min at RT.
6. Incubate cells, coverslip face down, on labeled microscope slides with 50 μL of primary antibody/1%BSA/PBS in a humid chamber for 1 h at RT. A range of antibody dilutions should be tested initially. The concentration of antibody usually needs to be much higher than that used for Western blot analysis. To label concurrently for desmin and CryAB, use mouse anti-desmin (1:50; Abcam, catalog number AB8592) and rabbit anti-CryAB (1:50; Vision, catalog number Abcrys512). Remember to include a PBS control without antibody.
7. Float coverslip off slide with 1.0% BSA/PBS, and using forceps, carefully return to a 6-well plate, keeping the orientation intact.
8. Rinse cells with PBS for 5 min 3 times at RT.

All remaining steps to be conducted in subdued lighting:

9. Incubate cells, coverslip face down, on labeled microscope slides with 50 μL of 1:12 diluted fluorescent secondary antibodies/1.0%BSA/PBS in a humid chamber for 30 min at RT. Use rhodamine-tagged goat anti-mouse IgG (Jackson ImmunoResearch, catalog number 115-025-003) to detect desmin and fluorescein-tagged goat anti-rabbit (Jackson Immuno Research, catalog number 111-095-045) to detect CryAB.
10. Float coverslip off slide with 1.0% BSA/PBS, and using forceps, carefully return to 6-well plate, keeping the orientation intact.
11. Rinse the coverslips with 1.0%BSA/PBS for 2 min.
12. Stain cells with Hoechst 33258 (Molecular Probes, catalog number H3570) diluted 1:50 in PBS for 2 min at RT.
13. Rinse cells with PBS for 2 min 3 times at RT.
14. Rinse cells with distilled water for 2 min twice at RT.
15. Mount coverslips on microscope slides with ~18 μL Vectashield mounting medium.

6.4. Preparation of PBS with Ca^{2+} and Mg^{2+} (PBS+)

To avoid precipitation, two 20x stock solutions should be prepared and autoclaved separately.

solution A : 348 mM Na_2HPO_4: 46.86 g
70 mM NaH_2PO_4: 9.66 g
Bring to 1 liter and autoclave

solution B : 18 mM $CaCl_2$: 2.6 g
70 mM KCl: 5.2 g
18 mM $MgCl_2$: 3.6 g
2.74 M NaCl: 160 g
Bring to 1 liter and autoclave

Store stocks at room temperature. Add 5 mL of each to 90 mL of water for working stock. Do not refrigerate.

6.5. Generation of a rabbit polyclonal anti-LC3 antibody

One of the initial steps in autophagy is conversion of LC3 from the cytosolic LC3-I isoform to the PE-conjugated membrane-bound LC3-II isoform. Changes in the abundance of LC3-II or the ratio of LC3-II to LC3-II + LC3-I have been used to track changes in autophagic flux (though the latter is discouraged because of differences in the amount of LC3-I in different cells/tissues and the detection of this form by many antibodies). Although commercial antibodies are available, we have had limited success using these on western analysis of heart extracts. We therefore generated our own affinity-purified antipeptide antibody. Briefly, amino acids 2–15 from the N-terminus of murine LC3 (PSEKTFKQRRSFEQC) were chemically coupled to keyhole limpet hemocyanin (KLH). New Zealand white rabbits were used to generate antisera to the LC3-KLH conjugate, and specific antibodies were affinity purified using the same peptide. This antibody works well for Western blot analysis but not immunohistochemistry.

7. Isolation of LC3 Proteins from NRVM in Culture

1. Aspirate the medium and wash cells (approximately 1.2 million) twice with ice-cold PBS.
2. Add RIPA buffer (150 mM NaCl, 50 mM Tris-HCl, pH7.4, 1 mM EDTA, 1% Triton X-100, 1% sodium deoxycholate, 0.1% SDS, plus protease inhibitors (Complete Mini-EDTA free, Roche 1836145) and scrape cells.

3. Transfer to a 1.5-mL microcentrifuge tube and incubate on ice, agitating every 2–3 min for 15 min.
4. Centrifuge at 16,000×*g* for 10 min.
5. Transfer the supernatant fraction to a fresh 1.5-mL microcentrifuge tube.
6. Quantify protein using coomassie blue reagent.

Note: Be sure to dilute protein extract in PBS to bring buffer components within the linear range of the protein assay being used.

8. Isolation of LC3 Protein from Heart or Skeletal Muscle Tissue

Detergent extraction of some proteins from striated muscle can be difficult presumably because they become trapped in the tightly packed sarcomere, which becomes an insoluble mass upon addition of most detergents. Extraction of cardiac myosins is carried out by homogenization in a buffer lacking detergent followed by incubation on ice. We have developed a similar technique for extracting LC3.

1. Use a 2-mL glass Dounce homogenizer to homogenize finely minced pieces of left ventricle in lysis buffer (150 m*M* NaCl, 20 m*M* Tris-HCl, pH7.4, 10 m*M* EDTA plus protease inhibitors) on ice using both the A and B pestles.
2. Centrifuge at 500×*g* for 10 min at 4 °C.
3. Transfer the supernatant fraction to a fresh microcentrifuge tube.
4. Quantify the protein using coomassie reagent.
5. Add an equal volume of 2xSDS sample buffer to the supernatant fraction.

Note: Components in the low-speed supernatant may settle with time. Be sure to resuspend before removing a sample for protein determination. If the extract is viscous due to the presence of genomic DNA, it can be passed through glass wool prior to loading for SDS-PAGE.

8.1. Western blot analysis using the LC3 antibody

1. Separate proteins on a 12.5% SDS-PAGE gel.
2. Transfer protein to a nitrocellulose membrane at 25V overnight at 4 °C.
3. Incubate membrane with blocking buffer (5% milk, 1xTBS, 0.1% Tween 20) for 1 h at RT.
4. Incubate with anti-LC3 polyclonal antibody (1: 2,000 dilution) in blocking buffer overnight at 4 °C or at RT for 1 h.
5. Wash in TBST (1xTBS, 0.1% Tween 20) for 10 min 4 times.
6. Incubate with HRP-conjugated goat antirabbit secondary antibody (1:5000 dilution) in blocking buffer for 1 h at RT.

7. Wash in TBST for 10 min 4 times.
8. Incubate with HRP substrate solutions (e.g., ECL PLUS).

8.2. Monitoring levels of other autophagy-related proteins

A change in the level of several other autophagy-related proteins may indicate altered flux through autophagic pathways. Beclin 1, the mammalian homolog of yeast Atg6 (Liang *et al.*, 1999), binds to class III phosphoinositides-3-kinase (PI3K) and recruits other autophagy molecules to the membrane at an early stage of autophagosome formation (Kihara *et al.*, 2001). In sTAB hearts, an increase in autophagy is accompanied by an increase in Beclin 1 protein abundance (Zhu *et al.*, 2007). Cardiomyocyte-restricted forced overexpression of the gene coding for Beclin 1 (α*MHC-beclin 1*) increases stress-induced cardiac autophagy (Zhu *et al.*, 2007). Conversely, autophagy is diminished in the hearts of heterozygous *beclin* $1^{+/-}$ mice (Zhu *et al.*, 2007). It is important to be aware that Beclin 1 expression does not always increase when autophagy is activated. For instance, during starvation-induced autophagy, Beclin 1 levels in the heart do not change (Zhu *et al.*, 2007). Furthermore, changes in Beclin 1 levels do not necessarily correlate with changes in processing of LC3-I (Matsui *et al.*, 2007).

Polyubiqitinated aggregates of misfolded or damaged proteins can be degraded via autophagy. A protein named p62, also called SQSTM1, binds both to LC3 and polyubiquitin and is thought to be involved in this process (Pankiv *et al.*, 2007). Consistent with this, disruption of autophagic activity in cardiomyocytes leads to accumulation of poly-ubiquitinated proteins and p62 (Nakai *et al.*, 2007; Tannous *et al.*, 2008b).

Western blot analysis of Beclin 1, p62, and protein ubiquitination can be carried out following the protocol described above for LC3. Incubate primary antibodies in blocking buffer overnight at 4 °C. Monoclonal Beclin 1 antibody (BD Bioscience, catalog number 612113) is used at a 1:500 dilution. Monoclonal p62 antibody (BD Bioscience, catalog number 610832) is used at a 1:500 dilution, and the polyclonal ubiquitin antibody (Zymed for immunohistochemistry (131600); Dako for western blot, catalog number Z0458) is used at 1:500 dilution.

8.3. Separation of soluble proteins from insoluble aggregates

Protein aggregation and increased autophagic activity are common features of many neurodegenerative disorders (Rubinsztein, 2006). We have observed an association between the accumulation of protein aggregates and increased autophagic activity in both the pressure overloaded heart (Tannous *et al.*, 2008b) and in a genetic model of desmin-related cardiomyopathy (Rothermel and Hill, 2008; Tannous *et al.*, 2008a). Protein aggregates can be both soluble and insoluble, and there is mounting evidence that the soluble preaggregate is the more toxic of the species (Arrasate *et al.*, 2004; Rubinsztein, 2006). The two protein extraction methods provided above for NRVM and heart were optimized for tracking LC3 and are not appropriate for analyzing soluble and insoluble proteins. In the case of the NRVM extraction, both soluble and insoluble proteins are present, whereas in the heart extraction, insoluble protein aggregates would be discarded. Thus, the following extractions are designed to separate soluble and detergent insoluble proteins. We have listed 1% Triton X-100 as the detergent for NRVM and 1% NP-40 in the heart extraction, however, the two give similar results in either protocol.

9. Soluble/Insoluble Fractionation of NRVM

1. Harvest cells by digestion in trypsin (0.25% dissolved in Hanks' Balanced Salt Solution containing 2.21 m*M* EDTA) at 37 °C for ~5 min. Pool 3 wells (from a 6 well-plate) to make 1 sample. Centrifuge at 3000×*g*, 5 min.
2. Aspirate off the trypsin. Wash cells once with PBS, centrifuging for 5 min at 3000×*g*.
3. Aspirate off the PBS. Resuspend the cell pellet in 200 μL of Triton Buffer (10 m*M* Tris, 150 m*M* NaCl, 1 m*M* EDTA, 1% Triton X-100, protease and phosphatase inhibitors (Complete Mini-EDTA free, Roche 1836145). Pass the lysate through an insulin syringe (Fisher Scientific 1482679) 5 times. Incubate on ice for (at least) 15 min.
4. Centrifuge at 12,000×*g*, 15 min, at 4 °C.
5. Collect the supernatant fraction as the soluble fraction. Wash the pellet fractions 3 times with cold PBS (add 1 mL PBS to each pellet, centrifuge at 12,000×*g* for 10 min). Try not to disturb the pellet during each wash.
6. Resuspend the washed pellet in 200 μL of 1xSDS loading buffer (dilute 2xSDS loading buffer in PBS).

9.1. Soluble/insoluble fractionation from heart or skeletal muscle

As discussed earlier, protein extraction from heart tissue can be difficult, because the sarcomere becomes insoluble when the tissue is homogenized in the presence of detergent. We have overcome this problem by first solubilizing myosins in a detergent-free extraction buffer and then adding detergent.

1. Dounce homogenize finely minced heart tissue in 500 μL to 1 mL (depending on size of tissue) of homogenization buffer (0.3 *M* KCl, 0.1 *M* KH_2PO_4, 50 m*M* K_2HPO_4,10 m*M* EDTA, 4 m*M* Na orthovanadate, 100 m*M* NaF, plus protease inhibitors).

2. Pass the homogenized sample through 100 micron mesh (Tetko, catalog number AC38842) on ice, collecting the lysate run-through.
3. Incubate the lysate on ice for 30 min.
4. Transfer a portion of the sample to another microcentrifuge tube, and add 10% NP-40 (Sigma, catalog number 18896) to achieve a final concentration of 1% NP-40. Incubate on ice for 15–30 min.
5. Centrifuge at 13,000 rpm for 15 min, 4 °C.
6. Collect the supernatant fraction as the soluble fraction (contains sarcomere).
7. With the pellets, wash 3 times with cold PBS (add 1mL PBS to each pellet, centrifuge at 13,000 rpm for 10 min).
8. Resuspend the washed pellet in 1% SDS, 10 m*M* Tris buffer, pH 8.0.

Note: Tissue disruption with a Dounce homogenizer is essential for this procedure, as it ensures that all cells are broken open despite the lack of detergent. The soluble fraction will contain all of the sarcomeric proteins and is therefore ideal for monitoring shifts in αMHC and *β*MHC content by Western blot analysis.

10. Perspective

Accumulating evidence links autophagy with the pathogenesis of multiple forms of heart disease, suggesting that autophagy may be a novel target of therapy. However, we and others have shown that the regulation and contribution of autophagy to heart disease is dose and context dependent, which poses new questions and special challenges (Rothermel and Hill, 2007). We believe that the mouse models of heart failure we have established and the methods we developed for monitoring autophagy in heart will be useful in dissecting the precise role(s) of cardiac autophagy in multiple forms of heart disease. Ultimately, it is our hope that insights gleaned using these tools will lead to the novel strategies of therapy for these devastating diseases.

ACKNOWLEDGMENTS

This work was supported by grants from the Donald W. Reynolds Cardiovascular Clinical Research Center (JAH), NIH (HL-72016, BAR; HL-075173, JAH; HL-080144, JAH), and AHA (0655202Y, BAR; 0640084N, JAH).

REFERENCES

Arrasate, M., Mitra, S., Schweitzer, E. S., Segal, M. R., and Finkbeiner, S. (2004). Inclusion body formation reduces levels of mutant huntingtin and the risk of neuronal death. *Nature* **431,** 805–810.

Baschong, W., Suetterlin, R., and Laeng, R. H. (2001). Control of autofluorescence of archival formaldehyde-fixed, paraffin-embedded tissue in confocal laser scanning microscopy (CLSM). *J. Histochem. Cytochem.* **49,** 1565–1572.

Brown, R. D., Ambler, S. K., Mitchell, M. D., and Long, C. S. (2005). The cardiac fibroblast: Therapeutic target in myocardial remodeling and failure. *Annu. Rev. Pharmacol. Toxicol.* **45,** 657–687.

Dammrich, J., and Pfeifer, U. (1983). Cardiac hypertrophy in rats after supravalvular aortic constriction. II. Inhibition of cellular autophagy in hypertrophying cardiomyocytes. *Virchows. Arch B. Cell Pathol. Incl. Mol. Pathol.* **43,** 287–307.

Decker, R. S., Decker, M. L., Herring, G. H., Morton, P. C., and Wildenthal, K. (1980). Lysosomal vacuolar apparatus of cardiac myocytes in heart of starved and refed rabbits. *J. Mol. Cell Cardiol.* **12,** 1175–1189.

Decker, R. S., and Wildenthal, K. (1980). Lysosomal alterations in hypoxic and reoxygenated hearts 1. Ultrastructural and cytochemical changes. *American Journal of Pathology* **98,** 425–444.

Elsasser, A., Vogt, A. M., Nef, H., Kostin, S., Mollmann, H., Skwara, W., Bode, C., Hamm, C., and Schaper, J. (2004). Human hibernating myocardium is jeopardized by apoptotic and autophagic cell death. *J. Am. Coll. Cardiol.* **43,** 2191–2199.

Frey, N., Katus, H. A., Olson, E. N., and Hill, J. A. (2004). Hypertrophy of the heart: A new therapeutic target? *Circulation* **109,** 1580–1589.

Frey, N., and Olson, E. N. (2003). Cardiac hypertrophy: The good, the bad, and the ugly. *Annu. Rev. Physiol.* **65,** 45–79.

Hein, S., Arnon, E., Kostin, S., Schonburg, M., Elsasser, A., Polyakova, V., Bauer, E. P., Klovekorn, W. P., and Schaper, J. (2003). Progression from compensated hypertrophy to failure in the pressure-overloaded human heart: Structural deterioration and compensatory mechanisms. *Circulation* **107,** 984–991.

Heineke, J., and Molkentin, J. D. (2006). Regulation of cardiac hypertrophy by intracellular signalling pathways. *Nat. Rev. Mol. Cell Biol.* **7,** 589–600.

Hill, J. A., Karimi, M., Kutschke, W., Davisson, R. L., Zimmerman, K., Wang, Z., Kerber, R. E., and Weiss, R. M. (2000). Cardiac hypertrophy is not a required compensatory response to short-term pressure overload. *Circulation* **101,** 2863–2869.

Hill, J. A., and Olson, E. N. (2008). Cardiac plasticity. *New. Engl. J. Med* **358,** 1370–1380.

Kabeya, Y., Mizushima, N., Ueno, T., Yamamoto, A., Kirisako, T., Noda, T., Kominami, E., Ohsumi, Y., and Yoshimori, T. (2000). LC3, a mammalian homologue of yeast Apg8p, is localized in autophagosome membranes after processing. *EMBO J.* **19,** 5720–5728.

Kihara, A., Kabeya, Y., Ohsumi, Y., and Yoshimori, T. (2001). Beclin-phosphatidylinositol 3-kinase complex functions at the trans-Golgi network. *EMBO Reports* **2,** 330–335.

Klionsky, D. J. (2007). Autophagy: From phenomenology to molecular understanding in less than a decade. *Nat. Rev. Mol. Cell Biol.* **8,** 931–937.

Knaapen, M. W. M., Davies, M. J., De Bie, M., Haven, A. J., Martinet, W., and Kockx, M. M. (2001). Apoptotic versus autophagic cell death in heart failure. *Cardiovascular Research* **51,** 304–312.

Kostin, S., Pool, L., Elsasser, A., Hein, S., Drexler, H. C., Arnon, E., Hayakawa, Y., Zimmermann, R., Bauer, E., Klovekorn, W. P., *et al.* (2003). Myocytes die by multiple mechanisms in failing human hearts. *Circ. Res.* **92,** 715–724.

Kundu, M., and Thompson, C. B. (2008). Autophagy: Basic principles and relevance to disease. *Annu. Rev. Pathol.* **3,** 427–455.

Levine, B., and Kroemer, G. (2008). Autophagy in the pathogenesis of disease. *Cell* **132,** 27–42.

Levy, D., Larson, M. G., Vasan, R. S., Kannel, W. B., and Ho, K. K. (1996). The progression from hypertension to congestive heart failure. *JAMA* **275,** 1557–1562.

Liang, X. H., Jackson, S., Seaman, M., Brown, K., Kempkes, B., Hibshoosh, H., and Levine, B. (1999). Induction of autophagy and inhibition of tumorigenesis by beclin 1. *Nature* **402,** 672–676.

Lockshin, R. A., and Zakeri, Z. (2002). Caspase-independent cell deaths. *Current Opinion in Cell Biology* **14,** 727–733.

Maass, A. H., and Buvoli, M. (2007). Cardiomyocyte preparation, culture, and gene transfer. *Methods. Mol. Biol.* **366,** 321–330.

Matsui, Y., Takagi, H., Qu, X., Abdellatif, M., Sakoda, H., Asano, T., Levine, B., and Sadoshima, J. (2007). Distinct roles of autophagy in the heart during ischemia and reperfusion: Roles of AMP-activated protein kinase and Beclin 1 in mediating autophagy. *Circ. Res.* **100,** 914–922.

Mizushima, N. (2007). Autophagy: Process and function. *Genes Dev.* **21,** 2861–2873.

Mizushima, N., Yamamoto, A., Matsui, M., Yoshimori, T., and Ohsumi, Y. (2004). *In vivo* analysis of autophagy in response to nutrient starvation using transgenic mice expressing a fluorescent autophagosome marker. *Mol. Biol. Cell* **15,** 1101–1111.

Nakai, A., Yamaguchi, O., Takeda, T., Higuchi, Y., Hikoso, S., Taniike, M., Omiya, S., Mizote, I., Matsumura, Y., Asahi, M., Nishida, K., Hori, M., *et al.* (2007). The role of autophagy in cardiomyocytes in the basal state and in response to hemodynamic stress. *Nat. Med* **13,** 619–624.

Pankiv, S., Clausen, T. H., Lamark, T., Brech, A., Bruun, J. A., Outzen, H., Øvervatn, A., Bjørkøy, G., and Johansen, T. (2007). p62/SQSTM1 binds directly to Atg8/LC3 to facilitate degradation of ubiquitinated protein aggregates by autophagy. *J. Biol. Chem.* **282,** 24131–24145.

Pfeifer, U., Fohr, J., Wilhelm, W., and Dammrich, J. (1987). Short-term inhibition of cardiac cellular autophagy by isoproterenol. *J. Mol. Cell Cardiol.* **19,** 1179–1184.

Rockman, H. A., Ross, R. S., Harris, A. N., Knowlton, K. U., Steinhelper, M. E., Field, L. J., Ross, J. Jr., and Chien, K. R. (1991). Segregation of atrial-specific and inducible expression of an atrial natriuretic factor transgene in an *in vivo* murine model of cardiac hypertrophy. *Proc. Natl. Acad. Sci. USA* **88,** 8277–8281.

Rosamond, W., Flegal, K., Furie, K., Go, A., Greenlund, K., Haase, N., Hailpern, S. M., Ho, M., Howard, V., Kissela, B., *et al.* (2008). Heart disease and stroke statistics–2008 update: A report from the American Heart Association Statistics Committee and Stroke Statistics Subcommittee. *Circulation* **117,** e25–e146.

Rothermel, B. A., Berenji, K., Tannous, P., Kutschke, W., Dey, A., Nolan, B., Yoo, K. D., Demetroulis, E., Gimbel, M., Cabuay, B., *et al.* (2005). Differential activation of stress-response signaling in load-induced cardiac hypertrophy and failure. *Physiol. Genomics* **23,** 18–27.

Rothermel, B. A., and Hill, J. A. (2007). Myocyte autophagy in heart disease: Friend or foe? *Autophagy* **3,** 632–634.

Rothermel, B. A., and Hill, J. A. (2008). The heart of autophagy: Deconstructing cardiac proteotoxicity. *Autophagy* **4(7),** 932–935.

Rubinsztein, D. C. (2006). The roles of intracellular protein-degradation pathways in neurodegeneration. *Nature* **443,** 780–786.

Shimomura, H., Terasaki, F., Hayashi, T., Kitaura, Y., Isomura, T., and Suma, H. (2001). Autophagic degeneration as a possible mechanism of myocardial cell death in dilated cardiomyopathy. *Jpn. Circ. J.* **65,** 965–968.

Shintani, T., and Klionsky, D. J. (2004). Autophagy in health and disease: A double-edged sword. *Science* **306,** 990–995.

Sybers, H. D., Ingwall, J., and DeLuca, M. (1967). Autophagy in cardiac myocytes. *Recent. Adv. Stud. Cardiac. Struct. Metab.* **12,** 453–463.

Tannous, P., Zhu, H., Johnstone, J. L., Shelton, J. M., Rajasekaran, N. S., Benjamin, I. J., Nguyen, L., Gerard, R. D., Levine, B., Rothermel, B. A., and Hill, J. A. (2008a). Autophagy is an adaptive response in desmin-related cardiomyopathy. *Proc. Natl. Acad. Sci. USA* **105,** 9745–9750.

Tannous, P., Zhu, H., Nemchenko, A., Berry, J. M., Johnstone, J. L., Shelton, J. M., Miller, F. J. Jr., Rothermel, B. A., and Hill, J. A. (2008b). Intracellular protein aggregation is a proximal trigger of cardiomyocyte autophagy. *Circulation* **117,** 3070–3078.

Yamamoto, S., Sawada, K., Shimomura, H., Kawamura, K., and James, T. N. (2000). On the nature of cell death during remodeling of hypertrophied human myocardium. *J. Mol. Cell Cardiol.* **32,** 161–175.

Zhu, H., Tannous, P., Johnstone, J. L., Kong, Y., Shelton, J. M., Richardson, J. A., Le, V., Levine, B., Rothermel, B. A., and Hill, J. A. (2007). Cardiac autophagy is a maladaptive response to hemodynamic stress. *J. Clin. Invest.* **117,** 1782–1793.

CHAPTER EIGHTEEN

Evaluation of Cell Death Markers in Severe Calcified Aortic Valves

Wilhelm Mistiaen* *and* Michiel Knaapen†

Contents

Abstract

Degenerative aortic valve disease is the most frequent acquired valve disease. Especially in the elderly, its prevalence is increasing. Once the disease becomes symptomatic, it is rapidly fatal. The disease cannot be considered a result of aging alone. The condition is an active process, which occurs with rapid progression, especially when calcification can be documented. This calcification can be the end result of cellular mechanisms involving cell death pathways (such as autophagy) and cellular matrix remodeling. These processes are beginning to be unraveled in the initiation and propagation of the disease. Autophagy could be the common step through which these mechanisms lead to this pathway of cell death in this disease. Autophagy can be detected by procedures described hereafter.

* Department of Healthcare Sciences, University College of Antwerp, Antwerp, Belgium
† ZNA Middelheim, Laboratory for Cardiovascular Pathology, Antwerp, Belgium

Methods in Enzymology, Volume 453
ISSN 0076-6879, DOI: 10.1016/S0076-6879(08)04018-4

1. Clinical Importance of Degenerative Aortic Valve Disease

In the Western community, degenerative aortic valve disease (AVD) is the most common valvular lesion. It presents itself usually as aortic valve stenosis. This condition is highly lethal (Hilton, 2000) and aortic valve replacement (AVR) is necessary for symptomatic and prognostic reasons (Otto, 1998). Although advanced age is not a contraindication to AVR (Mistiaen *et al.*, 2004) most of the difficulties for the optimal timing of AVR occur in this group of patients. Referral for surgery is increasing, especially in the elderly (Mistiaen, 2007). Although degenerative AVD occurs more in the elderly, it is not an inevitable consequence of ageing (Otto, 1998). Progression of this condition is not always easily estimated (Cosmi *et al.*, 2002) and requires detailed ultrasonography (Otto *et al.*, 1997).

2. Pathological Appearances of AVD

The normal aortic valve leaflets are macroscopically smooth, thin and opalescent. Three defined tissue layers can be distinguished: the fibrosa, the spongiosa, and ventricularis. They contain collagen and elastic fibers organized in different planes and orientations. These fibers are surrounded by glycosaminoglycans as the matrix component. The cellular components are a heterogeneous population of fibroblasts, smooth muscle cells and myofibroblasts. These cells renew the extracellular matrix by remodeling processes (Somers *et al.*, 2006; Merryman *et al.*, 2007). Degenerative aortic valve disease is considered as the end stage of a progressive and active process, including inflammation, oxidation, angiogenesis, and calcification with osteogenesis (Somers *et al.*, 2006).

With increasing age, nonspecific thickening of the tips of the aortic valve leaflets occurs with an increase in the number of adipocytes. In degenerative AVD, thickening of the leaflets is characterized by irregular nodular masses located in the bases of the cusp, deep within the sinuses of Valsalva, at the aortic surface. These subendothelial lesions displace the elastic lamina and extend into the adjacent fibrosa (Somers *et al.*, 2006). Irrespective of the severity of the condition, disruption of the endothelial layer and the valvular interstitial tissue and basement membrane is associated with thickening. Collagen fibers are disorganized, and inflammatory cells (macrophages and T lymphocytes secreting cytokines), lipoproteins, bone matrix, and bone minerals infiltrate. Proliferation of smooth muscle cells, which is common in atheromatosis, is absent in degenerative AVD. In contrast, calcification is

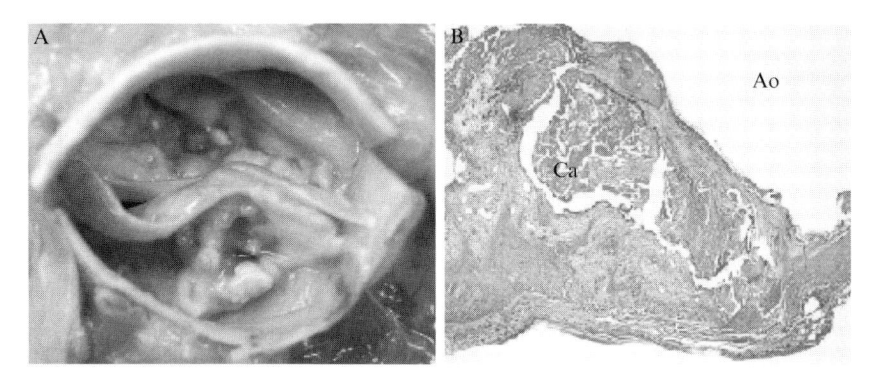

Figure 18.1 Calcified aortic valve. (A) Macroscopic presentation of a severe calcified bicuspid aortic valve. (B) Histological presentation of the calcified aortic valve (Ca = Calcium; Ao = Aortic side).

more dominant in degenerative AVD (Fig. 18.1), with hydroxyapatite similar to bone tissue and cells with up-regulated osteoblastic genes (O'Brien 1995; Somers *et al.*, 2006). In spite of the dissimilarities, aortic valve degeneration and coronary atheromatosis occur frequently simultaneously. Moreover, in the vicinity of calcified regions, apoptosis (Clark-Greuel, 2007; Jian *et al.*, 2003) and mechanisms of autophagic cell death are observed (Somers *et al.*, 2006b).

3. Mechanism of Progression of AVD

Two mechanisms can be held responsible for this progression: mechanical and biochemical.

1. Mechanical shear stress, which is responsible for high collagen turnover within the spongiosa of the aortic valve is a result of the normal blood flow and is required for the physiological modulation of the phenotype of aortic interstitial cells. *In vitro* stretch can, in the presence of transforming growth factor $\beta 1$ (TGF-$\beta 1$), activate a resting fibroblast to a contractile myofibroblast, which occurs in diseased and remodeled valves. This is due to their adaptive cytoskeletal remodeling and biosynthesis, which serves normally to preserve homeostasis within tissues (Merryman, 2007). Shear stress (in physiological conditions) has a protective effect by regulating oxidative and inflammatory genes in endothelial cells at the ventricular and vascular side of the valve. Disturbed flow could play a causal role through the initiation of endothelial activation in valvular pathology (Butcher, 2006). For these reasons,

one can assume that once an aortic valve becomes deformed, resulting turbulences will further damage the valve tissue.

2. Cytotoxic oxidized low-density lipoproteins, inflammatory T-lymphocytes and osteopontin-producing macrophages play an important biochemical role in formation of the calcium deposits in aortic valves (O'Brien *et al.*, 1995). Osteopontin binds readily to hydroxyapatite (Ortlepp *et al.*, 2004), and bone tissue is actively formed (Aubin, 1995; Rajamannan *et al.*, 2003). Macrophages and other inflammatory cells could be a source of cytokines, including TGF-β1. The latter induces osteoblast differentiation of aortic valve interstitial cells (Osman *et al.*, 2006). Although TGF-β1-induced apoptotic cell death in degenerative aortic valve disease is reported (Clark-Greuel *et al.*, 2007; Jian *et al.*, 2003), we cannot find apoptotic cell death by TUNEL labeling. Instead, our results suggest that cell death by autophagy could play a role in the release of matrix vesicles in AVD. This could induce inflammatory cells, which are responsible in regulating genes involving calcification (Somers *et al.*, 2006b). This possible chain of events is shown in Fig. 18.2.

4. Autophagy: Major Player in the Progression of Aortic Valve Disease?

Autophagy is a complex reparative and recycling mechanism (Klionsky *et al.*, 2008; Levine and Klionsky, 2004; Stromhaug and Klionsky, 2000). As in apoptosis, the mitochondrion can play a key role but also other organelles such as lysosomes and the endoplasmic reticulum have an important function in the release and activation of cathepsins, calpains, and other proteases (Broker *et al.*, 2005). Also, environmental conditions (including starvation), specific hormones, signaling pathways, calcium and protein synthesis play an important role in the regulation of autophagy (Wang and Klionsky, 2003). Autophagy involves the sequestration of intracellular components and their subsequent degradation in secondary lysosomes (Yoshimori, 2004).

Autophagy-related publications in cardiovascular research have increased considerably in the last 2 years, indicating that autophagy is becoming a topic of major importance. Nevertheless, the process is still an underestimated and highly neglected phenomenon in cardiovascular disease as compared with apoptosis. Transmission electron microscopy for the detection of autophagic vacuoles has become the gold standard. The demonstration of granular cytoplasmic-ubiquitin inclusions by immunohistochemistry is an attractive alternative technique (Elsässer *et al.*, 2004; Hein *et al.*, 2003; Martinet *et al.*, 2007). However, ubiquitinated aggregates also result from a malfunction in the autophagic pathway or from structural

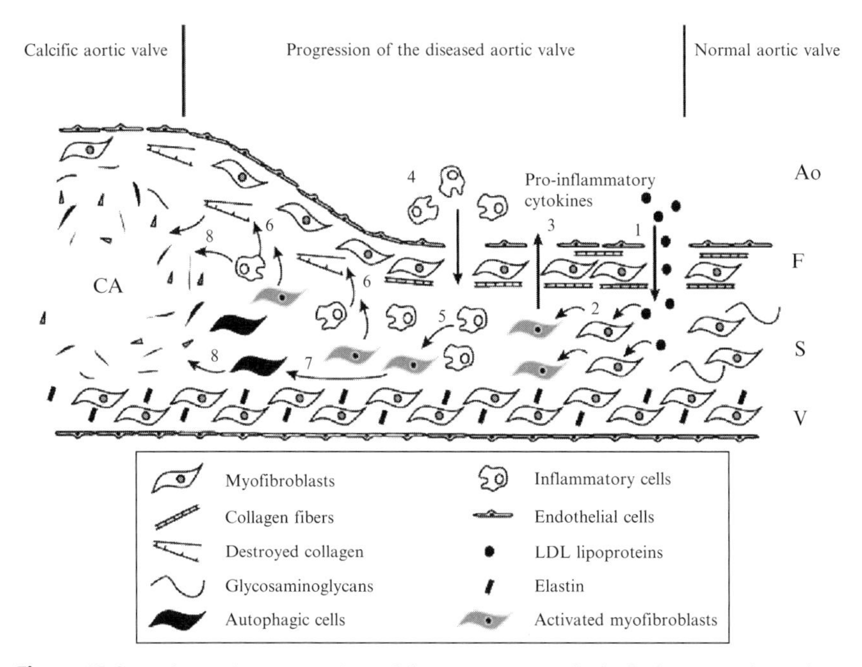

Figure 18.2 Schematic presentation of the progression of calcified aortic valve. This is a difficult and not fully understood pathway. However, the basis during the progression of the disease lies in an accumulation of lipids (LDLs) (1), triggering interstitial myofibroblasts (2) to secrete proinflammatory signals (3). Monocytes and macrophages enter the diseased aortic valve (4) to clean the inflammatory response of the aortic leaflet (5). Due to overproduction and over-stimulation, both myofibroblasts and macrophages produce matrix metalloproneinases destroying the collagen content of the aortic cellular matrix (6). Moreover, myofibroblasts or denditic cells cannot maintain the program of cleaning and become therefore phagocytotic (7). As a result both pathways lead to an increased up-regulation of genes (osteopontin, tenascin-C) responsible for calcification (8) of the diseased aortic valve, and subsequently result in a stiff calcified aortic valve.

changes in the protein substrates, halting their degradation. Caution therefore is warranted (Martinet *et al.*, 2007). Ubiquitin is used as marker for autophagy in several conditions such as heart failure (Elsässer *et al.*, 2004; Hein *et al.*, 2003; Knaapen *et al.*, 2001), neurodegenerative diseases (Bi *et al.*, 2007; McCray and Taylor, 2008), atherosclerosis (Martinet *et al.*, 2007), and degenerative AVD (Mistiaen *et al.*, 2006; Somers *et al.*, 2006b;). Ubiquitin-positive vacuoles are also positive for the autophagic marker LC3, suggesting that ubiquinated proteins are present in autolysosomes. Ubiquitination is required for the formation and maturation of autophagosomes, and ubiquitination of proteins could mark them for uptake by autophagy (Purdy and Russell, 2007).

5. Methods for Detection of Cell Death Markers in the Degenerated Aortic Valve

Transmission electron microscopy is the gold standard for the detection of cell death. However, because aortic valves are cellular poor this technique is not useful as well as time consuming for the detection of apoptotic or autophagic cells. Therefore, markers such as TUNEL for apoptosis and ubiquitin for autophagy are useful cell death markers. Decalcification is however, necessary for detection of cell death markers as well as for other immunohistochemistry stains. In general, Bouin fixative or treatment with formic acid or EDTA are protocols for decalcification. However, this could reduce epitope antigenicity resulting in poor staining for cell death markers as well as immunohistochemical stains. Recently, a decalcification procedure based on electrolysis has been introduced (Sakura Finetek, the Netherlands) to preserve antigenic properties of calcified tissue. For the detection of cell death markers and other immunohistochemical stains the following procedures were performed.

5.1. Subsampling of aortic valves

From every cusp, 2 regions, perpendicular to the edge of the cusp are sampled (Fig. 18.3). The sections are processed for histological examination by 48 hours fixation in 3.7% formalin at room temperature.

5.2. Decalcification

Decalcification is performed by bringing the calcified specimens into the decalcifier TDE30 system of Sakura (Sakura Finetek, The Netherlands, Cat Nr: 1427). The valves are placed in the middle of the equipment in approximately 750 ml of TDE 30 Decalcifier Solution (Sakura Finetek, The Netherlands, Cat Nr: 1428). Electrolysis is applied 1–4 h, depending on the size of the specimen. After decalcification, samples are embedded in paraffin.

5.3. Staining aortic valves

Staining of aortic valves is performed on paraffin embedded material.

1. Paraffin embedded tissues are sectioned (5 μm) and mounted on poly-L-lysine coated slides and dried over night at 58 °C.
2. After drying, the slides are deparaffinized by xylene (4 times, 10 min) followed by stepwise decreasing concentrations of ethanol (EtOH; 100%

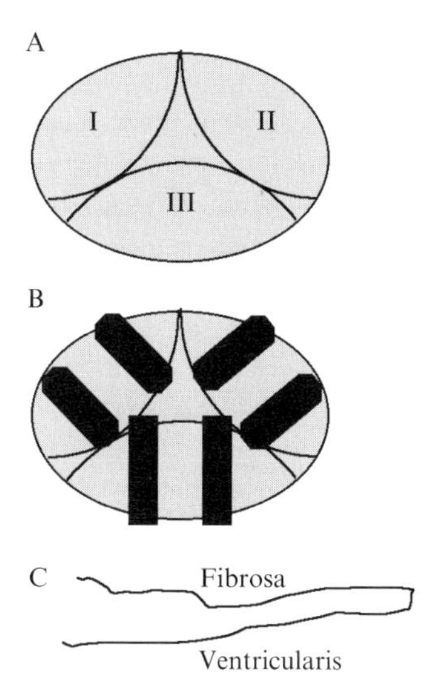

Figure 18.3 prelevation of aortic valve samples. (A) The aortic valve with the three cusps. (B) Where samples of valve tissue can be taken by sectioning. (C) The section of the valve cusp. The aortic side (fibrosa) is at the top, the ventricular side (ventricularis) is at the bottom.

Step	Reagent	Temperature	Vacuum	Time
1	Formalin	Room temp.		3–45 h
2	Formalin	39 ± 1°C	Yes	3 h
3	70% alcohol	39 ± 1°C	Yes	1 h
4	80% alcohol	39 ± 1°C	Yes	1 h
5	90% alcohol	39 ± 1°C	Yes	1 h
6	96% alcohol	39 ± 1°C	Yes	1 h
7	100% alcohol	39 ± 1°C	Yes	1 h
8	100% alcohol	39 ± 1°C	Yes	1 h
9	Xylene	39 ± 1°C	Yes	1:30 h
10	Xylene	39 ± 1°C	Yes	1:30 h
11	Paraffin	59 ± 1°C	Yes	0:50 h
12	Paraffin	59 ± 1°C	Yes	1:05 h

EtOH for 2 min, 95% EtOH for 2 min, 95% EtOH for 2 min, 80% EtOH for 2 min, 70% EtOH for 2 min).

3. Haematoxylin and eosin staining is performed for routine histological examination. The samples are washed for 5 min in tap water. This is followed by Hematoxylin (Harris, nucleus staining) for 3 min, and washing again with 0.1% HCl, 70% EtOH by repetitive dipping.

The reaction is neutralized by submersion for 10 min in tap water followed by repetitive dipping in ammonia water (2.52 ml of a 25% stock solution diluted in 1 L of water) for 5 min. Cytoplasm staining is performed with eosin for 1 min. The samples are washed with tap water for 30 s and the slides are dehydrated by stepwise increased concentrations of EtOH and finally cover-slipped.

4. Alcian Blue staining is performed for examination of mucoid degeneration. The de-paraffinized slides are placed in an alcian blue solution for 30 min. After staining, the sections are washed in running tap water for 2 min and rinsed in distilled water. Counterstaining occurs in nuclear fast red solution for 5 min following washing in running tap water for 1 min. Dehydration by a stepwise increasing concentration of EtOH gradient is followed by cover-slipping. The alcian blue (Sigma Aldrich, UK) solution contains 1 g of alcian blue dissolved in 100 ml of acetic acid, 3% solution. This solution is adjusted to a pH of 2.5 using acetic acid. The 0.1% nuclear fast red solution contains 0.1 g of nuclear fast red (Labconsult, Belgium) and 5 g of aluminium sulphate dissolved in 100 ml of distilled water. Acidic mucosubstances are stained blue, nuclei are stained pink to red, and cytoplasm is stained pale pink.
5. For immunohistochemical staining, a mouse polyclonal PGP9.5 antibody (DAKO, Denmark, CatNr: Z5116) is used. The deparaffinized slides are treated with 3% H_2O_2 for 15 min following staining with PGP 9.5 (dilution of 1:250) for 60 min. A secondary antibody of goat-anti-mouse is used for 45 min and the secondary antibody is visualized with 3-amino-9-ethylcarbazole (AEC) as the chromogen (SIGMA ALDRICH, USA, Cat. Nr. A5754). Twenty-five mg of AEC is dissolved in 2.5 ml of DMF and then into 47.5 ml of 50 m*M* acetate buffer, pH 5.0. Immediately before use, 25 μL of 30% hydrogen peroxide solution is added. This stains the areas indicated by immunohistochemistry. Counterstaining is performed with haemalun and sections are mounted in glycerin jelly.

5.4. Detection of cell death markers

Apoptosis is detected by the TUNEL technique, whereas autophagy is demonstrated with ubiquitin staining.

For the detection of apoptotic cell death, a stringent TUNEL (Terminal dUTP nick-end labeling) as described by Kockx *et al.* (1998) is used to avoid interference with cells with high RNA synthesis and/or splicing. This stringent TUNEL techniques is performed as follows:

1. The sections are deparaffinized three times for 3 min each in toluene.
2. Thereafter, they are immersed in ethanol three times for 3 min each and finally placed in distilled water.

3. The sections are treated with 3% H_2O_2 for 15 min.
4. This is followed by washing 3 times with distilled water.
5. The sections are treated with 3% citric acid for 1 h, at room temperature.

This is followed again by washing 3 times in distilled water.

6. The sections are treated with proteinase K working solution for 30 s.

(*Note*: be careful with this step. A nonspecific labeling could occur when section are treated too long with proteinase K). Proteinase K working solution is made by diluting 20 μL of a stock solution (10 mg of proteinase K in 10 mL of 0.05 *M* Tris-HCl, pH 7.6) in 1 mL of 0.05 *M* Tris-HCl, at pH 7.6.

7. This is followed by washing of at least 4 times in distilled water for 5 min each.
8. The sections are incubated with the TdT solution for 1 h at 37 °C. The TdT solution contains 750 μl of distilled water, 1 μl of 10 m*M* dATP (Sigma, Poole, UK), 200 μL of Tris-cacodylate buffer, 50 μL of 25 MM $CoCl_2$, 2.5 μL of 1 m*M* FITC (fluorescein-isothiocyanate)-dUTP (Amersham, Little Chalfont, UK) 50 U/ml TdT (Roche Diagnostics, Mannheim, Germany) at pH 6.6.
9. The sections are washed 4 times with distilled water for 5 min each.
10. This is followed by incubation with a blocking reagent for 1 h at room temperature. This blocking reagent contains 5 mg/ml blocking reagent (Amersham, Buckinghamshire, UK) in phosphate-buffered saline, pH 7.6.
11. Thereafter, they are incubated in sheep anti-FITC peroxidase conjugate (Roche Diagnostics, Mannheim, Germany) (dilution of 1:300) for 1 h at room temperature.
12. This is followed by washing with phosphate-buffered saline, pH 7.6, for 5 min.
13. Then, the sections are treated with 3-amino-9-ethylcarbazole (AEC) for 10 min.
14. They are washed twice in distilled water.
15. The sections are counterstained with haematoxylin.
16. Finally, they are washed in tap water and mounted in glycerin jelly.
17. Human tonsils are used as positive control.

For the detection of autophagic cell death, a ubiquitin stain is performed. This method is a robust and attractive method for the detection of autophagic cell death in tissue where transmission electron microscopy in not successful.

1. The sections are deparaffinized 3 times for 3 min each in toluene.
2. Then, they are immersed in ethanol 3 times for 3 min each and finally placed in distilled water.

3. Then, the sections are treated with 3% H_2O_2 for 15 min.
4. Thereafter, they are rinsed in distilled water.
5. The sections are pretreated with trypsin/citrate (100 mg trypsin in 100 ml Tris-HCl, pH 7.8; the citrate solution contains 12.35 g disodium citrate, 1.54 g citric acid in 5 L water).
6. Thereafter, they are rinsed in distilled water.
7. The sections are incubated in the primary antibody polyclonal mouse Anti-ubiquitin (DAKO, Denmark, P64) (1:100) for 60 min at room temperature.
8. This is followed by rinsing in phosphate-buffered saline, pH 7.6, for 5 min.
9. The sections are incubated in the secondary antibody EPR (DAKO, Denmark) (1:9) for 45 min at room temperature.
10. This is followed by rinsing in phosphate-buffered saline, pH 7.6, for 5 min
11. The sections are treated with 3-amino-9-ethylcarbazole (AEC) for 10 min.
12. Then they are washed twice in distilled water.
13. This is followed by counterstaining with haematoxylin.
14. Finally, the sections are washed in tap water and mounted in glycerin jelly.

Colocalization studies are performed on adjacent sections of the ubiquitin- and TUNEL-stained sections with sections stained for Alizarin red (Sigma-Aldrich, Cat Nr: 11996-2). Calcium forms an alizarin red S-calcium complex in a chelation process. The solution consists of 2 g of alizarin red S in 100 ml of distilled water, which is adjusted to pH 4.1–4.3 using 0.5% ammonium hydroxide. The adjacent sections of ubiquitin- and/or TUNEL-labeled sections are placed in the preceding solution. Colocalization is performed by imaging the tissue section by a digital coordinating system (Märzhauser, Wetzlar, Germany) mounted on an Olympus BX40 microscope (Hamburg, Germany). This allows the study of the calcified aortic valve, including a semi-quantitative analysis with scoring of several components.

6. Methods for Quantification of Calcified Aortic Valves

Semiquantitative analysis of calcified aortic valves is a useful and quick method of evaluating the severity of the diseased aortic valve. In general the percents of endothelization, inflammation, autophagic cell death and calcification of the individual valves are scored. The following methodology for quantification calcified aortic valves can be used.

Score	Endothelization
0	Complete endothelization of the aortic valve (endothelial cells covered more than 99% of the aortic surface)
1	Focal spots of inadequate endothelization at the cusp or tears of the aortic valve (endothelial cells covered between >80 % and <99% of the aortic surface)
2	Severe spots of inadequate endothelization at the cusp and tears at both sites of the aortic valve (endothelial cells covering the aortic surface between > 50% and < 80%)
3	Incomplete endothelization of aortic valve at both sides (endothelial cells covering less than 50% of the aortic surface)

Score	Inflammation
0	No inflammation (inflammation less than 1%)
1	Focal inflammation in the cusp or tears of the aortic leaflet (inflammation between 20% and 1%)
2	Focal inflammation in the cusp as well as in the tears of the aortic leaflet at both sites of the aortic leaflet in fibrosa or ventricularis (inflammation between 20% and 50%)
3	Total inflammation in the cusp and tears of the aortic leaflet both in fibrosa, spongiosa and *ventricularis* (inflammation more than 50%)

Score	Autophagic Cell Death
0	No signs for autophagic cell death
1	Focal autophagic cell death in fibrosa or ventricularis (1% to 20%)
2	Mild autophagic cell death in fibrosa and/or ventricularis (between 20% and 50%)
3	Severe autophagic cell death in the aortic valves (more than 50% of cells showing autophagic cell death)

Score	Calcification
0	No calcification
1	Focal calcification in fibrosa and/or ventricularis (less than 20%)
2	Calcification in the fibrosa and ventricularis (between 20 to 50%)
3	Severe calcification in fibrosa and ventricularis (more than 50%)

7. Discussion

Autophagy is a process that involves the sequestration of intracellular components and their subsequent degradation in secondary lysosomes (Yoshimori, 2004). This indicates that autophagic cell death might play a role in the release of matrix vesicles in degenerative aortic valves, and possibly trigger inflammation and calcification. Nevertheless, the nature of these triggers for the enhancement of inflammatory cells is unclear. Neither is the nature of the target molecules of increased ubiquitin production clear, or how ubiquitination interacts with the apoptotic machinery, although evidence does exist that ubiquitination inhibits the caspase-dependent apoptotic pathway (Yang *et al.*, 2003). The findings explain the identification of ubiquitinated cells in severely calcified aortic valves. Ubiquitination, together with Light Chain 3 processing was also found in a related disease, namely atheromatosis (Martinet *et al.*, 2004). The overall conclusion is that autophagic cell death, rather than apoptosis, is the most important trigger for the release of matrix vesicles in calcific aortic stenosis.

8. Prospects and Concluding Remarks

Autophagy plays an important role in degenerative aortic valve disease. The presence of a sparse number of autophagic cells within a calcified valve could be compared with a few burning coals within the ashes of a slow burning fire (Mistiaen *et al.*, 2006). However, the effect of autophagy on the rate of progression needs more investigation. This requires a prospective comparison of explanted valves from a range of between slow and rapid progressors of aortic valve degeneration. This might be done in patients with moderate degenerative AVD, needing coronary bypass surgery. Whether autophagy is an early cause or a terminal event associated with the end stage of valve disease needs also to be established. Recognition of inducing mechanisms could open ways for the modulation of autophagy in cardiovascular disease and possibly define targets for future drug design. However, there are few inhibitors for autophagy available and these are toxic. The central player in the autophagy signaling complexes and pathways is the mammalian target of rapamycin (mTOR). Inhibition of mTOR activity leads to autophagic cell death and, in some conditions, to apoptosis. Rapamycin and its analogues have already been used as immunosuppressants after transplantation by blocking proliferation of activated T lymphocytes, and after stenting by blocking proliferation of smooth muscle cells in arteries (Martinet *et al.*, 2007). Whether this is useful in degenerative aortic valve disease remains still to be seen.

REFERENCES

Aubin, J. E., Liu, F., Malaval, L., and Gupta, A. K. (1995). Osteoblast and chondroblast differentiation. *Bone* **17,** 77S–S83.

Bi, X. N., and Liao, G. H. (2007). Autophagic-lysosomal dysfunction and neurodegeneration in Niemann-Pick type C mice: Lipid starvation or indigestion? *Autophagy* **3**(6), 646–648.

Broker, L. E., Kruyt, F. A. E., and Giaccone, G. (2005). Cell death independent of caspases: A review. *Clin. Cancer Res.* **11**(9), 3155–3162.

Butcher, J. T., Tressel, S., Johnson, T., Turner, D., Sorescu, G., Jo, H., and Nerem, R. M. (2006). Transcriptional profiles of valvular and vascular endothelial cells reveal phenotypic differences. *Arterioscler. Thromb. Vasc. Biol.* **26,** 69–77.

Clark-Greuel, J. N., Conolly, J. M., Sorichillo, E., Narula, N. R., Rapoport, H. S., Mohler, E. R. III, Gorman, J. H. III, Gorman, R. C., and Levy, R. J. (2007). Transforming growth factor-β1 mechanisms in aortic valve calcification: Increased alkaline phosphatase and related events. *Ann. Thor. Surg.* **83**(3), 946–953.

Cosmi, J. E., Kort, S., Tunick, P. A., Rosenzweig, B. P., Freedberg, R. S., Katz, E. S., Applebaum, R. M., and Kronzon, I. (2002). The risk of the development of aortic stenosis in patients with "benign" aortic valve thickening. *Arch. Intern. Med.* **162,** 2345–2347.

Elsässer, A., Voigt, A. M., Nef, H., Kostin, S., Mollmann, H., Skwara, W., Bode, C., Hamm, C., and Schaper, J. (2004). Human hibernating myocardium is jeopardized by apoptotic and autophagic cell death. *JACC* **43**(12), 2191–2199.

Hein, S., Arnon, E., Kostin, S., Schonburg, M., Elsässer, A., Polyakova, V., Bauer, E. P., Kloverkorn, W. P., and Schaper, J. (2003). Progression from compensated hypertrophy to failure in the pressure-overloaded human heart: Structural deterioration and compensatory mechanisms. *Circulation* **107,** 984–991.

Hilton, T. C. (2000). Aortic valve replacement for patients with mild to moderate aortic stenosis undergoing coronary artery bypass surgery. *Clin. Cardiol.* **23,** 141–147.

Jian, B., Narula, N., Li, Q-y., Mohler, E. R. II I., and Levy, R. J. (2003). Progression of aortic valve stenosis: TGF-β1 is present in calcified aortic valve cusps and promotes aortic valve interstitial cell calcification via apoptosis. *Ann. Thor. Surg.* **75**(2), 457–465.

Klionsky, D., *et al.* (2008).Guidelines for the use and interpretation of assays for monitoring autophagy in higher eukaryotes*Autophagy* **4**(2), 1–25.

Knaapen, M. W. M., Davies, M. J., De Bie, M., Haven, A. J., Martinet, W., and Kockx, M. M. (2001). Apoptotic versus autophagic cell death in heart failure. *Cardiovasc. Res.* **51,** 304–312.

Kockx, M. M., Muhring, J., Knaapen, M. W., and de Meyer, G. R. (1998). RNA synthesis and splicing interferes with DNA *in situ* end labeling techniques used to detect apoptosis. *Am. J. Pathol.* **152,** 885–888.

Levine, B., and Klionsky, D. J. (2004). Development by self-digestion: Molecular mechanisms and biological functions of autophagy. *Dev. Cell* **4,** 463–477.

Martinet, W., De Bie, M., Schrijvers, D. M., De Meyer, G. R. Y., Herman, A. G., and Kockx, M. M. (2004). 7-Ketocholesterol induces protein ubiquitination, myelin figure formation, and light chain 3 processing in vascular smooth muscle cells. *Arterioscler. Thromb. Vasc. Biol.* **24,** 2296–2301.

Martinet, W., De Meyer, G. R. Y., Andries, L., Herman, A. G., and Kockx, M. M. (2006). Detection of autophagy in tissue by standard immunohistochemistry. Possibilities and limitations. *Autophagy* **2,** 55–57.

Martinet, W., Knaapen, M. W. M., Kockx, M. M., and De Meyer, G. R. Y. (2007). Autophagy in cardiovascular disease. *Trends in molecular medicine* **13**(11), 482–491.

McCray, B. A., and Taylor, J. P. (2008). The role of autophagy in age-related neurodegeneration. *Neurosignals* **16**(1), 75–84.

Merryman, W. D., Lukoff, H. D., Long, R. A., Engelmayr, G. C. Jr, Hopkins, R. A., and Sacks, M. S. (2007). Synergistic effects of cyclic tension and transforming growth factor-β1 on the aortic valve myofibroblast. *Cardiovascular pathology* **16,** 268–276.
Mistiaen, W., Van Cauwelaert, P. H., Muylaert, P. H., Wuyts, F. L., Harrisson, F., and Bortier, H. (2004). Risk factors and survival after aortic valve replacement in octogenarians. *J. Heart Valve Dis.* **13,** 538–544.
Mistiaen, W. P., Somers, P., Knaapen, M. W. M., and Kockx, M. M. (2006). Autophagy as mechanism for cell death in degenerative aortic valve disease: An underestimated phenomenon in cardiovascular diseases. *Autophagy* **2**(3), 221–223.
Mistiaen, W., Van Cauwelaert, Ph., Muylaert, Ph., and De Worm, E. (2007). One thousand Carpentier-Edwards pericardial valves in the aortic position: What has changed in the past 20 years, and what are the effects on hospital complications? *J. Heart Valve Dis.* **16**(4), 417–422.
O'Brien, K. D., Kuusisto, J., Reichenbach, D. D., Ferguson, M., Giachelli, C., Alpers, C. E., and Otto, C. M. (1995). Osteopontin is expressed in human aortic valvular lesions. *Circulation* **92,** 216–218.
Ortlepp, J. R., Schmitz, F., Mevissen, V., Weiss, S., Huster, J., Dronskowski, R., Langebartels, G., Autschbach, R., Zerres, K., Weber, C., Hanrath, P., and Hoffmann, R. (2004). The amount of calcium-deficient hexagonal hydroxyapatite in aortic valves is influenced by gender and associated with genetic polymorphisms in patients with severe calcific aortic stenosis. *Eur. Heart J.* **25,** 514–522.
Osman, L., Yacoub, M. H., Latif, N., Amrani, M., and Chester, A. H. (2006). Role of human valve interstitial cells in valve calcification and their response to atorvastatin. *Circulation* **114,** 1547–1552.
Otto, C. M., Burwash, I. G., Legget, M. E., Munt, B. I., Fujioka, M., Healy, N. L., Kraft, C. D., Miyake-Hull, C. Y., and Schwaegler, R. G. (1997). Prospective study of asymptomatic valvular aortic stenosis. Clinical, echocardiographic, and exercise predictors of outcome. *Circulation* **95,** 2262–2270.
Otto, C. M. (1998). Aortic stenosis. Clinical evaluation and optimal timing of surgery. *Cardiol. Clin.* **16,** 353–373.
Purdy, P. E., and Russell, D. G. (2007). Ubiquitin trafficking to the lysosome. *Autophagy.* **3** (4), 399–401.
Rajamannan, N. M., Subramaniam, M., Rickard, D., Stock, S. R., Donovan, J., Spingett, M., Orszulak, T., Fullerton, D. A., Tajik, A. J., Bonow, R. O., and Spelsberg, T. (2003). Human aortic valve calcification is associated with an osteoblast phenotype. *Circulation* **107,** 2181–2184.
Somers, P., Van Cauwelaert, P., Knaapen, M., Kockx, M., Bortier, H., and Mistiaen, W. (2006). Histological evaluation of autophagic cell death in calcified aortic valve stenosis. *J. Heart Valve Dis.* **15,** 43–48.
Somers, P., Knaapen, M., and Mistiaen, W. (2006). Histopathology of calcific aortic valve stenosis. *Acta Cardiologica* **61,** 557–562.
Stromhaug, P. E., and Klionsky, D. J. (2000). Approaching the molecular mechanism of autophagy. *Traffic* **2,** 524–531.
Takemura, G., Miyata, S., Kawase, Y., Okada, H., Maruyama, R., and Fujiwara, H. (2006). Autophagic degeneration and death of cardiomyocytes in heart failure. *Autophagy* **2,** 212–214.
Wang, C- W., and Klionsky, D. J. (2003). The molecular mechanism of autophagy. *Molecular Medicine* **9,** 65–76.
Yang, Y., and Yu, X. (2003). Regulation of apoptosis; the ubiquitous way. *FASEB J.* **17,** 790–799.
Yoshimori, T. (2004). Autophagy: A regulated bulk degradation process inside cells. *Biochem. Biophys. Res. Commun.* **313,** 453–458.

CHAPTER NINETEEN

Monitoring Autophagy in Muscle Diseases

May Christine V. Malicdan,* Satoru Noguchi,* *and* Ichizo Nishino*

Contents

* Department of Neuromuscular Research, National Institute of Neurosciences, National Center of Neurology and Psychiatry, Tokyo, Japan

Methods in Enzymology, Volume 453
ISSN 0076-6879, DOI: 10.1016/S0076-6879(08)04019-6

Abstract

Autophagy is a tightly regulated pathway for the degradation and recycling of proteins delivered to lysosomes, and is an important process in maintaining cellular homeostasis. Whereas a basal level of autophagy can be detected in skeletal muscles, its perturbation can be seen in a variety of conditions affecting the muscle. In certain muscle diseases, moreover, autophagy seems to be a characteristic feature, although the exact role of autophagy in these disorders is just starting to be understood. As autophagy is not only an index of disease progression but also a potential target for treatment in certain disease conditions, its characterization is indeed of relevance. Thus, in this chapter, methods applicable to both human and murine skeletal muscle preparation for the analysis and monitoring of autophagy are presented.

1. Introduction

The skeletal muscle is one of the organs where there is a high level of constitutive autophagy and where autophagy can be physiologically enhanced (Mizushima *et al.*, 2004; Lunenmann *et al.*, 2007). The accumulation of autophagic vacuoles is documented in many neuromuscular disorders (Fujita *et al.*, 2007; Jongen *et al.*, 1995; Kaneda *et al.*, 2003; Lünemann *et al.*, 2007; Nascimbeni *et al.*, 2008; Nishino *et al.*, 2005; Raben, 2007 *et al.*; Shintani and Klionsky, 2004), and thus could be thought to be a nonspecific response to various signals. Hereditary muscle diseases in which autophagy is a remarkable feature can be classified broadly into two categories: rimmed vacuolar myopathies; and autophagic vacuolar myopathies (Nishino, 2006). Rimmed vacuolar myopathies are most likely secondary lysosomal myopathies as autophagy is activated secondarily to cellular events such as protein misfolding or aggregation, and because so far the identified causative genes encode extralysosomal proteins; this category includes distal myopathy with rimmed vacuoles (DMRV), otherwise known as hereditary inclusion body myopathy (hIBM), inclusion body myositis, oculopharyngeal muscular dystrophy (OPMD), and others. The second category primarily involves the lysosomal system, where there is buildup of autophagosomes most likely because of a block or inhibition of the autophagic process; this group includes, but is not limited to, acid maltase deficiency or Pompe disease, X-linked myopathy with excessive autophagy (XMEA), and a distinct group of myopathies characterized by the presence of a unique feature called autophagic vacuoles with sarcolemmal features (AVSF) (Nishino, 2006). The latter is comprised of myopathies that appear to be genetically heterogeneous, but is best exemplified by Danon disease, which is due to primary deficiency of a lysosomal membrane protein, LAMP-2 (Nishino *et al.*, 2000).

The relevance of autophagy in muscle disorders is highlighted by different murine models, which resemble the disease phenotype in humans (Bijvoet *et al.*, 1998; Malicdan *et al.*, 2007; Tanaka *et al.*, 2000).

Studies concerning autophagy in skeletal muscles have been evolving in recent years. But despite these recent advances, the precise role of autophagy in these neuromuscular disorders is not fully clarified, making the analysis of autophagy in skeletal muscles relevant. In general, methods to analyze the autophagic process in skeletal muscles do not differ from those that are employed in other tissues. However, the isolation and preparation of skeletal muscles for these analyses need specific techniques when compared to those used in other tissues. In skeletal muscles, necessary steps for sample preparation are very important to obtain optimal conditions for study. Hence, this chapter focuses on the preparation of skeletal muscle cells and certain methodologies that are specific to skeletal muscle cells.

2. Histological Observation of Skeletal Muscles

2.1. Isolation and fixation of frozen muscle biopsy specimen

Isolation of muscle by biopsy is done following standard protocols. To obtain an ideal sample, at least three pieces consisting of 20 × 10 mm samples should be harvested. In addition, some cautionary notes could help in optimizing sample collection. During the biopsy procedure, maintain normal tautness of muscle and avoid over-stretching, which could give rise to artifacts. Unnecessary forceps marks on the muscle should be avoided. If the specimen cannot be processed for fixation immediately, wrap it in saline-moistened gauze and keep it on ice, maximally for 2 h. The same cautionary notes are maintained when obtaining skeletal muscle tissues from murine animals, except that usually the whole muscle (*i.e.*, muscle from the point of origin to the point of attachment) is isolated. Moreover, careful surgical excision will allow identification and exclusion of unnecessary parts of the limb muscles, including fatty tissues and excess fascia, which should not be included for fixation.

Studying muscle pathology is best done in snap frozen muscle specimens. The method of freezing muscles that produces the least artifacts is by using isopentane, which is cooled in liquid nitrogen.

1. Trim the biopsy specimen, ideally into a length of 15–20 mm, and a width of 5 mm. Remove areas that have sutures and forceps marks.
2. Make tragacanth gum by mixing tragacanth powder (WAKO Chemical Industries, 200-02245) and enough water to achieve the consistency of

whipped cream. Apply tragacanth gum on a 2-cm chunk of block, creating a small mound.

3. Embed the muscle specimen in tragacanth gum, orienting the muscle in an upright cylindrical position.
4. Pour isopentane into a small beaker suspended over a larger container with liquid nitrogen. Lower the beaker with isopentane into the liquid nitrogen slowly to allow the temperature of isopentane to drop to its melting point (−160 °C). Continuously stir the isopentane to make the temperature homogenous (the optimal temperature is reached when isopentane shows white crystals on the base of the beaker).
5. Using long-handled forceps to hold the block with the biopsy specimen, immerse the block with tissue into the precooled isopentane and shake the specimen rapidly (vigorously) for 1 min.
6. Quickly transfer the frozen biopsy onto dry ice to evaporate the isopentane and then place into a storage container prechilled with dry ice. The specimen can be stored indefinitely at less than −80 °C.

2.2. Routine histological staining

This section describes protocols for the commonly used methods used to describe pathological changes in autophagic myopathies. Perform routine histological staining using unfixed, 10-μm thick cryosections placed on a non-coated glass slide. After placing cryosections on the glass, leave it to dry at room temperature for 2–3 min. These cryosections can be stored indefinitely at −80 °C.

Hematoxylin and eosin staining is commonly used to describe the general morphology of myofibers, including fiber size and the presence of internalized nuclei. In some myopathies, prominent lysosomes are stained blue; interpretation, however, can be difficult, because other structures such as mitochondria can also be stained blue.

Clues in light microscopy that certain structures are autophagic vacuoles are usually obtained by doing selected histochemical studies (Fig. 19.1). Modified Gomori trichrome demonstrates rimmed vacuoles (RVs) and cytoplasmic bodies, among other findings; RVs appear reddish in color. Acid phosphatase highlights lysosomal structures in skeletal muscles, phagosomes in macrophages, and other structures/compartments that have low pH; positive staining appears red. Nonspecific esterase also stains lysosomes, in addition to macrophages, neuromuscular junctions, and myotendinous junctions; highlighted structures appear brown to brick red. Acetylcholinesterase stains the neuromuscular and myotendinous junctions. However, it also highlights the vacuoles that are seen in autophagic myopathies with AVSFs.

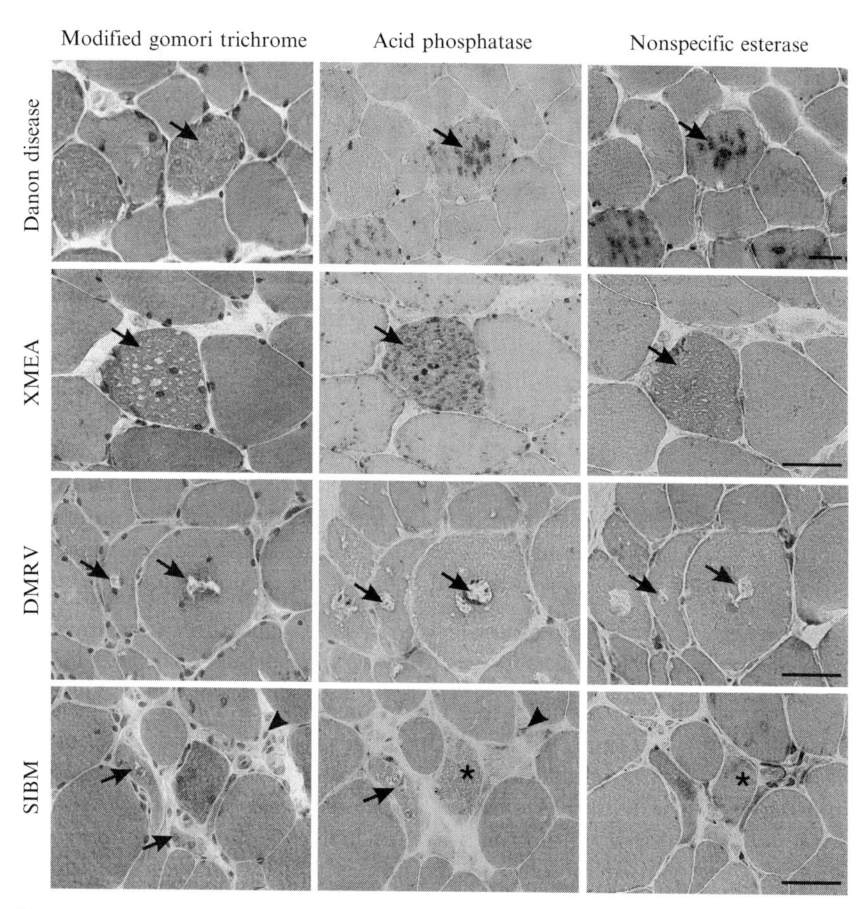

Figure 19.1 Common histochemical stains for analyzing autophagy in skeletal muscle. Sections were taken from patients with Danon disease, X-linked myopathy with excessive autophagy, DMRV/hIBM, sporadic inclusion body myositis (sIBM). In Danon disease, modified Gomori trichrome (mGT) staining shows tiny intracytoplasmic vacuoles (arrow) that occupy most of the fiber. These vacuoles have high acid phosphatase (ACP) activity, and also show intense nonspecific esterase activity. Similar tiny intracytoplasmic vacuoles (arrow) on mGT are seen in XMEA patients. The vacuoles have rather a more remarkable increase in ACP activity but usually do not exhibit an increase in nonspecific esterase activity. Rimmed vacuoles (RVs) (arrow), one of the common pathologic features of DMRV, are evident with mGT staining; note the rimming of the vacuoles (in red color). The vacuoles also show increased ACP activity but do not consistently exhibit an increase in nonspecific esterase activity. The same RVs (arrow) are characteristic in sIBM, except that the RVs are accompanied by other findings, such as ragged red fiber (asterisk), and inflammatory cell infiltrates (arrowhead, with increased ACP and nonspecific esterase activities). Bars represent 50 μm. (See Color Insert.)

2.3. Modified Gomori trichrome staining

1. Put glass slides in coplin jars. Stain with Harris hematoxylin for 10 min, shake coplin jars occasionally. To make Harris hematoxylin solution:
   a. Make Harris hematoxylin stock solution. Mix 300 ml of 50% ethanol (EtOH), 1.5 g of hematoxylin (Kanto Chemicals, 6835) and 0.9 g of aluminum ammonium sulfate (Sigma, 202568). Boil the mixture and slowly add 1.8 g of mercuric red oxide (in no less than 20 minutes). Cool solution and keep at room temperature for 16–20 h; this is stable at 4 °C for 12 months.
   b. In another beaker, dissolve 200 g of aluminum ammonium sulfate powder in 500 ml deionized water by heating at 35 °C to 37 °C for 16 h; store at room temperature.
   c. To make hematoxylin working solution, dilute Harris hematoxylin 5x with aluminum ammonium sulfate solution. Filter with Whatman filter paper (No. 2) before use.
2. Wash in running tap water for 5–10 min.
3. Stain with Gomori's reagent for 30 min. To make Gomori's reagent, mix the following in a glass beaker: 0.6 g of chromotrope 2R (Sigma, C3143), 0.3 g of fast green FCF (Sigma, F7258), 0.6 g of phosphotungstic acid (Kanto Chemicals, 32200-30), 1 ml of glacial acetic acid, 100 ml of deionized water; adjust to pH 3.4 (this solution is good for 3 weeks stored at room temperature).
4. Briefly rinse twice with 0.2% acetic acid.
5. Dehydrate in ascending EtOH (50% → 70% → 90% → 100% → absolute EtOH:xylene in 1:1 volume ratio) 10 s each, and dip in xylene for another 10 s.
6. Mount in Permount (Fisher 537-16271) or Canada balsam (Wako, C1795) (these mounting media can be used directly).

2.4. Acid phosphatase

1. Fix slides in a mixture of 12 ml of Veronal acetate solution and 18 ml of 100% acetone at 4 °C or –20° for 10 min. To make Veronal acetate solution, mix 1.94 g of sodium acetate (Kanto Chemicals, 37092-00), 2.94 g of sodium barbital (WAKO, 021-0032), 280 ml of deionized water, and 180 ml of 0.1 *N* HCl; adjust to pH 5.0.
2. Wash in tap water for 4–5 min.
3. Incubate in ACP solution at 37 °C in a water bath for 60 min. To make ACP solution, make Solution A, which is a mixture of 0.6 of ml 4% sodium nitrite in deionized water, 0.6 ml of 4% pararosaniline (Acros organics, 569-61-9) in 2 *N* HCl, 19.5 ml of deionized water, and 7.5 ml of Veronal acetate solution. In a separate beaker, dissolve 15 mg of naphthol AS-B1 phosphate (Sigma, N2125) in 1.5 ml *N,N*-dimethiformamide

(WAKO, 045-02916); when the naphthol is completely dissolved, add to Solution A. Adjust to pH 4.7–5.0 using 2 *N* NaOH.

4. Wash in tap water for 1 min.
5. Counterstain the nuclei by immersing the glass slides in 0.3% methyl green (Acros Organics, 1485576-6) for 15–20 s.
6. Wash briefly in tap water.
7. Mount in prewarmed glycerin jelly. This is made by dissolving 5 g of gelatin (Merck, Art., 4078) in 30 ml of water under low heat; when the solution is completely dissolved, continue stirring and add 0.5 ml of TE-saturated phenol (Nippon Gene, 319-90093). Cool and use directly. Store at 4 °C and prewarm before use.

2.5. Nonspecific esterase

1. Incubate slides in nonspecific esterase solution at 37 °C in a water bath for 60 min. To make nonspecific esterase solution, mix 1.2 ml of 4% sodium nitrite, 30 ml of 0.1 *M* sodium phosphate buffer, pH 6.5, and 0.75 ml of 1% alpha-naphtyl acetate (Sigma, N8505) dissolved in acetone.
2. Wash in running tap water for 10 min.
3. Dehydrate in ascending EtOH, as in the procedure for the modified Gomori trichrome staining, and clear in xylene.
4. Mount in Permount.

2.6. Acetylcholine esterase

1. Prepare acetylcholine esterase solution by mixing the following: 12.5 mg acetylthiocholine iodide (Sigma, A5751), 2.5 ml of deionized water, 15.8 ml of 0.8% sodium acetate, 0.5 ml of 0.6% acetic acid, 1.5 ml of 2.94% sodium citrate, 2.5 ml of 0.5% copper sulfate, and 2.5 ml of 0.17% potassium ferricyanide.
2. Incubate the slides in acetylcholine esterase solution at 37 °C in a water bath for 15 min, then wash briefly in tap water.
3. Dehydrate in ascending EtOH, clear in xylene, and mount in Permount as described earlier.

2.7. Immunohistochemical staining

For immunohistochemical analysis, 5- to 6-μm-thick frozen sections are placed on Matsunami adhesive silane (MAS) (Matsunami, S094410) or poly-L-lysine (PLL) (Matsunami, S074410) coated glass slides, and are air-dried for 15–30 min. Fixation depends on the type of antibody used, but in general, probing for proteins associated with membranous compartments and that are cytoskeleton-associated is best observed on sections fixed with acetone. For proteins in cytosolic compartments, other cross-linking fixatives are used, including paraformaldehyde. In characterizing autophagy in neuromuscular disorders, we commonly use acetone for fixation.

1. Immerse slides in acetone for 5 min at room temperature, or in cold acetone for 20 s at −20 °C.
2. Remove slides from acetone and air-dry for a few seconds.
3. Proceed with blocking and immunostaining following standard protocols (see the chapter by Kimura *et al.*, in volume 452).

Methods for immunohistochemistry are not different from other protocols, so these will not be presented in this chapter. However, there are some issues regarding the use of fluorescence in analyzing specimens from autophagic myopathies, as will be discussed in the succeeding paragraph, so some modification of these methods is done. Alternatively, enzyme detection (3,3′ diaminobenzidine or alkaline phosphatase staining) is sometimes preferred.

2.8. Special considerations in various autophagic myopathies

In skeletal muscles, the presence of autophagy is usually suspected when RVs are seen at the light microscopy level. In addition to these RVs, various pathological findings, including intracellular protein accumulations and aggregations, can also be seen. From our experience, these fibers with RVs and intracellular accumulations can exhibit autofluorescence, which is not related to the use of fixatives, and thus can make precise interpretation of results difficult. This observation is more remarkable in mice models of autophagic myopathies.

Autofluorescence is primarily due to the presence of lipopigments such as lipofuscin and ceroid. Lipofuscin is a nondegradable lysosomal byproduct of organelle and protein degradation, which contains sugars and metals (mercury, aluminum, iron, copper, and zinc), in addition to its large lipid composition. Ceroid has been defined as a lipofuscin-like lipopigment that arises from pathological conditions such as disease, malnutrition, and cell stress (Yin, 1996). The protein composition of ceroid and lipofuscin also differs in that ceroid composition in most neuronal ceroid lipofuscinoses (NCLs) primarily contains subunit c of the mitochondrial ATP synthase, whereas the protein components of lipofuscin appear to be more heterogeneous, with only amyloid-β currently identified (Seehafer and Pearce, 2006). These lipopigments have broad emission and excitation patterns, and are thus visible under fluorescent filters for UV, fluorescein, rhodamine, and Cy5. Disturbances in the process of transporting macromolecules and organelles to the lysosomes may give rise to the accumulation of such lipopigments. However, it is beyond the scope of this chapter to discuss pathomechanistic theories on how autofluorescence could occur in skeletal muscles with autophagy, in addition to the identification of the precise molecule responsible for the autofluoresence in these myopathies. Thus, this section limits its discussion to protocols on how to quench this phenomenon.

2.9. Methods to quench autofluorescence

In the literature, several techniques are described for reducing autofluorescence, and the most convenient methods include the use of dyes (Sudan Black, Pontamine Sky Blue, Tryptan Blue), chemicals (sodium borohydide, cupric sulfate), heavy metals (osmium tetroxide, ferric chloride, or mercuric chloride), photobleaching, or mathematical models that attempt to subtract the background fluorescence because of the broader autofluorescent excitation spectra compared to the spectra of the fluorescent label. There is usually, however, a trade-off between the attempt to quench the autofluoresence and the maintenance of the desired signal intensity.

1. The use of the Autofluorescence Eliminator Reagent (Chemicon, 2160) is effective to decrease autofluorescence in muscle cryosections. After the last PBS wash following immunohistochemistry, rinse slides briefly in deionized water and dehydrate in 70% EtOH for 3 min. Incubate in Autofluorescence Eliminator Reagent for 6 min, and rinse three times with 70% EtOH. Mount with mounting medium following standard protocols.
2. Cupric sulfate ($CuSO_4$) (Schnell *et al.*, 1999). After standard immunofluorescence, using secondary antibodies labeled with fluorophores of interest, immerse the sections in 0.1% $CuSO_4$ in 50 m*M* ammonium acetate buffer, pH 5.0, for 1 h. Wash briefly with deionized water then PBS. Mount slides following standard procedures.
3. Sudan Black (Schnell *et al.*, 1999). After immunohistochemistry, immerse slides in 1% Sudan Black B (Sigma, S2380) in 70% methanol for 10 min. Dip in three washes of 70% EtOH. Mount slides following standard procedures.

3. Measuring Autophagy by Protein Quantification

3.1. Preparation of total protein lysates and subfractionation of samples from skeletal muscle

The skeletal muscle is an organ that contains the largest amounts of cytosolic proteins. For analysis of autophagic markers, most of which are membranous proteins, the larger amounts of cytosolic proteins can interfere with the solubilization of such membrane-associated proteins. For total protein extraction, a high concentration of detergent is necessary. Nevertheless, preparation of microsome fractions is effective for analyzing minor proteins.

3.2. Extraction of total proteins

1. Cut 20–30 10-μm thick cryosections in a cryostat and collect into a precooled 1.5-ml microcentrifuge tube. Weigh the pile of sections.
2. Mix with SDS buffer (10% SDS, 10 m*M* EDTA, 5% β-mercaptoethanol, 70 m*M* Tris-HCl, pH 6.7) to make a 10% (weight/volume) suspension. Immediately boil in a 95 °C block incubator for 5 min and then sonicate the solution for 2–3 min to reduce its viscosity. Centrifuge at 5000×*g* for 10 min at 4 °C; the supernatant can be used for SDS-PAGE.

3.3. Specific analysis of membrane fractions

1. To prepare membranous (microsomal) fractions from skeletal muscles, cut 50 10-μm thick cryosections and collect into a precooled test tube.
2. Homogenize by using a polytron (Kinematica AG, PT1200) homogenizer in 1 ml of microsome-preparation buffer consisting of 10% sucrose, 20 m*M* sodium pyrophosphate, 20 m*M* sodium phosphate, 0.5 m*M* EGTA and 1 m*M* magnesium chloride, with Complete, Mini, EDTA-free proteinase inhibitors (Roche, 11836170001) (adjust buffer to pH 7.1).
3. Centrifuge the solution at 5000×*g* for 10 min at 4 °C and recover the supernatant fraction. Centrifuge the supernatant at 100,000×*g* for 60 min at 4 °C to recover the microsomal fraction in the pellet.
4. Wash the microsomal pellet with 0.6 *M* potassium chloride to remove actomyosin. Solubilize the residual pellet in SDS sample buffer for SDS-PAGE analysis.

4. Monitoring Autophagy in Cultured Skeletal Myocytes

4.1. Muscle primary culture

The skeletal muscle is a multinucleated organ that contains postmitotic nuclei. It undergoes degeneration and regeneration, which can result from injury or disease. The skeletal muscle has a remarkable capacity for regeneration when it is injured, and this is predominantly due to the presence of mononucleated satellite cells, which, when activated, give rise to muscle precursor cells that express myogenic regulatory factors. By taking advantage of these progenitor cells present in skeletal muscles, primary cultures of myocyte can be established.

Preparation of primary culture derived from skeletal muscles has a problem due to contamination from many fibroblasts. The methodology to isolate skeletal muscle primary culture cells is based on the difference in the ability of attachment to the surface of dishes between fibroblasts and myoblasts. During the initial week or two of culturing, selective growth and passaging conditions allow the myoblasts to become the dominant cell type. Primary cells can be isolated from mice of any age, but isolation from neonatal mice (1–5 days) gives a greater yield of myogenic cells.

4.2. Isolation and growth of murine primary muscle cells

1. Warm proliferating medium (DMEM with 20% heat-inactivated FBS [Sigma F4135] and 1% chick embryo, penicillin-streptomycin [Nacalai tesque, 26253-84]) in a 37 °C water bath.
2. After euthanizing 1–2 mice by CO_2 asphyxiation, disinfect the skin by wiping with 70% EtOH. Decapitate and cut off forepaws and hindpaws with sterile surgical scissors. Gently remove the skin completely. Make a longitudinal section on the anterior abdominal region and remove all internal organs. Remove all fatty tissues carefully, which are predominantly over the dorsal neck and over the thighs. Harvest all limb muscles, intercostal muscles with ribs and paravertebral muscles (excluding the vertebra). Dissection is easier if done under a stereo dissecting microscope. Keep the specimen moistened with PBS(–) (PBS without Ca^{2+} and Mg^{2+}) on a sterile non-coated culture dish.
3. Move the specimen to new dishes and wash out the blood twice with Hanks solution.
4. Move washed tissues into a 50-ml tube and enzymatically remove the cells attached to the outside of the tissues (fibroblasts, adipocytes) by incubating in 10 ml of 0.25% trypsin, 0.05%. DNase I, 1 m*M* EDTA, in Hank's Balanced Salt Solution (HBSS) (GIBCO BRL, 14185-052).
5. Incubate at 37 °C with constant agitation for 10 min.
6. Stand tube for a couple of minutes to allow the tissues to settle by gravity. Carefully pipette out the supernatant fraction and discard.
7. Mechanically dissociate tissues by mincing them into a slurry using surgical scissors and a razor blade and enzymatically dissociate the tissues in 10 ml of collagenase/trypsin solution (0.4% collagenase type II (GIBCO BRL, 17101-015) in 0.25% trypsin, 1 m*M* EDTA, pH 8).
8. Incubate at 37 °C with constant agitation for 40 min.
9. Stand tube for a couple of minutes to allow the tissues to settle by gravity. Carefully pipette out the supernatant fraction and save the pellet.

10. Transfer the supernatant fraction (approximately 8 ml) into a new 50-ml tube and add 10 ml of proliferating medium. Mix by pipetting, and filter the cells through a 70-μm cell strainer (BD Falcon, 352350).
11. Resuspend the pellet (obtained in step 9) in 10 ml of collagenase/trypsin solution (to subject it to another round of digestion) by hand pipetting.
    a. Incubate at 37 °C in a water bath for an additional 15 min.
    b. Strain through a 70-μm cell strainer and transfer the pass-through to a 50-ml tube.
    c. Combine with the supernatant fraction from step 10.
12. Cell separation by differential attachment
    a. After washing the cells with proliferating medium by centrifugation at 1,500 rpm for 5 min, plate into 2–3 100-mm diameter nontreated dishes (do not use the dishes for cell culture).
    b. Incubate at 37 °C in a humidified chamber with 95% air, 5% CO_2, for 40 min.
    c. Collect the supernatant and transfer to a new 50-ml tube (label this "sup-1"). Wash the surface of a dish with 5 ml of medium; collect this washing medium and combine with sup-1. Add 10 ml of medium to the washed dish and incubate at 37 °C in a humidified chamber (label this as "bound fraction").
    d. Reconstitute the sup-1 fraction to 30 ml with medium, and plate it into three 100-mm type I-coated dishes (these are now referred to as "unbound fraction"), and incubate in a humidified CO_2 chamber at 37 °C.
13. Check the plates the next day; if you find layering of the cells, transfer the debris (outermost layer, supernatant) to another dish.
14. Change the medium every 2 days. Primary myoblasts grow best when dense, but avoid more than 80% confluence, or else they may start to differentiate (contact induction leads to differentiation) or die. Split cells at no more than a 1:4 dilution.
15. You can measure the population of myoblasts and fibroblasts by staining with mouse monoclonal desmin antibody (ICN Biomedicals, 69-181; myoblasts are positive for this antibody).
16. Unbound fractions can be immediately used. It is, however, possible to increase the population of myoblasts when desired. When the cells are ready to be split, remove the cells from the dish using PBS(–) with no trypsin or EDTA. Aspirate off the medium and rinse the dish with PBS (–). Add 1 ml PBS into the dish, and hit the dish very firmly in a sideways fashion against the edge of a tabletop to dislodge the cells. In this step, most of the myoblasts will be detached while the fibroblasts will remain attached to the dish. Transfer the suspension of cells to a new collagen-coated dish. Repeat these steps during the initial week of culture expansion or until most of the fibroblasts are gone from the culture.

17. Alternatively, unbound fractions can be mixed and stored together using standard cell culture techniques. Likewise, bound fractions can also be stored together for cryopreservation.

4.3. Differentiating myoblasts into myotubes

1. Seed unbound cells on a type-I-collagen-coated plastic dish with proliferating medium (as defined above) in a humidified CO_2 chamber at 37 °C.
2. Allow the cells to grow to at least 90% confluence.
3. Change the medium to differentiating medium that consists of 5% heat-inactivated horse serum (GIBCO BRL, 26050-070, 0.1 *M* hydroxyurea and penicillin-streptomycin in DMEM (hydroxyurea suppresses growth of the mitotic cells such as fibroblasts and induces them to die, while postmitotic cells such as myotubes survive in this condition).
4. Change the medium every 3 days. Indices which support differentiation are usually observed 5–7 days after switching to differentiating medium, and these include elongation of cells, presence of multiple nuclei in a cell, expression of muscle-specific proteins and spontaneous contraction of skeletal myotubes.

4.4. Analysis using LysoTracker

Weakly basic amines selectively accumulate in cellular compartments with low internal pH and can be used to investigate the biosynthesis and pathogenesis of lysosomes. The use of acidotrophic dyes including LysoTracker Red can contribute to the quantification of lysosomes in muscle cell culture. However, one caveat that should be remembered is that these dyes stain autolysosomes and not the early autophagosome. When used together with early markers of autophagy, these can, however, be of benefit for monitoring autophagy in skeletal muscle cells.

1. Culture myoblasts overnight in proliferating medium on type-I collagen-coated cover slips replaced in the center part of 35-mm diameter culture dishes overnight.
2. Prepare LysoTracker (Molecular Probes/Invitrogen, L7528) working solution by dilution of the 0.1 *M* stock solution in culture medium to a final concentration of 50 n*M* and keep it in a 37 °C water bath, protected from light.
3. Remove the medium from the dish and replace with prewarmed medium containing LysoTracker. Swirl immediately to distribute the dye evenly. Put the dishes back in the humidified chamber for 30 min to 1 h.

4. After incubation, briefly rinse three times with prewarmed 10% FBS in HBSS and then replace with culture medium.
5. The dish can be directly viewed under an inverted fluorescence microscope.

5. Electron Microscopy Observation of Skeletal Muscles

5.1. Routine electron microscopy

Electron microscopy is a valid method to determine the quantitative and qualitative characteristics of the phagophore and prelysosomal autophagosome. After obtaining biopsy specimens, immediate processing for fixation is preferred (also see the chapter by Anttila *et al.*, in volume 452).

5.2. Specimen preparation from muscles

1. Orient the muscle in such a way that the longitudinal direction is identified. Using a razor blade, cut small pieces of muscle 15 × 4 × 4 mm, the longest dimension being the length of the muscle. Pick up these pieces carefully and with a light grip using forceps and place them on the frosted edge of a glass slide to keep the tissues from bending.
2. Immerse the glass slide with the specimen in a solution with 2.5% glutaraldehyde in 100 m*M* sodium cacodylate buffer, pH 7.0, 5m*M* $CaCl_2$. The volume should be at least 10x that of the specimen. Let the slide stand at room temperature for 2 h before proceeding to the next step. For murine skeletal muscles, fixation may be prolonged from hereon to overnight at 4 °C. For murine cardiac muscles, use 4% glutaraldehyde; in addition, fixation is better when a perfusion method is done to deliver fixative into the cardiac muscle.
3. Decant the glutaraldehyde solution or aspirate with a pipette and collect the waste into a separate bottle.
4. Transfer the muscle specimen into a glass vial from the glass slides. Wash the specimen twice with washing buffer, which consists of 0.1 *M* sodium cacodylate buffer in 0.1 *M* PBS, pH 7.0. The specimen can be kept at 4 °C for several weeks.
5. Trim off the edges of the specimen and cut into smaller pieces (5 × 1.5 × 1.5 mm). Avoid drying of the specimen by applying drops of washing buffer on the specimen while trimming. Using a toothpick or small applicator stick, pick up pieces of the tissues and transfer them into a new glass tube with fresh washing buffer.

6. Remove the washing buffer and postfix the specimen in a mixture containing 4% osmium tetroxide (OsO_4) (Sigma, 201030) in 0.1 *M* phosphate buffer:0.2 *M* s-Collidine (TAAB Laboratories Equipment, C013):3% lanthanum nitrate (TAAB Laboratories Equipment, L001) (1:1:1 in volume) for 2 h on ice. The volume should be at least 10x the sample.

Note: OsO_4 is very toxic and the vapors are extremely noxious. Perform these steps in a fume hood.

7. Wash the specimen at least 3 times in deionized water.
8. Dehydrate in ascending concentrations of EtOH (50%, 70%, 2 times of 80%, 95%, 3 times of 100%), for 5 min per change.
9. Immerse in 3 changes of absolute EtOH, for 5 min each.
10. Wash 3 times in absolute EtOH:N-butyl-glycidyl ether (QY-1) (Nisshin EM Co., 310) (1:1 in volume), for 5 min per change.
11. Wash 3 times in 100% QY-1, 5 min per change.
12. Prepare EPON resin. In sequential order, mix 22.8 ml of EPON 812 (TAAB Laboratories Equipment, T002), 11.8 ml of MNA (TAAB Laboratories Equipment, M010), 15.4 ml of DDSA (TAAB Laboratories Equipment, D025). Add 1.5 ml of DMP30 (TAAB Laboratories Equipment, D035).
13. Carefully decant the QY-2. Preembed in 1–2 ml QY-2:EPON resin (1:1 in volume), and incubate for 1 h at 37 °C.
14. Change the preembedding solution to QY-2:EPON (1:2 in volume) and incubate overnight (16 h).
15. Embed specimens in labeled embedding capsules with freshly prepared EPON resin. Polymerize the resin by heating at 45° for 8 h and 60° for 3 days.
16. Proceed with cutting semi-thin sections (1-μm thick) using microtome for toluidine blue staining to screen areas for further trimming.
    a. Make toluidine blue solution: Dissolve 1 g of Borax (Sigma, B9876) in 100 ml of deionized water. Add 1 g of toluidine blue (Sigma, T3260) and stir until dissolved. Filter solution before use.
    b. Obtain semithin sections and transfer these to a drop of deionized water on a glass slide. Place the glass slide on a slide warmer (40 °C–50 °C) to dry the sections.
    c. When the sections are completely dried, apply toluidine solution to cover the sections completely. Continue heating the slides until the sections turn blue (duration of heating depends on the darkness of staining desired, but usually 1 min is enough).
    d. Wash gently with deionized water. Air-dry the slides.
17. Obtain thin sections (10-nm thick) and place them on 200- or 300-mesh, copper grids.
18. Stain with 1% uranyl acetate for 15–30 min and wash well with deionized water. Incubate in Reynolds lead citrate buffer (dissolve

1.33 g of lead citrate (Sigma, 15326) in 30 ml of deionized water. Add 1.76 g of tribasic sodium citrate (Sigma, S1804) and continue mixing. Add 1 *N* NaOH, then bring the solution to 50 ml with deionized water. Keep the solution at 4 °C, properly sealed. Discard when precipitates appear) for 3–5 min and then wash with deionized water 3 times. Dry grids and observe under EM.

5.3. Specimen preparation from skeletal myoblasts and myotubes

1. Seed cells on biocoat slides (Nalge Nunc, Inc., 564350) using the appropriate medium (for myoblasts: DMEM with 20% heat inactivated FBS and 1% penicillin-streptomycin; for myotubes: DMEM with 5% heat-inactivated horse serum and 1% penicillin-streptomycin). For myoblasts, allow to grow to near confluence. For myotubes, wait until the cells are fully differentiated and contracting.
2. Wash the cells 3 times with washing buffer (0.1 *M* sucrose in 0.1 *M* phosphate buffer, pH 7.4), 1 min per wash.
3. Fix the cells with 1.2% glutaraldehyde in 0.1 *M* phosphate buffer, pH 7.4, at room temperature for 30 min.
4. Aspirate the glutaraldehyde solution and wash the cells 5 times with washing buffer, 1 min per wash.
5. Postfix in 1% OsO_4/0.1 *M* phosphate buffer, pH 7.4, for 1 h on ice.
6. Dehydrate in ascending concentration of EtOH (50%, 70%, 80%, 90%, 95%, 3 times of 100%, each for 1 min). Incubate in 100% absolute EtOH for 1 min.
7. Prepare EPON resin by mixing the following in sequential order: 9.1 ml of Epon 812, 4.7 ml of MNA (22%), 6.2 ml of DDSA. Add 0.3 ml of DMP30.
8. Preembed with the following schedule: 100% EtOH:EPON resin (2:1 in volume) for 30 min; 100% EtOH:EPON resin (1:1 in volume) for 30 min; 100% EtOH:EPON resin (1:2 in volume) for 1 h; and then EPON resin for 2 h.
9. Polymerize in freshly prepared EPON resin at 45° for 8 h and 60° for 3 days.
10. Process for microtome sectioning and EM observation.

6. Immunoelectron Microscopy

For the optimal study of autophagy, the ability to identify the double-membraned autophagosomal structure is a prerequisite. Even among experts, however, this precise identification of the autophagosome is complicated by the fact that single-membraned structures (e.g., the autolysosome) are also seen in the process of autophagosomal maturation. When such difficulties arise, the use immunoelectron microscopy may help clarify the nature of such structures. The readers are referred to published protocols for such methods (Askanas *et al.*, 1994; Mayhew *et al.*, 2007; Mayhew, 2002).

7. Conclusion

This chapter emphasizes the importance of sample preparation as a prerequisite for analyzing autophagy in muscles. When done according to these protocols, the fixation of frozen muscle will give rise to a sample that is almost devoid of artifacts, the most common of which would be the presence of the so-called freezing artifacts, appearing like empty vacuoles within the myofibers; these empty vacuoles could easily be misinterpreted as a pathologic finding.

Another approach that has been employed to study the role of autophagy in the pathomechanism of muscle diseases is isolation of single fibers. The group of Raben *et al.* (2007; see also the chapter by Raben *et al.*, in this volume), has used this technique in analyzing autophagic buildup in the muscles. This method, however, is not covered in this chapter; hence the reader is referred to previous publications.

REFERENCES

Askanas, V., Engel, W. K., Bilak, M., Alvarez, R. B., and Selkoe, D. J. (1994). Twisted tubulofilaments of inclusion body myositis muscle resemble paired helical filaments of Alzheimer brain and contain hyperphosphorylated tau. *Am. J. Pathol.* **14,** 177–187.

Bijvoet, A. G., van de Kamp, E. H., Kroos, M. A., Ding, J. H., Yang, B. Z., Visser, P., Bakker, C. E., Verbeet, M. P., Oostra, B. A., Reuser, A. J., and van der Ploeg, A. T. (1998). Generalized glycogen storage and cardiomegaly in a knockout mouse model of Pompe disease. *Hum. Mol. Genet.* **7,** 53–62.

Fujita, E., Kouroku, Y., Isoai, A., Kumagai, H., Misutani, A., Matsuda, C., Hayashi, Y. K., and Momoi, T. (2007). Two endoplasmic reticulum-associated degradation (ERAD) systems for the novel variant of the mutant dysferlin: ubiquitin/proteosome ERAD(I) and autophagy/lysosome ERAD(II). *Hum. Mol. Genet.* **16,** 618–629.

Jongen, P. J., Ter Laak, H. J., and Stadhouders, A. M. (1995). Rimmed basophilic vacuoles and filamentous inclusions in neuromuscular disorders. *Neuromuscul. Disord.* **5,** 31–38.

Kaneda, D., Sugie, K., Yamamoto, A., Matsumoto, H., Kato, T., Nonaka, I., and Nishino, I. (2003). A novel form of autophagic vacuolar myopathy with late-onset and multiorgan involvement. *Neurology* **61,** 128–131.

Lünemann, J. D., Schmidt, J., Dalakas, M. C., and Münz, C. (2007). Macroautophagy as a pathomechanism in sporadic inclusion body myositis. *Autophagy* **3,** 384–386.

Malicdan, M. C., Noguchi, S., and Nishino, I. (2007). Autophagy in a mouse model of distal myopathy with rimmed vacuoles or hereditary inclusion body myopathy. *Autophagy* **3,** 396–398.

Mayhew, T. M. (2007). Quantitative immunoelectron microscopy: Alternative ways of assessing subcellular patterns of gold labeling. *Methods Mol. Biol.* **369,** 309–329.

Mayhew, T. M., Lucocq, J. M., and Griffiths, G. (2002). Relative labelling index: A novel stereological approach to test for non-random immunogold labelling of organelles and membranes on transmission electron microscopy thin sections. *J. Microsc.* **205,** 153–164.

Mizushima, N., Yamamoto, A., Matsui, M., Yoshimori, T., and Ohsumi, Y. (2004). *In vivo* analysis of autophagy in response to nutrient starvation using transgenic mice expressing a fluorescent autophagosome marker. *Mol. Biol. Cell* **15,** 1101–1111.

Nascimbeni, A. C., Fanin, M., Tasca, E., and Angelini, C. (2008). Molecular pathology and enzyme processing in various phenotypes of acid maltase deficiency. *Neurology* **70,** 617–626.

Nishino, I. (2003). Autophagic vacuolar myopathies. *Curr. Neurol. Neurosci. Rep.* **3,** 64–69.

Nishino, I. (2006). Autophagic vacuolar myopathy. *Semin. Pediatr. Neurol.* **13,** 90–95.

Nishino, I., Fu, J., Tanji, K., Yamada, T., Shimojo, S., Koori, T., Mora, M., Riggs, J. E., Oh, S. J., Koga, Y., Sue, C. M., Yamamoto, A., *et al.* (2000). Primary LAMP-2 deficiency causes X-linked vacuolar cardiomyopathy and myopathy (Danon disease). *Nature* **406,** 906–910.

Nishino, I., Malicdan, M. C., Murayama, K., Nonaka, I., Hayashi, Y. K., and Noguchi, S. (2005). Molecular pathomechanism of distal myopathy with rimmed vacuoles. *Acta Myol.* **24,** 80–83.

Raben, N., Takikita, S., Pittis, M. G., Bembi, B., Marie, S. K., Roberts, A., Page, L., Kishnani, P. S., Schoser, B. G., Chien, Y. H., Ralston, E., and Nagaraju, K. (2007). Deconstructing Pompe disease by analyzing single muscle fibers: To see a world in a grain of sand. *Autophagy* **3,** 546–552.

Schnell, S. A., Staines, W. A., and Wessendorf, M. W. (1999). Reduction of lipofuscin-like autofluorescence in fluorescently labeled tissue. *J. Histochem. Cytochem.* **47,** 719–730.

Seehafer, S. S., and Pearce, D. A. (2006). You say lipofuscin, we say ceroid: Defining autofluorescent storage material. *Neurobiol. Aging* **27,** 576–588.

Shintani, T., and Klionsky, D. J. (2004). Autophagy in health and disease: A double-edged sword. *Science* **306,** 990–995.

Tanaka, Y., Guhde, G., Suter, A., Eskelinen, E.-L., Hartmann, D., Lüllmann-Rauch, R., Janssen, P. M., Blanz, J., von Figura, K., and Saftig, P. (2000). Accumulation of autophagic vacuoles and cardiomyopathy in LAMP-2-deficient mice. *Nature* **406,** 902–906.

Yin, D. (1996). Biochemical basis of lipofuscin, ceroid, and age pigment-like fluorophores. *Free Radic. Biol. Med.* **21,** 871–888.

CHAPTER TWENTY

Analyzing Macroautophagy in Hepatocytes and the Liver

Wen-Xing Ding* *and* Xiao-Ming Yin*

Contents

Abstract

Mammalian autophagy has been well characterized in the liver and in hepatocytes. Autophagy plays important roles in the normal physiology of the liver and in the pathogenesis of several liver diseases. This chapter will discuss the commonly used methods for analysis of autophagy in hepatocytes and in the liver.

* Department of Pathology, University of Pittsburgh School of Medicine, Pittsburgh, Pennsylvania, USA

Methods in Enzymology, Volume 453
ISSN 0076-6879, DOI: 10.1016/S0076-6879(08)04020-2

1. The Pathophysiological Relevance of Macroautophagy in the Liver

Three modes of autophagy have been defined, macroautophagy, microautophagy, and chaperone-mediated autophagy, which differ in how the cytoplasmic materials are delivered to the lysosome. In macroautophagy cellular contents are sequestered in double-membraned structures, called autophagosomes, which fuse with lysosomes to form autolysosomes to degrade the sequestered materials. This chapter will focus on the analysis of macroautophagy (hereafter referred to as *autophagy*) in the liver and in hepatocytes.

Autophagy plays important roles in liver physiology and pathology (Yin *et al.*, 2008). In fact, most of the early work on autophagy was conducted with the liver (Deter and De Duve, 1967). One of the important roles of hepatic autophagy is to provide the body under starvation conditions with necessary nutrients through the degradation of intracellular materials. This activity can be regulated by the plasma levels of amino acids, insulin, and glucagon (Deter and De Duve, 1967; Mortimore and Poso, 1987; Pfeifer, 1978). The ATP level could also affect hepatic autophagy via AMP-activated protein kinase (AMPK), which is activated by a high AMP/ATP ratio (Meijer and Codogno, 2007).

Hepatic autophagy is also required for the homeostasis of subcellular organelles, including mitochondria, endoplasmic reticulum and peroxisomes (Yin *et al.*, 2008). Mitophagy, selective autophagic degradation of mitochondria, leads to the removal of damaged mitochondria that are associated with increased generation of reactive oxygen species, and can thus serve to protect against cell death, aging and tumorigenesis (Jin, 2006).

A number of liver-related diseases involve the generation of misfolded proteins, such as the Z mutant of alpha-1 antitrypsin (ATZ), which causes a significant deficiency of this enzyme (Perlmutter, 2006), and mutant fibrinogen γ chain (Aguadilla γD), which causes hypofibrinogenemia (Kruse *et al.*, 2006). In addition, alcoholic liver disease and nonalcoholic steatohepatitis are characterized by the presence of Mallory bodies (Ku *et al.*, 2007). Mallory bodies are cytoplasmic inclusions, which contain keratin 8, keratin 18, polyubiquitinated proteins and p62/SQSTM1. Autophagy is an important mechanism for the clearance of these misfolded proteins.

2. Analysis of Autophagy in Isolated Hepatocytes

Primary hepatocytes are critically important for analyzing the physiological and pathological functions of the liver *in vitro*. Primary culture offers the general benefits of pharmacological and molecular manipulations and the ability to determine the contribution of individual cellular components. We generally use 2- to 4-month-old mice to isolate hepatocytes for most experiments *in vitro*, as it has been found that older hepatocytes may have decreased autophagic proteolysis (Donati *et al.*, 2001).

2.1. Preparation of primary mouse hepatocyte culture

Detailed protocols to prepare human and rodent hepatocytes for primary culture have been well developed (Klaunig *et al.*, 1981a,b; Ryan *et al.*, 1993; Seglen, 1976). Following is a protocol that we have used most frequently for preparing mouse hepatocytes (Ding *et al.*, 2004; Zhao *et al.*, 2003). It is adapted from Klaunig and colleagues (Klaunig *et al.*, 1981a,b) and describes a collagenase-based reverse perfusion method.

2.1.1. Instruments and reagents

A peristaltic pump (e.g., Gilson minipuls 3) and a regular water bath are required for the perfusion. Other small instruments include surgical clamp, scissors and forceps. The following perfusion solutions should be freshly prepared and placed in the 37 °C water bath before and during perfusion.

Solution A (150 mL): 142.5 mL of calcium- and magnesium-free Hank's buffer (Thermo Scientific, SH30588.02), 0.75 mL of 0.1*M* EGTA, and 7.5 mL of HEPES buffer, pH 7.2.

Solution B (100 mL): 99.3 mL of L15 medium (Cambrex Bio Scinence, 12-700Q), 0.748 mL of 0.5 *M* $CaCl_2$, 50 mg of Type IV collagenase (Sigma, C5138).

2.1.2. Procedures

i. Anesthetize the mouse with a proper anesthetic and ensure that the mouse remains alive until the perfusion starts. Position the mouse on a platform with the abdomen facing up. Stretch the limbs and secure them on the platform with a piece of laboratory labeling tape. Sterilize the abdominal skin with a 70% ethanol solution.
ii. Open the abdomen and carefully isolate the vena cava. Insert a 20-G catheter into the vein, tighten the suture proximal to the liver and fix the catheter with a surgical clamp.
iii. Connect the catheter to the warmed Solution A through the peristaltic pump, cut the portal vein and start the perfusion by turning on the peristaltic pump at the rate of 12 ml/min. After the completion of Solution A, perfuse with Solution B at the rate of 8 ml/min. After the completion of the latter perfusion, the liver becomes "softened" because of the effect of collagenase.
iv. Remove the liver and place the liver in a 100-mm Petri dish with 10 mL of Solution B. Incubate the liver at 37 °C for 10 min. By this time, the

liver is quite fragile and can be easily dispersed in the Petri dish with a pair of tweezers, which will lead to the dissociation of individual hepatocytes from the matrix. Add another 10 mL of phosphate buffered saline (PBS, pH 7.2) to the dish. Pass the cell suspension, which contains the released hepatocytes, through a sterile metal filter (40 mesh) (Sigma, S0770-5EA) into a 50-mL centrifuge tube.

v. Centrifuge at 600 rpm for 3 min at 4 °C. The parenchymal cells (*i.e.*, the hepatocytes) are in the pellet fraction. Discard the supernatant fraction and wash the cell pellet with phosphate balanced saline (PBS, pH 7.2) 3 times and resuspend the cells in Williams' Medium E (Sigma, W4128) supplemented with 10% fetal bovine serum, 2 m*M* L-Glutamine, 10 m*M* Hepes, and 100 units/mL penicillin and 0.1 mg/mL streptomycin. In general, a successful perfusion should yield approximately 10–30 million hepatocytes from 1 mouse liver with more than 90% viability.

vi. The hepatocytes are then allowed to attach in a culture dish, flask or plate for 2 h at 37 °C with 5% CO_2. The cells are then washed with PBS before incubated in Williams' Medium E supplemented with 2 mmol/L L-glutamine and antibiotics, but no serum. The cells are usually cultured overnight before experimental manipulations are conducted, but they may be treated soon after the attachment is established.

2.2. General methods to monitor autophagy in isolated hepatocytes

Autophagy may be induced in hepatocytes by different means depending on the nature of the issues being examined. A common means to induce autophagy is amino acid deprivation (starvation). Examples of autophagy illustrated in this chapter are all based on starvation. For the isolated hepatocytes, starvation can be easily initiated by removing the Williams' Medium E from overnight cultured cells, washing the cells with PBS 3 times, and then culturing the cells in Earle's Balanced Salt Solution (EBSS) (Hyclone, SH30029.02). Autophagy induction can be detected in the next 2–12 hours depending on the method used.

2.2.1. Monitoring autophagy using GFP-LC3

The microtubule-associated protein 1 light chain 3B (LC3) is a homologue of yeast Atg8. LC3 is proteolytically converted to a short form (LC3-I) by a cysteine protease, Atg4B, which becomes conjugated to phosphatidylethanolamine (PE) to form LC3-II upon autophagy induction. LC3-II is located at the autophagosomal membrane. The translocation of LC3-I to the membrane (LC3-II) can be tracked by immunofluorescence staining. For real-time assessment in viable cells, the use of fluorescently labeled LC3, such as GFP-LC3, is a sensitive method to monitor autophagy status (see the chapter by Kimura *et al.*, in volume 452). The translocation of GFP-LC3 from the cytosol to the membranes during autophagy induction is manifested as the formation of puncta that can be easily visualized and quantified. GFP-LC3 can be introduced into cells via transient transfection. However, this is not recommended, as the background level of GFP-LC3 puncta could be high due to the transfection procedure. This is particularly true for hepatocytes. For many routinely used cell lines, clones stably expressing GFP-LC3 can be established and analyzed. For primary cells, it is possible to use GFP-LC3 transgenic mice (Mizushima *et al.*, 2004) as the source (also see the chapter by N. Mizushima, in this volume). Alternatively, for hepatocytes isolated from wild-type mice an adenoviral vector can be used to carry the GFP-LC3 construct into the cell.

For this latter purpose, the adenoviral GFP-LC3 can be added to hepatocytes in Williams' Medium E with 10% fetal bovine serum soon after the isolation. Cells can then be plated for attachment. After the 2-h incubation, the cells are washed with PBS and cultured in serum-free Williams' Medium E. Strong GFP signals can usually be detected in 16–24 h. The cells are then ready for experimental manipulations, such as starvation. Although the expression is transient, it usually lasts for several days, long enough for most of the *in vitro* work with the primary hepatocytes.

In the example shown in Fig. 20.1, hepatocytes are starved for 12 h and then fixed with 4% paraformaldehyde in PBS for 30 min before being examined under a confocal fluorescence microscope. Most of the untreated hepatocytes display diffuse fluorescence of GFP-LC3 in the cytosol with few puncta. In contrast, there is a significant increase of punctated GFP-LC3 in the starved hepatocytes.

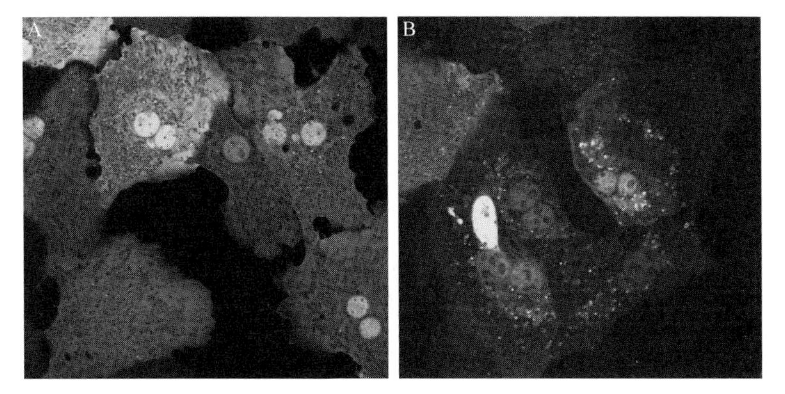

Figure 20.1 Starvation induces GFP-LC3 puncta in cultured hepatocyte. Isolated murine hepatocytes were first infected with adenoviral GFP-LC3 overnight, then either cultured in Williams' Medium E (A) or in EBSS (B) for 12 h. Cells were fixed in 4% paraformaldehyde and examined by confocal microscopy. GFP-LC3 puncta can be observed in starved (B), but not in control hepatocytes (A).

The extent of GFP-LC3 punctation can be quantified. Briefly, the number of GFP-LC3 puncta can be enumerated and compared between the control and treated cells. If the background of GFP-LC3 puncta in the control cells is low, autophagy may be estimated by determining the percentage of cells positive for GFP-LC3 puncta. Under less ideal conditions, it may be necessary to predetermine the threshold of the number of dots per cell to qualify for the positive response. For example, if most control cells contain a low degree of GFP-LC3 puncta (<2 dots/cell), then only cells with more than 2 dots can be considered positive after the proper stimulation.

2.2.2. Western blot analysis for LC-I and LC3-II conversion

The change from LC3-I to LC3-II can be also monitored by SDS-PAGE followed by immunoblot analysis. When separated by SDS-PAGE, the LC3-II form migrates faster than the LC3-I form, and thus the increase in the LC3-II form can be correlated with autophagic turnover. However, it is necessary to carry out this analysis in the presence and absence of lysosomal protease inhibitors (or inhibitors that block fusion of autophagosomes with the lysosome) to distinguish between an increase in LC3-II due to induction of autophagy and blockage of turnover, versus complete autophagic flux as discussed in section 2.3. We use the following protocol.

2.2.2.1. Procedures

i. Hepatocytes can be seeded at a density of 4×10^5 cells/well in a 6-well plate.
ii. After treatment, the cells are collected using a cell scraper and are centrifuged at 1000 rpm for 5 min. The cell pellets are resuspended in 50 μL of RIPA buffer (50 m*M* Tris-HCl, pH 7.4, 1% NP-40, 0.25% sodium deoxycholate, 150 m*M* NaCl) with protease inhibitors including 10 μg/mL pepstatin, 10 μg/mL leupeptin, and 5 μg/mL aprotinin. Pipette the pellets up and down several times and incubate on ice for 30 min.
iii. The lysates are centrifuged at 13,000 rpm for 10 min.
iv. The supernatant fractions are removed to new tubes and the protein concentration is determined by using the BCA protein assay (Thermo Scientific, 23228).
v. Thirty to fifty μg of proteins are separated on a mini SDS-PAGE gel and transferred onto a PVDF membrane (also see the chapter by Sarkar *et al.*, in this volume).
vi. After blocking for 1 h in 5% milk in TBST buffer (10 m*M* Tris HCl, pH 7.5, 10 m*M* NaCl, and 0.1% Tween 20), the membrane is incubated with an anti-LC3 antibody for 1 h.

vii. The membrane is washed 3 times in TBST, and then incubated with a secondary antibody conjugated with horseradish peroxidase. LC3 can be detected by the enhanced chemiluminescence method.

The conversion of LC3-I to LC3-II also occurs for the exogenously introduced GFP-LC3 molecule. It is sometimes easier to detect GFP-LC3 conversion using an anti-GFP antibody. Figure 20.2 shows an example of such conversion in hepatocytes following starvation.

2.2.3. Electron microscopy analysis

Autophagy was first defined by electron microscopy (EM). EM is a valid and reliable method for analyzing autophagy in hepatocytes.

2.2.3.1. Procedures

i. Primary hepatocytes can be seeded at a density of 4×10^5 cells/well in a 6-well plate.
ii. The cells are cultured in EBSS for 6 h.
iii. The cells are fixed with 2% paraformaldehyde and 2% glutaraldehyde in 0.1 *M* phosphate buffer, pH 7.4, for 2 h, followed by 1% OsO_4 in 0.1 *M* phosphate buffer for another hour.
iv. After dehydration in an ethanol gradient (70% ethanol for 20 min, 96% ethanol for 20 min, and 100% ethanol for 2 × 20 min), samples are embedded in Epon-812 (Electron Microscopy Sciences, RT 13940).

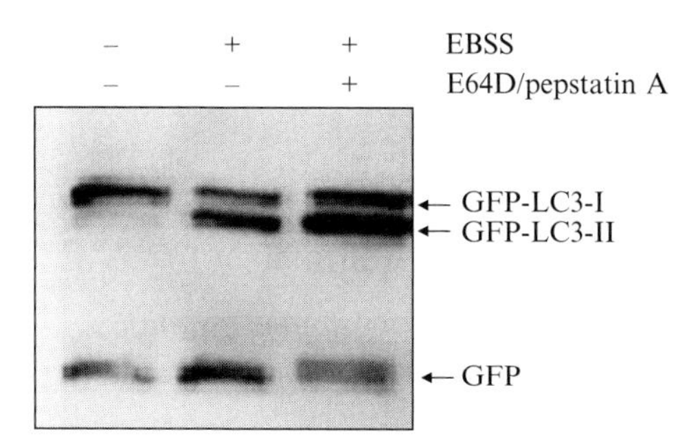

Figure 20.2 Starvation induces autophagy flux in cultured hepatocytes. Primary hepatocytes isolated from a GFP-LC3 transgenic mouse were either left untreated, or cultured in EBSS in the presence or absence of E64D (10 μM) plus pepstatin A (10 μM) for 12 h. Total cell lysates were subjected to Western blot analysis using an anti-GFP antibody. Starvation induced the conversion of GFP-LC3-I to GFP-LC3-II, which is membrane associated. The enhanced appearance of GFP suggested an increased autophagic flux, which can be confirmed by the use of lysosomal inhibitors. In the latter, the degradation of GFP-LC3 is suppressed, accompanied with a reduced GFP level and an increased GFP-LC3-II level.

v. Thin sections are prepared and stained with 2% uranyl acetate and 2.5% lead citrate in water before examination by electron microscopy.

Autophagy is a dynamic process, in which autophagic vesicles at different developmental stages can be observed even in one cell. Figure 20.3 shows the electron microscopy detection of the autophagosomes in starved hepatocytes. The early stage autophagosomes can be identified with the clear presence of the double membrane and cellular content, such as cytoplasm or subcellular organelles. These are usually termed as initial autophagic compartments (autophagosomes). The more matured autophagosomes are those that have completed the fusion with the lysosomes. They are delimited by a single membrane and may contain partially degraded cytoplasmic materials, which typically show signs of disintegration and increased electron density. These autolysosomes are also known as degradative autophagic compartments.

The extent of autophagy may be determined by morphometric analysis. The number of autophagosomes, autolysosomes, or total autophagic vesicles can be determined in a given cell section, or area. The size or the area of the autophagic vesicles, relative to the size or area of the cell, can be also determined. A detailed description of the EM detection of autophagy can be found in the chapter by Anttila *et al.*, in volume 452.

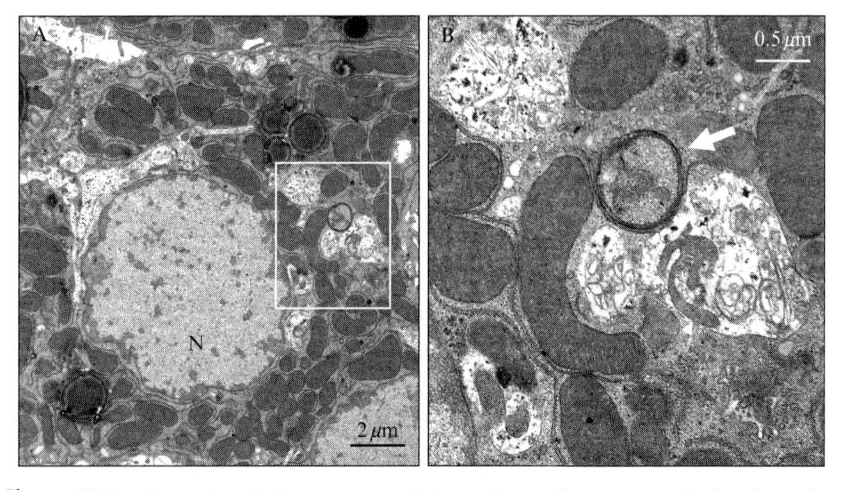

Figure 20.3 Starvation induces accumulation of autophagosomes. Isolated murine hepatocytes were cultured in EBSS for 12 h. Cells were fixed in 2% paraformaldehyde and 2% glutaraldehyde, and processed for electron microscopy (A). Boxed area in panel A is shown in panel B. The arrow indicates a typical autophagosome with double membrane and cytosolic contents. N: nuclei. Scale bars of 2 μm and 0.5 μm are shown.

2.3. Examination of autophagy flux

Newly generated autophagosomes are rapidly degraded in the lysosome. Early studies suggested that the half-life of an autophagosome in the liver is about 10 min (Pfeifer, 1978). It is thus important to bear in mind that the level of autophagosomes or autophagosomal markers detected at a given time is a reflection of the turnover status of the autophagosome at that time, determined not only by the generation, but also the degradation of the autophagosomes. Consequently, it is necessary to measure the level of autophagy flux to understand whether the experimental manipulations affect the generation or the degradation of the autophagosomes. In particular, it is necessary to differentiate whether the accumulation of autophagosomes or the increase of autophagosomal markers is due to increased production or reduced degradation as mentioned earlier.

2.3.1. Flux analysis based on electron microscopy

The quantification of autophagosomes and autolysosomes by morphometric analysis may help to differentiate whether a stimulus can induce or inhibit autophagy. If the amount of autophagosomes is not different between the control and the treated, but the amount of autolysosomes is increased after treatment, it may suggest the blockage of the lysosomal functions (e.g., following the treatment with chloroquine). In contrast, if only the autophagosomes are significantly accumulated, it may suggest that the fusion of the autophagosome with the lysosome is affected (e.g., in LAMP-2-deficient hepatocytes or following treatment with vinblastine, a microtubule inhibitor (Fengsrud *et al.*, 1995; Tanaka *et al.*, 2000).

2.3.2. Analysis of LC3 puncta or LC3-II with lysosome or microtubule inhibitors

If the increase in LC3 puncta or the LC3-II form is due to induction of autophagy, suppression of autophagosome fusion with microtubule inhibitors or suppression of autophagosome degradation with lysosomal protease inhibitors may lead to further elevations in the amount of LC3-II. If the increase is not due to induction but to the blockage of degradation, then further use of these inhibitors should not lead to a higher level of accumulation.

Commonly used lysosomal inhibitors include protease inhibitors, such as E64D, leupeptin and pepstatin A, and agents that disturb the lysosome acidic environment, such as chloroquine and NH_4Cl (Seglen *et al.*, 1979). Agents that prevent the fusion of the autophagosome with the lysosome include microtubule inhibitors, such as vinblastine. Fig. 20.2 shows that in starved hepatocytes treatment of E64D plus pepstatin further increases the level of GFP-LC3-II, suggesting that the original increase is due to induction, not due to reduced degradation.

2.3.3. Analysis of the release of GFP moiety from GFP-LC3

When GFP-LC3 is delivered to the lysosome, the LC3 part of the fusion molecule is sensitive to degradation, whereas the GFP moiety is relatively resistant to hydrolysis (Hosokawa *et al.*, 2006). Therefore, it is possible to examine the presence of the free GFP moiety, which would indicate the progression of autophagy through the degradation stage. This assay can be easily applied to hepatocytes expressing GFP-LC3 (Fig. 20.2). The appearance of the GFP moiety can be clearly observed on immunoblot assay following starvation, which can be blocked by lysosomal inhibitors. Concomitantly, treatment with lysosomal inhibitors will cause the accumulation of the full-length GFP-LC3-II.

2.3.4. Western blot analysis of the level of p62/SQSTM1

p62/SQSTM1 is a molecule that can bind to misfolded protein. It can be recruited to the autophagosome via an interaction with LC3 (Bjorkoy *et al.*, 2005). The p62 protein is largely degraded in the autophagosomes and its level is increased when autophagy is defective (Yue, 2007). Its level, as detected by Western blot, can thus suggest whether the autophagic degradation is increased or inhibited. Thus one may expect a reduction of p62/SQSTM1 when autophagy is induced, as its degradation is increased. Conversely, suppression of autophagy may result in an accumulation of this protein. It should be noted, however, that p62 is also degraded via the ubiquitin-proteasome system.

2.4. Monitoring autophagic protein turnover in isolated hepatocytes

Degradation of long-lived proteins is mainly mediated by the autophagic machinery. Whereas the preceding analyses of autophagy are more or less static, measuring long-lived protein degradation is a functional analysis of dynamic autophagy. Although complicated, this method had been successfully used in isolated hepatocytes. In fact, a number of important findings were made using this assay, such as the revelation of 3-methyladenine (3-MA) as an inhibitor of autophagy (Seglen and Gordon, 1982). Detailed protocols can be found in the chapter by Bauvy *et al.*, in volume 452.

2.5. Monitoring autophagic mitochondrial turnover in isolated hepatocytes

Autophagic degradation of subcellular organelles is common in the liver and in hepatocytes (Pfeifer, 1978; Yin *et al.*, 2008). Mitochondrial autophagy can be relatively easy to study in hepatocytes due to the abundance and the larger size of the mitochondria. Mitochondrial degradation by autophagy can be induced by starvation; this is not referred to as mitophagy, because it is not a selective process. As in other scenarios, the determination of mitochondria removal by the autophagy machinery requires evidence from different assays, some of which are described here.

2.5.1. Double labeling of the mitochondria and the autophagosomal compartment

The autophagosomal compartment may be labeled with GFP-LC3. Hepatocytes can be isolated from GFP-LC3 transgenic mice or they can be infected with Ad-GFP-LC3 as described earlier. In the latter, after the GFP-LC3 is expressed, the cells can be loaded with MitoTracker Red (MTR, 50 n*M*, Molecular Probe) at 37 °C for 15 min. This dye is taken into the functional mitochondria in a potential-dependent manner. It will react with mitochondrial proteins to form stable complexes and will not be released even in depolarized or damaged mitochondria.

Hepatocytes are then subjected to autophagic stimulations, such as amino acid deprivation. The change in GFP-LC3 in relationship to MTR can be monitored in live cells or in cells fixed with 4% paraformaldehyde by fluorescence microscopy. The colocalization of GFP-LC3 puncta with MTR can be observed during starvation. Frequently, GFP-LC3 forms ring-shaped puncta surrounding the MTR-labeled mitochondria

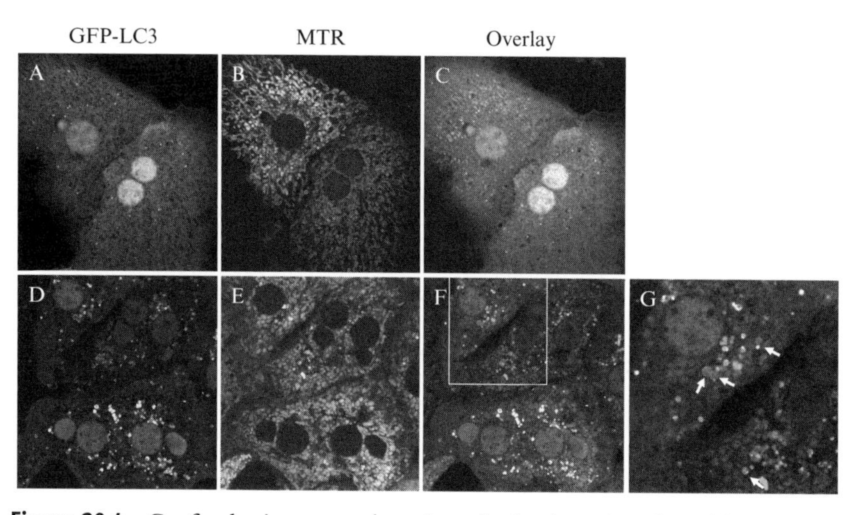

Figure 20.4 Confocal microscopy detection of mitophagy in cultured hepatocytes. Isolated murine hepatocytes were first infected with adenoviral GFP-LC3 overnight, and then loaded with MitoTracker Red (50 n*M*) for 15 min. Cells were then cultured in Williams's Medium E (A–C) or in EBSS (D-G) for 6 h. Boxed area in panel F is enlarged in panel G. Arrows indicate GFP-LC3-positive structures (presumably autophagosomes) that contain the mitochondria. Increased GFP-LC3 puncta are seen in starved cells and many to enwrap the mitochondria (arrows). (See Color Insert.)

(Fig. 20.4). Although not formally proved, this colocalization may indicate an on-going process of mitochondrial autophagy.

2.5.2. Double labeling of the mitochondria and the lysosomal compartment

Further evidence of mitochondrial autophagy may be sought by examining whether the MTR-labeled mitochondrial compartment can be colocalized with the lysosomal compartment. If so, this will strongly suggest the uptake of the mitochondria into the lysosome. The lysosome can be labeled by several methods, including the use of LysoTracker (Elmore *et al.*, 2001).

LysoTracker Red (LTR, 50 n*M*, Molecular Probe) and MitoTracker Green (MTG, 50 n*M*, Molecular Probe) can be added to the cultured hepatocytes at 37 °C for 15 min. Autophagic stimulus, such as starvation, can then be applied. Cells are observed at different times by fluorescence microscopy. Colocalization of the two dyes may suggest mitochondrial autophagy. One problem with LysoTracker is that other acidic compartments, not just the lysosomes, can be also labeled. Thus a more specific approach is to locate the lysosome using antibodies against LAMP-1 and/or LAMP-2. This can be done on fixed cells that have been loaded with MitoTracker and treated. Alternatively, immunostaining for cathepsin D can also help with the identification of the lysosomal compartment.

2.5.3. Electron microscopy analysis of mitochondrial autophagy

Mitochondrial autophagy in hepatocytes can be observed by EM, which would provide an even more direct piece of evidence. In the example given in Fig. 20.5, mitochondria were found to be taken up by autophagosomes in hepatocytes undergoing starvation. Here the sample was processed as described in section 2.2.3.

2.5.4. Determination of the loss of mitochondria

Whereas the colocalization of mitochondria with LC3 may suggest the engulfment of mitochondria by the autophagosome, which can be further confirmed by the colocalization with the lysosomal marker and EM examination, the actual loss of the mitochondria would have to be assessed by other methods (see the chapter by Zhang *et al.*, in volume 452 for an additional discussion of this topic). This involves the commonly used methodologies of assessing mitochondrial mass. Thus the fluorescence level of MitoTracker in cells undergoing mitochondrial autophagy can be assessed by flow cytometry. This approach has been used to monitor mitochondrial turnover in other types of cells (Sandoval *et al.*, 2008; Schweers *et al.*, 2007).

Alternatively, immunoblot analysis can be applied to determine the levels of mitochondrial proteins, such as cytochrome c, cytochrome c oxidase subunits, or voltage dependent anion channel (VDAC). It is recommended that multiple proteins are assessed. The reduction of these proteins

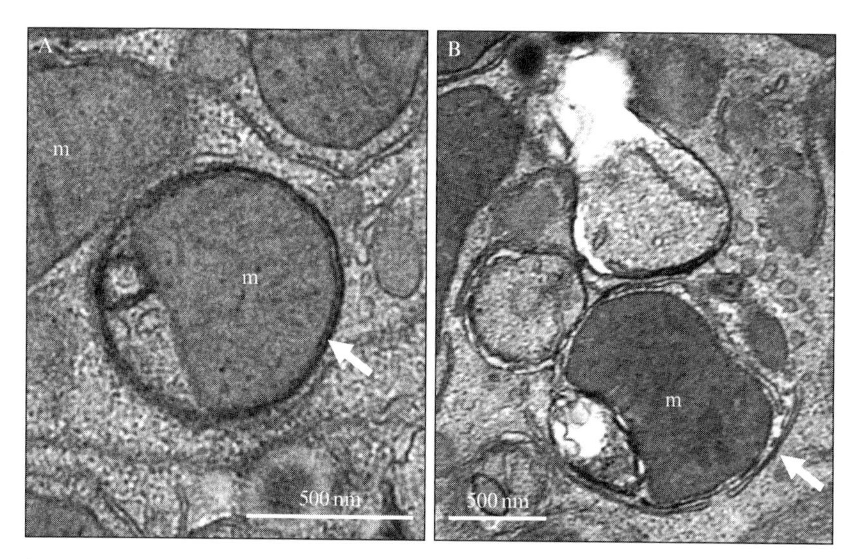

Figure 20.5 Electron microscopy detection of mitochondrial autophagy in cultured hepatocytes. Isolated murine hepatocytes were cultured in EBSS for 12 h. Cells were fixed in 2% paraformaldehyde and 2% glutaraldehyde and processed for electron microscopy. Panels A and B show autophagosomes (arrows) with engulfed mitochondria (indicated by m). Scale bars of 500 nm are shown.

can suggest a reduction of the mitochondrial mass. If this reduction can be inhibited by interfering with the autophagy machinery, it will indicate the mitochondria turnover by the autophagy machinery.

2.6. Modulation of autophagy in hepatocytes

Primary hepatocytes are amenable to in vitro pharmacological treatment and molecular manipulations. Several pharmacological agents such as 3-MA have been used successfully to inhibit autophagy in isolated hepatocytes. Subsequent studies confirm that similar to 3-MA, other PI-3 kinase inhibitors, such as wortmannin and LY294002, can also inhibit autophagy in isolated rat hepatocytes (Blommaart *et al.*, 1995). In addition, vinblastine (a microtubule inhibitor), leupeptin (a lysosomal protease inhibitor), and propylamine (a lysosomotropic amine) all are able to inhibit autophagy in hepatocytes (Kovacs *et al.*, 1982). As shown in Fig. 20.2, we use E64 plus pepstatin to suppress the lysosomal protease activity and therefore autophagic degradation. We have also determined that chloroquine, which has been used in other scenarios to suppress autophagy (Amaravadi *et al.*, 2007), can successfully inhibit starvation-induced autophagy flux in hepatocytes (unpublished observations).

Genetic manipulations of hepatocytes can be achieved using mice deficient or over-expressing certain genes in hepatocytes. Alternatively, if wild-type mouse cells are used, adenoviral vector is a preferred way to introduce nucleic acids into hepatocytes (as described above for the use of GFP-LC3, section 2.2.1). Transient transfection can be also conducted, although the efficiency can vary greatly. Following is a protocol that was designed for introducing siRNA into primary hepatocytes.

2.6.1. Procedures

i. Hepatocytes are prepared as described in section 2.1.
ii. Cells are then plated at a density of 4×10^5 cells per well in a 6-well plate overnight.
iii. Cells in duplicate wells will be transfected with a particular siRNA(s), such as that against Atg5, at a final concentration of 120 n*M* using Oligofectamine (Invitrogen) according to the manufacturer's protocol.
iv. Hepatocytes can be subjected to experimental treatment 36–48 h later. The effects of the knockdown should be also verified by immunoblot assay.

3. Analysis of Autophagy in the Liver

Liver is a complicated organ involved in multiple physiological functions and is susceptible to various pathological insults. Autophagy has been demonstrated in the liver under various conditions (Yin *et al.*, 2008). A commonly used scheme is to subject mice to overnight fasting. Upon starvation, the liver is one of the most active organs to rapidly turn on the autophagy machinery to degrade proteins, and glycogen in the neonates, in an attempt to supply the body with amino acids and glucose, respectively (Kuma *et al.*, 2004; Mortimore *et al.*, 1989; Schiaffino *et al.*, 2008). In the past, analysis of autophagy in the liver had been limited to the electron microscopy and protein degradation assays. With the recent identification of a number of key molecules in autophagy, the approach to study liver autophagy has been expanded significantly.

3.1. Animal models

Mice deficient in ULK1/Atg1, Beclin 1/Atg6, Atg5, Atg7, Atg4C, or LC3B/Atg8 have been established (Cann *et al.*, 2008; Hara *et al.*, 2006; Komatsu *et al.*, 2005; Kundu *et al.*, 2008; Marino *et al.*, 2007; Qu *et al.*, 2003). Mice lacking Beclin 1 are embryonic lethal, but the heterozygous Beclin $1^{+/-}$ mice show decreased autophagic activity and increased tumorigenesis including the development of hepatocellular carcinoma, suggesting that Beclin 1 is a haploinsufficient tumor-suppressor (Qu *et al.*, 2003). Mice deficient in either Atg5 or Atg7 are born alive but die within one day after birth (Komatsu *et al.*, 2005; Kuma *et al.*, 2004). Mice deficient in ULK1/Atg1 have a deficiency in autophagic removal of mitochondria and ribosomes during erythrocyte maturation although they seem to be normal in other aspects of autophagy (Kundu *et al.*, 2008). In contrast, mice lacking Atg4C or LC3B are viable and do not display any obvious abnormality (Cann *et al.*, 2008; Hara *et al.*, 2006; Komatsu *et al.*, 2005; Marino *et al.*, 2007; Qu *et al.*, 2003). These observations indicate that compensatory mechanisms are present for some autophagy molecules, such as ULK1, Atg4C and LC3, but not present for others, such as Beclin 1, Atg5, and Atg7, in mammalian cells.

To bypass the lethal phenotype, liver-specific conditional deletion is preferred to analyze the role of autophagy in the liver under both physiological and pathological conditions. Toward this end, liver-specific conditional Atg7-deficient mice have been developed by Dr. Komatsu and colleagues. These mice possess low autophagy activities and accumulate deformed mitochondria and ubiquitin-positive aggregates in hepatocytes (Komatsu *et al.*, 2005). This strain would be thus useful to analyze the impact and the role of liver autophagy under various physiological and pathological conditions.

Another potentially valuable strain is the GFP-LC3 transgenic mice (Mizushima *et al.*, 2004). This strain may be particularly useful to monitor autophagy in vivo, although the expression level of GFP-LC3 varies in different tissues. The expression in the liver seems adequate for analysis (Mizushima *et al.*, 2004) (Fig. 20.2). To detect the change in GFP-LC3, mice can be subjected to a proper autophagic treatment, followed by cryosection of the liver tissue using a standard protocol (Mizushima *et al.*, 2004) as follows in brief.

3.1.1. Procedures

i. Liver tissue can be first fixed in 4% paraformaldehyde at room temperature for 4 h.
ii. The fixed tissue is submerged in 15% sucrose in PBS for 4 h and then 30% sucrose overnight at 4 °C.
iii. The tissue is then embedded in OCT (SAKURA Finetechnical Co., 4583) compound.
iv. The samples are stored at −80 °C. Sectioning is conducted at −20 °C in 5- to 7-μm thickness with a cryostat. Sample is air-dried for 30 min and stored at −20 °C until use.
v. The liver samples can be stained with Hoechst 33342 (5 μg/mL) for 15 min at room temperature followed by washing with PBS twice.

vi. The samples are then examined by fluorescence microscopy. Formation of GFP-LC3 punctation is determined to assess the autophagy status (Mizushima *et al.*, 2004). Alternatively, immunoblot analysis can be conducted to determine the formation of GFP-LC3-II as shown in Fig. 20.2.

Without the use of GFP-LC3 mice, Ad-GFP-LC3 can be introduced into mice by a hydrodynamic injection through the tail vein (Knapp and Liu, 2004), which could enforce the take-up of the virus by the liver and expression in this organ.

3.2. Histological examination of autophagy

Regular hematoxylin and eosin staining of tissue sections will not reveal autophagic changes. However, immunostaining with an anti-LC3 antibody or antibodies against other autophagy molecules may be helpful to study autophagy in the liver. Cryosection slides can be used for immunostaining to detect the puncta formation of the endogenous LC3 or to determine the expression levels of other autophagy molecules.

3.3. Electron microscopy examination of liver autophagy

Liver is frequently studied with EM, and this approach is still one of the most effective methodologies to detect autophagy in the liver. Liver can be fixed in 2.5% glutaraldehyde solution for 2 h and then processed for EM as described above for hepatocytes. Additional comments for EM may be found in the chapter by Anttila *et al.*, in volume 452.

3.4. Immunoblot analysis of LC3 change in the liver

As described for hepatocytes, the conversion of LC3-I to LC3-II can be also analyzed using liver tissues to assess autophagy.

3.4.1. Procedures

i. Approximately 50 mg of liver tissue is minced with a pair of scissors.
ii. The minced liver is suspended in 0.5 mL RIPA buffer and sonicated for a total of 4 times (each time for 5 s) on ice at a setting of 9 watts output power.
iii. The lysate is centrifuged at 12,000 rpm for 15 min at 4 °C.
iv. The supernatant fractions are collected into new tubes and are used for SDS-PAGE and immunoblot analysis using an anti-LC3 antibody.

3.5. Analysis of autophagic protein degradation in the liver

Dr. Mortimore and his colleagues have extensively used an *in situ* liver perfusion technique to study autophagy. Using this technique, they determined the kinetics of protein degradation in the liver and developed the concept that long-lived protein degradation is mediated by autophagy (Mortimore and Poso, 1987). In addition, they also identified several key autophagy inhibitors such as amino acids, insulin, cycloheximide and 3-MA. Although this technique has not been widely used in recent studies, it may offer unique advantages to study the functions of autophagy in protein degradation at the organ level (Gohla *et al.*, 2007). The readers are advised to consult previously published papers for details in this technique (Mortimore and Mondon, 1970; Mortimore and Schworer, 1977; Mortimore *et al.*, 1972; Woodside and Mortimore, 1972).

3.6. Proteomic analysis of autophagy in the liver

Liver is one of the most metabolically active organs. Proteomic analysis may help to identify new regulatory molecules or autophagic substrates by comparing starved livers with normal livers. This approach can be applied to the total liver (Matsumoto *et al.*, 2008) or to the autophagosomes isolated from the liver (Overbye *et al.*, 2007). For the latter, it is relatively easier to obtain an enriched autophagosome fraction from the liver due to the large size of this organ (Stromhaug *et al.*, 1998). For this purpose, mice or rats could be given chemicals that block the fusion of autophagosomes with the lysosome, such as vinblastine, or chemicals that block lysosomal degradation, such as leupeptin, so that the autophagosomal compartment can be enriched. The availability of the tools for isolation of the autophagosome and for proteomic analysis will definitely advance the study on hepatic autophagy.

4. Summary

Autophagy plays important roles in the pathophysiology of the liver and liver diseases. Besides providing nutrients during starvation, autophagy is involved in diseases related to misfolded proteins, organelle turnover, liver injury, and liver cancer (Yin *et al.*, 2008). As one of the most metabolically active organs in mammals, the liver can serve as a good model for autophagy studies. A wide array of analytical methods and tools has been developed in recent years and many have been adapted to the studies of hepatic autophagy. We hope that the methods discussed in this chapter will further stimulate studies of autophagy in the area of hepatology.

ACKNOWLEDGMENTS

We would like to thank Dr. Donna Stolz for expert assistance in electron microscopy and Dr. Noboru Mizushima (Tokyo Metropolitan Institute of Medical Science, Japan) for the transgenic GFP-LC3 mice. Xiao-Ming Yin was in part supported by the NIH funds (CA83817, CA111456). Wen-Xing Ding was a recipient of the American Liver Foundation/Alpha-1 Foundation scholar award.

REFERENCES

Amaravadi, R. K., Yu, D., Lum, J. J., Bui, T., Christophorou, M. A., Evan, G. I., Thomas-Tikhonenko, A., and Thompson, C. B. (2007). Autophagy inhibition enhances therapy-induced apoptosis in a Myc-induced model of lymphoma. *J. Clin. Invest.* **117,** 326–336.

Bjorkoy, G., Lamark, T., Brech, A., Outzen, H., Perander, M., Overvatn, A., Stenmark, H., and Johansen, T. (2005). p62/SQSTM1 forms protein aggregates degraded by autophagy and has a protective effect on huntingtin-induced cell death. *J. Cell Biol.* **171,** 603–614.

Blommaart, E. F., Luiken, J. J., Blommaart, P. J., van Woerkom, G. M., and Meijer, A. J. (1995). Phosphorylation of ribosomal protein S6 is inhibitory for autophagy in isolated rat hepatocytes. *J. Biol. Chem.* **270,** 2320–2326.

Cann, G. M., Guignabert, C., Ying, L., Deshpande, N., Bekker, J. M., Wang, L., Zhou, B., and Rabinovitch, M. (2008). Developmental expression of LC3alpha and beta: Absence of fibronectin or autophagy phenotype in LC3beta knockout mice. *Dev. Dyn.* **237,** 187–195.

Deter, R. L., and De Duve, C. (1967). Influence of glucagon, an inducer of cellular autophagy, on some physical properties of rat liver lysosomes. *J. Cell Biol.* **33,** 437–449.

Ding, W. X., Ni, H. M., DiFrancesca, D., Stolz, D. B., and Yin, X. M. (2004). Bid-dependent generation of oxygen radicals promotes death receptor activation-induced apoptosis in murine hepatocytes. *Hepatology* **40,** 403–413.

Elmore, S. P., Qian, T., Grissom, S. F., and Lemasters, J. J. (2001). The mitochondrial permeability transition initiates autophagy in rat hepatocytes. *Faseb J.* **15,** 2286–2287.

Fengsrud, M., Roos, N., Berg, T., Liou, W., Slot, J. W., and Seglen, P. O. (1995). Ultrastructural and immunocytochemical characterization of autophagic vacuoles in isolated hepatocytes: Effects of vinblastine and asparagine on vacuole distributions. *Exp. Cell Res.* **221,** 504–519.

Gohla, A., Klement, K., Piekorz, R. P., Pexa, K., vom Dahl, S., Spicher, K., Dreval, V., Haussinger, D., Birnbaumer, L., and Nurnberg, B. (2007). An obligatory requirement for the heterotrimeric G protein Gi3 in the antiautophagic action of insulin in the liver. *Proc. Natl. Acad. Sci. USA* **104,** 3003–3008.

Hara, T., Nakamura, K., Matsui, M., Yamamoto, A., Nakahara, Y., Suzuki-Migishima, R., Yokoyama, M., Mishima, K., Saito, I., Okano, H., and Mizushima, N. (2006). Suppression of basal autophagy in neural cells causes neurodegenerative disease in mice. *Nature* **441,** 885–889.

Hosokawa, N., Hara, Y., and Mizushima, N. (2006). Generation of cell lines with tetracycline-regulated autophagy and a role for autophagy in controlling cell size. *FEBS Lett.* **580,** 2623–2629.

Jin, S. (2006). Autophagy, mitochondrial quality control, and oncogenesis. *Autophagy* **2,** 80–84.

Klaunig, J. E., Goldblatt, P. J., Hinton, D. E., Lipsky, M. M., Chacko, J., and Trump, B. F. (1981a). Mouse liver cell culture. I. Hepatocyte isolation. *In Vitro* **17,** 913–925.

Klaunig, J. E., Goldblatt, P. J., Hinton, D. E., Lipsky, M. M., and Trump, B. F. (1981b). Mouse liver cell culture. II. Primary culture. *In Vitro* **17,** 926–934.

Knapp, J. E., and Liu, D. (2004). Hydrodynamic delivery of DNA. *Methods Mol. Biol.* **245,** 245–250.

Komatsu, M., Waguri, S., Ueno, T., Iwata, J., Murata, S., Tanida, I., Ezaki, J., Mizushima, N., Ohsumi, Y., Uchiyama, Y., Kominami, E., Tanaka, K., *et al.* (2005). Impairment of starvation-induced and constitutive autophagy in Atg7-deficient mice. *J. Cell Biol.* **169,** 425–434.

Kovacs, A. L., Reith, A., and Seglen, P. O. (1982). Accumulation of autophagosomes after inhibition of hepatocytic protein degradation by vinblastine, leupeptin or a lysosomotropic amine. *Exp. Cell Res.* **137,** 191–201.

Kruse, K. B., Dear, A., Kaltenbrun, E. R., Crum, B. E., George, P. M., Brennan, S. O., and McCracken, A. A. (2006). Mutant fibrinogen cleared from the endoplasmic reticulum via endoplasmic reticulum-associated protein degradation and autophagy: An explanation for liver disease. *Am. J. Pathol.* **168,** 1299–1308; quiz 1404-1405.

Ku, N. O., Strnad, P., Zhong, B. H., Tao, G. Z., and Omary, M. B. (2007). Keratins let liver live: Mutations predispose to liver disease and crosslinking generates Mallory-Denk bodies. *Hepatology* **46,** 1639–1649.

Kuma, A., Hatano, M., Matsui, M., Yamamoto, A., Nakaya, H., Yoshimori, T., Ohsumi, Y., Tokuhisa, T., and Mizushima, N. (2004). The role of autophagy during the early neonatal starvation period. *Nature* **432,** 1032–1036.

Kundu, M., Lindsten, T., Yang, C. Y., Wu, J., Zhao, F., Zhang, J., Selak, M. A., Ney, P. A., and Thompson, C. B. (2008). Ulk1 plays a critical role in the autophagic clearance of mitochondria and ribosomes during reticulocyte maturation. *Blood* **112,** 1493–1502.

Marino, G., Salvador-Montoliu, N., Fueyo, A., Knecht, E., Mizushima, N., and Lopez-Otin, C. (2007). Tissue-specific autophagy alterations and increased tumorigenesis in mice deficient in Atg4C/autophagin-3. *J. Biol. Chem.* **282,** 18573–18583.

Matsumoto, N., Ezaki, J., Komatsu, M., Takahashi, K., Mineki, R., Taka, H., Kikkawa, M., Fujimura, T., Takeda-Ezaki, M., Ueno, T., Tanaka, K., and Kominami, E. (2008). Comprehensive proteomics analysis of autophagy-deficient mouse liver. *Biochem. Biophys. Res. Commun.* **368,** 643–649.

Meijer, A. J., and Codogno, P. (2007). AMP-activated protein kinase and autophagy. *Autophagy* **3,** 238–240.

Mizushima, N., Yamamoto, A., Matsui, M., Yoshimori, T., and Ohsumi, Y. (2004). *In vivo* analysis of autophagy in response to nutrient starvation using transgenic mice expressing a fluorescent autophagosome marker. *Mol. Biol. Cell* **15,** 1101–1111.

Mortimore, G. E., and Mondon, C. E. (1970). Inhibition by insulin of valine turnover in liver. Evidence for a general control of proteolysis. *J. Biol. Chem.* **245,** 2375–2383.

Mortimore, G. E., and Poso, A. R. (1987). Intracellular protein catabolism and its control during nutrient deprivation and supply. *Annu. Rev. Nutr.* **7,** 539–564.

Mortimore, G. E., Poso, A. R., and Lardeux, B. R. (1989). Mechanism and regulation of protein degradation in liver. *Diabetes Metab. Rev.* **5,** 49–70.

Mortimore, G. E., and Schworer, C. M. (1977). Induction of autophagy by amino-acid deprivation in perfused rat liver. *Nature* **270,** 174–176.

Mortimore, G. E., Woodside, K. H., and Henry, J. E. (1972). Compartmentation of free valine and its relation to protein turnover in perfused rat liver. *J. Biol. Chem.* **247,** 2776–2784.

Overbye, A., Fengsrud, M., and Seglen, P. O. (2007). Proteomic analysis of membrane-associated proteins from rat liver autophagosomes. *Autophagy* **3,** 300–322.

Perlmutter, D. H. (2006). The role of autophagy in alpha-1-antitrypsin deficiency: A specific cellular response in genetic diseases associated with aggregation-prone proteins. *Autophagy* **2,** 258–263.

Pfeifer, U. (1978). Inhibition by insulin of the formation of autophagic vacuoles in rat liver. A morphometric approach to the kinetics of intracellular degradation by autophagy. *J. Cell Biol.* **78,** 152–167.

Qu, X., Yu, J., Bhagat, G., Furuya, N., Hibshoosh, H., Troxel, A., Rosen, J., Eskelinen, E. L., Mizushima, N., Ohsumi, Y., Cattoretti, G., and Levine, B. (2003). Promotion of tumorigenesis by heterozygous disruption of the beclin 1 autophagy gene. *J. Clin. Invest.* **112,** 1809–1820.

Ryan, C. M., Carter, E. A., Jenkins, R. L., Sterling, L. M., Yarmush, M. L., Malt, R. A., and Tompkins, R. G. (1993). Isolation and long-term culture of human hepatocytes. *Surgery* **113,** 48–54.

Sandoval, H., Thiagarajan, P., Dasgupta, S. K., Schumacher, A., Prchal, J. T., Chen, M., and Wang, J. (2008). Essential role for Nix in autophagic maturation of erythroid cells. *Nature* **454,** 232–235.

Schiaffino, S., Mammucari, C., and Sandri, M. (2008). The role of autophagy in neonatal tissues: Just a response to amino acid starvation? *Autophagy* **4,** 727–730.

Schweers, R. L., Zhang, J., Randall, M. S., Loyd, M. R., Li, W., Dorsey, F. C., Kundu, M., Opferman, J. T., Cleveland, J. L., Miller, J. L., and Ney, P. A. (2007). NIX is required for programmed mitochondrial clearance during reticulocyte maturation. *Proc. Natl. Acad. Sci. USA* **104,** 19500–19505.

Seglen, P. O. (1976). Preparation of isolated rat liver cells. *Methods Cell Biol.* **13,** 29–83.

Seglen, P. O., and Gordon, P. B. (1982). 3-Methyladenine: Specific inhibitor of autophagic/ lysosomal protein degradation in isolated rat hepatocytes. *Proc. Natl. Acad. Sci. USA* **79,** 1889–1892.

Seglen, P. O., Grinde, B., and Solheim, A. E. (1979). Inhibition of the lysosomal pathway of protein degradation in isolated rat hepatocytes by ammonia, methylamine, chloroquine and leupeptin. *Eur. J. Biochem.* **95,** 215–225.

Stromhaug, P. E., Berg, T. O., Fengsrud, M., and Seglen, P. O. (1998). Purification and characterization of autophagosomes from rat hepatocytes. *Biochem. J.* **335**(Pt 2), 217–224.

Tanaka, Y., Guhde, G., Suter, A., Eskelinen, E. L., Hartmann, D., Lullmann-Rauch, R., Janssen, P. M., Blanz, J., von Figura, K., and Saftig, P. (2000). Accumulation of autophagic vacuoles and cardiomyopathy in LAMP-2-deficient mice. *Nature* **406,** 902–906.

Woodside, K. H., and Mortimore, G. E. (1972). Suppression of protein turnover by amino acids in the perfused rat liver. *J. Biol. Chem.* **247,** 6474–6481.

Yamamoto, A., Tagawa, Y., Yoshimori, T., Moriyama, Y., Masaki, R., and Tashiro, Y. (1998). Bafilomycin A1 prevents maturation of autophagic vacuoles by inhibiting fusion between autophagosomes and lysosomes in rat hepatoma cell line, H-4-II-E cells. *Cell Struct. Funct.* **23,** 33–42.

Yin, X. M., Ding, W. X., and Gao, W. (2008). Autophagy in the liver. *Hepatology* **47,** 1773–1785.

Yue, Z. (2007). Regulation of neuronal autophagy in axon: Implication of autophagy in axonal function and dysfunction/degeneration. *Autophagy* **3,** 139–141.

Zhao, Y., Ding, W. X., Qian, T., Watkins, S., Lemasters, J. J., and Yin, X. M. (2003). Bid activates multiple mitochondrial apoptotic mechanisms in primary hepatocytes after death receptor engagement. *Gastroenterology* **125,** 854–867.

CHAPTER TWENTY-ONE

Monitoring Autophagy in Lysosomal Storage Disorders

Nina Raben,* Lauren Shea,* Victoria Hill,* *and* Paul Plotz*

Contents

Abstract

Lysosomes are the final destination of the autophagic pathway. It is in the acidic milieu of the lysosomes that autophagic cargo is metabolized and recycled. One would expect that diseases with primary lysosomal defects would be among the first systems in which autophagy would be studied. In reality, this is not the case. Lysosomal storage diseases, a group of more than 60 diverse inherited disorders, have only recently become a focus of autophagic research. Studies of these clinically severe conditions promise not only to clarify pathogenic mechanisms, but also to expand our knowledge of autophagy itself. In this chapter, we will describe the lysosomal storage diseases in which autophagy has been explored, and present the approaches used to evaluate this essential cellular pathway.

* The Arthritis and Rheumatism Branch, NIAMS, National Institutes of Health, Bethesda, Maryland, USA

Methods in Enzymology, Volume 453
ISSN 0076-6879, DOI: 10.1016/S0076-6879(08)04021-4

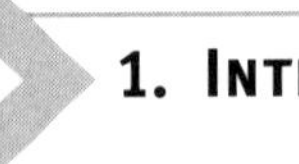

1. Introduction

Lysosomal storage disorders (LSDs) are an extremely diverse group of more than 60 genetic diseases. Individual LSDs are rare disorders, but, as a group, their incidence is estimated to be 1 in 5000–8000 live births (Hodges and Cheng, 2006; Wenger *et al.*, 2003). The majority of LSDs are caused by the deficiency of a single lysosomal hydrolase, leading to the accumulation of the corresponding substrate. LSDs can also result from mutations in proteins involved in the intracellular trafficking of lysosomal enzymes. Two such diseases, mucolipidoses II and III, are caused by a deficiency of the enzyme that catalyzes the addition of mannose-6-phosphate (required for lysosomal targeting) to newly synthesized lysosomal hydrolases. Another group of LSDs results from defective transport of the products of lysosomal catabolism across the lysosomal membrane. Examples include cystinosis and sialic acid storage disease. Finally, there are several atypical LSDs, which involve mutations in integral membrane proteins of poorly defined function. These disorders include Niemann–Pick Type C and juvenile neuronal ceroid lipofuscinosis.

A great deal of progress has been made in the field of LSDs in elucidating genetic defects, developing therapeutic approaches, improving patient care, generating animal models etc. However, the biological pathways from lysosomal storage to cellular dysfunction and death remain largely unknown. In the not so distant past, lysosomes were viewed as simple garbage disposal organelles. Recently, an attractive new view of lysosomes has been proposed by Steven Walkley (Walkley, 2007). According to this view, lysosomes are credited with a prominent role as essential mediators of cellular metabolism. The well-being of lysosomes has broad-reaching effects on many cellular systems, including the multiple vesicular pathways that terminate at these organelles.

One of the pathways destined for the lysosome, macroautophagy (hereafter referred to as autophagy), is a catabolic pathway responsible for the turnover of long-lived cytosolic proteins and organelles, such as mitochondria. Autophagic cargo is sequestered by double-membrane vesicles (autophagosomes) and is ultimately degraded after autophagosome-lysosome fusion. Functional lysosomes are critical for the maturation of autophagosomes and for the degradation of their content. Thus, autophagy seemed an attractive target for studies in LSDs, in which lysosomal abnormalities are the primary defects. These studies became possible due to the great progress in, and fascination with, the field of autophagy.

The case for examining autophagy is bolstered by similarities between LSDs, most of which involve severe neurodegeneration, and a number of age-related neurodegenerative diseases, such as Alzheimer's disease,

Parkinson's disease, and Huntington's disease. Although the etiology of LSDs is different, namely the accumulation of undegraded compounds in lysosomes, the mechanisms of neurodegeneration may be similar, including the major role of autophagy. Furthermore, inactivation of crucial autophagy genes in mice has been shown to lead to severe neurodegeneration and the accumulation of undegraded polyubiquitinated proteins (Hara *et al.*, 2006; Komatsu *et al.*, 2006). Accumulation of undegraded material may be particularly harmful for post-mitotic cells such as neurons or muscle cells as these cells cannot dilute the accumulated material by cell division.

The precise mechanisms leading to changes in or malfunction of the autophagic pathway in LSDs remain elusive. Furthermore, the link between abnormal autophagy and cell death, in particular neuronal cell death, still needs to be established. It is important to emphasize that the study of autophagy in LSDs is just emerging. The role of autophagy has been investigated in only a handful of LSDs, but the approach is certain to be extended to a broader range of LSDs in the near future.

The LSDs in which autophagy has been studied to date include Multiple Sulfatase Deficiency (MSD), Mucopolysaccharidosis Type IIIA (MPS IIIA), G_{M1}-Gangliosidosis, Pompe Disease (PD), Niemann-Pick Disease Type C (NPC), Neuronal Ceriod Lipofuscinosis (NCL), Mucolipidosis Type IV, and Danon Disease. A variety of available methods have been adapted to monitor autophagy in LSDs, which now compose an arsenal of general techniques. We will first describe these general techniques, providing references for their application to LSDs. We will then focus on the diseases themselves and highlight some of the less commonly used methods (referred to as specific techniques) employed to study them.

2. General Techniques to Monitor Autophagy In LSDs

2.1. Lysosomes and associated vesicular pathways

2.1.1. Vesicular markers

Several membrane proteins are particularly useful in the study of vesicle morphology and distribution. Microtubule-associated protein 1 light chain 3 (LC3) exists in the cytoplasm as a soluble form (LC3-I). A phosphatidylethanolamine (PE)-conjugated form of LC3 (LC3-II) is associated with the phagophore and the inner autophagosomal membrane and is a highly specific marker for these structures (Kabeya *et al.*, 2000). Upon fusion with the lysosome, LC3-II on the inner autophagosomal membrane is degraded (Xie and Klionsky, 2007). LC3-I and II can be distinguished based on their migration in an SDS-PAGE gel (see the chapter by Kimura *et al*,, in this volume). The LC3-II:LC3-I ratio, as determined by western blotting, is a commonly used measure of autophagosome levels (Cao *et al.*, 2006; Koike *et al.*, 2005). Recently it has been noted that a more accurate standardization may be achieved by comparing LC3-II levels to other internal controls (Klionsky *et al.*, 2008; Liao *et al.*, 2007; Settembre *et al.*, 2008a). Staining cells or tissues with anti-LC3 antibodies indicates the steady state level of autophagosomes. Additional methods include transfection of diseased cells with GFP-LC3 (Cao *et al.*, 2006; Gonzalez-Polo *et al.*, 2005), as well as generation of GFP-LC3 transgenic mice on an LSD background (Koike *et al.*, 2005) (also see the chapter by N. Mizushima in volume 452).

Other vesicular markers have also proven useful in studying LSD pathology. Cathepsins are lysosomal proteases that can serve as markers for the lysosomal lumen (Tanaka *et al.*, 2000), and LAMP-1 and LAMP-2 are transmembrane proteins localized specifically to lysosomal membranes. Double labeling with LAMP-1 and LC3 antibodies has been used in fibroblasts (Raben *et al.*, 2003; Settembre *et al.*, 2008a) and brain (Koike *et al.*, 2005) to quantify autolysosomes as a surrogate measure of the efficiency of autophagosome/lysosome fusion. Late endosomes stain positive for both LAMP-1 and the cation-independent mannose-6-phosphate receptor (CI-MPR), whereas early endosomes are marked by both rab5 and early endosome antigen 1 (EEA1) (Fukuda *et al.*, 2006b). Structural information obtained with these markers complements functional studies of endocytosis (see subsequently).

As an alternative to immunological methods, autophagosomes have been visualized with the autofluorescent drug monodansylcadaverine (MDC). Cells are incubated with MDC and examined via fluorescence microscopy (Jennings *et al.*, 2006) (also see the chapter by Vázquez and Colombo in volume 452). In addition, MDC-labeled cells can be lysed, and a solution of ethidium bromide added to the lysate to normalize for cell number by measuring DNA fluorescence (Pacheco *et al.*, 2007). Concerns have been raised, however, that MDC is not completely specific for autophagosomes; this technique is thus best employed in combination with other methods of autophagosome detection (Klionsky *et al.*, 2008).

LysoTracker is a fluorescent dye that preferentially accumulates in vesicles with an acidic pH. LysoTracker staining can be used to detect abnormalities in vesicular pH (Eskelinen *et al.*, 2004) and to examine the efficiency of autophagosome/lysosome fusion in live cells (Gonzalez-Polo *et al.*, 2005). It is important to note that as a weakly basic amine, LysoTracker may cause lysosome alkalinization. Thus, in studies of live cells, images should be taken shortly after staining to reduce this effect. LysoTracker-stained cells can also be fixed in paraformaldehyde followed by mounting in Vectashield mounting medium without significant loss of fluorescent signal (Cao *et al.*, 2006).

2.1.2. Electron microscopy (EM)

Autophagosomes are typically identified with EM by the presence of a double membrane, and are often classified as AVi (immature vesicles, containing undegraded cellular material) and AVd (mature vesicles, containing partially degraded identifiable cellular material) (Eskelinen *et al.*, 2002, 2004; Koike *et al.*, 2005; Tanaka *et al.*, 2000). This classification has recently drawn criticism, however, as autophagosome maturation is now seen as a gradual process characterized by numerous intermediate states (Klionsky *et al.*, 2008) (also see the chapter by Ylä-Anttila and Eskelinen in volume 452). Morphometric analysis offers a quantitative interpretation of EM images. With this technique, several EM image frames are scored for the number of various vesicular structures present (double membrane, multivesicular, multilamellar, and multidense bodies were categories used in one study of neuronal ceroid lipofuscinosis (Cao *et al.*, 2006)). The combination of EM and immunogold labeling can facilitate identification of vesicles that are difficult to distinguish in conventional EM. In a mouse model of Pompe disease, for example, this technique was used to characterize non-contractile inclusions in muscle as LAMP-1-positive (Fukuda *et al.*, 2006b).

2.1.3. Subcellular fractionation

Many of the techniques described previously can also be applied to subcellular fractions isolated from tissues, allowing for more specific and detailed analysis of autophagosomes and other vesicle types (for example, see the chapter by Kaushik and Cuervo in volume 452). In studying LSDs, fractionation has been performed on brain (Liao *et al.*, 2007) and liver (Cao *et al.*, 2006) samples from mice, as well as on cultured fibroblasts (Eskelinen *et al.*, 2004).

Brain:

1. Mouse brains from wild-type and $Npc1^{-/-}$ (a model of Niemann-Pick type C disease) mice are dissected in ice-cold artificial cerebrospinal fluid (124 mM NaCl, 3 mM KCl, 1.25 mM KH_2PO_4, 3.4 mM $CaCl_2$, 2.5 mM $MgSO_4$, 26 mM $NaHCO_3$, 10 mM D-glucose (Kramar *et al.*, 2004)), followed by homogenization in buffer containing 3 mmol/L imidazole, 250 mmol/L sucrose, 1 mmol/L EDTA, pH 7.4, and protease inhibitors (Sigma, St. Louis, MO).
2. Homogenates are centrifuged for 10 min at $1500\times g$.
3. The sucrose concentration of the supernatant fraction (postnuclear lysates) is adjusted to 40.6% by the slow addition of 62% sucrose in homogenization buffer-EDTA. The supernatant is then overloaded with 35% and 25% sucrose in the same buffer, and the samples are centrifuged in an SW 55 rotor (Beckman Instruments, Palo Alto, CA) at $14{,}000\times g$ for 90 min at 4 °C.

4. Following centrifugation, 3 fractions are collected: a late-endosome/lysosome enriched fraction (25% sucrose), an early endosomes-enriched fraction (25%–35% sucrose) and a fraction enriched in plasma membranes (35%–40.6% sucrose). The distribution of particular proteins in these fractions is then determined by western blotting (Liao *et al.*, 2007). The details of the fractionation protocol can be found in de Araujo *et al.*, (2006).

Liver:

Cellular fractionation and isolation of autophagic vacuoles have been performed in livers from a mouse model of Juvenile Neuronal Ceroid Lipofuscinosis (Cao *et al.*, 2006). The isolation protocol is a modification (Yu *et al.*, 2005) of the original method described by Marzella *et al.*, (1982). Briefly, the procedure involves the following steps:

1. Mice are starved for 12 h and injected intraperitoneally with 5 mg/100 g of body weight of vinblastine in normal saline (0.9% NaCl) 3 h before sacrificing.
2. Livers are harvested (livers from 4 mice are pooled), homogenized (in 0.25 *M* sucrose (1:2 wt/vol)), and separated by differential centrifugation to produce a low-speed pellet containing the nuclear fraction and unbroken cells, and a high-speed pellet that is enriched in AVs, lysosomes, and mitochondria.
3. A discontinuous metrizamide gradient is then used to separate the lysosomal fraction, two AV fractions (light AV fraction and heavy AV fraction), and the mitochondrial fraction. Fractions are collected and examined by Western blotting and EM (Cao *et al.*, 2006).

Fibroblasts:

Subcellular fractionation has been performed in mouse embryonic fibroblasts obtained from LAMP-1 and LAMP-2 double-deficient mice (Eskelinen *et al.*, 2004). In this case, the lysosome-enriched Percoll gradient fraction is isolated and analyzed with lysosomal markers.

2.1.4. Pharmacological modulation

Pharmacological agents have been used to determine the functional capacity of autophagy in LSDs. Bafilomycin, for example, blocks autophagosome/lysosome fusion by causing lysosomal alkalinization. If bafilomycin-treated cells accumulate more LC3-II than untreated cells, this indicates that autophagosome turnover through fusion with lysosomes is occurring to some extent (Settembre *et al.*, 2008a). Similarly, LC3-II accumulation in response to treatment with lysosomal protease inhibitors such as E64 and pepstatin relative to the untreated control indicates that any block in autophagy is not complete (Pacheco *et al.*, 2007) (see the chapter by Kimura *et al.*, in volume 452).

2.1.5. Pathway analysis

Pathway analysis has been used to determine whether an observed increase in the number of autophagosomes is due to an induction of autophagy. The Akt/mTOR pathway, downstream of class I PI3K, and the ERK pathway are known to regulate autophagic activity (Corcelle *et al.*, 2006; Lum *et al.*, 2005; Tassa *et al.*, 2003). Class I PI3K-independent regulation of autophagy has also been observed; in this case, the class III PI3K-interacting protein Beclin 1 appears to play a major role (Tassa *et al.*, 2003). Western blot quantification of the levels of these proteins has been used in LSDs where abnormal autophagic activity is observed (Cao *et al.*, 2006; Pacheco *et al.*, 2007; Takamura *et al.*, 2008).

2.1.6. Vesicular pH and endocytic trafficking

Several techniques are available for measuring late endosomal/lysosomal pH, which is another key parameter of vesicular functionality.

i. The assay we utilize in our lab (to measure the lysosomal pH in myoblasts derived from Pompe mice) is based on measuring the ratio of pH-sensitive ORG488 (Oregon Green) or FL (fluorescein) to pH-insensitive TMR (tetramethylrhodamine) fluorescence emissions (Fukuda *et al.*, 2006b).

Procedure:

1. Myoblasts are grown in Dulbecco's modified Eagle's medium (DMEM) supplemented with 20% fetal calf serum, 10% horse serum, 1% chick embryo extract, 100 IU/ml penicillin, and 100 μg of streptomycin (Invitrogen, Carlsbad, CA).
2. Cells are split 24 h before the experiments on collagen-coated (0.4%) or poly-D-Lysine chamber slides [Lab-Tek® Chamber Slide™ System, Nunc, Naperville, IL, Cat. No. 154941 (8 well glass slide)] at low density (2.0×10^3 cells/cm^2).
3. Myoblasts are incubated with either a mixture of ORG488 (dextran, Oregon Green® 488; Cat. No. D7170) and TMR-conjugated dextrans (which allows pH detection in a range from 4.0 to 5.5; Cat. No. D1868) (1mg/ml each) or FL/TMR-double conjugated dextran (for pH 4.8–6.0) (2 mg/ml) (Cat. No. D1950) (all from Molecular Probes/Invitrogen) for 24 h.
4. Cells are washed and incubated in dextran-free complete medium (chase) for 2 and 36 h.
5. Confocal images of cells in serum-free DMEM (Biosource, Rockville, MD) are recorded using the appropriate emission ranges.

6. The FL/ORG488 over TMR ratio of individual dextran-containing vesicles is used to determine lysosomal pH; TMR acts as an internal control for dye uptake (Maxfield 1989). The calculated ratio is then converted to a pH value by using a calibration curve, which is generated by treating the dextran labeled cells with 1 m*M* monensin (Cat. No. M5273) and 1 m*M* nigericin (Cat. No. N7143) (both from Sigma) for 5 min in buffers of pH 4.8 to 6.0 in increments of 0.1–0.2 pH units. Monensin and nigericin are ionophores that neutralize the acidic interior of endosomes/lysosomes without affecting the membrane potential. Acetate buffer (25 m*M*) is used for pH 4.8 to 5.7, and MES 2-(N-morpholinoethanesulfonic acid) buffer (25 m*M*) is used for pH 5.7 to 6.0. Both buffer solutions contain 5 m*M* NaCl, 115 m*M* KCl, and 1.2 m*M* $MgSO_4$.

ii. Acridine orange (AO) (solution in water) (Molecular Probes/Invitrogen, Cat. No. A3568), a fluorescent weak base that accumulates in acidic compartments, can be used to visualize lysosomes and to evaluate gross abnormalities in lysosomal pH. Upon entering acidic compartments such as lysosomes, AO becomes protonated and sequestered; AO shows red fluorescence in an acidic environment and green fluorescence in a neutral environment.
We use AO to demonstrate the presence of large alkalinized lysosomes in cultured myotubes from Pompe mice: the cells are loaded with AO (2.5 μg/ml) in DMEM medium for 10 min at 37 °C and analyzed by confocal microscopy (our unpublished data).
AO can also be used to demonstrate a higher than normal acidic environment of lysosomes, as shown in fibroblasts derived from patients with Mucolipidosis type IV (Soyombo *et al.*, 2006). In this case, the cells are incubated with AO (1 μM for 10 min.) in the standard bath solution, containing 140 m*M* NaCl, 5m*M* KCl, 1m*M* $CaCl_2$, 1 m*M* $MgCl_2$, 10 m*M* Hepes, pH7.4. Increased AO accumulation in the lysosomes resulting in a brighter AO fluorescence indicates that the lysosomes in the diseased cells are over-acidified (Soyombo *et al.*, 2006).

iii. To evaluate the pH of enlarged lysoSomes in myotubes derived from Pompe mice we have also used two LysoSensors: DND-189 (p*K*a = 5.2) (Cat. No. L-7535), which fluoresces in an acidic environment (pH $\leq$ 5.2), and DND-153 (p*K*a = 7.5) (Cat. No. L-7534), which fluoresces in both an acidic and in a neutral environment (pH $\leq$ 7.5) (both lysoSensors are from Molecular Probes). The cells are incubated with DND-153 or DND-189 (in DMEM medium) at 1 μM for 60 minutes, washed several times with medium and analyzed by confocal microscopy.

iv. DAMP (N-(3-((2,4-dinitrophenyl)amino) propyl)-N-(3-aminopropyl) methylamine), a nonfluorescent, weakly basic amine that traffics to acidic organelles, can be visualized by staining with anti-dinitrophenol antibodies. This approach, developed by Anderson and Orci (1988) has been used to study vesicular pH in LAMP-2-deficient mice (Tanaka *et al.*, 2000) and in LAMP-1/LAMP-2-deficient fibroblasts (Eskelinen *et al.*, 2004).

To monitor endocytic trafficking to lysosomes, labeled dextran, a fluid-phase endocytic marker, can be used in combination with LysoTracker Red. This approach has been used, for example, in studying autophagy in a mouse model of Juvenile Neuronal Ceroid Lipofuscinosis (JNCL) (Cao *et al.*, 2006; Fossale *et al.*, 2004).

Procedure (Fossale *et al.*, 2004):

1. Neuron-derived JNCL cells are seeded at a density of 3–5 × 10^4 cells/well in 4-well chamber slides and grown overnight at 33 °C.
2. Growth medium (DMEM, Gibco BRL) supplemented with 10% fetal bovine serum and 24 m*M* KCl is exchanged for fresh, prewarmed medium containing 500 n*M* LysoTracker Red DND-99 (Cat. No. L-7528) or 1 mg/ml dextran-FITC (Cat. No.D-7178) (both from Molecular Probes/Invitrogen).
3. Cells are incubated at 33 °C for 45 min with LysoTracker or for 15 min with labeled dextran.
4. Cells are then placed on ice, washed for 10 min in ice-cold dye-free medium, and fixed with 4% formaldehyde in PBS for 20 min on ice.
5. Slides are prepared with Vectashield mounting medium (Cat. No. H-1000; Vector Laboratories; Burlingame, CA), and cells are analyzed by confocal microscopy.

2.2. Consequences of Impaired Autophagy in LSDs

2.2.1. Protein turnover

Turnover of long-lived proteins, known substrates of autophagy, has been quantified in a radio-labeled amino acid release assay (Pacheco *et al.*, 2007; Tanaka *et al.*, 2000) (also see the chapter by Bauvy *et al*,, in volume 452) In a variation on this procedure, cells are treated with ammonium chloride (20 m*M*) and leupeptin (0.1 m*M*) to inhibit all lysosomal proteolysis, allowing evaluation of the contribution of lysosomal pathways. Similarly, treating cells with the macroautophagy inhibitor 3-methyladenine (3-MA, 10 m*M*) reveals the contribution of macroautophagic proteolysis. Subtracting the macroautophagic proteolysis from the lysosomal proteolysis yields the contribution of nonmacroautophagic lysosomal pathways (chaperone-mediated autophagy and microautophagy). In these experiments, proteolysis measurements are initiated 1 h after the inhibitors are added and are continued for only 3 additional hours, to avoid possible side effects from lengthy inhibitor treatment (Eskelinen *et al.*, 2004).

2.2.2. Ubiquitinated proteins

Autophagy plays a major role in the degradation of ubiquitinated proteins (Hara *et al.*, 2006; Komatsu *et al.*, 2005, 2006). Western blotting of tissue lysates and subcellular fractions with antiubiquitin antibodies reveals defects in autophagic removal of ubiquitinated substrates in several LSDs (Liao *et al.*, 2007; Settembre *et al.*, 2008a). Accumulation of p62/SQSTM1, a protein that binds both ubiquitinated substrates and LC3 (Pankiv *et al.*, 2007), is an indicator of defective autophagy (Settembre *et al.*, 2008a).

2.2.3. Cell viability

In vitro models of LSDs are particularly well suited for cell viability assays. Multiple methods are available to assess cell viability. For example, viability can be measured using the vital dye, propidium iodide (PI; Sigma, Cat. No. P4170), which is added to cultured cells at a concentration of 1 μg/ml. The cells are then incubated for 10–15 min at room temperature (protected from light), washed and examined by fluorescence microscopy or flow cytometry (PI is excluded from viable cells) (Gonzalez-Polo *et al.*, 2005). Cell viability in a cerebellar cell model of juvenile neuronal ceroid lipofuscinosis was measured using the Cell-Titer-96 AQueous Non-Radioactive Cell Proliferation Assay (Promega, Cat. No.G5421), according to the manufacturer's specifications (Cao *et al.*, 2006; Fossale *et al.*, 2004).

The viability of wild-type and LSD cells can be measured after induction of autophagy by starvation (Gonzalez-Polo *et al.*, 2005) or by oxidative stress following the addition of varying concentrations of hydrogen peroxide as a stressor (Fossale *et al.*, 2004). The effects of autophagic inhibitors (3-MA) and stimulators (rapamycin) on cell survival can be examined (Cao *et al.*, 2006). These compounds may also be administered in combination with a stress-inducing agent (for example, paraquat for oxidative stress (Takamura *et al.*, 2008)) to further elucidate the role of autophagy in LSD pathology.

2.3. Mitochondrial Abnormalities

The mitochondrial-lysosomal axis theory of aging postulates that oxidized material accumulates in lysosomes as cells age, which results in decreased degradative capacity of lysosomes (Terman and Brunk 2006). The autophagic/lysosomal pathway is the major route of destruction for damaged mitochondria (potent generators of reactive oxygen species); therefore, mitochondria accumulate and subject cells to increasing oxidative stress when autophagy is dysfunctional. Investigating the condition of mitochondria is thus highly relevant in LSDs where autophagy may be impaired (Kiselyov *et al.*, 2007).

2.3.1. Mitochondrial markers

Mitochondria can be visualized with antibodies against grp75, a chaperone protein that localizes to the inner mitochondrial membrane (Fossale *et al.*, 2004), as well as with the fluorescent dye 123 rhodamine (Jennings, Jr. *et al.*, 2006). Immunostaining with grp75 antibody is performed on fixed cells, whereas fluorescence experiments are done on live cells, and in both cases the standard procedures can be followed. MitoTracker Red is also used in studies of LSDs (Takamura *et al.*, 2008). This compound offers the advantage of being retained in mitochondria even after membrane potential is lost during fixation, although initial labeling does require intact membrane potential (Poot *et al.*, 1996). Several probes, such as MitoTracker Red CMXRos or MitoTracker JC-1 (Molecular Probes) can be used for mitochondrial labeling of cells according to standard techniques (Takamura *et al.*, 2008).

EM and transfection with the mitochondrial marker mt-DsRed may also be employed to visualize mitochondria (Fossale *et al.*, 2004; Gonzalez-Polo *et al.*, 2005; Settembre *et al.*, 2008a). Mitochondrial fragmentation is observed in LSD models (Jennings, Jr. *et al.*, 2006; Takamura *et al.*, 2008). Cytochrome *c* oxidase subunit IV (cox4) is an abundant protein in mitochondria, and Western blotting for cox4 can be used to detect mitochondrial aberrations (Settembre *et al.*, 2008a).

2.3.2. Membrane potential

To assess the functional condition of mitochondria, the mitochondria-specific, voltage-dependent dye $DiOC_6$ can be employed. This positively charged fluorescent probe accumulates only in mitochondria with a strongly negative membrane potential. The approach has been used, for example, to demonstrate the presence of dysfunctional mitochondria in Multiple Sulfatase Deficiency (MDS); the experiments have been done *in vitro*, in mouse embryonic fibroblasts (MEFs) derived from wild type and MDS mice (Settembre *et al.*, 2008a). The procedure includes treatment of PBS-washed cells with $DiOC_6$ dye (Sigma; 40nM) and 1μg/ml propidium iodine for 15 min at 37 °C followed by flow cytometry, which identifies the percentage of cells in which mitochondrial membrane potential has been lost (Gonzalez-Polo *et al.*, 2005; Settembre *et al.*, 2008a). MitoTracker JC-1, another probe that accumulates in mitochondria, is also used to measure mitochondrial membrane potential. The procedure involves incubation of cultured cells with the MitoTracker probe followed by imaging using confocal microscopy: for example, primary cultured astrocytes from β-galactosidase knockout mice (a model of G_{M1}-gangliosidosis) grown on coverslips in Hanks' balanced solution were loaded with 3μM MitoTracker JC-1 (Molecular Probes) for 20 min at 37 °C. In mitochondria with weak membrane potential, JC-1 monomers give off green fluorescence. In mitochondria with high membrane potential, however, the probe accumulates and forms aggregates that give off red fluorescence (Takamura *et al.*, 2008).

2.3.3. ATP synthase subunit *c* accumulation

Autophagy is the major cellular pathway responsible for recycling aging mitochondria. Thus, it is expected that mitochondrial proteins will accumulate in autophagy-deficient cells. Accumulation of the mitochondrial ATP synthase subunit *c* has been detected by Western blotting and/or immunostaining in patients' samples and in animal models in several LSDs (Cao *et al.*, 2006; Elleder *et al.*, 1997; Koike *et al.*, 2005). Both immunoblot and immunostaining are performed according to standard procedures.

3. LSDs Analyzed for Autophagic Involvement

3.1. Lysosomal hydrolase deficiencies

3.1.1. Multiple sulfatase deficiency (MSD) and mucopolysaccharidosis Type IIIA (MPSIIIA, Sanfilippo A)

Multiple sulfatase deficiency (MSD) is caused by defects in sulfatase modifying factor 1 (SUMF1), the only enzyme responsible for an essential post-translational modification of sulfatases (Cosma *et al.*, 2003; Dierks *et al.*, 2003). The deficiency results in a profound reduction in the activity of all sulfatases, leading to massive lysosomal storage of sulfated glycosaminoglycans and sulfolipids (Hopwood and Ballabio, 2001). Deficiencies of individual sulfatases cause six separate lysosomal storage diseases (LSDs), including MPS IIIA [sulfamidase (Heparan N-sulfatase) deficiency]. The phenotype of MSD patients combines all features observed in individual sulfatase deficiencies, including progressive neurodegeneration, developmental delay, visceromegaly, and skeletal involvement.

Model systems

Sumf1 knockout (KO) mice were developed as an animal model of MSD (Settembre *et al.*, 2007). Two natural murine models are available for MPSIIIA (Bhattacharyya *et al.*, 2001; Bhaumik *et al.*, 1999). Mouse embryonic fibroblasts (MEF) have been derived from both MSD and MPSIIIA mice and embryonic liver macrophages (ELM) from the MSD model. Both cell types demonstrate lysosomal storage pathology (Settembre *et al.*, 2008a,b).

General techniques

Several general techniques have been applied to detect an increase in the number of autophagosomes *in vivo*, specifically in the brains of MSD mice (Settembre *et al.*, 2008a,b). Immunoblotting and immunofluorescence reveal elevated LC3-II levels and an increased number of LC3-positive vesicles in brain. EM shows abundant early/immature autophagic vacuoles in cerebellum and cerebral cortex sections, suggesting a defect in autophagosome maturation. The development of adequate tissue culture models, MEFs and ELMs derived from MSD mice, facilitates analysis of the autophagic defect. Reduced co-localization of LAMP-1 and LC3, observed by confocal microscopy in starved and non-starved MEFs, is an indication that a block in autophagosomal-lysosomal fusion, rather than induction of autophagy, may be responsible for the increased number of autophagosomes. The *in vitro* system also makes possible assessment of the degree of the autophagic block: treatment of MEFs with bafilomycin A_1, a known inhibitor of autophagosomal-lysosomal fusion or lysosomal hydrolase activity, increases LC3-II levels. This finding indicates that the defect in fusion is partial rather than complete (Settembre *et al.*, 2008a,b).

Specific techniques

The functional consequences of the autophagic block in MSD and MPSIIIA are analyzed by several approaches (Settembre *et al.*, 2008a,b).

a. The ability of MEFs and ELMs to degrade exogenous substrates of autophagy was tested. Mutant huntingtin (involved in Huntington disease) and alpha-synuclein (involved in familial Parkinson's disease) are two such substrates. Mutant Gln74Q74 huntingtin (which encodes the first exon of huntingtin with 74 glutamine repeats) or A53T alpha-synuclein were fused with GFP and transiently expressed in MEFs and ELMs. The increased accumulation of GFP-fused mutant proteins in MSD cells was examined by western blot analysis with anti-GFP antibody, and an increase in the number of GFP-Q74 aggregates was monitored by immunofluorescence.
b. The accumulation of endogenous substrates, such as ubiquitinated proteins, p62/SQSTM1, and mitochondria was monitored. Immunohistochemical staining, immunofluoresence, and western analyses were used to demonstrate the progressive age-dependent accumulation of ubiquitin-positive inclusions in the cerebral cortex and other regions of the MSD brain. Double-staining with NeuN, a neuronal marker, was used to show that these inclusions are located in MDS neurons. Immunofluorescence with anti-ubiquitin and anti-p62/SQSTM1 antibodies showed significant colocalization of p62/SQSTM1 with polyubiquitinated proteins in the diseased brain.

c. Mitochondrial number and function were analyzed. EM of brain sections and MEFs as well as immunoblotting with anti-cox4 demonstrated an increased number of mitochondria. A significant reduction in mitochondrial membrane potential in MSD, measured by using a mitochondria specific voltage-dependent dye ($DiOC_6$) was observed in both the normal and starved condition.

Thus, lysosomal dysfunction in MSD and MPSIIIA is associated with impairment in autophagososmal-lysosomal fusion and a consequent block in autophagy. The reduced capacity of lysosomes to degrade the autophagic cargo results in accumulation of damaged mitochondria and polyubiquitinated protein aggregates, the putative mediators of neuronal cell death in these two disorders and possibly in other LSDs (Settembre *et al.*, 2008a).

3.1.2. G_{M1}-gangliosidosis

G_{M1}-gangliosidosis (Norman-Landing Disease) is caused by a deficiency of the lysosomal enzyme acid beta-galactosidase-1 (β-gal), which hydrolyzes the terminal beta-galactosidic residue from G_{M1} ganglioside and other glycoconjugates. The deficiency results in the generalized accumulation of G_{M1} ganglioside and other enzyme substrates, including glycolipid G_{A1}, oligosaccharides, and intermediates of keratan sulfate degradation. The clinical manifestations of G_{M1}-gangliosidosis include visceromegaly, skeletal dysplasias, and severe progressive central nervous system damage. A deficiency in the same beta-galactosidase is also the primary defect in mucopolysaccharidosis type IVB (MPSIVB, Morquio disease), in which keratan sulfate is the predominant storage material. No neurological involvement is seen in MPSIVB.

The data on autophagy in G_{M1}-Gangliosidosis have been presented in one publication so far (Takamura *et al.*, 2008).

Model systems

A knockout mouse model with the neurological phenotype of human G_{M1}-gangliosidosis is available (Itoh *et al.*, 2001; Matsuda *et al.*, 1997). Primary culture of astrocytes from the murine model has also been utilized.

General techniques

Several general methods were used to demonstrate the increase in autophagy in G_{M1}-gangliosidosis (Takamura *et al.*, 2008). Immunoblotting and immunoflourescence revealed an increase in the LC3-II level in brains of β-gal$^{-/-}$ mice, particularly in the sites of neuronal death, the cerebellum and brain stem. Immunofluorescence also revealed the colocalization of LC3 and G_{M1} ganglioside. Induction of autophagy was associated with increased expression of Beclin 1 in both brain lysates and cultured astrocytes from beta-gal$^{-/-}$ mice.

Specific techniques

A number of specific methods were used to address the question of mitochondrial involvement.

a. Measurement of cytochrome *c* oxidase activity in brains from β-gal$^{-/-}$ mice revealed decreasing mitochondrial functionality, paralleling disease progression. Mitochondrial isolation kit and mitochondrial activity kit (both from BioChain Institute, Inc., Hayward, CA USA) are used to isolate mitochondria from the mouse brain and cultured astrocytes and to determine the activity of cytochrome *c* oxidase (Takamura *et al.*, 2008). Confocal microscopy and mitochondrial membrane potential measurement with MitoTracker JC-1 staining showed abnormally small fragmented or circulated mitochondria with a reduced membrane potential in beta-gal$^{-/-}$ primary astrocytes.

b. The functional relevance of the observed mitochondrial dysfunction to cell death was addressed by subjecting primary astrocytes to oxidative stress (the cells were cultured with 250 μM paraquat (Cat. No. 160-08871; Wako Pure Chemical Industries, Ltd, Tokyo, Japan) for 24 h. Beta-gal$^{-/-}$ astrocytes die at an increased rate compared to wild-type cells as shown by lactate dehydrogenase (LDH) cytotoxicity assay. LDH is a soluble cytosolic enzyme that is released into the culture medium if the integrity of the cellular membrane is lost. LDH activity (LDH Cytotoxicity Assay Kit; Cayman Chemical, Ann Arbor, MI, Cat. No. 10008882) in the medium can be used for evaluation of cytotoxicity caused by chemical compounds. Addition of ATP and 3-MA, inhibitors of autophagy, to paraquat treated β-gal$^{-/-}$ astrocytes results in decreased cell death. The addition of rapamycin, an inducer of autophagy, has no effect on cell survival.

Thus, in G_{M1}-gangliosidosis there is evidence of mitochondrial abnormalities and activation of autophagy, both being age-dependent, and paralleling the degree of neurodegeneration.

3.1.3. Pompe disease (PD)

Pompe disease (Glycogen Storage Disease type II) is caused by a deficiency of acid alpha-glucosidase (GAA), the enzyme responsible for degradation of glycogen to glucose in the lysosome. Lysosomal retention of glycogen occurs in all tissues. Unlike many other LSDs, PD is not known to be a neurodegenerative disease, and the major pathology is observed in cardiac and skeletal muscle. In the most severe cases, any possible neurological involvement may not be detected due to early mortality, which occurs within 1 or 2 years of life. Studies in animals and autopsy results clearly indicate that lysosomal glycogen does accumulate in the CNS. Thus, latent neurological manifestations may eventually come to light as enzyme replacement therapy prolongs the lives of infantile patients.

Since our group has studied the role of autophagy in PD, more detailed descriptions of methods we have used are included.

Model systems

$GAA^{-/-}$ mice and myoblasts derived from these mice have been studied (Fukuda *et al.*, 2006b; Raben *et al.*, 1998). In addition, myoblasts and myotubes, as well as muscle biopsies, from patients with Pompe disease were analyzed (Raben *et al.*, 2007).

General techniques

Conventional transmission electron microscopy remains the most reliable technique to visualize autophagic vacuolization in skeletal muscle in Pompe disease as well as in other disorders affecting muscle. One of the intrinsic pitfalls of EM in Pompe disease is that lysosomal glycogen is often lost during processing leaving large empty spaces in the EM images. The following fixation method can be used to preserve lysosomal glycogen for EM (as recommended by Dr. Andrew Engel). We have used this protocol for EM of $GAA^{-/-}$ mice, but it can also be used to analyze human muscle biopsy.

Procedures

i. The mouse is euthanized, placed on a platform, and dissected to expose muscle. In mice white gastrocnemius, externum digitorum longus (EDL), tibialis anterior (TA) and psoas muscles are a good source of fast-twitch glycolytic type II fibers, whereas soleus muscle is a good source of slow-twitch oxidative type I fibers.
ii. Immediately after dissection, the clamped muscles are injected with ice-cold 5% glutaraldehyde (Cat. No. 16220; Electron Microscopy Sciences, Hatfield, PA) buffered with 0.1*M* sodium phosphate, using a 27-gauge hypodermic needle. (A commonly used fixation in 3% EM grade glutaraldehyde buffered with 0.1*M* sodium cacodylate buffer results in a significant loss of lysosomal glycogen.) Clamps suitable for mouse muscle can be obtained from Fine Science Tools, Inc. (Cat. No. 00396-01; North Vancouver, British Columbia, Canada). Ice-cold buffered glutaraldehyde can also be injected into the selected muscle *in situ* so that the muscle is first fixed for at least several minutes in its natural setting.
iii. The clamped muscles are then removed, pinned at rest length (or slightly extended) on a Sylgard-coated plate (preparation of Sylgard-coated plates are described at the end of this protocol), and kept immersed in the same fixative for 2 h. The clamps can be removed once the muscle is pinned.
iv. The muscles are divided into small blocks, rinsed and kept in ice-cold 0.1*M* sodium phosphate buffer overnight.

v. Small blocks of muscle are then cut, osmicated and dehydrated before Epon embedding according to standard procedures.

To preserve microtubules, room temperature fixative should be used, but some glycogen may be lost. EM reveals the presence of large autophagic areas in predominantly Type II-rich muscle derived from $GAA^{-/-}$ mice (Fukuda *et al.*, 2006b).

Preparation of Sylgard dissection plates Sylgard 182 Silicone Elastomer kit; Robert McKeown Company, Branchburg, NJ, Cat. No. 1821.1:

a. Mix 1 vol of curing agent in 10 vol of silicone elastomer in a plastic disposable container; stir slowly to avoid bubbles.
b. Pour the solution into small tissue culture dishes (Cat. No. 353002; Becton Dickinson Labware, Franklin Lakes, NJ); approximately 5 ml of the solution is needed for one dish.
c. The plates are then allowed to sit un-covered for 48 h for the Sylgard to polymerize.
d. Note, that if needed (for example, for culturing of live single muscle fibers) the procedure can be done under sterile conditions in a tissue culture hood.

Specific techniques

Autophagy in skeletal muscle can be detected by immunostaining of single fixed muscle fibers. The fixation for immunostaining is quite different from that for EM; the protocol was originally described by Ploug *et al.* (1998) and Ralston *et al.* (1999).

Muscle fixation

Reagents

1. 2% paraformaldehyde in 0.1 *M* sodium phosphate buffer:
   a. 4% paraformaldehyde: mix 10ml of 16% paraformaldehyde (Cat. No. 15710; Electron Microscopy Sciences, Fort Washington, PA) and 30 ml of H_20.
   b. 0.2 *M* phosphate buffer, pH 7.2: mix 13.6 ml 1 *M* Na_2HPO4 and 6.32 ml 1 *M* NaH_2PO4, add H_20 up to 100 ml.
   c. Mix equal volumes of 4% paraformaldehyde and 0.2 *M* phosphate buffer, pH 7.2.
2. Sylgard Silicone Elastomer kit (see above).
3. Methanol (stored at −20 °C).
4. PBS (1X).

Procedures

1. Muscle samples are removed immediately after sacrifice from wild-type and $GAA^{-/-}$ mice, pinned to a Sylgard-coated dish, and submerged in 2%

paraformaldehyde in 0.1 *M* phosphate buffer for 1 h for fixation at room temperature.
2. Samples are gently blotted using a paper towel, rinsed in PBS, and cleaned of any extraneous material under a dissecting microscope.
3. After cleaning each sample, muscles are blotted again, transferred into tubes containing cold methanol, and placed at −20 °C for 6 min. We recommend using a 24-well plate when multiple muscles from several mice are analyzed.
4. Samples are again rinsed with PBS; make sure that the muscle is fully submerged in PBS.
5. Samples may be used immediately or stored. For storage, samples are placed in 50% glycerol at 4 °C for 1 day and then moved to −20 °C. For immediate use, samples are placed in PBS at 4 °C.

The methanol step is important for immunostaining with some antibodies (for example, anti-alpha-tubulin for staining of the microtubules). However, LAMP and LC3 immunostaining works well on single muscle fibers fixed in paraformaldehyde alone.

Isolation of single muscle fibers and immunostaining

Reagents

1. 4% saponin stock solution (100x); keep at −20 °C in aliquots; once defrosted the solution is stable for approximately 1 week at 4 °C.
2. Sylgard-coated plates.
3. 10% Triton X-100 in PBS stock solution.
4. Vector M.O.M. Immunodetection Kit (Vector Laboratories, Burlingame, CA). This kit is specifically designed for immunodetection using mouse monoclonal or polyclonal primary antibodies on mouse tissues (including muscle). We routinely use the reagents contained in the kit (blocking buffer and diluent for primary and secondary antibodies) for all immunostaining of single muscle fibers.
5. Hoechst 33342 (Cat. No. H21492; Invitrogen, Carlsbad, CA). The solid dye may be dissolved in PBS to make concentrated stock solutions of 0.1 mg/ml. Stock solutions may be stored frozen, protected from light.

Procedures

1. Fixed muscle stored at −20 °C in a 50% solution of glycerol in PBS must be gradually transitioned to lesser concentrations of glycerol (first 25% glycerol then 12% glycerol for at least 10 min each). When transitioning and for all other steps, muscle bundles should be held at the ends with tweezers.
2. Muscle bundles are placed in a puddle of 0.04% saponin in PBS on a Sylgard-coated plate under a dissecting microscope. Fibers are then

gently pulled apart from muscle bundles with the finest available tweezers (we use tweezers from Dumont, Switzerland, Cat. No. RS-4905). Make sure the fibers stay wet at all times.

3. Once isolated, fibers are transferred to a well of a 24-well plate (approximately 20 fibers per well) containing blocking solution (prepared according to the manufacturer's instructions). We add 0.2% Triton X-100 to the blocking solution to make the myofiber membranes more permeable. The effect of Triton X-100 is irreversible, so there is no need to have it in any subsequent solutions. The fibers tend to stick to the tweezers, so be wary of that when placing them in the well.
4. Fibers are incubated in the blocking solution for at least an hour.
5. Under the microscope, the blocking solution is removed from the wells using a 200-μL pipette.
6. 500 μL of diluent solution (M.O.M. kit) is added immediately.
7. Fibers are incubated at 4 °C on a shaker overnight with primary antibody in diluent (dilution is antibody-specific). More than one antibody may be used.
8. Fibers are washed twice for 10–15 min in PBS on a shaker.
9. PBS is removed and 500 μL of diluent are added to each well. Samples are allowed to incubate in secondary antibody and diluent at room temperature for 2 h. The fibers should be protected from light to avoid a loss of fluorescence due to photobleaching.
10. Fibers are washed twice with PBS for 10–15 min. Hoechst staining for nuclei can be added to the first wash to a final concentration of 0.5 μg/ml (a 200x dilution of the stock).
11. Fibers are mounted on a slide in Vectashield mounting medium (Vector Laboratories, Burlingame, CA; catalog No. H-1000). Using the microscope to view the well, fibers are transferred to a drop of Vectashield on a slide. The slide is placed under the microscope and fibers are dragged out of the drop and lined up next to each other. A coverslip is gently and evenly placed over the fibers and the Vectashield, so that the Vectashield spreads evenly under the cover slip. Excess Vectashield is removed (remove as much of the mounting medium as possible) and the cover slip is sealed in place with nail polish.
12. Fibers are analyzed by confocal microscopy.

Immunostaining of muscle fibers isolated from $GAA^{-/-}$ mice with LAMP-1 (rat anti-mouse LAMP-1; BD Pharmingen, San Diego, CA; Cat. No. 553792) and LC3 (rabbit anti-LC3; Sigma, Cat. No. L7543) antibodies reveals lysosomal enlargement in all fibers and massive autophagic vacuolization in glycolytic fast muscle. The integrity of late endosomal/lysosomal and autophagosomal vesicles in the affected fibers is lost as the disease progresses in older animals. The autophagic buildup appears to disrupt muscle architecture, indicating that the failure of autophagy contributes to the pathology of Pompe disease along with enlargement of glycogen-filled lysosomes (Fukuda *et al.*, 2006a,b; Raben *et al.*, 2007). Functional consequences of the autophagic failure in skeletal muscle can be evaluated by immunostaining of the fibers with antibodies specific for ubiquitin-protein conjugates [mouse monoclonal FK1 (Cat. No. PW8805) and FK2 (Cat. No. PW8810) antibodies; BIOMOL International, L.P., Philadelphia, PA)] as well as for p62/SQSTM1 (mouse monoclonal, Santa Cruz Biotechnology, Inc., Santa Cruz, Cat. No. sc-28359).

3.2. Defects in Transmembrane Proteins

3.2.1. Niemann-Pick type C (NPC)

NPC is a fatal early-onset progressive neurodegenerative disorder that mainly affects children. The vast majority of cases are caused by mutations in the *NPC1* gene (Carstea *et al.*, 1997), which encodes a membrane-spanning protein that functions in late endosomes to promote lipid sorting and vesicular trafficking. The remaining cases are due to mutations in the *NPC2* gene (Naureckiene *et al.*, 2000), which encodes a cholesterol-binding late endosomal lumenal protein. NPC is an example of a disease in which both the accumulation of undegraded compounds (cholesterol and other lipids) and the deficiency of metabolic products normally derived from the stored material, such as oxysterols, play a key role in the pathogenesis (Frolov *et al.*, 2003; Walkley, 2007). The major pathology occurs in liver, spleen, and the CNS.

Model systems

Two spontaneous mouse models of NPC are available: C57BLKS/J *spm* (Miyawaki *et al.*, 1982) and BALB/c $npc1^{nih}$ (Pentchev *et al.*, 1980). Both contain mutations in the *NPC1* gene and show progressive loss of cerebellar Purkinje cells (PC) and early decline in glia in the corpus callosum (Loftus *et al.*, 1997). Autophagy was studied in a different NPC mouse model, termed aggregation chimeras, which was generated by fusing the embryos of NPC KO mice and transgenic mice with ubiquitous GFP expression. In these chimeric NPC mutant mice, wild-type cells were GFP positive, whereas mutant NPC cells were not. The contribution from wild-type cells ranged from 11–61%. The ongoing death of mutant NPC cells surrounded by wild-type cells established the cell-autonomous nature of neuronal loss, pointing to the involvement of intracellular pathways, such as autophagy (Ko *et al.*, 2005). Human fibroblasts from patients with the disease have also been used (Pacheco *et al.*, 2007). The importance of choosing an adequate model system was emphasized by studies in NPC-deficient mutant CT60 Chinese hamster ovary cells. Although these cells were capable of responding to autophagic stimuli, they did not faithfully replicate the disease pathology in terms of autophagic abnormalities (Ko *et al.*, 2005).

General techniques

An increase in autophagy in chimeric mice and primary human fibroblasts was shown by ultrastructural detection of autophagic vacuoles, immunoblotting and immunofluorescence analysis of LC3, measurement of protein degradation, and analysis of the mTOR signaling pathway (Bi and Liao 2007; Ko *et al.*, 2005; Pacheco *et al.*, 2007). Numerous inclusion bodies and membranous vacuoles with double-membranes or multilamellar electron-dense material were observed in the affected brain by EM (Bi and Liao 2007). The fact that LC3-II levels in human fibroblasts are increased by treatment with lysosomal protease inhibitors E64D and pepstatin A indicates that autophagosomal-lysosomal fusion is occurring (Pacheco and Lieberman, 2007; Pacheco *et al.*, 2007). The progression of neuronal degeneration was associated with a dramatic increase in LC3-II levels in brains of chimeric mice (Ko *et al.*, 2005), enhanced protein turnover in NPC fibroblasts, and upregulation of Beclin1 (but no change in activity of Akt, mTOR, or p70 S6K) in mouse tissues and human fibroblasts (Pacheco and Lieberman, 2007; Pacheco *et al.*, 2007). SiRNA down-regulation of Beclin 1 reversed the increased protein turnover in NPC-deficient cells (Pacheco and Lieberman, 2007; Pacheco *et al.*, 2007). LC3-labeled autophagosomes accumulated within filipin-labeled cholesterol clusters inside Purkinje cells (Bi and Liao, 2007).

Specific techniques

Particular aspects of this pathogenetically complex disease are addressed by a number of tailored approaches. These approaches include:

a. Identification of the substrate responsible for autophagic induction by pharmacological means. This is done by exposing the NPC1-deficient cells to *N*-butyl-deoxygalactonojirimycin (*N*B-DGJ, Toronto Research Chemicals, Cat. No. B690500), a compound used in substrate deprivation therapy for glycosphingolipid disorders, and by exposing control fibroblasts to the U18666A (Sigma, Cat. No. U3633), a compound known to inhibit cholesterol transport and to induce an accumulation of unesterified cholesterol (reviewed in (Koh and Cheung, 2006)). A decrease in glycosphingolipid level in the NPC1-deficient fibroblasts by treatment with *N*B-DGJ (Vruchte *et al.*, 2004) did not affect the levels of LC3 or Beclin 1. In contrast, U18666A, increased both LC3 and Beclin 1 in control fibroblasts, indicating that elevated autophagy in NPC cells is a result of altered lipid trafficking and not glycosphingolipid accumulation (Pacheco and Lieberman, 2007; Pacheco *et al.*, 2007).

b. Investigation of the role of Beclin 1 in the induction of autophagy by analyzing a panel of primary fibroblasts derived from patients with several sphingolipid disorders. An increase in Beclin 1 is found in the diseases with abnormal trafficking of sphingolipids (NPC1, NPC2, and Sandhoff disease, a deficiency of both lysosomal hexosaminidase A and B), but not in Gaucher cells, which have proper trafficking (Pacheco and Lieberman, 2007; Pacheco *et al.*, 2007).
c. Analysis of cathepsin D and B levels in NPC1 KO mice. Western blot analysis, immunohistochemistry, and enzymatic assays revealed increased levels of cathepsin D and B throughout the brains of NPC1 KO mice with the highest increase in the regions that exhibit early and profound neurodegeneration. Immunoblotting of fractionated brain homogenates revealed high levels of ubiquitinated proteins, cathepsin D and rab7 (a protein involved in vesicle maturation) in the late endosomal/lysosomal fraction (Bi and Liao, 2007; Gutierrez *et al.*, 2004; Jager *et al.*, 2004; Liao *et al.*, 2007).

Thus, unlike other LSDs, NPC is characterized not by a block but rather an induction of autophagy, which is associated with increased levels of Beclin 1 expression.

3.2.2. Neuronal ceroid lipofuscinosis (Batten disease)

Batten disease belongs to a group of neurodegenerative disorders called the neuronal ceroid lipofuscinoses (NCLs), the most common causes of neurodegeneration among children. There are at least seven known genes that, when mutated, cause human NCL, including Cathepsin D (CD) (Mole, 2004; Shacka and Roth 2007; Siintola *et al.*, 2006; Steinfeld *et al.*, 2006). A major pathological hallmark is the accumulation of lipofuscin and ceroid, a lipofuscin-like lipopigment, in the lysosomes of neurons and other cell types. Most of the NCLs result in the lysosomal accumulation of subunit *c* of the mitochondrial ATP synthase, which requires cathepsin D for turnover in the lysosome (Ezaki *et al.*, 2000). Mutations in the *CLN3* gene encoding battenin cause juvenile onset NCL (JNCL, Batten disease), which is the most common among NCLs. Battenin is a late endosomal/lysosomal multi-pass transmembrane protein implicated in membrane trafficking and mitochondrial function. The majority of JNCL patients share a founder mutation: a deletion of exons 7–8 in *CLN3*. Clinical manifestations include visual failure, seizures, and progressive physical and mental decline associated with massive loss of cortical neurons.

Model systems

Autophagy was studied in cathepsin D knockout and cathepsin B and L double-knockout mice (Koike *et al.*, 2005), which were generated from pre-existing heterozygous lines. The neuronal pathology in mouse brains near the terminal stage (less than one month) is highly reminiscent of NCL pathology: progressive accumulation of undigested lipopigment and proteins including subunit *c* of mitochondrial ATP synthase in the affected neurons (Koike *et al.*, 2000, 2005). Transgenic GFP-LC3 mice on cathepsin D knockout background were also developed (Koike *et al.*, 2005). A knock-in mouse model (*Cln3*$^{\Delta ex7/8}$ mice) has also been generated, which replicates the common JNCL mutation in the murine *Cln3* gene (Cotman *et al.*, 2002).

An immortalized cerebellar cell culture model (Cb*Cln3*$^{\Delta ex7/8}$ cells) has been derived from the *Cln3*$^{\Delta ex7/8}$ mice (Fossale *et al.*, 2004). These cells express low levels of mutant battenin and can differentiate into MAP-2 and NeuN-positive, neuron-like cells. At sub-confluent growth, the cells exhibit abnormalities in cathepsin D transport and processing, altered lysosomal size and distribution, abnormally elongated mitochondria, reduced endocytosis, reduction in cellular ATP levels, and decreased survival following oxidative stress. Only when aged at confluency do Cb*Cln3*$^{\Delta ex7/8}$ cells accumulate ATP synthase subunit *c*.

General techniques

For analysis of cathepsin-deficient mice (Koike *et al.*, 2005): Immunoblot revealed a gradual age-dependent increase in LC3-II in brain, and immunohistochemistry showed intense granular staining of LC3 in cortical neurons, as well as strong autofluorescence and immunoreactivity for subunit *c* in neurons and microglia. Double staining with LAMP-1 and LC3 showed some co-localization, but a considerable number of LC3-positive structures were free from LAMP-1, suggesting a defect in autophagosomal maturation. A combination of EM and morphometry was used to evaluate the volume density of autophagosome-like structures, including various types of AVs, dense bodies, granular osmiophilic deposit (GROD)-like inclusions, and fingerprint profiles.

For analysis of knock-in Cln3$^{\Delta ex7/8}$*mice* (Cao *et al.*, 2006): Increased autophagy in brains of homozygous *Cln3*$^{\Delta ex7/8}$ mice, particularly in regions where ATP synthase subunit *c* accumulates, was shown by immunohistochemistry with anti-LC3 antibody. Western analysis revealed elevated LC3-II/LC3-I ratios in brain from these mice. Reduced levels of phospho-mTOR and phospho-p70 S6 kinase were detected in brains of homozygous *Cln3*$^{\Delta ex7/8}$ mice.

For analysis of CbCln3$^{\Delta ex7/8}$ *cells* (Cao *et al.*, 2006): As in brain, elevated LC3-II/LC3-I ratios and reduced levels of phospho-mTOR and phospho-p70 S6 kinase were detected in the mutant cells. Cb*Cln3*$^{\Delta ex7/8}$ cells accumulate subunit *c* only when grown at confluency (as stated above) or when treated with rapamycin (Cao *et al.*, 2006). 3-MA treatment (5–10 m*M*) resulted in significant cell loss in both wild type and Cb*Cln3*$^{\Delta ex7/8}$

cells; the effect was more pronounced in the mutant cells. 3-MA treatment produced similar results in JNCL patient lymphoblasts.

Specific techniques

Several specific methods were used to demonstrate the disruption of autophagy in knock-in *Cln3*$^{\Delta ex7/8}$ mice and Cb*Cln3*$^{\Delta ex7/8}$ cells (Cao *et al.*, 2006).

a. Fractions corresponding to autophagosomes, autophagolysosomes, and lysosomes are isolated from liver of *Cln3*$^{\Delta ex7/8}$ and wild-type mice (see previously). These fractions are then analyzed by western blotting and transmission EM. These methods reveal that the autophagosomal and lysosomal fractions from *Cln3*$^{\Delta ex7/8}$ liver contain a significant amount of undigested material including multilamellar membranes and fingerprint-like storage material, neither of which is seen in control livers. In addition, the recovery of subunit *c* is significantly enhanced in the light AV fraction and in the lysosomal fraction of *Cln3*$^{\Delta ex7/8}$ liver. Increases in LC3-II, LAMP1, and cathepsin D are also detected in the light AV fraction from *Cln3*$^{\Delta ex7/8}$ liver.
b. A combined endocytic uptake and LysoTracker (labels lysosomes and autolysosomes) staining assay was performed in wild-type and *Cln3*$^{\Delta ex7/8}$ cells. Two endocytic dyes were used: the lysine-fixable dextran BODIPY FL (Cat. No. D7168) or dextran Alexa Fluor 488 (Cat. No. D22910) (both from Molecular Probes/Invitrogen). The procedure involves incubation of the cells (seeded at a density of 3–5 $\times$ 10^4 /well in 4-well chamber slides in DMEM medium supplemented with 10% fetal bovine serum and 24 m*M* KCl) with both Lysotracker and the dextran dye (1 mg/ml) at 33 °C for 20–50 min. The labeled cells are then placed on ice, washed for 10 min in ice-cold dye-free medium, fixed in 4% formaldehyde in PBS for 20 min on ice, and analyzed by confocal microscopy (Cao *et al.*, 2006; Fossale *et al.*, 2004).

Stimulation of autophagy with rapamycin (100 n*M*, 4 h) in these cells up-regulated both endocytic uptake and Lisotracker stain and promoted their co-localization in both wild-type and *Cln3*$^{\Delta ex7/8}$ cells, emphasizing the communication between the autophagic and endocytic pathways. Co-localization, however, was much lower in *Cln3*$^{\Delta ex7/8}$ cells, suggesting impaired trafficking between the endocytic and autophagosomal-lysosomal pathways (Cao *et al.*, 2006).

These experiments led to the proposal of a hypothetical model of the disruption of autophagy in NCL. According to this model, battenin resides on the membranes of late endosomes, lysosomes, and autophagosomes, where it plays a role in autophagosomal maturation and vesicular fusion. Thus, defective battenin would result in nutrient deficiency, up-regulation of autophagy, and accumulation of undegraded substrates, including subunit *c* of ATP synthase. The upregulation of autophagy, however, is unlikely to rescue the cells, since autophagosomal maturation is profoundly impaired (Cao *et al.*, 2006).

3.2.3. Mucolipidosis (ML)

Mucolipidosis type IV (MLIV) belongs to a group of lysosomal storage diseases characterized by accumulation of membranous lipid inclusions in lysosomes (Bach 2001). The disease is caused by mutations in the gene encoding mucolipin 1 (TRP-ML1), a member of the TRP family of ion channels. The pathological hallmark is neuromotor retardation. TRP-ML1 is ubiquitously expressed, and the degenerative processes are not limited to the brain but are also documented in cornea, retina, skeletal muscles, pituitary gland, and other tissues. The exact role of TRP-ML1 is still unclear. The protein is thought to be involved in vesicle fusion in the endocytic pathway, and recently TRP-ML1 was shown to control lysosomal pH (Soyombo *et al.*, 2006).

Model systems

Cultured skin fibroblast from MLIV patients have been used to investigate the mitochondrial-autophagosomal axis in mucolipidoses. Heterozygous (phenotypically normal) fibroblasts from patients' parents served as controls along with MLII and MLIII fibroblasts. In addition to their ready availability, fibroblasts are an attractive model because corneal opacification is a common clinical manifestation of ML, and corneal keratinocytes are similar to fibroblasts in many respects (Jennings, Jr. *et al.*, 2006). *MCOLN1* knockout mice have been recently created, which replicate human MLIV. These mice may be used to study autophagy in the future (Venugopal *et al.*, 2007).

Specific approaches

a. Analysis of mitochondrial appearance in MLIV skin fibroblasts. Fibroblasts from a MLIV patient and a heterozygous control were loaded with MitoTracker Red (Cat. No. M-22425) (1 μm) or 123-rhodamine (Cat. No. R-22420) (1 μm) (both from Molecular Probes/Invitrogen) according to standard procedures. Confocal microscopy revealed mitochondrial fragmentation in MLIV fibroblasts. The significance of this finding is underlined by the fact that the same results were obtained in MLII, MLIII, and NCL2 fibroblasts (Jennings *et al.*, 2006).
b. Pharmacological inhibition of autophagosomal or lysosomal function in control fibroblasts by treatment with bafilomycin (an inhibitor of the lysosomal H^+ pump), 3-MA, nigericin (an H^+ ionophore), or ammonium chloride (a lysosome alkalinizing agent). These compounds induce mitochondrial fragmentation in control cells. By extension, this suggests that in ML and perhaps other LSDs, a block in the lysosomal/

autophagosomal pathway leads to accumulation of fragmented mitochondria (Jennings *et al.*, 2006).

c. Measurement of mitochondrial Ca^{2+} buffering capacity. MLIV and control fibroblasts are treated with bradykinin (Bk), a Ca^{2+}-mobilizing agonist, to increase the level of cytoplasmic Ca^{2+}. A fraction of the cytoplasmic Ca^{2+} is taken up by functional mitochondria. By exposing cells to FCCP (the mitochondrial uncoupler, which collapses mitochondrial membrane potential and causes mitochondria to release Ca^{2+}), it is possible to quantify the uptake. Cytoplasmic Ca^{2+} is then measured by incubation of the cells with Fura-2AM, a dye that undergoes a shift in excitation maximum upon binding calcium (Jennings, Jr. *et al.*, 2006; Soyombo *et al.*, 2006). Unlike in wild-type cells, no Ca^{2+} release was seen in MLIV or III cells, indicating impaired mitochondrial buffering capacity. Consistent with these data, Rhod2, a fluorescent marker for mitochondrial Ca^{2+}, accumulated in control but not MLIV fibroblasts upon bradykinin treatment. FCCP-mediated Ca^{2+} release was eliminated by pretreatment of unaffected cells with 3-MA or bafilomycin, suggesting that accumulation of mitochondria with reduced Ca^{2+} buffering capacity in ML cells is due to defects in the lysosomal/autophagosomal system (Jennings *et al.*, 2006).
d. Analysis of the susceptibility of MLIV fibroblasts to Ca^{2+}-induced apoptosis. Mitochondrial buffering is thought to play a protective role against Ca^{2+}-mediated cytochrome *c* release and subsequent apoptosis. Treatment of the MLIV and control fibroblasts with bradykinin (1 μM for 1-3 hours) induced apoptosis in a greater fraction of mutant than control cells (Jennings *et al.*, 2006). Thus, impaired lysosomal function in MLIV affects autophagic degradation of mitochondria, which in turn results in accumulation of abnormal mitochondria with reduced Ca^{2+} buffering capacity, leaving cells more susceptible to apoptosis.

3.2.4. Danon disease

Danon disease is an X-linked disorder caused by mutations in the gene encoding lysosome-associated membrane protein 2, LAMP-2 (Nishino *et al.*, 2000). The major pathological features of the disease are vacuolar myopathy and cardiomyopathy. Originally the disease was described as a glycogen storage disease similar to Pompe syndrome, but with normal acid alpha-glucosidase activity (Danon *et al.*, 1981). However, it later became clear that not all patients accumulate lysosomal glycogen (Nishino, 2003; Nishino *et al.*, 2000). Severe cardiomyopathy is a feature in all affected patients, while skeletal muscle involvement and mental retardation are seen in most (Yang and Vatta, 2007).

Model systems

LAMP-2 knockout mice replicate some of the major pathological findings in patients with Danon disease, including the accumulation of autophagic vacuoles in cardiac and skeletal muscle. In these mice, however, neutrophilic leukocytes, hepatocytes, and acinar gland cells are also affected, which is not the case in patients (Tanaka *et al.*, 2000).

Hepatocytes (Tanaka *et al.*, 2000; Eskelinen *et al.*, 2002) and fibroblasts (Eskelinen *et al.*, 2004) derived from LAMP-2 knockout mice have also been analyzed. In addition, HeLa cells transfected with LAMP-2 siRNA have been studied (Gonzalez-Polo *et al.*, 2005).

General techniques

EM reveals accumulation of autophagic vacuoles in many tissues from LAMP-$2^{-/-}$ mice, including liver, kidney, pancreas, cardiac muscle and skeletal muscle. Cultured LAMP$2^{-/-}$ hepatocytes show increased density of early autophagic vacuoles relative to control cells (Tanaka *et al.*, 2000). DAMP labeling of cultured hepatocytes indicated that LAMP-2-deficiency did not affect autophagosomal pH (Tanaka *et al.*, 2000). Western blotting revealed lower levels of mature cathepsin D in LAMP-$2^{-/-}$ hepatocytes, and metabolic labeling of these cells showed increased secretion of cathepsin D as compared to control cells (Eskelinen *et al.*, 2002). Metabolic labeling also demonstrated lower baseline and starvation-induced proteolysis in LAMP-$2^{-/-}$ hepatocytes as compared to controls (Tanaka *et al.*, 2000). Knocking down LAMP-2 in HeLa cells with siRNA decreased starvation-induced co-localization of GFP-LC3 and LysoTracker Red, suggesting a role for LAMP-2 in autophagosomal-lysosomal fusion. $DiOC_6$ staining in these cells revealed a loss of mitochondrial membrane potential after starvation (Gonzalez-Polo *et al.*, 2005).

Specific techniques

a. A combination of several traditional methods is used to study the maturation of autophagosomes in LAMP-$2^{-/-}$ hepatocytes. Cells are cultured in starvation medium (without serum and amino acids) for 5 h to induce autophagy, followed by a 1- or 3-h chase in complete medium with 3-MA (which prevents *de novo* autophagosome formation). Cells are harvested before starvation, after starvation, and after chase and analyzed by EM. In wild-type cells, starvation-induced autophagosomes are mostly resolved after a 3-hour chase. In contrast, in LAMP-$2^{-/-}$ cells, autophagosomes persist after a 3-h chase, suggesting a defect in autophagosome maturation and resolution (Eskelinen *et al.*, 2002).
b. Western blotting shows a lower density of cation-dependent mannose-6-phosphate receptor (MPR46, a receptor involved in delivery of

endogenous lysosomal enzymes) in LAMP-2$^{-/-}$ hepatocytes (Eskelinen *et al.*, 2002). Radiolabeling and immunoprecipitation of MPR46 in these cells shows that the receptor has a decreased half-life, which could be restored by inhibition of lysosomal proteases. Immunogold EM of LAMP-2$^{-/-}$ hepatocytes revealed increased amounts of MPR46 within autophagosomes and loss of MPR46 from multivesicular endosomes, which are involved in receptor recycling. Thus, aberrant targeting of MPR46 to the autophagosomal/lysosomal pathway may lead to down-regulation of the receptor in LAMP-2$^{-/-}$ cells (Eskelinen *et al.*, 2002).

c. Hela cells are transfected with LAMP-2 siRNA, mt-dsRed (a mitochondrial marker), and SytVII-GFP (a lysosomal marker). Confocal microscopy of cells reveals decreased co-localization of mitochondrial and lysosomal markers in LAMP-2-deficient cells, indicating decreased mitochondrial turnover (Gonzalez-Polo *et al.*, 2005).

Thus, the role of LAMP-2, previously thought to be limited to protecting the cytoplasm from the contents of the lysosome, is now seen to include lysosomal biogenesis and the late steps of the autophagic pathway (Eskelinen *et al.*, 2004).

4. Conclusion

Lysosomal storage disorders are not new nosologic entities; they were described decades, and in some cases even a century ago. The discovery of the lysosome and the connection between a particular enzyme deficiency and a corresponding storage material came much later in the mid to late part of the last century. These discoveries paved the way for the identification of the genes involved and the mutations found in patients. Exploration of the primary defects of these diseases and the development of enzyme replacement therapy for some of these disorders followed. The new chapter in LSDs is marked by extending research beyond the lysosome and focusing on the secondary events following lysosomal dysfunction. Autophagy is clearly one of the interrelated pathways affected. One consequence of autophagic abnormalities is the impaired degradation of mitochondria. The similar range of pathological manifestations in many LSDs, in particular those with neurological involvement, suggested that a generalized mechanism of degeneration may exist. Autophagy, indeed, appears to be a common factor in these diseases, although the mechanisms by which it contributes to pathology and pathogenesis are variable. The study of autophagy may yet lead to the elucidation of the mechanisms of cellular death and the development of new tailored therapies for these devastating disorders.

REFERENCES

Anderson, R. G., and Orci, L. (1988). A view of acidic intracellular compartments. *J.Cell Biol.* **106,** 539–543.

Bach, G. (2001). Mucolipidosis type IV. *Mol. Genet. Metab.* **73,** 197–203.

Bhattacharyya, R., Gliddon, B., Beccari, T., Hopwood, J. J., and Stanley, P. (2001). A novel missense mutation in lysosomal sulfamidase is the basis of MPS III A in a spontaneous mouse mutant. *Glycobiology* **11,** 99–103.

Bhaumik, M., Muller, V. J., Rozaklis, T., Johnson, L., Dobrenis, K., Bhattacharyya, R., Wurzelmann, S., Finamore, P., Hopwood, J. J., Walkley, S. U., and Stanley, P. (1999). A mouse model for mharidosis type III A (Sanfilippo syndrome). *Glycobiology* **9,** 1389–1396.

Bi, X., and Liao, G. (2007). Autophagic-lysosomal dysfunction and neurodegeneration in Niemann-Pick Type C mice: Lipid starvation or indigestion? *Autophagy* **3,** 646–648.

Cao, Y., Espinola, J. A., Fossale, E., Massey, A. C., Cuervo, A. M., Macdonald, M. E., and Cotman, S. L. (2006). Autophagy is disrupted in a knock-in mouse model of juvenile neuronal ceroid lipofuscinosis. *J. Biol. Chem.* **29,** 20483–20493.

Carstea, E. D., Morris, J. A., Coleman, K. G., Loftus, S. K., Zhang, D., Cummings, C., Gu, J., Rosenfeld, M. A., Pavon, W. J., Krman, D. B., *et al.* (1997). Niemann-Pick C1 disease gene: Homology to mediators of cholesterol homeostasis. *Science* **277,** 228–231.

Corcelle, E., Nebout, M., Bekri, S., Gauthier, N., Hofman, P., Poujeol, P., Fenichel, P., and Mograbi, B. (2006). Disruption of autophagy at the maturation step by the carcinogen lindane is associated with the sustained mitogen-activated protein kinase/extracellular signal-regulated kinase activity. *Cancer Res.* **66,** 6861–6870.

Cosma, M. P., Pepe, S., Annunziata, I., Newbold, R. F., Grompe, M., Parenti, G., and Ballabio, A. (2003). The multiple sulfatase deficiency gene encodes an essential and limiting factor for the activity of sulfatases. *Cell* **113,** 445–456.

Cotman, S. L., Vrbanac, V., Lebel, L. A., Lee, R. L., Johnson, R. L., Donahue, L. R., Teed, A. M., Antonellis, K., Bronson, R. T., Lerner, T. J., and Macdonald, M. E. (2002). Cln3 (Delta ex7/8) knock-in mice with the common JNCL mutation exhibit progressive neurologic disease that begins before birth. Hum. *Mol. Genet.* **11,** 2709–2721.

Danon, M. J., Oh, S. J., DiMauro, S., Manaligod, J. R., Eastwood, A., Naidu, S., and Schliselfeld, L. H. (1981). Lysosomal glycogen storage disease with normal acid maltase. *Neurology* **31,** 51–57.

de Araujo, M. E. G., Huber, L. A., and Stasyk, T. (2006). Isolation of endocitic organelles by density gradient centrifugation. *In* "2D PAGE: Sample Preparation and Fractionation" (A. Posch, ed.), pp. 317–331. Humana Press, Totowa, NJ.

Dierks, T., Schmidt, B., Borissenko, L. V., Peng, J., Preusser, A., Mariappan, M., and von Figura, K. (2003). Multiple sulfatase deficiency is caused by mutations in the gene encoding the human C(alpha)-formylglycine generating enzyme. *Cell* **113,** 435–444.

Elleder, M., Sokolova, J., and Hrebicek, M. (1997). Follow-up study of subunit c of mitochondrial ATP synthase (SCMAS) in Batten disease and in unrelated lysosomal disorders. *Acta Neuropathol.* **93,** 379–390.

Eskelinen, E. L., Illert, A. L., Tanaka, Y., Schwarzmann, G., Blanz, J., von Figura, K., and Saftig, P. (2002). Role of LAMP-2 in lysosome biogenesis and autophagy. *Mol. Biol. Cell* **13,** 3355–3368.

Eskelinen, E. L., Schmidt, C. K., Neu, S., Willenborg, M., Fuertes, G., Salvador, N., Tanaka, Y., Lullmann-Rauch, R., Hartmann, D., Heeren, J., von Figura, K., Knecht, E., *et al.* (2004). Disturbed cholesterol traffic but normal proteolytic function in LAMP-1/LAMP-2 double-deficient fibroblasts. *Mol. Biol. Cell* **15,** 3132–3145.

Ezaki, J., Takeda-Ezaki, M., and Kominami, E. (2000). Tripeptidyl peptidase I, the late infantile neuronal ceroid lipofuscinosis gene product, initiates the lysosomal degradation of subunit c of ATP synthase. *J. Biochem.* **128,** 509–516.

Fossale, E., Wolf, P., Espinola, J. A., Lubicz-Nawrocka, T., Teed, A. M., Gao, H., Rigamonti, D., Cattaneo, E., Macdonald, M. E., and Cotman, S. L. (2004). Membrane trafficking and mitochondrial abnormalities precede subunit c deposition in a cerebellar cell model of juvenile neuronal ceroid lipofuscinosis. *BMC. Neurosci.* **5,** 57.

Frolov, A., Zielinski, S. E., Crowley, J. R., Dudley-Rucker, N., Schaffer, J. E., and Ory, D. S. (2003). NPC1 and NPC2 regulate cellular cholesterol homeostasis through generation of low density lipoprotein cholesterol-derived oxysterols. *J. Biol. Chem.* **278,** 25517–25525.

Fukuda, T., Ahearn, M., Roberts, A., Mattaliano, R. J., Zaal, K., Ralston, E., Plotz, P. H., and Raben, N. (2006a). Autophagy and mistargeting of therapeutic enzyme in skeletal muscle in pompe disease. *Mol. Ther.* **14,** 831–839.

Fukuda, T., Ewan, L., Bauer, M., Mattaliano, R. J., Zaal, K., Ralston, E., Plotz, P. H., and Raben, N. (2006b). Dysfunction of endocytic and autophagic pathways in a lysosomal storage disease. *Ann. Neurol.* **59,** 700–708.

Gonzalez-Polo, R. A., Boya, P., Pauleau, A. L., Jalil, A., Larochette, N., Souquere, S., Eskelinen, E. L., Pierron, G., Saftig, P., and Kroemer, G. (2005). The apoptosis/autophagy paradox: Autophagic vacuolization before apoptotic death. *Journal of Cell Science* **118,** 3091–3102.

Gutierrez, M. G., Munafo, D. B., Beron, W., and Colombo, M. I. (2004). Rab7 is required for the normal progression of the autophagic pathway in mammalian cells. *J. Cell Sci.* **117,** 2687–2697.

Hara, T., Nakamura, K., Matsui, M., Yamamoto, A., Nakahara, Y., Suzuki-Migishima, R., Yoyama, M., Mishima, K., Saito, I., Okano, H., and Mizushima, N. (2006). Suppression of basal autophagy in neural cells causes neurodegenerative disease in mice. *Nature* **441,** 885–889.

Hodges, B. L., and Cheng, S. H. (2006). Cell and gene-based therapies for the lysosomal storage diseases. *Curr. Gene Ther.* **6,** 227–241.

Hopwood, J. J., and Ballabio, A. (2001). *In* "Multiple Sulfatase Deficiency and the Nature of the Sulfatase family," (C. Scriver, A. Beaudet, D. Valle, and W. Sly, eds.), pp. 3725–3732. McGraw-Hill, New York.

Itoh, M., Matsuda, J., Suzuki, O., Ogura, A., Oshima, A., Tai, T., Suzuki, Y., and Takashima, S. (2001). Development of lysosomal storage in mice with targeted dis ruption of the beta-galactosidase gene: A model of human G(M1)-gangliosidosis. *Brain Dev.* **23,** 379–384.

Jager, S., Bucci, C., Tanida, I., Ueno, T., Kominami, E., Saftig, P., and Eskelinen, E. L. (2004). Role for Rab7 in maturation of late autophagic vacuoles. *J. Cell Sci.* **117,** 4837–4848.

Jennings, J. J. Jr,, Zhu, J. H., Rbaibi, Y., Luo, X., Chu, C. T., and Kiselyov, K. (2006). Mitochondrial aberrations in mucolipidosis Type IV. *J. Biol. Chem.* **281,** 39041–39050.

Kabeya, Y., Mizushima, N., Ueno, T., Yamamoto, A., Kirisako, T., Noda, T., Kominami, E., Ohsumi, Y., and Yoshimori, T. (2000). LC3, a mammalian homologue of yeast Apg8p, is localized in autophagosome membranes after processing. *EMBO J.* **19,** 5720–5728.

Kiselyov, K., Jennigs, J. J. Jr, Rbaibi, Y., and Chu, C. T. (2007). Autophagy, mitochondria and cell death in lysosomal storage diseases. *Autophagy* **3,** 259–262.

Klionsky, D. J., *et al.* (2008). Guidelines for the use and interpretation of assays for monitoring autophagy in higher eukaryotes. *Autophagy* **4,** 151–175.

Ko, D. C., Milenkovic, L., Beier, S. M., Manuel, H., Buchanan, J., and Scott, M. P. (2005). Cell-autonomous death of cerebellar purkinje neurons with autophagy in Niemann-Pick type C disease. *PLoS. Genet.* **1,** 81–95.

Koh, C. H., and Cheung, N. S. (2006). Cellular mechanism of U18666A-mediated apoptosis in cultured murine cortical neurons: Bridging Niemann-Pick disease type C and Alzheimer's disease. *Cell Signal* **18,** 1844–1853.

Koike, M., Nakanishi, H., Saftig, P., Ezaki, J., Isahara, K., Ohsawa, Y., Schulz-Schaeffer, W., Watanabe, T., Waguri, S., Kametaka, S., Shibata, M., Yamamoto, K., *et al.* (2000). Cathepsin D deficiency induces lysosomal storage with ceroid lipofuscin in mouse CNS neurons. *J. Neurosci.* **20,** 6898–6906.

Koike, M., Shibata, M., Waguri, S., Yoshimura, K., Tanida, I., Kominami, E., Gotow, T., Peters, C., von Figura, K., Mizushima, N., Saftig, P., and Uchiyama, Y. (2005). Participation of autophagy in storage of lysosomes in neurons from mouse models of neuronal ceroid-lipofuscinoses (Batten disease). *Am. J. Pathol.* **167,** 1713–1728.

Komatsu, M., Waguri, S., Chiba, T., Murata, S., Iwata, J., Tanida, I., Ueno, T., Koike, M., Uchiyama, Y., Kominami, E., and Tanaka, K. (2006). Loss of autophagy in the central nervous system causes neurodegeneration in mice. *Nature* **441,** 880–884.

Komatsu, M., Waguri, S., Ueno, T., Iwata, J., Murata, S., Tanida, I., Ezaki, J., Mizushima, N., Ohsumi, Y., Uchiyama, Y., Kominami, E., Tanaka, K., *et al.* (2005). Impairment of starvation-induced and constitutive autophagy in Atg7-deficient mice. *J. Cell Biol.* **169,** 425–434.

Kramar, E. A., Lin, B., Lin, C. Y., Arai, A. C., Gall, C. M., and Lynch, G. (2004). A novel mechanism for the facilitation of theta-induced long-term potentiation by brain-derived neurotrophic factor. *J. Neurosci.* **24,** 5151–5161.

Liao, G., Yao, Y., Liu, J., Yu, Z., Cheung, S., Xie, A., Liang, X., and Bi, X. (2007). Cholesterol accumulation is associated with lysosomal dysfunction and autophagic stress in Npc1 -/- mouse brain. *Am. J. Pathol.* **171,** 962–975.

Loftus, S. K., Morris, J. A., Carstea, E. D., Gu, J. Z., Cummings, C., Brown, A., Ellison, J., Ohno, K., Rosenfeld, M. A., Tagle, D. A., Pentchev, P. G., and Pavan, W. J. (1997). Murine model of Niemann-Pick C disease: Mutation in a cholesterol homeostasis gene. *Science* **277,** 232–235.

Lum, J. J., DeBerardinis, R. J., and Thompson, C. B. (2005). Autophagy in metazoans: Cell survival in the land of plenty. *Nat. Rev. Mol. Cell Biol.* **6,** 439–448.

Marzella, L., Ahlberg, J., and Glaumann, H. (1982). Isolation of autophagic vacuoles from rat liver: Morphological and biochemical characterization. *J. Cell Biol.* **93,** 144–154.

Matsuda, J., Suzuki, O., Oshima, A., Ogura, A., Noguchi, Y., Yamamoto, Y., Asano, T., Takimoto, K., Sukegawa, K., Suzuki, Y., and Naiki, M. (1997). Beta-galactosidase-deficient mouse as an animal model for GM1-gangliosidosis. *Glycoconj. J.* **14,** 729–736.

Maxfield, F. R. (1989). Measurement of vacuolar pH and cytoplasmic calcium in living cells using fluorescence microscopy. *Methods Enzymol.* **173,** 745–771.

Miyawaki, S., Mitsuoka, S., Sakiyama, T., and Kitagawa, T. (1982). Sphingomyelinosis, a new mutation in the mouse: A model of Niemann-Pick disease in humans. *J. Hered* **73,** 257–263.

Mole, S. E. (2004). The genetic spectrum of human neuronal ceroid-lipofuscinoses. *Brain Pathol.* **14,** 70–76.

Naureckiene, S., Sleat, D. E., Lackland, H., Fensom, A., Vanier, M. T., Wattiaux, R., Jadot, M., and Lobel, P. (2000). Identification of HE1 as the second gene of Niemann-Pick C disease. *Science* **290,** 2298–2301.

Nishino, I. (2003). Autophagic vacuolar myopathies. *Curr. Neurol. Neurosci. Rep.* **3,** 64–69.

Nishino, I., Fu, J., Tanji, K., Yamada, T., Shimojo, S., Koori, T., Mora, M., Riggs, J. E., Oh, S. J., Koga, Y., *et al.* (2000). Primary LAMP-2 deficiency causes X-linked vacuolar cardiomyopathy and myopathy (Danon disease). *Nature* **406,** 906–910.

Pacheco, C. D., Kunkel, R., and Lieberman, A. P. (2007). Autophagy in Niemann-Pick C disease is dependent upon Beclin-1 and responsive to lipid trafficking defects. *Hum. Mol. Genet.* **16,** 1495–1503.

Pacheco, C. D., and Lieberman, A. P. (2007). Lipid trafficking defects increase Beclin-1 and activate autophagy in Niemann-Pick type C disease. *Autophagy* **3,** 487–489.

Pankiv, S., Clausen, T. H., Lamark, T., Brech, A., Bruun, J. A., Outzen, H., Overvatn, A., Bjorkoy, G., and Johansen, T. (2007). p62/SQSTM1 binds directly to Atg8/LC3 to facilitate degradation of ubiquitinated protein aggregates by autophagy. *J. Biol. Chem.* **282,** 24131–24145.

Pentchev, P. G., Gal, A. E., Booth, A. D., Omodeo-Sale, F., Fouks, J., Neumeyer, B. A., Quirk, J. M., Dawson, G., and Brady, R. O. (1980). A lysosomal storage disorder in mice characterized by a dual deficiency of sphingomyelinase and glucocerebrosidase. Biochim. *Biophys. Acta* **619,** 669–679.

Ploug, T., van Deurs, B., Ai, H., Cushman, S. W., and Ralston, E. (1998). Analysis of GLUT4 distribution in whole skeletal muscle fibers: Identification of distinct storage compartments that are recruited by insulin and muscle contractions. *J. Cell Biol.* **142,** 1429–1446.

Poot, M., Zhang, Y. Z., Kramer, J. A., Wells, K. S., Jones, L. J., Hanzel, D. K., Lugade, A. G., Singer, V. L., and Haugland, R. P. (1996). Analysis of mitochondrial morphology and function with novel fixable fluorescent stains. *J. Histochem. Cytochem.* **44,** 1363–1372.

Raben, N., Danon, M., Gilbert, A. L., Dwivedi, S., Collins, B., Thurberg, B. L., Mattaliano, R. J., Nagaraju, K., and Plotz, P. H. (2003). Enzyme replacement therapy in the mouse model of Pompe disease. *Mol. Genet. Metab.* **80,** 159–169.

Raben, N., Nagaraju, K., Lee, E., Kessler, P., Byrne, B., Lee, L., LaMarca, M., King, C., Ward, J., Sauer, B., and Plotz, P. (1998). Targeted disruption of the acid alpha-glucosidase gene in mice causes an illness with critical features of both infantile and adult human glycogen storage disease type II. *J. Biol. Chem.* **273,** 19086–19092.

Raben, N., Takikita, S., Pittis, M. G., Bembi, B., Marie, S. K.N, Roberts, A., Page, L., Kishnani, P. S., Schoser, B. G.H, Chien, Y..H, Ralston, E., Nagaraju, K., *et al.* (2007). Deconstructing Pompe disease by analyzing single muscle fibers. *Autophagy* **3,** 546–552.

Ralston, E., Lu, Z., and Ploug, T. (1999). The organization of the Golgi complex and microtubules in skeletal muscle is fiber type-dependent. *J. Neurosci.* **19,** 10694–10705.

Settembre, C., Annunziata, I., Spampanato, C., Zarcone, D., Cobellis, G., Nusco, E., Zito, E., Tacchetti, C., Cosma, M. P., and Ballabio, A. (2007). Systemic inflammation and neurodegeneration in a mouse model of multiple sulfatase deficiency. *Proc. Natl. Acad. Sci. USA* **104,** 4506–4511.

Settembre, C., Fraldi, A., Jahreiss, L., Spampanato, C., Venturi, C., Medina, D., de Pablo, R., Tacchetti, C., Rubinsztein, D. C., and Ballabio, A. (2008a). A block of autophagy in lysosomal storage disorders. *Hum. Mol. Genet.* **17,** 119–129.

Settembre, C., Fraldi, A., Rubinsztein, D. C., and Ballabio, A. (2008b). Lysosomal storage diseases as disorders of autophagy. *Autophagy* **4,** 113–114.

Shacka, J. J., and Roth, K. A. (2007). Cathepsin D deficiency and NCL/Batten disease: There's more to death than apoptosis. *Autophagy* **3,** 474–476.

Siintola, E., Partanen, S., Stromme, P., Haapanen, A., Haltia, M., Maehlen, J., Lehesjoki, A. E., and Tyynela, J. (2006). Cathepsin D deficiency underlies congenital human neuronal ceroid-lipofuscinosis. *Brain* **129,** 1438–1445.

Soyombo, A. A., Tjon-Kon-Sang, S., Rbaibi, Y., Bashllari, E., Bisceglia, J., Muallem, S., and Kiselyov, K. (2006). TRP-ML1 regulates lysosomal pH and acidic lysosomal lipid hydrolytic activity. *J. Biol. Chem.* **281,** 7294–7301.

Steinfeld, R., Reinhardt, K., Schreiber, K., Hillebrand, M., Kraetzner, R., Bruck, W., Saftig, P., and Gartner, J. (2006). Cathepsin D deficiency is associated with a human neurodegenerative disorder. *Am. J. Hum. Genet.* **78,** 988–998.

Takamura, A., Higaki, K., Kajimaki, K., Otsuka, S., Ninomiya, H., Matsuda, J., Ohno, K., Suzuki, Y., and Nanba, E. (2008). Enhanced autophagy and mitochondrial aberrations in murine G(M1)-gangliosidosis. Biochem. Biophys. *Res. Commun.*

Tanaka, Y., Guhde, G., Suter, A., Eskelinen, E. L., Hartmann, D., Lullmann-Rauch, R., Janssen, P. M.L, Blanz, J., von Figura, K., and Saftig, P. (2000). Accumulation of autophagic vacuoles and cardiomyopathy in LAMP-2-deficient mice. *Nature* **406,** 902–906.

Tassa, A., Roux, M. P., Attaix, D., and Bechet, D. M. (2003). Class III phosphoinositide 3-kinase–Beclin1 complex mediates the amino acid-dependent regulation of autophagy in C2C12 myotubes. *Biochem. J.* **376,** 577–586.

Terman, A., and Brunk, U. T. (2006). Oxidative stress, accumulation of biologicasl "garbage", and aging. *Antioxid. Redox. Signal.* **8,** 197–204.

Venugopal, B., Browning, M. F., Curcio-Morelli, C., Varro, A., Michaud, N., Nanthakumar, N., Walkley, S. U., Pickel, J., and Slaugenhaupt, S. A. (2007). Neurologic, gastric, and opthalmologic pathologies in a murine model of mucolipidosis type IV. *Am. J. Hum. Genet.* **81,** 1070–1083.

Vruchte, D., Lloyd-Evans, E., Veldman, R. J., Neville, D. C., Dwek, R. A., Platt, F. M., van Blitterswijk, W. J., and Sillence, D. J. (2004). Accumulation of glycosphingolipids in Niemann-Pick C disease disrupts endosomal transport. *J. Biol. Chem.* **279,** 26167–26175.

Walkley, S. U. (2007). Pathogenic mechanisms in lysosomal disease: A reappraisal of the role of the lysosome. *Acta Paediatr.Suppl* **96,** 26–32.

Wenger, DA., Coppola, S., and Liu, SL. (2003). Insights into the diagnosis and treatment of lysosomal storage diseases. *Arch. Neurol.* **60,** 322–328.

Xie, Z., and Klionsky, D. J. (2007). Autophagosome formation: Core machinery and adaptations. *Nat. Cell Biol.* **9,** 1102–1109.

Yang, Z., and Vatta, M. (2007). Danon disease as a cause of autophagic vacuolar myopathy. Congenit. *Heart Dis.* **2,** 404–409.

Yu, W. H., Cuervo, A. M., Kumar, A., Peterhoff, C. M., Schmidt, S. D., Lee, J. H., Mohan, P. S., Mercken, M., Farmery, M. R., Tjernberg, L. O., Jiang, Y., Duff, K., *et al.* (2005). Macroautophagy: A novel -amyloid peptide-generating pathway activated in Alzheimer's disease. *J. Cell Biol.* **171,** 87–98.

Author Index

B

D

L

M

P

U

V

Z

Subject Index

A

B

D

E

P

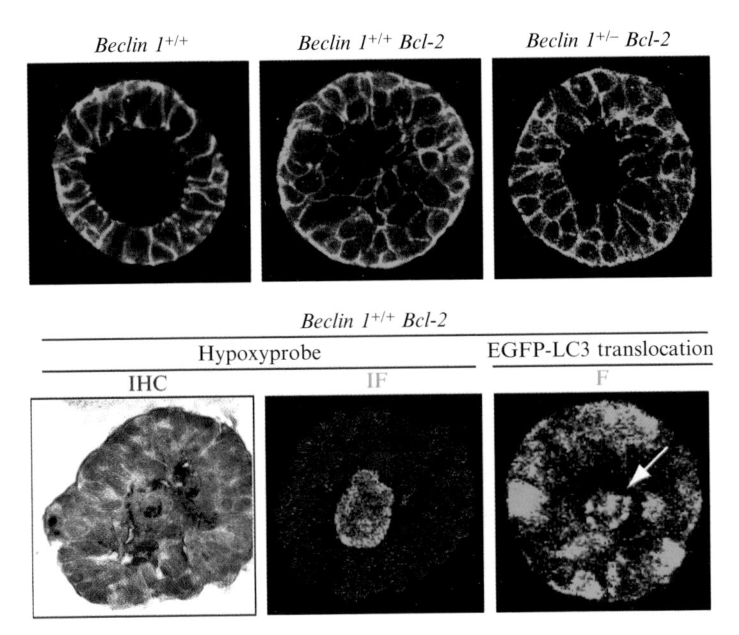

Robin Mathew *et al.*, Figure 4.2 Please refer to the legend in the text.

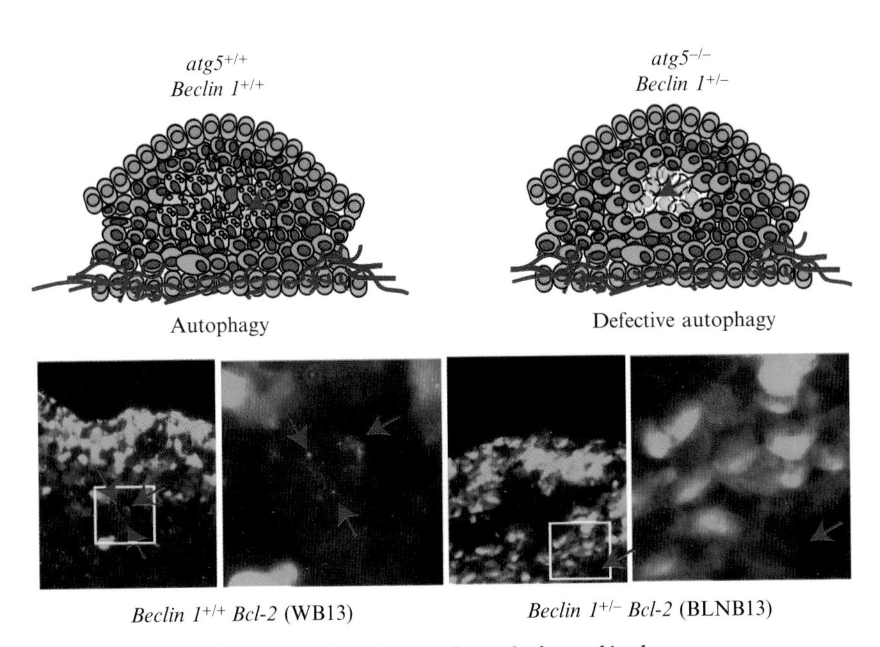

Robin Mathew *et al.*, Figure 4.3 Please refer to the legend in the text.

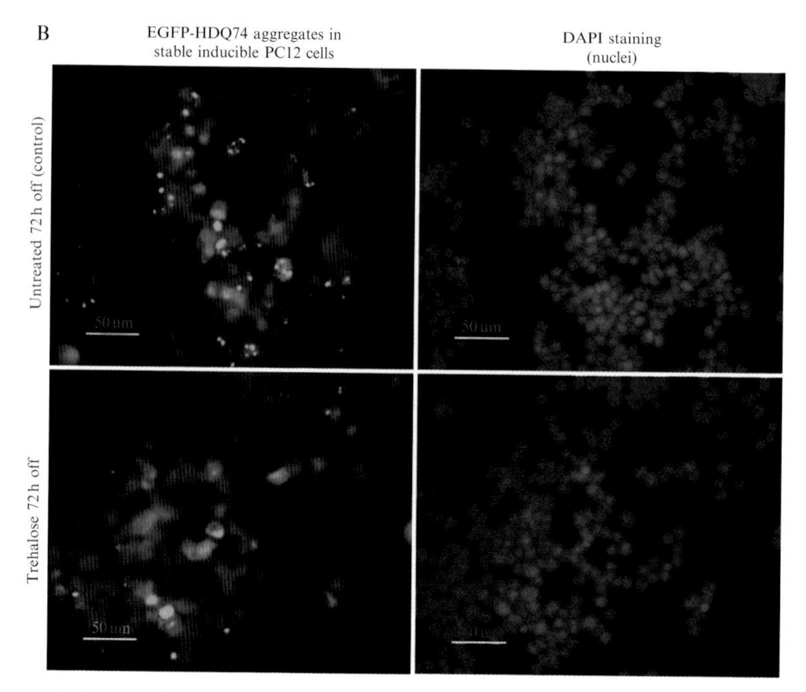

Sovan Sarkar *et al.*, Figure 5.2 Please refer to the legend in the text.

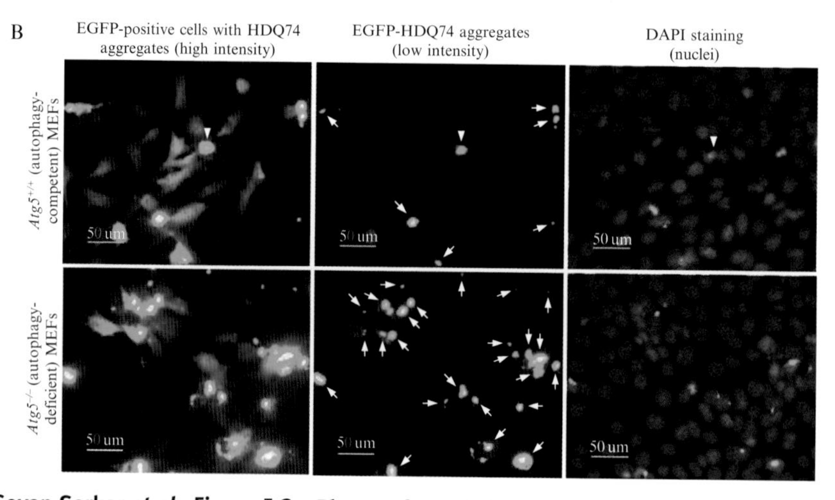

Sovan Sarkar *et al.*, Figure 5.3 Please refer to the legend in the text.

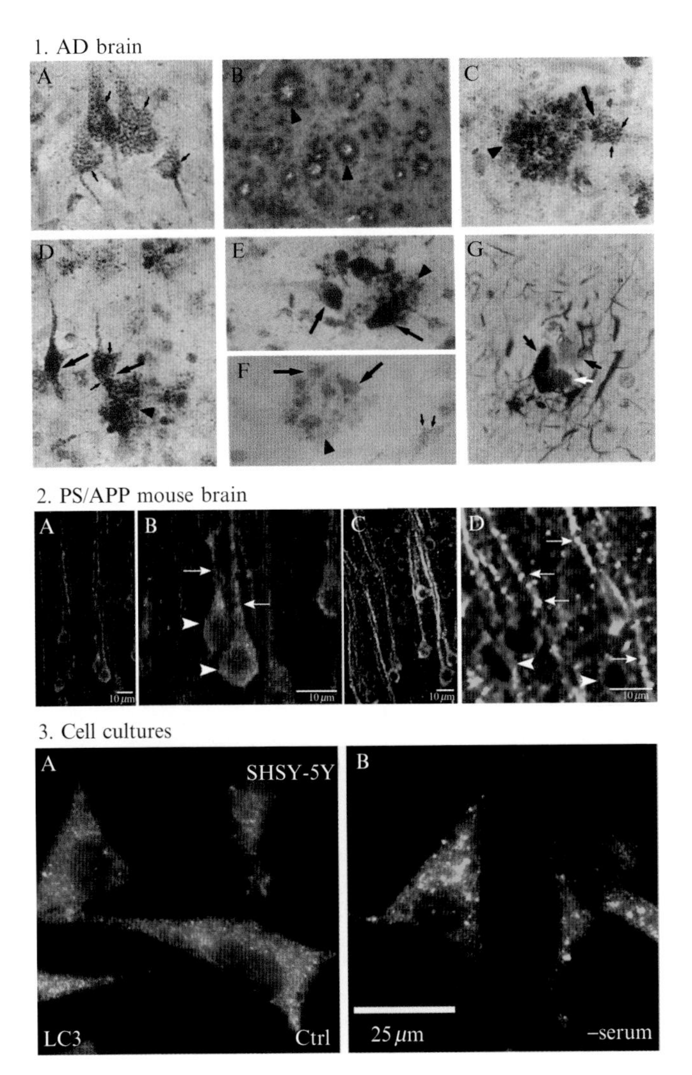

Dun-Sheng Yang *et al.*, Figure 6.1 Please refer to the legend in the text.

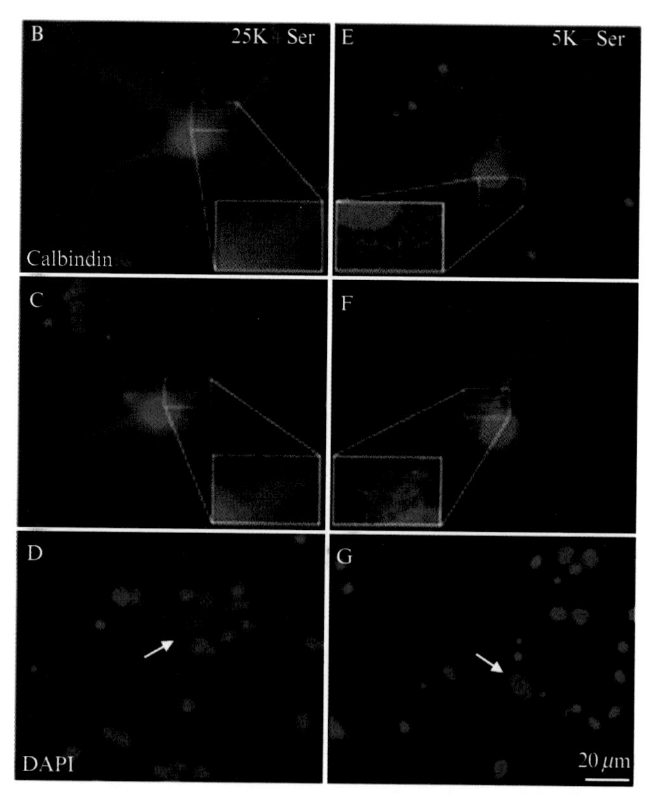

Mona Bains and Kim A. Heindenreich, Figure 7.1 Please refer to the legend in the text.

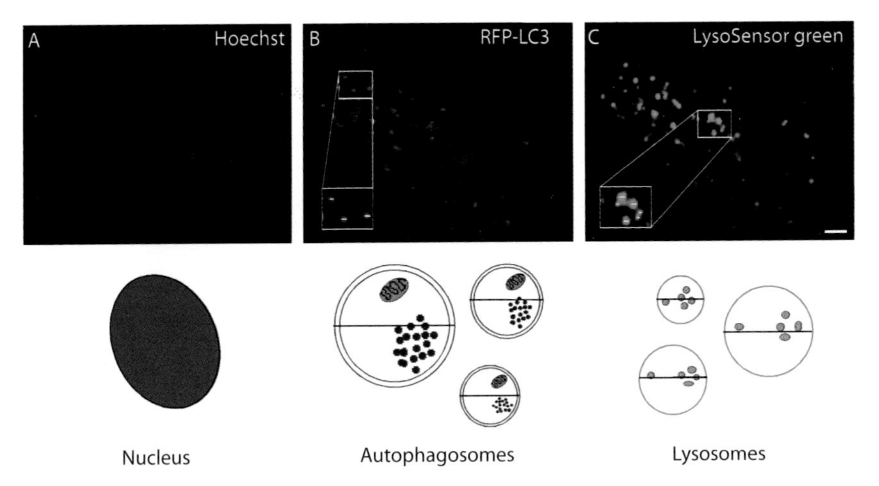

Mona Bains and Kim A. Heindenreich, Figure 7.2 Please refer to the legend in the text.

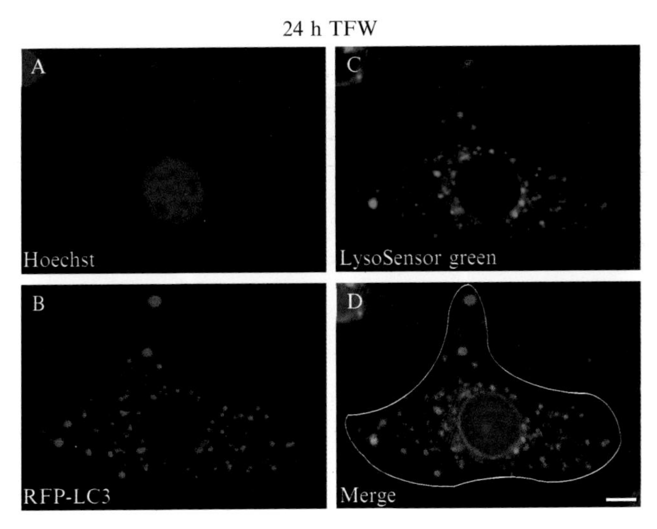

Mona Bains and Kim A. Heindenreich, Figure 7.4 Please refer to the legend in the text.

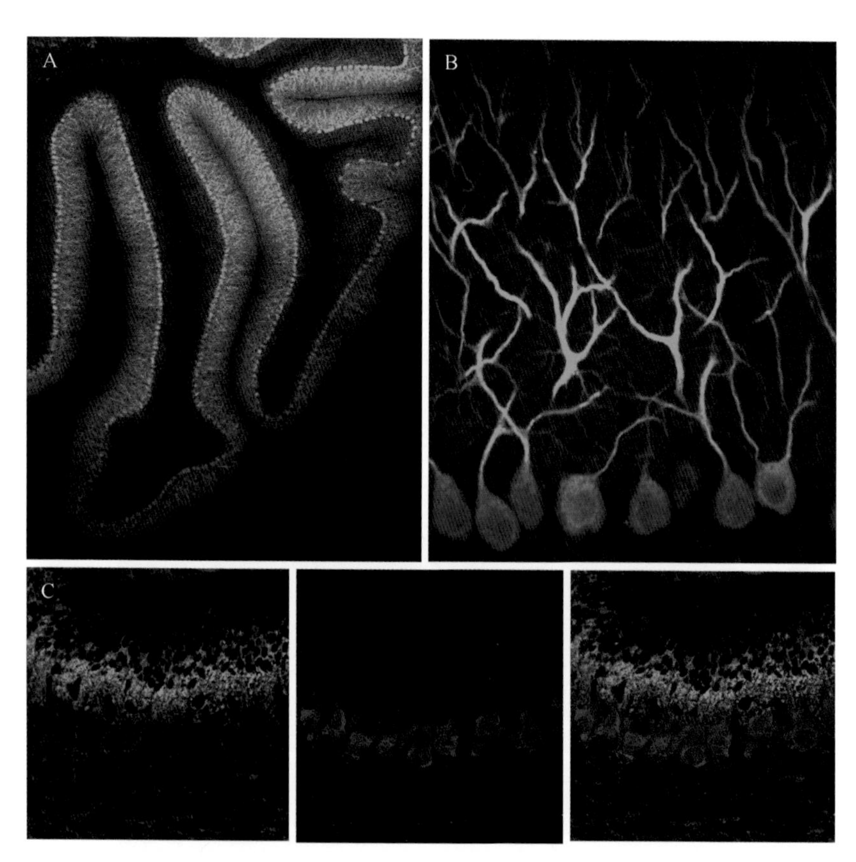

Zhenyu Yue *et al.*, Figure 8.1 Please refer to the legend in the text.

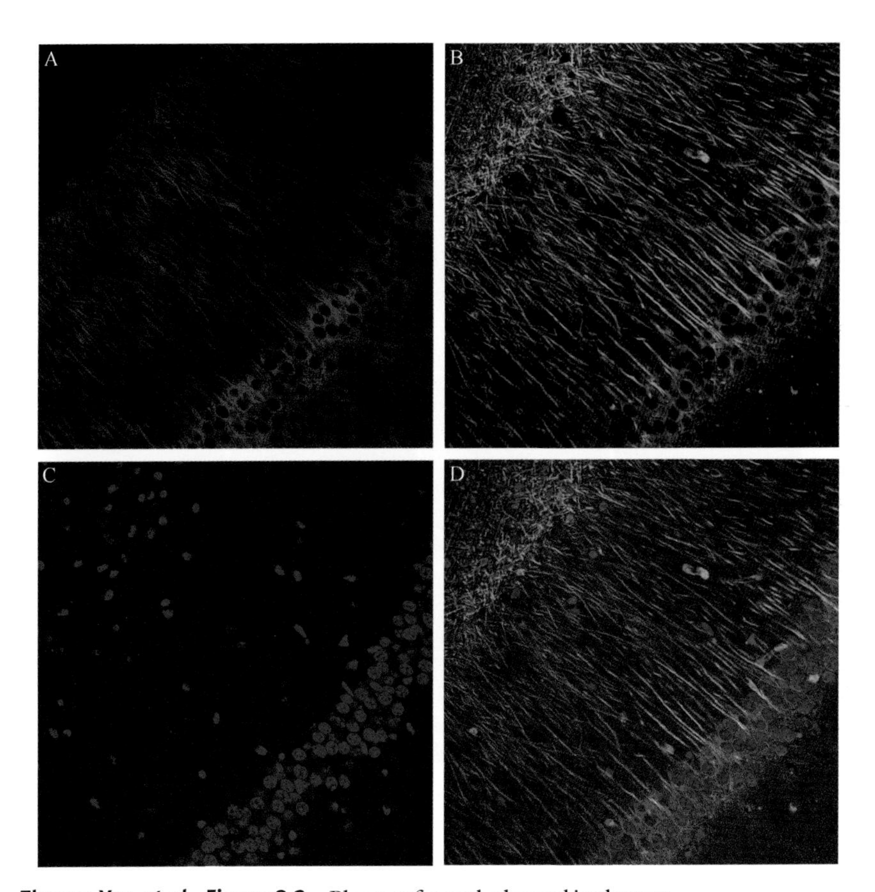

Zhenyu Yue *et al.*, Figure 8.2 Please refer to the legend in the text.

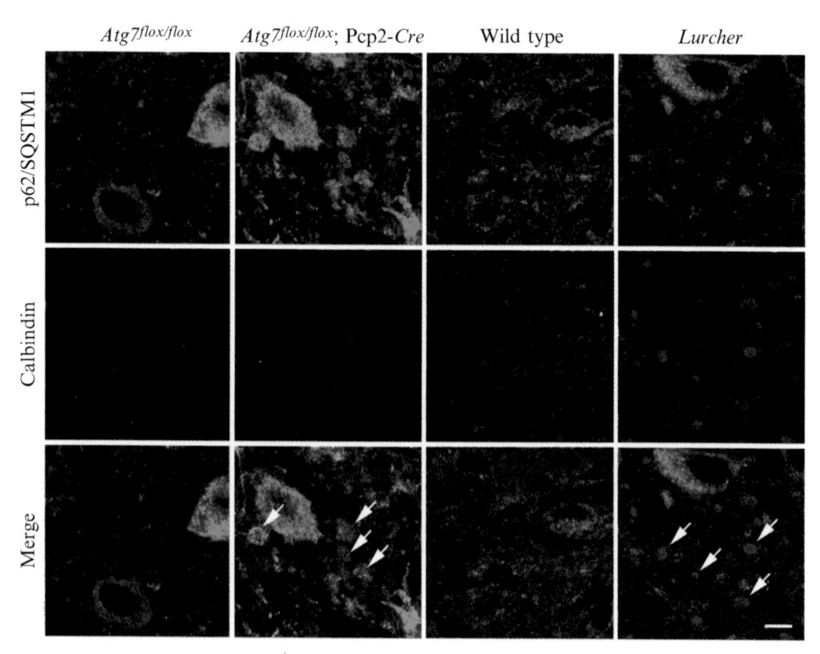

Zhenyu Yue *et al.*, Figure 8.3 Please refer to the legend in the text.

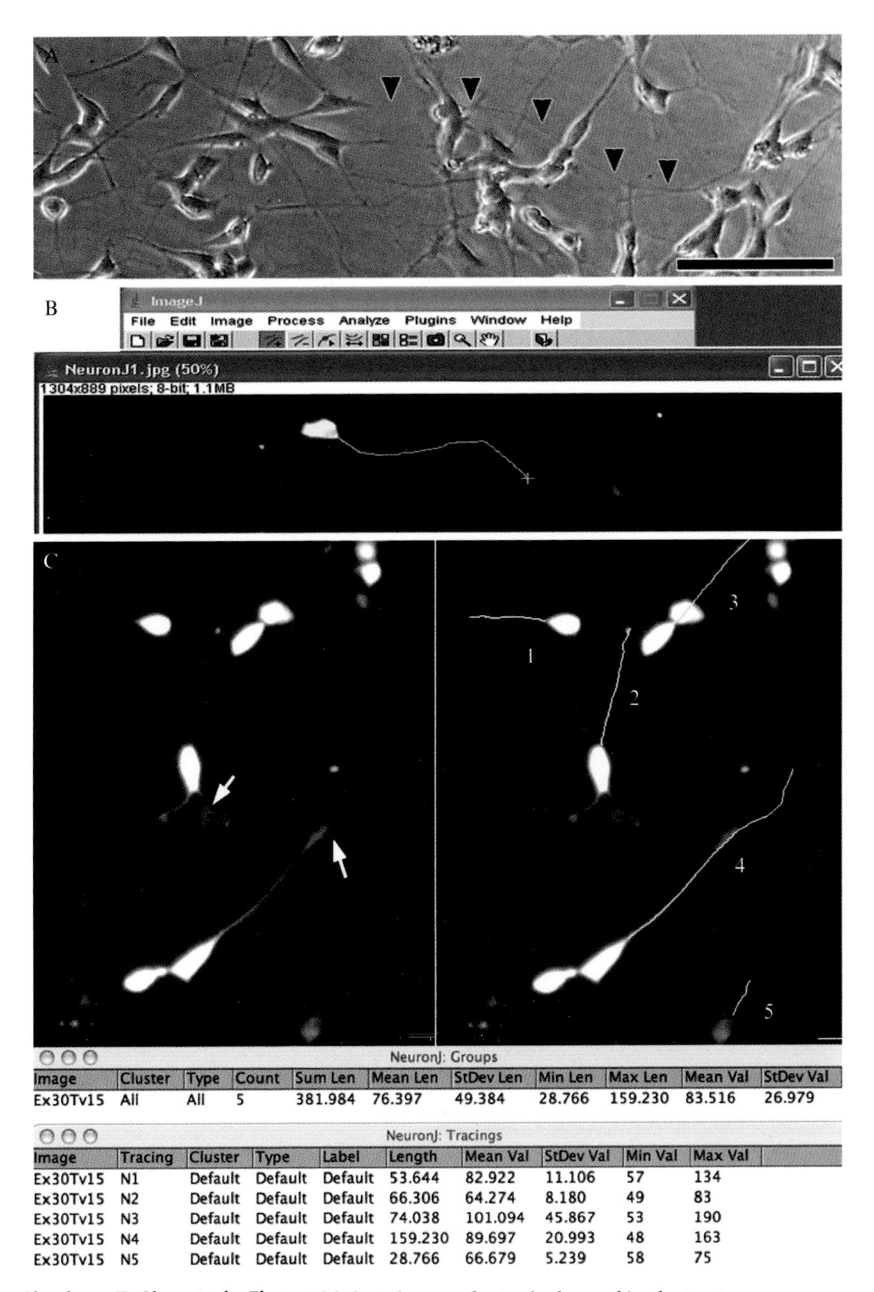

NeuronJ: Groups

Image	Cluster	Type	Count	Sum Len	Mean Len	StDev Len	Min Len	Max Len	Mean Val	StDev Val
Ex30Tv15	All	All	5	381.984	76.397	49.384	28.766	159.230	83.516	26.979

NeuronJ: Tracings

Image	Tracing	Cluster	Type	Label	Length	Mean Val	StDev Val	Min Val	Max Val
Ex30Tv15	N1	Default	Default	Default	53.644	82.922	11.106	57	134
Ex30Tv15	N2	Default	Default	Default	66.306	64.274	8.180	49	83
Ex30Tv15	N3	Default	Default	Default	74.038	101.094	45.867	53	190
Ex30Tv15	N4	Default	Default	Default	159.230	89.697	20.993	48	163
Ex30Tv15	N5	Default	Default	Default	28.766	66.679	5.239	58	75

Charleen T. Chu *et al.*, Figure 11.1 Please refer to the legend in the text.

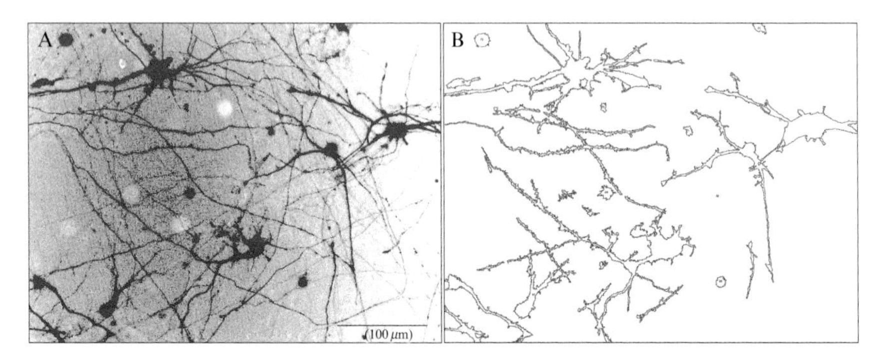

Charleen T. Chu *et al.*, Figure 11.2 Please refer to the legend in the text.

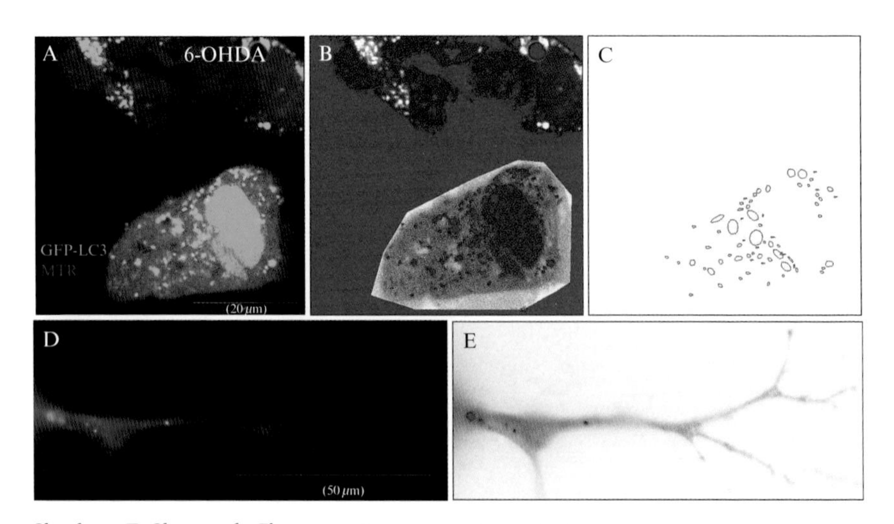

Charleen T. Chu *et al.*, Figure 11.3 Please refer to the legend in the text.

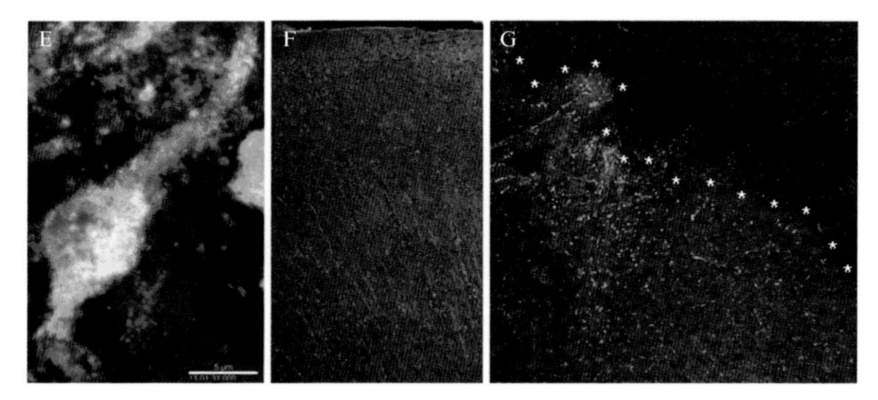

Charleen T. Chu *et al.*, Figure 11.5 Please refer to the legend in the text.

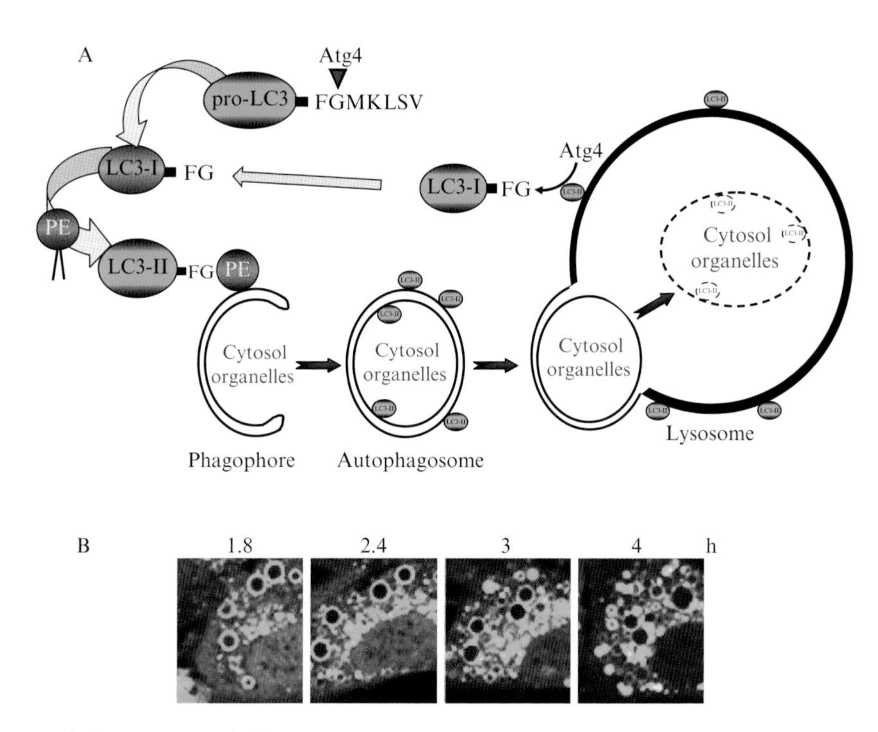

Frank C. Dorsey *et al.*, Figure 12.1 Please refer to the legend in the text.

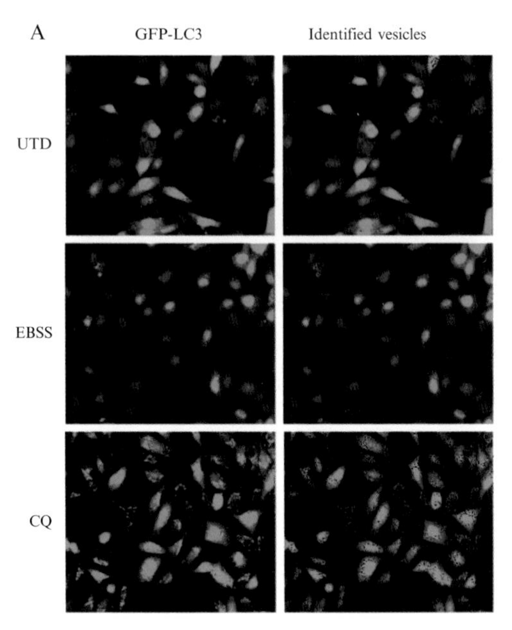

Frank C. Dorsey *et al.*, Figure 12.2 Please refer to the legend in the text.

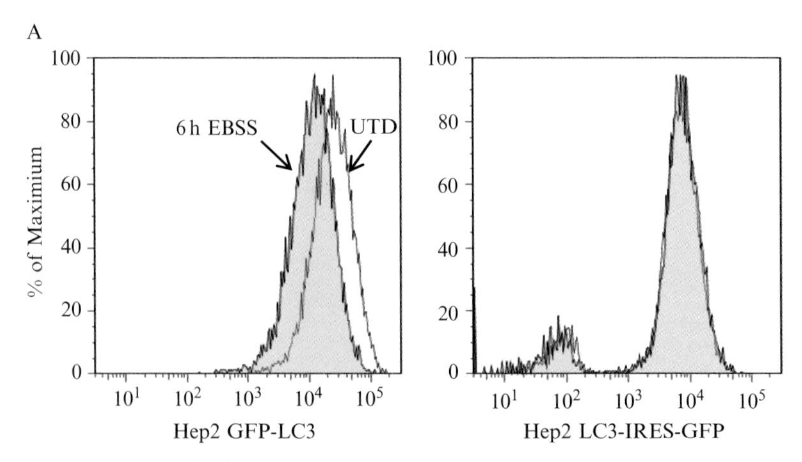

Frank C. Dorsey *et al.*, Figure 12.3 Please refer to the legend in the text.

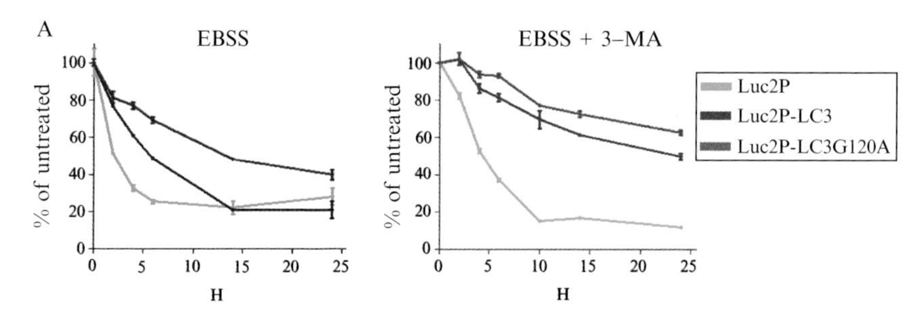

Frank C. Dorsey *et al.*, Figure 12.4 Please refer to the legend in the text.

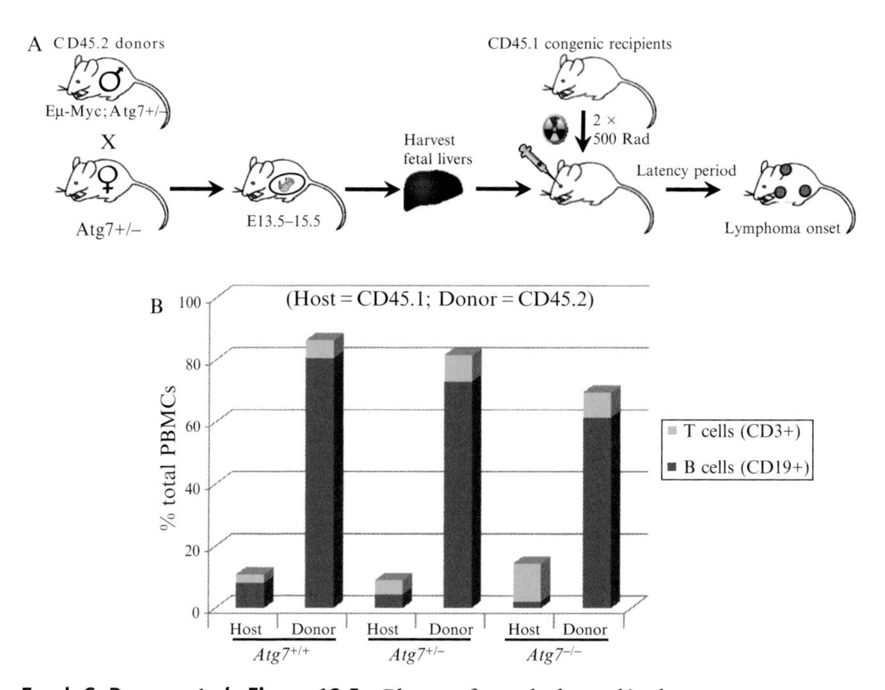

Frank C. Dorsey *et al.*, Figure 12.5 Please refer to the legend in the text.

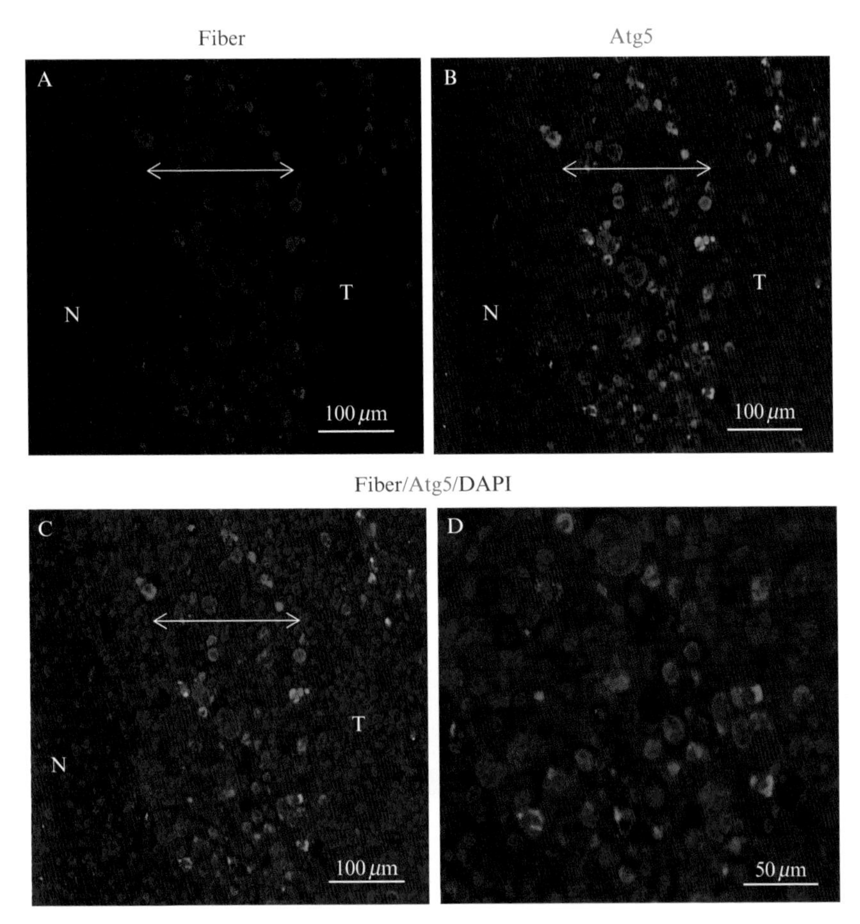

Hong Jiang *et al.*, Figure 13.5 Please refer to the legend in the text.

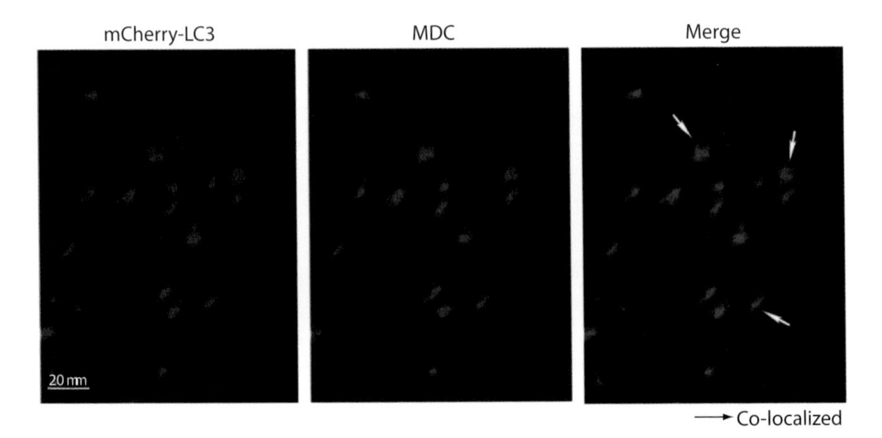

Cynthia N. Perry *et al.*, Figure 16.5 Please refer to the legend in the text.

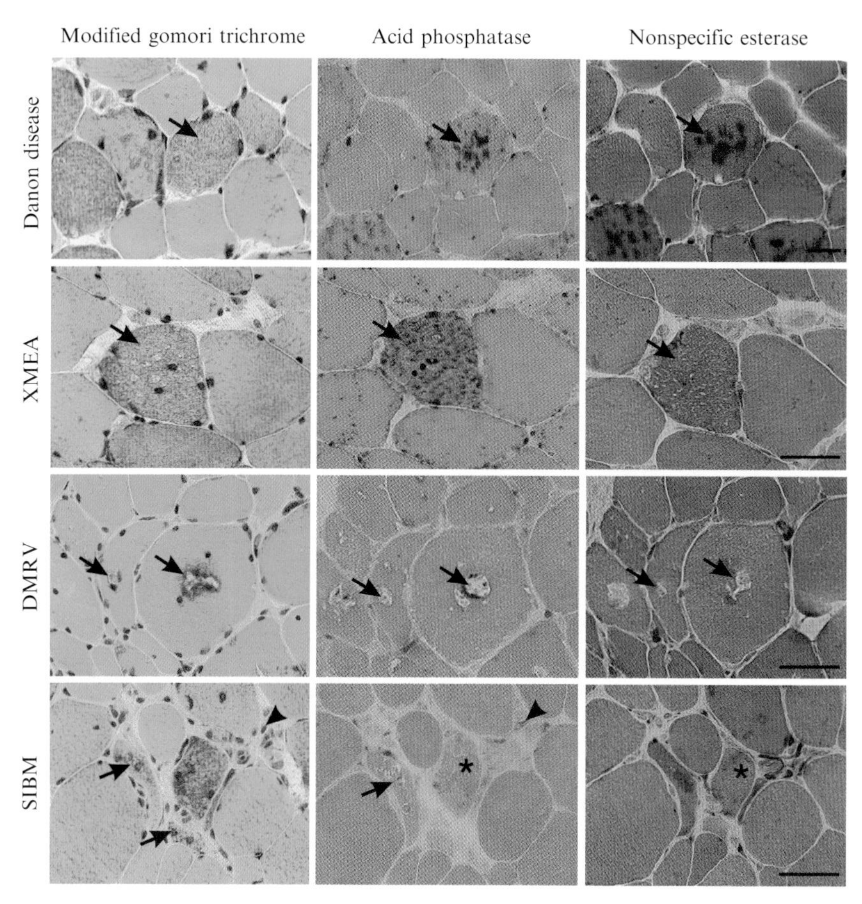

May Christine V. Malicdan *et al.*, Figure 19.1 Please refer to the legend in the text.

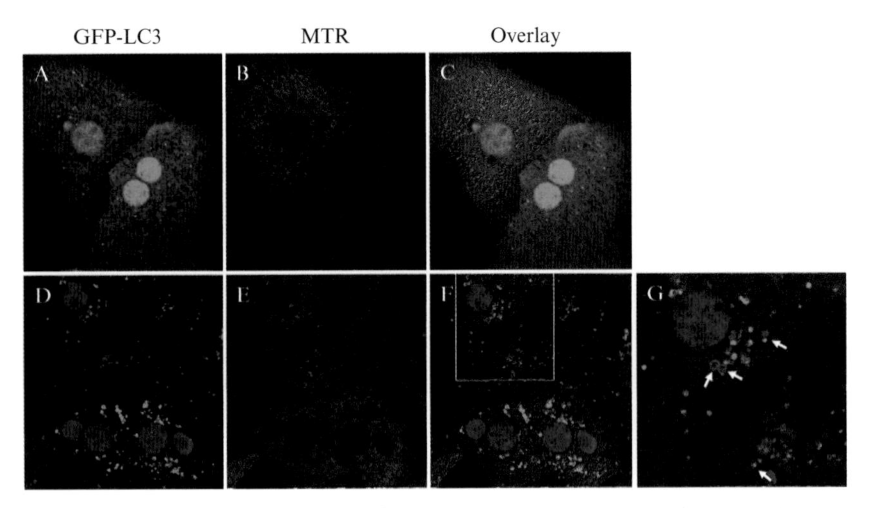

Wen-Xing Ding and Xiao-Ming Yin, Figure 20.4 Please refer to the legend in the text.